Clinical Neuropsychology

Behavioral and Brain Science

Clinical

Neuropsychology

Behavioral and Brain Science

John L. Bradshaw

&

Jason B. Mattingley

DEPARTMENT OF PSYCHOLOGY
MONASH UNIVERSITY
CLAYTON, VICTORIA, AUSTRALIA

ACADEMIC PRESS

SAN DIEGO   BOSTON   NEW YORK
LONDON   SYDNEY   TOKYO   TORONTO

This book is printed on acid-free paper. ∞

Academic Press, Inc.
A Division of Harcourt Brace & Company
525 B Street, Suite 1900, San Diego, California 92101-4495

United Kingdom Edition published by
Academic Press Limited
24-28 Oval Road, London NW1 7DX

Library of Congress Cataloging-in-Publication Data

Bradshaw, John L., date.
    Clinical neuropsychology : behavioral and brain science / by John
  L. Bradshaw, Jason B. Mattingley.
        p.   cm.
    Includes bibliographical references and index.
    ISBN 0-12-124545-4
    1. Clinical neuropsychology.   2. Neuropsychiatric disorders.
  I. Mattingley, Jason B.   II. Title.
    [DNLM: 1. Neuropsychology.   2. Nervous System Diseases.   3. Mental
  Disorders.   WL 103.5 B812c   1995]
  RC386.B72   1995
  616.8--dc20
  DNLM/DLC
  for Library of Congress   94-49187
                                                             CIP

PRINTED IN THE UNITED STATES OF AMERICA
95   96   97   98   99   00   EB   9   8   7   6   5   4   3   2   1

Contents

Preface

*I*f psychology is the scientific study of behavior, neuropsychology places that study firmly in the domain of its underlying neural and neuronal determinants. It has long been a truism that if one wishes to understand how a system works, a careful examination of all its potentialities for malfunction may provide unique insights. This volume will concentrate on the clinical aspects of brain malfunction, but will not ignore those experimental techniques, especially the newer ones of neuroimaging, which throw a complementary light on the general issue of brain–behavior interrelationships. While we can learn much from unfortunate experiments of nature, new minimally invasive neuroimaging techniques seem set to illuminate the essentially "mental" contributions of, for example, thought and imagination. For the first time *healthy* individuals can be scanned while mentally rehearsing all manner of material, and engaging in complex *everyday* tasks like viewing a video, composing music, or even knitting. The last, a project of our own, aims to determine those brain areas involved in complex, highly overlearned bimanually integrative skills that do not require vision.

We hope in this book to go some way toward simultaneously achieving several different aims: an account of the major neuropsychological and neurological syndromes in their own right; a description of those rare and intriguing anomalies of behavior, where relevant, which have long fascinated the general public from the reports of such gifted medical writers as Oliver Sacks and Harold Klawans; and an enhanced understanding of normal brain processes in language, speech, object recognition, attention, memory, movement control, and thought.

While psychology's historical roots lie within the twin fields of physiology and philosophy, neurology and psychiatry sprang from a common medical stem and only now, after a period of separation, are beginning once again to grow together. The driving force behind their reunion is an increasing awareness of the organicity of so many disorders. The behavioral neurologist is of course comfortable with the aphasias, agnosias, amnesias, apraxias, and other movement disorders, and the clinical neurologist treats the patient with Parkinson's disease

(PD); however, the Huntington's disease (HD) patient, with very similar disabilities, may be managed by the psychiatrist, who of course also treats patients with Tourette's syndrome (TS), schizophrenia, and mood disorders. The discipline of neuropsychiatry is doing much to blur many of these old lines of demarcation.

Lost or altered function can arise from destruction (by stroke, injury, or tumor) of the brain tissue directly involved in mediating that function, as with aphasia, apraxia, or agnosia; it can also stem from loss of activation due to damage elsewhere in the system, as in PD, or from changes in neurotransmitter activity or balance, as with the affective disorders. Clearly there are more degrees of freedom for treating disorders of the last type than is the case with the first, though cell transplantation is being employed or actively considered in the treatment of Parkinson's, Huntington's, and even Alzheimer's disease (AD).

Apart from practical handbooks of neuropsychological testing procedures, textbooks of neuropsychology have tended to adopt either a cognitive or a more traditional biological emphasis. The former owes much to modern computational and information-processing theory, and more or less explicitly rejects biological substrates in its emphasis upon software and "architecture." Indeed, for related ideological reasons it tends to eschew group studies, arguing that every individual, in sickness and health, is unique. The other, more traditional, biological approach, which we have adopted, seeks to ground function within structure, and to discuss commonalities of deficit consequent upon similarities of lesion. While we admit to finding this latter approach more satisfying, due presumably to our biological biases, and suspect that it may lend itself more readily to practical applications or therapeutic interventions, we would emphasize the potential complementarity of the two traditions.

The genesis of this volume arose from our teaching (senior undergraduate and graduate) and research. At one time or another we have empirically investigated aspects of many of the syndromes described herein, and we would emphasize the reality of the two-way nexus between teaching and research; both stimulate each other in unforseen ways. A single convenient, integrative volume could not, we felt, religiously cover all primary sources, so we have selectively built upon existing major integrative reviews, where available, and fleshed them out with new, major, provocative, or intriguing material from primary sources published within the last two or three years. Similarly, rather than meticulously referencing every statement, we have tended to refer the reader to the major reviews listed as Further Readings at the close of each chapter, though occasionally, as in Chapter 6, in the absence of suitable material, we have adopted more traditional referencing procedures. The material in Further Readings has been carefully selected to provide an up-to-date, balanced, and comprehensive overview of the issues addressed. Some items in the Further Readings and not previously cited within the chapter may be of major integrative significance. The provision of a Summary and Conclusions section to each chapter aims to provide both a general survey and a balanced judgment of any controversial aspects previously discussed in detail. Not all chapters are of similar length, as their designation has been largely a matter of content. Thus controversial or unresolved issues have been treated at greater length, whereas short bridging chapters (e.g., Chapter 11) provide the necessary (largely nonclinical) background to those that follow. We have drawn together and reviewed many diverse areas of information, but if at times we may appear uncritical, it is because so much of the information is often so new and controversial that it cannot yet be properly assessed and evaluated. While we have attempted to address most major areas of relevance, and hope to have included most integrative reviews up to 1994, we note that the field is far too dynamic for us to be able to offer anything completely free of personal bias, or totally comprehensive or exhaustive.

The book is aimed at senior undergraduates, postgraduates, and researchers in the behavioral and medical sciences. We are extremely grateful to the many individuals who so readily provided us with illustrative material, and hope we have been able to acknowledge everyone appropriately. Our grateful thanks go to our partners, who are also our colleagues, Judy Bradshaw and Ada Kritikos, to our drafting artist and photographer Rosemary Williams, and to our research colleagues in psychology, neurology, and psychiatry, especially Jim Phillips, Bob Iansek, and Ed Chiu.

Chapter 1

An Introduction to Clinical Neuropsychology

THE INSCRUTABLE ARCANUM OF THE ORGANIZATION OF THE BRAIN . . . ATTRACTED US IRRESISTIBLY. WE SAW THAT AN EXACT KNOWLEDGE OF THE STRUCTURE OF THE BRAIN WAS OF SUPREME INTEREST FOR THE BUILDING UP OF A RATIONAL PSYCHOLOGY. TO KNOW THE BRAIN, WE SAID, IS EQUIVALENT TO ASCERTAINING THE MATERIAL COURSE OF THOUGHT AND WILL.

—SANTIAGO RAMÓN Y CAJAL, *Recollections of My Life*

The basic precept of neuroanatomists such as Cajal, that explanations of human behavior can develop from an understanding of the structural organization of the brain, represents an approach to brain–behavior relationships that lies at the opposite end of the spectrum to that of neuropsychology. From its narrowest interpretation, the primary goal of neuropsychology is to use detailed behavioral analyses to infer underlying structural and functional properties of the brain. Thus, the kinds of data yielded by neuroanatomical and neuropsychological approaches are likely to be quite different, even though their ultimate goals are essentially the same. Moreover, although the levels of explanation offered by the two approaches are often different, the insights provided by one may be useful in refining and constraining the interpretations of the other. In this volume, although our view of brain–behavior relationships is taken principally from a neuropsychological perspective, we shall also draw on insights obtained from neuroanatomical and neurophysiological studies. This reflects our basic philosophy that cognitive processes do not exist in a vacuum; they are determined by the inherent structure and function of the neural substrate.

In this introductory chapter, we provide an overview of some of the important conceptual and methodological issues in neuropsychology. Our intention is to familiarize the reader with the assumptions and principles upon which hypotheses about brain–behavior relationships are based, and to explain the different approaches to data acquisition and interpretation in neuropsychological research. In addition, because we place considerable emphasis on studies of brain-lesioned patients, we have included sections on the cerebral vasculature and on the behavioral consequences of pathologies involving specific arterial territories. Finally, we consider two modern techniques, neuroimaging and artificial neural network modeling, both of which have had a considerable impact on contemporary models of the structural and functional organization of the brain.

Neuropsychology, Neuropsychiatry, and Behavioral Neurology

The discipline of neuropsychology began in the late 19th century with the patient studies of Broca, Wernicke, Lichtheim, and Hughlings Jackson. These workers adopted the principle of functional localization, in which specific cognitive functions are assumed to be subserved by discrete brain regions. Using the method of anatomoclinical correlation, these workers established associations between damage to particular brain regions and distinct manifestations of behavioral anomalies such as aphasia. The localizationist approach, which developed from Gall's original, though misguided, attempts to localize such various complex behaviors as "philoprogenetiveness" according to the topography of an individual's skull (Spurzheim, 1815; cited in Farah, 1994), became the cornerstone of the developing discipline of neuropsychology. However, whereas the early anatomoclinical studies persisted in assigning complex functions to specific brain regions, it was not until Hughlings Jackson established his famous dictum, that localizing a *lesion* and localizing a *function* are conceptually distinct, that real progress was made in elucidating the nature of human brain–behavior relationships.

Despite its relative youth, neuropsychology has already found a distinct niche within the fields of medicine and psychology. There are, however, several other disciplines that are clearly related; thus, *neuropsychiatry* involves "understanding the neurobiological basis, optimal assessment, natural history, and most efficacious treatment of disorders of the central nervous system with behavioral manifestations" (Cummings & Hegarty, 1994, p. 209). One subtle distinction between neuropsychology and neuropsychiatry is that the latter is often motivated by a desire to understand disordered behavior in terms of its implications for diagnosis and treatment. Similarly, the discipline of *behavioral neurology* (Devinsky, 1992) involves the use of data concerning the anatomy and physiology of the central nervous system (CNS) to guide interpretations of disordered behavior consequent upon neural damage. Although neuropsychology shares these aims, at least insofar as it too is concerned with patient diagnosis and management, it is ultimately concerned with elucidating the mechanisms underlying abnormal and *normal* behavior. Indeed, modern neuropsychology is based on data obtained not only from brain-damaged patients, but also from normal individuals; it exploits over 100 years of research in experimental psychology to help explain the patterns of disordered perceptual, cognitive, and motor processes seen in patients with neural damage.

There is also a sense in which neuropsychology itself can be further divided into distinct subspecialties. Whereas *clinical* neuropsychology is concerned with explaining how specific patterns of disordered human behavior may arise from disruption of particular brain processes, *cognitive* neuropsychology is focused primarily on explaining disordered behavior in terms of disruption to particular information-processing units, with (at best) only peripheral concern for their anatomical or physiological instantiation. Thus, the aim of cognitive neuropsychology is to understand information processing independently of the neural hardware. The reason for this approach, it has been argued (e.g., Caramazza, 1988), is that we know too little about normal brain processes and behavior to begin relating the two; this limitation becomes even more obvious when we attempt to understand the behavioral consequences of brain damage. An alternative view is that despite the obvious gaps in our knowledge, we nevertheless have a sufficiently detailed database from which to begin relating results of behavioral analyses with those obtained from studies of brain anatomy and physiology (e.g., Kosslyn, 1994). It is this latter view that we have adopted in this book; although in some cases the links we have proposed between particular behavioral anomalies and underlying brain processes remain tentative, we believe such endeavors facilitate the iterative process of developing and refining models of normal and disordered behavior.

Important Assumptions in Neuropsychology

Modularity

The prevailing assumption in neuropsychology is that cognitive processes are organized into distinct processing units or *modules*. In one conceptualization, modules are hardwired, autonomous, computational mechanisms that perform highly specific functions, such as feature detection (Fodor, 1983). The outputs from several modules may be combined in the service of more complex cognitive operations, though the content of any given module is not available to consciousness. The concept of modularity, or some variant thereof, underpins most theorizing in modern neuropsychology (Shallice, 1988). It is appealing from both a computational and an evolutionary perspective; dividing a computational system into smaller subunits permits rapid, parallel processing of many different inputs, and it permits changes in specific subunits without necessarily affecting the operations of others. In addition, local damage may impair only a subset of specialized functions, rather than causing a global reduction in processing efficiency.

Although the general assumption of modularity as outlined above relates to *functional* processing units, it may be extended to provide a potential organizing principle for the brain itself. In the visual system, for example, anatomically distinct modules exist for processing color, form, location, and motion, to name but a few (DeYeo, Felleman, Van Essen, & McClendon, 1994). These regions are themselves structured in a modular fashion; consider, for example, the organization of the visual cortex in which neurons responsible for processing edges of a specific orientation are clustered into vertical columns (Livingstone & Hubel, 1988, and see Chapter 5). The anatomical notion of modular organization has also been extended to include widely distributed neural circuits subserving, for example, spatial attention, language, and memory (Mesulam, 1990). We will not consider here the philosophical arguments for and against any particular model of modularity (see Chomsky, 1980; Fodor, 1983; Marr, 1982); rather, our intention is to point out the importance of the modularity assumption in neuropsychology (i.e., that a discrete brain lesion will disrupt one or more functional modules, producing a restricted set of deficits in the context of otherwise normal behavior).

Locality

We have already suggested that one consequence of modularity is that local brain damage should produce exclusively local effects. In other words, provided there is some degree of correspondence between anatomical and cognitive architectures, a discrete lesion should result in a specific set of behavioral deficits in the context of otherwise normal functioning; this is the *locality assumption* in neuropsychology (Farah, 1994). Provided the locality assumption is true, conclusions about the organization of normal cognitive processes can be drawn from studies of brain-damaged patients; thus, the finding of a deficit in a particular ability following damage to a given module can be taken to imply that the module is necessary for the ability in question. However, despite the evidence for modularity, it is also clear that the brain is a highly interactive system; thus, even relatively small lesions may modify the functioning of distant anatomical regions, which may in turn affect apparently unrelated cognitive domains. Such notions have in fact been implicit in neuropsychology since the time of Ferrier and Hughlings Jackson. Indeed, von Monakow (1914) hypothesized that a circumscribed lesion may disrupt the functioning of brain regions distant from the site of primary damage (diaschisis). This concept has in fact received support from recent physiological measures of neural blood flow and metabolism, which show that damage to discrete cortical or subcortical regions may reduce physiological activity in remote regions, probably via interruption of efferent connections to such regions from the area of primary damage (Vallar et al., 1988).

The existence of diaschisis poses problems for the locality assumption, at least insofar as one wishes to base inferences about the organization of normal cognitive processes on the performances of brain-damaged patients. How do we know whether the behavioral deficit being observed is a consequence of primary damage to a particular module or set of modules, or whether it also reflects the anomalous activity of undamaged modules? In other words, to what extent does damage to one set of modules induce changes in the normal functioning of other modules? One possible consequence of local damage is that certain functions are released from inhibition; such a mechanism was suggested by Hughlings Jackson to underlie the "positive symptoms," such as primitive reflexes, recurrent utterances, tremors, and hallucinations, that are often encountered in cases of neurological illness.

Another potential problem for the locality assumption is raised by recent studies showing remarkable plasticity of structure and function in the mature brains of nonhuman species (e.g., Darian-Smith & Gilbert, 1994; Rajan, Irvine, Wise, & Heil, 1993). In human lesion studies, the potential for reorganization is likely to vary as a function of etiology, patient age, and time since onset of the pathological process (Kertesz, 1991). Certain premorbid factors such as handedness, which is itself related to structural and functional hemispheric asymmetries (see Bradshaw, 1989; Bradshaw & Nettleton, 1983), may also play a role in determining the nature and extent of reorganization following brain damage. Finally, humans in particular may adopt alternative cognitive strategies to compensate for perceived deficits, so that their performance on any given task may mislead interpretations of normal functioning.

In view of the caveats outlined above, can we derive any meaningful conclusions from human lesion studies? Our answer is an emphatic yes. The first step in overcoming the potential difficulties associated with the modularity and locality assumptions is to recognize the situations in which they are most likely to lead to erroneous conclusions. For example, diaschisis often resolves in the weeks and months following acute cerebral damage (Vallar et al., 1988), so that distance effects may be less of a problem when considering the performances of patients in the chronic stages after damage. Moreover, functional neuroimaging techniques such as positron emission tomography (PET) may be used to verify the presence or absence of physiological abnormalities in nonlesioned areas (Vallar et al., 1988).

The fact that clinical neuropsychology has revealed systematic and consistent patterns of behavioral impairment following discrete brain lesions suggests that the assumptions of modularity and locality have some validity (McCarthy, 1994). Many contemporary explanations of disordered functioning after brain damage correspond with models derived from studies of normal performance (e.g., Posner, 1994). Moreover, hypotheses regarding the neural instantiation of certain cognitive functions based upon anatomoclinical correlations have in several instances been verified by studies of normal humans using functional neuroimaging techniques (Posner, 1993). Of course, as in any branch of science, particular hypotheses are best supported by converging lines of evidence; in this book, therefore, we have drawn wherever possible upon evidence from multiple sources, including cognitive and neuroimaging studies of normal and brain-lesioned patients, animal studies, and occasionally upon artificial neural network modeling.

Conceptual Issues in Neuropsychology

The Syndrome Approach to Neuropsychology

One long-established tradition in medical diagnosis is for the practitioner to elicit signs and record symptoms from individual patients in the hope of establishing recognizable patterns of clinical features, or *syndromes*. The identification of a specific syndrome assists in diagnosis and management by implicating the underlying pathological process and suggesting an appropriate form of treatment. Because clinical neuropsychology grew out of (or at least was strongly influenced by) clinical neurology, it is perhaps not surprising that contemporary neuropsychologists have also adopted the syndrome approach in dealing with brain-damaged patients. This approach has proved useful in helping to establish anatomoclinical correlations. For example, it has been demonstrated repeatedly that the syndrome of Broca's aphasia occurs after damage to the left frontal operculum (see Chapter 2), and that the syndrome of left unilateral neglect occurs after right parietal damage (Vallar & Perani, 1986, and see Chapter 6). However, with the now widespread availability of sophisticated neuroimaging techniques, clinical neuropsychologists are rarely called upon to predict patterns of structural damage on the basis of observed behavioral anomalies. For this and other more theoretically motivated reasons, contemporary theorists have questioned the validity of the syndrome approach in neuropsychology (e.g., Caramazza & Badecker, 1991). Indeed, it has been suggested that neuropsychological syndromes are no more than "theoretically arbitrary categories reflecting only pretheoretic intuitions about the nature of language and cognition" (Caramazza & Badecker, 1991, p. 215). Such provocative statements continue to fuel a vigorous debate on the value of syndrome-based neuropsychology (Caramazza & Badecker, 1991; McCloskey, 1993; L. C. Robertson, Knight, Rafal, & Shimamura, 1993; Zurif, Swinney, & Fodor, 1991). Before addressing some of the central issues in this debate, we shall consider in more detail the rationale behind syndrome-based neuropsychology. Vallar (1991) has distinguished between three types of syndrome in neuropsychology, which are elaborated below.

Anatomical syndromes involve conglomerations of behavioral anomalies that reflect damage to adjacent or contiguous brain regions. Thus, given their probabilistic nature, partial associations and dissociations of symptoms are frequently encountered in anatomical syndromes; a relatively large lesion may produce all of the deficits associated with a particular anatomical syndrome, whereas a small, circumscribed lesion may result in only a subset of these anomalies. Crucially, the anatomical syndrome has no value in determining the functional architecture of the normal cognitive system. Consider, for example, Gerstmann's syndrome, which comprises finger agnosia, left–right disorientation, acalculia, and agraphia (see Chapter 4); if anything, this should be considered an anatomical syndrome, because although the specific association of deficits often indicates left parietal damage, it is unlikely to reflect dysfunction of a coherent functional module.

Functional syndromes, the second type of syndrome distinguished by Vallar, involve associations of deficits that arise from damage to a specific functional module. If this functional module is subserved by a circumscribed anatomical area, then damage to this area should produce all the characteristic symptoms of the syndrome. If instead the module is subserved by a widely distributed network of structures, the resulting syndrome will have no strictly localizable neural correlate. Unlike anatomical syndromes, functional syndromes do not generally allow for partial associations between observed deficits, except in terms of preexisting differences between the cognitive architectures of particular individuals (e.g., those of left- vs. right-handers).

The third type of syndrome distinguished by Vallar (1991) involves a combination, within the same individual, of anatomical and functional syndromes, the so-called *mixed syndrome.* Mixed syndromes comprise symptom complexes that arise from damage to discrete functional units that happen to be anatomically adjacent.

In clinical practice, the syndrome approach to neuropsychology has substantial heuristic value. It provides a useful shorthand for communicating the essential behavioral features of a particular patient's deficits, and often carries a strong implication as to the likely side and site of neural damage. Historically, the syndrome approach has also provided a convenient means of classifying patient groups for research. Dividing patients into homogeneous groups on the basis of observed behavioral deficits provides an a priori strategy for generating hypotheses about specific aspects of normal and disordered brain functioning. These can then be tested empirically, and the adequacy of the classification system itself can be modified accordingly (Zurif et al., 1991).

In contrast, there are those who believe that the syndrome-based approach to neuropsychology exerts a pernicious influence on theories of normal and disordered cognitive processes. For example, Caramazza and Badecker (1991) have argued that syndrome-based research cannot provide valid data on the likely mechanisms underlying normal cognition. The objections of these and other authors (see also McCloskey, 1993) revolve around the premise that, despite the apparent homogeneity of lesion sites or behavioral deficits or both among patients exhibiting a given syndrome, there will always be some variation between individuals. Thus, data derived from group studies typically reflect the averaged results of a range of different performances, rather than the specific pattern of normal and impaired processes of any given individual. This in turn may lead to erroneous conclusions about the nature of the underlying cognitive architecture (i.e., the so-called averaging artifact) (Shallice, 1988). Antagonists of the syndrome-based approach have suggested that the only convincing way of avoiding the averaging artifact is to restrict inferences about normal cognition to investigations of single patients; they assert that it is only at the level of the individual that we can infer facts about the underlying functional architecture. Indeed, much of the controversy surrounding the syndrome-based approach to neuropsychology can be reduced to a consideration of the relative merits of group versus single-case studies.

The single-case approach to neuropsychology had its origins in 19th-century descriptions of patients with focal brain lesions by workers such as Broca, Wernicke, and Hughlings Jackson. The behaviors of an individual, or of a small group of individuals, were described in a more or less qualitative fashion, and the particular pattern of impairments was associated with specific lesion sites. With the development of a more empirical approach to patient assessment, however, group studies were increasingly the rule in neuropsychological research, though single-case studies of rare or anomalous disorders continued unabated. The perceived failure of group studies to address issues relating to normal cognitive processes, coupled with the practical difficulties of assessing large, homogeneous groups of brain-damaged patients, has led to the revival of the single-case design (see Caramazza, 1986, 1988). Unlike early qualitative studies, however, the contemporary single-case approach uses properly controlled experimental paradigms, matched control groups, and accepted methods of statistical analysis. As we outlined earlier, advocates of the single-case approach consider its principal advantage to be that it avoids the averaging artifact to which group studies are susceptible, thereby permitting valid conclusions about the underlying architecture of normal cognitive functions.

Predictably, there are also those who point to weaknesses in the single-case-only approach to neuropsychology (e.g., L. C. Robertson et al., 1993). Perhaps the most critical of these weaknesses is that the single-case-only approach violates a central assumption of the scientific method, namely, that of replicability. Because in the single-case design each patient is considered to be unique, the possibility of replicating a particular finding in another patient is rendered impossible. The prerequisite of replicability is particularly important in clinical neuropsychology, because the performances of individual patients may differ as a consequence of factors other than those arising directly from neural damage. For example, individual differences in etiology, chronicity, premorbid functioning, and compensatory strategies may all affect the particular pattern of anomalies exhibited by an individual patient. Another obvious consequence of single-case-only research is that anatomoclinical correlations become impossible; it is only through repeated observations on patients with discrete lesions that specific functional losses can be related to damage of distinct brain regions (e.g., Vallar & Perani, 1986). Although, to some cognitive neuropsychologists, such a limitation is of little consequence, we consider that the exercise of relating brain structure and function is of critical importance. Finally, just as data from group studies may provide misleading information about the organization of normal cognitive functions, so those from single-case studies may be used incorrectly to infer general principles about the nature of disordered behavior in groups of patients with superficially similar symptoms.

The reader will note from the chapter titles in this book that we are sympathetic to the syndrome-based approach to clinical neuropsychology, at least insofar as it facilitates the process of dividing the vast array of cognitive deficits seen after brain damage into broadly related domains. We, along with other authors (e.g., Goodglass, 1993; L. C. Robertson et al., 1993), view neuropsychological syndromes as representing probabilistic tendencies of particular symptom complexes to reveal the brain's functional organization. Given the inherent capacity for individual variation in patterns of neural wiring as a function of, for example, maturational influences, the acknowledged heterogeneity between patients with a given syndrome is not surprising. Nonetheless, we believe there is a sufficient degree of consistency in the various neuropsychological syndromes to justify their treatment as theoretically coherent entities; such an approach is particularly valuable for students of clinical neuropsychology who may be searching for a conceptual framework on which to build a broad understanding of brain–behavior relationships. Having said this, however, the contents of this book are testimony to the fact that clinical neuropsychology is full of dissociations both within and between individuals and groups of patients. We do not view these two circumstances as being mutually exclusive; syndromes provide an organizing principle with which to conceptualize disorders within different cognitive domains, whereas the occurrence of dissociations emphasizes their inherent heterogeneity and provides clues as to the functional architecture of the normal brain.

We are also sympathetic to both group and single-case studies, provided the conclusions drawn from each are weighed with respect to the caveats outlined above. In particular, we view group studies as an attempt to sample disordered behavior, thereby providing a range over which performances may vary and a set of core features to be used in patient classification. By an iterative process, the group may subsequently be subdivided into smaller clusters that separately are more homogeneous than the original with respect to a particular cognitive domain. Hence, it is possible to funnel down to progressively more realistic explanations of underlying pathological mechanisms. Single-case studies permit conclusions about the organization of normal cognitive functions, though they tend to preempt the distillation process described above; nonetheless, the recent emergence of multiple single-case designs (e.g., Behrmann, Moscovitch, Black, & Mozer, 1990) suggests that even this potential limitation may be overcome.

Associations and Dissociations in Neuropsychology

In considering the syndrome approach to neuropsychology, we have emphasized the potential importance of associations of symptoms. As we have seen, these may arise from damage to a specific functional module, from damage to contiguous anatomical regions, or from a combination of the two. However, *dissociations* of symptoms are perhaps even more important for understanding the functional organization of normal cognitive processes. In general terms, a dissociation is said to occur when a patient performs poorly on one task, or a small number of tasks, but is normal or at least relatively unimpaired on another set of tasks. The implications of such a set of results for understanding the organization of normal cognitive processes is clear; dissociations reveal the "cleavage lines" between discrete functional modules (Shallice, 1988). It has been suggested that the primary aspects of a particular disorder are best described in terms of a set of dissociations, whereas its secondary aspects may be characterized in terms of associations (Shallice, 1988). Whereas the former are critical in drawing inferences about the architecture of normal cognitive processes, the latter exert a more tenuous grasp on models of normal functioning; subsequent cases may reveal further dissociations between secondary deficits initially thought to be associated. Throughout this book, we will describe examples of how particular disorders have come to be fractionated on the basis of dissociated performances exhibited by individual patients. We will also demonstrate how the emergence of such dissociations shapes our understanding of normal cognitive processes, and how in some instances it leads researchers to refine the taxonomy assigned to disorders in particular cognitive domains.

One potential difficulty with the kind of dissociation described above (known as a *simple dissociation*) is that its existence need not necessarily reflect a true discontinuity between preserved and impaired functions; a patient may perform normally on one task and poorly on another simply because the latter is relatively more difficult. An almost ubiquitous finding is that brain damage of any kind tends to reduce an individual's information-processing capacity; thus, differences in task performance may simply reflect the level of difficulty of the tests used. (The notion of such "resource" artifacts stems from the work of Norman & Bobrow, 1975.) This potential ambiguity may, however, may be resolved by seeking a *double dissociation* (Teuber, 1955). In straightforward instances, this type of dissociation is considered to occur when one patient (or one group of patients) is impaired on task X but normal on task Y, and another patient (or group of patients) is normal on task X and impaired on task Y. Under such circumstances, an interpretation of the dissociation in terms of differential task difficulty (or sensitivity) can now be rejected, and an alternative explanation in terms of damage to discrete functional units becomes more likely.

Indeed, the same general conclusion can be drawn when the dissociated performances exhibited by such patients are all below the normal range. In these instances, the double dissociation may be conceptualized as a crossover interaction of the kind often encountered in formal statistical analyses (G. Jones, 1983). Thus, contrary to traditional views (e.g., Teuber, 1955), the finding of a double dissociation does not require that patients exhibit deficient performances on one task and normal performances on the other; the only requirement is that their performance levels form a crossover interaction, irrespective of the possible existence of significant main effects between individuals or tasks.

Whatever their specific manifestations, it is important to bear in mind that, ultimately, we are seeking double dissociations between cognitive processes rather than between tasks. The finding of a double dissociation between the performances of two patients on individual tests of a particular function does not necessarily imply that distinct or independent cognitive processes must underlie such performances. The most convincing evidence in favor of the existence of independent cognitive processes is obtained from experiments in which patients exhibit a double dissociation between *sets* of *functionally related* tasks. Thus, for example, a double-dissociation between cognitive processes may be claimed when one patient is impaired on tasks $X1, X2, X3, \ldots Xn$ and normal on tasks $Y1, Y2, Y3, \ldots Yn$, where X and Y represent putative tests of a given cognitive function, and a second patient is normal on the set of tasks denoted X and impaired on those denoted Y. Unfortunately, many of the double dissociations reported in the neuropsychological literature are not usually of this latter kind.

Cerebral Circulation and Stroke

Although the clinical neuropsychologist has many tools with which to investigate the cognitive and behavioral manifestations of brain damage, the nature and locus of cerebral dysfunction is invariably beyond his or her control. With the obvious exceptions of surgical resection (e.g., lobectomy) and functional deactivation (e.g., sodium amobarbital suppression, magnetic stimulation), clinical neuropsychologists are restricted to working with patients in whom brain damage has already been determined by specific pathological mechanisms. The most common of these mechanisms is *cerebral stroke*. The World Health Organization (WHO) has defined stroke as "rapidly developed clinical signs of focal or global disturbance of cerebral function lasting more than 24 hours or until death, with no apparent nonvascular cause" (WHO MONICA Project, 1988, p. 108). Unlike many of the cerebral pathologies to be considered in this book, stroke typically causes focal brain damage that can be delineated with considerable accuracy using modern imaging techniques (see below). Our current understanding of normal and disordered brain function is based largely upon anatomoclinical studies of stroke patients. Therefore, in view of the special status of stroke in clinical neuropsychology, we provide here an introduction to the major categories of cerebrovascular disease, and consider the cognitive and behavioral impairments associated with involvement of specific arterial territories.

In view of the high metabolic demands of neural tissue, it is perhaps not surprising that even relatively brief departures from the normal parameters of regional cerebral blood flow (rCBF) may lead to cell death or *necrosis*. Destruction of many nerve cells in a localized region is known as cerebral *infarction*. There are several types of vascular pathology that may lead to cerebral infarction. Narrowing or occlusion of arteries due to thickening of the vessel wall, partial or total occlusion arising from an *embolus* (foreign material) or *thrombus* (obstructing clot), and abnormalities of the vessel wall due to deposition of *atheromatous plaques,* are usually classified as *ischemic* strokes. In certain circumstances blood vessels may rupture (*hemorrhagic* stroke), either as a consequence of hypertension, or from vessel wall abnormalities including *aneurysms* (outpouchings of the vessel wall) and *arteriovenous malformations* (local conglomerations of abnormal cerebral vessels). This leads to extravasation of blood into the subarachnoid space (*subarachnoid hemorrhage*) or into the brain parenchyma (*intracerebral hemorrhage*). In addition to these hemodynamic factors, alterations in the normal composition of the blood itself may result in tissue damage (Zülch & Hossmann, 1988).

Stroke pathogenesis determines the distribution and extent of infarcted tissue, and is therefore an important consideration in anatomoclinical investigations. Cerebral infarction after acute ischemia arises initially from local changes in rCBF followed by ischemia-induced cellular abnormalities in neurons and glial cells. Mild ischemia, such as that arising in patients resuscitated from cardiac arrest, may only affect "vulnerable" neurons such as those in specific zones of the hippocampus (Pulsinelli, 1992). Even complete occlusion of an artery will rarely abolish delivery of oxygen and glucose to the affected vascular territory. Compensatory flow from *collaterals* (parallel circulatory pathways) and *anastomoses* (reticular intercommunications between supply systems) may act to reduce the extent of damage. During a prolonged ischemic episode, the area of infarction begins in the region of lowest blood flow and spreads out toward its maximum extent. A narrow zone at the interface between the evolving infarct and surrounding normal tissue (the *ischemic penumbra*) contains neurons whose activity, although temporarily suppressed, may be returned to a normal level of functioning if the patient is treated with certain pharmacological agents within a few hours of symptom onset (Pulsinelli, 1992). Brain *edema,* which involves an accumulation of water and proteins in the damaged tissue, creates swelling in the affected area and surrounding structures. Edema is most marked in the days and weeks immediately after stroke, but resolves completely after two months in 70% of cases.

Although the pathogenesis of hemorrhagic strokes also involves ischemia, its effects are often less predictable. In subarachnoid hemorrhage, infarction tends to be insidious and diffuse (van Gijn, 1992). Intracerebral hemorrhages may dissect along major fiber tracts, decompressing surrounding tissue and raising intracranial pressure; herniation and distortion of brain structures (*mass effects*) are frequent concomitants. Local collections of intracranial blood (*hematomas*) may be surgically drained, but are often spontaneously resorbed over time. Residual cognitive and behavioral deficits tend to be less severe in patients with hematomas than in those with ischemic infarcts, because the former tend to disconnect or displace normal tissue, whereas the latter cause tissue destruction (Caplan, 1992).

It is critically important for the clinical neuropsychologist to be aware of the mechanisms underlying stroke, and their various consequences. In the early stages postinjury, a patient's clinical presentation is likely to reflect the combined effects of permanent tissue damage in the region of infarction, in addition to temporary dysfunction of tissue in the ischemic penumbra. The presence of widespread edema, raised intracranial pressure, and mass effects must also be considered when observing the signs and symptoms exhibited by patients shortly after stroke. Such acute effects may cause generalized deficits in arousal, attention, concentration, and speed of information processing. In many individuals, an unambiguous interpretation of cognitive and behavioral anomalies can only be made once these acute impairments have resolved.

Stroke Syndromes Associated with Lesions in Major Arterial Fields

The locus and extent of an infarct arising from localized ischemia is determined by the arterial territory (or territories) involved. The anterior (ACA), middle (MCA), and posterior (PCA) cerebral arteries, which supply blood to the entirety of the cerebrum, arise from the circle of Willis. In cases of severe arterial stenosis or occlusion, the size and locus of the resulting infarct is determined by the particular arterial branch involved, and by the distribution of collaterals and anastomoses in and around the affected region. Occlusion of the proximal portion of a major cerebral artery will cut supply to its entire field, resulting in an extensive infarct. However, if collateral supply from other arterial systems is available, the infarct will be restricted to the proximal part of the affected territory adjacent to the main stem. Vascular insufficiency in two or more arterial systems can lead to *watershed* or *border zone* infarcts, so-named because they occur at the junctions of arterial territories. A relatively common watershed infarct occurs in the occipitoparietal region at the junction of the ACA, MCA, and PCA territories. Other watershed infarcts may develop between the MCA and either the ACA or the PCA alone, or between the superficial and deep branches of the MCA, in the region of the putamen, caudate nucleus, and internal capsule. *Lacunar* infarcts (10–15 mm in diameter) occur in the most distant fields of small subcortical vessels, particularly around the centrum semiovale and at the periphery of the putamen and caudate nucleus (Zülch & Hossman, 1988).

Anterior Cerebral Artery

The ACA supplies a large part of the mesial surface of the cerebral hemisphere, including the frontal and parietal lobes, the corpus callosum, the head of the caudate nucleus, and the anterior limb of the internal capsule. Compared with the other major cerebral arteries, infarction involving the ACA territory is relatively uncommon (Hung & Ryu, 1988). Moreover, a patent anterior communicating artery tends to prevent infarction of the entire ACA territory, so that the majority of lesions arising from ACA involvement are restricted to only a portion of its territory.

Total ACA infarcts produce severe hemiplegia and some sensory loss on the contralesional side, affecting predominantly the lower limb. Bilateral and unilateral lesions of the entire ACA territory may produce *akinetic mutism,* in which the patient is apparently alert but does not move or speak, whereas unilateral damage of the dominant hemisphere may also result in *aphasia* (Chapter 2) and left limb *apraxia* (Chapter 10). Damage to the supplementary motor area (SMA), which occupies the mesial surface of the superior frontal gyrus, causes deficits in initiating and controlling voluntary movements. Dominant hemisphere SMA damage may result in a language disturbance characterized by a paucity of spontaneous speech, but with relatively preserved repetition and comprehension (*transcortical motor aphasia* or *dynamic aphasia*; see Chapter 2). Dominant hemisphere SMA damage has also been associated with bilateral *ideomotor apraxia* and a right *alien-hand* sign (Stuss & Benson, 1986). Damage to the anterior cingulate region of the nondominant hemisphere has been reported to produce attentional deficits such as *unilateral neglect* (Heilman & Valenstein, 1972; Posner & Petersen, 1990; see also Chapter 6). The pericallosal branches of the ACA supply the anterior four-fifths of the corpus callosum. Damage to this structure produces behavioral anomalies associated with disconnection of the cerebral hemispheres (see Chapter 4), including left-sided ideomotor apraxia, alien-hand syndrome, *agraphia,* and *tactile anomia.* Anterior cerebral artery infarctions involving the mesial and orbital frontal regions may result in deficits including personality change, disinhibition, and apathy (Cummings, 1993; see also Chapter 10). Infarction of hypothalamic and septal regions following rupture and repair of anterior communicating artery aneurysms often produces an impairment similar to *Korsakoff's psychosis,* including anterograde and retrograde amnesia, and confabulation (Volpe & Hirst, 1983; see also Chapter 8). Dominant and nondominant lesions of the caudate nuclei may produce, respectively, aphasia and unilateral neglect (see Chapters 2 and 6).

Middle Cerebral Artery

The MCA supplies most of the lateral convexity of the hemisphere, including the frontal, temporal, and parietal lobes, via its superior and inferior cortical branches. The superior division of the MCA supplies the inferior and middle frontal gyri, the pre- and postcentral gyri, and the anterior portion of the parietal lobe. The inferior division supplies the superior and middle temporal gyri, and the inferior parietal lobule (supramarginal and angular gyri). The penetrating (lenticulostriate) branches of the MCA supply the basal ganglia (BG) (putamen, caudate, and lateral aspect of the globus pallidus), corona radiata, and posterior limb of the internal capsule.

The MCA and its branches are the most commonly occluded in cases of major cerebral artery thrombosis and embolism. Occlusion of the main stem of the MCA without collateral flow will produce damage to structures supplied by both cortical and penetrating branches, resulting in disrupted functioning of almost an entire hemisphere. There is usually a dense contralesional hemiplegia and hemianesthesia involving the face, arm, and leg, in addition to contralesional visual field loss (*homonymous hemianopia*) and ipsilesional ocular deviation. Dominant hemisphere lesions will typically produce *global aphasia* (Chapter 2), whereas nondominant lesions produce unilateral neglect (Chapter 6).

Complete infarction of the superior division MCA territory will typically include the frontal convexity, pre- and postcentral gyri, and the anterior portion of the parietal lobe. There is a prominent hemiparesis that affects the face and arm more than the leg, with a corresponding sensory deficit. A contralesional visual field defect is usually absent because the optic radiations lie inferior to the lower reaches of this arterial field. As with MCA stem infarcts, those arising from superior division involvement in the dominant hemisphere may produce a predominantly "expressive" (Broca's type) aphasia and ideomotor apraxia (see Chapters 2 and 10), with corresponding damage to the nondominant hemisphere producing impaired prosody and unilateral neglect (Chapters 4 and 6). Infarction in the inferior division MCA territory will usually involve the inferior parietal lobule and the middle and superior temporal gyri. Visual field defects are characteristic; complete involvement of the optic radiation will result in homonymous hemianopia, whereas partial involvement may produce inferior (parietal) or superior (temporal) quadrantanopic defects. In contrast to superior division MCA lesions, primary motor deficits are either mild or absent in inferior division damage; if hemiparesis does occur, it may involve only transient weakness of the contralesional face and hand. Inferior division MCA damage of the dominant hemisphere is associated with language deficits such as *Wernicke's aphasia* and *conduction aphasia* (see Chapter 2). Other deficits associated with such dominant hemisphere lesions include agraphia, left–right disorientation, acalculia, and finger agnosia (*Gerstmann's syndrome*; see Chapter 4). Corresponding damage of the nondominant hemisphere is frequently associated with unilateral neglect, and occasionally with *anosognosia*, constructional and dressing apraxia, and agitated delirium (see Chapters 5 and 6).

Infarctions in the territory of the lenticulostriate branches of the MCA are typically of the lacunar form and affect the BG, posterior limb of the internal capsule, and corona radiata (*striatocapsular infarcts*; Bladin & Berkovic, 1984). Lesions restricted to the internal capsule produce "pure motor hemiparesis." This may be accompanied by dysarthria, but there are typically no other sensory or cognitive deficits. Aphasia and unilateral neglect are often associated, respectively, with dominant and nondominant striatocapsular infarcts. However, recent evidence suggests that there is no specific BG–internal capsule site associated with their occurrence, and that disruption of distant cortical structures may in fact be responsible (Weiller et al., 1993).

Posterior Cerebral Artery

Unlike the ACA and MCA, which arise from the internal carotid artery, the PCA is part of the vertebrobasilar system. Each PCA arises from the basilar artery at the junction of the pons and mesencephalon, its branches supplying lateral portions of the thalamus, the inferior surface of the temporal lobes (including the hippocampus, parahippocampal, and fusiform gyri), most of the occipital lobe, and the splenium of the corpus callosum. Whereas bilateral infarction is relatively rare in the ACA and MCA territories, it occurs somewhat more frequently in the PCA territories because of their common (basilar) origin. Because the PCA and its branches supply the lateral geniculate nucleus, the geniculocalcarine tract, and the striate cortex, visual deficits are frequent concomitants of PCA territory infarcts. Homonymous visual field defects occur without unilateral neglect, unless the lesion encroaches upon the parietal lobe. Such patients are acutely aware of their visual loss, and tend to make compensatory head and eye movements toward the hemianopic side.

Cortical damage sparing the geniculostriate pathways may result in impaired perception of shape, size, and color, with preserved motion perception. Visual perseverations are occasionally reported by patients, in which an object initially in central vision is projected into the hemianopic field. Visual images may also persist after the stimulus has left the patient's visual field (*palinopsia*, see Chapter 5). Occipitotemporal lesions may cause *metamorphopsia* (distortion of visual images); objects may appear larger (*macropsia*) or smaller (*micropsia*) than they really are, and distortions may be limited to one visual field (e.g., *hemimicropsia*, L. Cohen, Gray, Meyrignac, Dehaene, & Degos, 1994; see also Chapter 5). Abnormalities of color perception may occur following bilateral or unilateral damage of the calcarine cortex or lingual gyrus; impaired color naming in the context of normal perception may also occur, and is invariably associated with *alexia* (Caplan, 1988; see also Chapters 3 and 5).

Dominant hemisphere PCA infarcts involving the posterior forceps of the corpus callosum may cause *alexia without agraphia* ("pure alexia"), where reading is impaired despite intact expressive language (see Chapter 3). This disorder may or may not be accompanied by *visual agnosia*, in which the patient cannot name objects or pictures by sight, even though a normal verbal description of the object can be provided in response to the object name. Patients with visual agnosia can also recognize objects by touch and sound (see Chapter 5). Lesions of the thalamic or mesial temporal regions of the dominant hemisphere may result in severe *anterograde amnesia* (Chapter 8), whereas lesions involving the occipitotemporal region are associated with *transcortical sensory aphasia*, in which speech is fluent and paraphasic, comprehension is poor, but repetition is intact (see Chapter 2). Nondominant hemisphere PCA damage has been associated with unilateral neglect (when lesions encroach on the parietal area), constructional apraxia, and agitated delirium, the latter having been related to infarcts of the lingual and fusiform gyri (Caplan, 1988).

Extensive bilateral PCA damage is associated with the characteristic triad of cortical blindness, anterograde amnesia, and agitated delirium; this pattern typically occurs after occlusion of the basilar artery from a cardiac embolism. Such patients will occasionally deny their blindness. *Balint's syndrome* also results from bilateral infarction, though in this case the damage is distributed in patches throughout the calcarine and occipitoparietal regions (Caplan, 1988). Again, the syndrome is characterized by a triad of symptoms including *simultanagnosia* (narrowed or restricted attention), *optic ataxia* (impaired hand–eye coordination), and *gaze apraxia* (impaired voluntary eye movements). *Prosopagnosia* (an inability to recognize familiar faces) may occur after bilateral lesions of the mesial surfaces of the occipital lobes, though the critical lesion may involve the nondominant hemisphere (see Chapter 5).

Neuroimaging Techniques

One of the principal roles of early clinical neuropsychologists was to predict the likely location of brain pathology on the basis of behavioral data. However, contemporary clinicians and researchers have access to a variety of advanced neuroimaging techniques that usually render irrelevant the need for behavioral assessment solely for the purpose of lesion localization. These neuroimaging methods, which we shall summarize here, can be subdivided according to whether they map brain *structure* or brain *function*.

Structural Techniques

The most well-established and widely used of the modern neuroimaging techniques is *computerized tomography* (CT). CT scanning provides a visual reconstruction of body (in this case cranial) morphology by measuring tissue density. The technique involves passing a highly focused X-ray beam through the head and measuring its strength as it emerges from the other side. The denser the tissue through which the X-ray beam passes, the more energy it absorbs. By passing X-ray beams through the head from many different angles on a single plane, a "slice" of brain tissue can be reconstructed. A series of horizontal slices can be obtained at regular intervals, thereby imaging successively the morphology of the entire brain. In a standard CT image, hyperdense structures such as bone appear white, hypodense structures such as the ventricles appear black, and the various neural tissues appear different shades of gray between these extremes. Cerebral pathology is detected as a change in the normal density of specific brain structures, and by local and diffuse anomalies in the size and shape of these structures. The sensitivity of CT scanning can be augmented by intravenous injection of an X-ray opaque contrast medium.

Although the technology for CT scanning has been available since the early 1970s, another technique used to map brain morphology, *magnetic resonance imaging* (MRI), is a relatively recent development. Unlike CT, MRI involves detecting the radiowaves emitted by nuclei (particularly the proton nucleus of the hydrogen atom) that have been made to resonate at their characteristic frequency by exposing them to a strong magnetic field. A brief radio-frequency electromagnetic pulse is applied to perturb these nuclei; detectors then measure the radiowaves emitted by these perturbed nuclei as they "relax" back to a steady state, thereby allowing a computer to reconstruct a visual image. Conventional MRI detects the resonance of protons present in water molecules, which themselves occur in different densities throughout different anatomical regions of the brain. The advantage of MRI over CT lies in its high spatial resolution; standard MRI is able to resolve structures 1–2 mm in diameter, though some authors have suggested that its resolution may improve to 100 μm (M. Cohen & Bookheimer, 1994). Another advantage of MRI is that it can detect ischemic changes in brain tissue within 45 min post-onset, whereas CT can only detect the full extent of tissue damage after several days (Donnan, 1992). Finally, with MRI the development of new technologies has permitted dynamic imaging of *functional* activity with high spatial resolution (see below).

Functional Techniques

In studying brain–behavior relationships, we are ultimately concerned with documenting the neural mechanisms underlying both simple and high-level behavior. Although structural neuroimaging techniques provide information on brain morphology and the extent of tissue damage in patients, they contribute little to our understanding of the *functioning* of the normal or damaged brain. Fortunately, there are several alternative neuroimaging techniques that provide a window on the complex changes in neural activity that accompany human behavior. One of these techniques is PET. As with other measures of functional brain activity, PET is based on the assumption that an increase in the level of neuronal activity in a particular region will be accompanied by a concomitant increase in physiological indices, such as rCBF, regional glucose metabolism, and regional oxygen consumption. PET is reliant on radioactive isotopes (e.g., ^{15}O,

^{11}C, ^{13}N) that emit positrons as they decay. These isotopes are administered intravenously as radioactively labeled water; the amount of radioactive material accumulated in various regions of the brain increases as a function of the level of local activity. Positrons emitted by the decaying isotope collide with electrons; the resultant "annihilation" produces two gamma rays that project in nearly opposite directions. These gamma rays are detected by devices located in an array surrounding the imaged structure, from which a computer determines their point of origin and constructs a visual image.

Measures of rCBF have been suggested to provide the most reliable indication of brain function (Raichle, 1994). However, PET is also able to measure the transport and metabolism of specific substances (e.g., amino acids, free fatty acids); specific receptor ligands labeled with positron-emitting isotopes have also been used to map the distribution of binding sites of neurotransmitters including dopamine, serotonin, and benzodiazepine (Prichard & Brass, 1992). PET is also useful in documenting regions of abnormal functioning in patients with focal neurological disorders such as stroke, epilepsy, and tumor, and in those with more diffuse pathology, such as dementia and affective disorders. Several recent studies have used PET in healthy individuals to measure regional cerebral activity during specific cognitive tasks. These images are compared with control scans taken in a nonactive or "baseline" state. Blood flow measures from the latter are then subtracted from those obtained in the "activated" state, thereby showing the regions of activation associated with the particular cognitive task (Raichle, 1994). Subtractive methodologies of this kind have been used in studies of perception, attention, memory, motor control, language, and visual imagery (see Chapters 3, 5, 6, 8, and 10).

One of the many advantages of PET is that it uses radioactive isotopes that have a relatively short half-life (commonly around 2 min), allowing several images to be obtained from a single individual. The disadvantage of such a short half-life is that there must be ready access to a cyclotron, making the total cost of PET prohibitive. A more affordable alternative for imaging rCBF is *single photon emission computed tomography* (SPECT). This method uses photon-emitting isotopes that have a relatively long half-life and can therefore be produced commercially at distant sites. Another advantage of SPECT is that once the radioactive isotopes enter brain tissue, their metabolites remain trapped for several hours. Thus, images reflecting rCBF at the time of injection can be taken up to several hours later (Prichard & Brass, 1992). The uses of SPECT are similar to those of PET, although the long half-life of SPECT isotopes means that this technique is generally unsuitable for activation studies. In addition, the spatial resolution of SPECT is somewhat poorer than that of PET.

Finally, recent technological advances have permitted researchers to obtain images of *functional* activity using MRI. Functional MRI (fMRI), like PET and SPECT, is based on the measurement of changes in rCBF that accompany local variations in the level of neural activity. Increases in rCBF exceed the oxygen demands of cells in active areas, thereby resulting in a net increase in local blood oxygen levels. The amount of oxygen bound to hemoglobin affects the magnetic properties of the latter, and rapid MRI methods are sensitive to these hemodynamic changes. There are several advantages of fMRI. First, the signal is obtained from the blood, which acts as an endogenous contrast agent; injection of radioactive isotopes is therefore unnecessary. Second, fMRI provides functional images that can be used in conjunction with structural images from static MRI, thereby permitting extremely accurate localization of activity. Finally, fMRI has a relatively high spatial resolution (distinguishing structures approximately 1–2 mm in diameter), though this is unlikely to be as high as that obtained using static MRI techniques.

Because the fMRI signal is dependent upon changes in rCBF, its temporal resolution is comparable to that obtained using electroencephalographic (EEG) recording, and inferior to that obtained from single-unit recordings. It has been predicted that fMRI methods will eventually be able to resolve temporal signal changes on the order of a few tens of milliseconds (M. Cohen & Bookheimer, 1994). Despite its relatively recent arrival, fMRI has already been used to study patterns of neural activity accompanying primary sensory and motor functions, language, and visual imagery (M. Cohen & Bookheimer, 1994).

Before concluding this section on functional neuroimaging, it seems pertinent to examine some of the assumptions underlying such techniques and to identify their possible limitations. Perhaps the most critical assumption made by researchers using PET, SPECT, and fMRI is that changes in rCBF reflect corresponding changes in the level of neural activity. Although this assumption is probably valid, increases in neural activity may correspond with either excitation or *inhibition* of local activity. Thus, increases in rCBF do not necessarily imply the direct contribution of a given area to a specific function. On the other hand, the absence of changes in rCBF does not necessarily imply that a region is not involved in mediating a particular function; a combination of increases and decreases in rCBF during task performance may cancel each other out, resulting in no overall change (Sergent, 1994).

So-called activation studies involve an additional set of caveats. First, researchers must ensure that apparently trivial procedural variables (e.g., the wording of instructions given to subjects) are held constant: even minor variations may lead to subjects adopting different cognitive strategies. Second, subjects' affective states also require careful monitoring, because transient mood changes (e.g., anxiety) have been shown to produce characteristic alterations in rCBF (Sergent, 1994). Finally, because the volume of blood supplying the brain at any given moment is finite, any task-induced increase in blood flow to one region must induce a concomitant reduction in blood flow to other regions (Sergent, 1994).

To conclude, although structural and functional neuroimaging techniques such as CT, MRI, PET, SPECT, and fMRI make a unique contribution to our understanding of brain–behavior relationships, this contribution must be considered in the context of acknowledged limitations. Any conclusions derived from neuroimaging studies should ideally be considered in the context of behavioral data obtained both from normal healthy subjects, and from brain-damaged patients. In the chapters that follow, we will endeavor to fulfill this ideal.

Neural Networks

One recent development in clinical neuropsychology is the use of computer models to simulate the operations of the normal and damaged human brain. *Neural network modeling* involves the use of computer programs designed to simulate the activity of networks of neurons, in which the number of individual processing units, their firing thresholds, and their connections with other units, are programmed to perform a particular task. The theoretical background to neural network modeling can be traced back to the work of William James (1890), who described how cognitive processes might be connected or associated in terms of their temporal contiguity, and to the work of Donald Hebb (1949), whose neurophysiological concept of reverberatory activity suggested a means by which communications between individual nerve cells might be modified by prior patterns of activation.

Neural network modeling is based on the premise that cognitive processes are mediated by widely distributed networks of cells operating in parallel (so-called *parallel distributed processing*), with considerable interaction between the inputs and outputs of each component. This approach is therefore conceptually distinct from earlier models of cognitive functioning, in which information was considered to move from one discrete processing module to another in a serial or hierarchical fashion, with minimal interaction between modules. Parallel distributed processing models come much closer to simulating the activity of the brain than serial models, because they are able to handle extremely complex operations very quickly (Parks et al., 1991). Thus, researchers can readily compare the data from their neural network models with those obtained from, for example, in vivo neurophysiological recordings. In certain applications, specific receptor and synaptic properties, in addition to the pattern of connections between neurons, may be specified (for review, see Parks et al., 1991).

Although a detailed account of neural network modeling is beyond the scope of this chapter, a brief explanation of how such models are developed should illustrate its essential components. Neural networks are composed of "units" that represent individual neurons. The simplest networks are composed of three interconnected layers of such units. Units in the *input layer* represent information fed into the network. Units in the *hidden layer* represent the combined output from those input units with which they are connected, as a function of the "weights" attached to these connections. Weights specify the connection strength between individual units, thereby accounting for excitatory and inhibitory interactions between neurons. Similarly, units in the *output layer* represent the activity of the hidden units with which they are connected, and their associated weights. By modifying the activity of input and output units and connection weights, researchers can examine how the hidden units react. Training of neural networks in this way may proceed in one of two ways: in *supervised learning,* an external agent adjusts the various connection weights to optimize the match between the actual and desired outputs. This kind of learning is unlikely to resemble the functioning of the brain, in which learning occurs spontaneously, without the need for external monitoring. A more realistic alternative is *unsupervised learning,* where connection strengths are adjusted according to prespecified rules.

The applications of neural network models to clinical neuropsychology are manifold. They have been applied to investigations of cognitive processes such as pattern recognition, attention, learning, and memory (J. Cohen, Romero, Farah & Servan-Schreiber, 1994; Mozer & Behrmann, 1990; Parks et al., 1991; Zipser & Andersen, 1988). Some recent investigations have also examined the effects of "lesioning" neural networks by deleting specific modules from the computer program (e.g., J. Cohen et al., 1994). The output from such lesioned networks can then be compared with the performances of real patients. It may even be possible to use the results of such simulation studies to guide the development and implementation of rehabilitative techniques (Margolin, 1991).

Summary and Conclusions

The purpose of this introductory chapter was to familiarize the reader with an overview of some important conceptual and methodological issues in clinical neuropsychology. We began by distinguishing between clinical neuropsychology, which we believe is best conceptualized as being concerned with elucidating the brain mechanisms underlying both abnormal and normal behavior, from the related disciplines of neuropsychiatry and behavioral neurology, which are perhaps motivated more by a desire to understand disordered behavior in terms of its implications for diagnosis and treatment. We then considered the assumptions of modularity and locality that underlie most contemporary theorizing in clinical neuropsychology, and suggested ways in which their potential shortcomings may be minimized.

We also explored the different approaches to collecting and interpreting data in clinical neuropsychology; the syndrome approach, it was argued, is valuable not only as a convenient shorthand for communicating the nature of patients' deficits, but also as a heuristic for classifying patient groups for research and for generating hypotheses regarding patterns of normal and disordered behavior. On the other hand, there are those who suggest that syndromes are merely theoretically arbitrary categories. Indeed, it is unlikely that any two patients, however similar they may appear clinically, will have exactly the same functional lesions; thus, group studies of patients exhibiting superficially similar deficits are likely to obscure important underlying differences between individuals, and perhaps even lead to erroneous conclusions about the likely organization of normal cognitive processes.

Single-case studies are more likely to avoid such artifactual conclusions, though they have their own limitations in terms of replicability and in terms of their capacity to inform us about the brain regions involved in specific cognitive functions. We believe that neuropsychological syndromes are useful in providing a conceptual framework within which to understand disordered human behavior, whereas careful examination of individual patients may help to elucidate the functional architecture of the normal brain. In seeking to understand the organization of normal cognitive processes, clinical neuropsychologists are guided by specific patterns of association and dissociation between observed deficits. Double dissociations in particular provide the crucial data upon which our understanding of the organization of the normal cognitive system is ultimately based.

Just as knowledge of these basic principles is an important prerequisite for conceptualizing the behavioral anomalies shown by brain-lesioned patients, so is an understanding of the nature of the causative pathologies and their likely distribution in the brain. A substantial proportion of what we know about brain–behavior relationships stems from studies of patients with lesions caused by cerebrovascular accidents. The distribution of damage in such stroke patients is determined by the cerebral arterial vasculature, with distinct behavioral syndromes arising from involvement of each of the major arterial fields. Structural neuroimaging techniques such as CT and MRI allow mapping of the locus and extent of neural tissue necrosis, whereas functional techniques such as PET, SPECT, and fMRI document the distribution and extent of physiological changes induced by a given pathological process. Although functional neuroimaging techniques are constrained by their own methodological limitations, their combined application with traditional lesion studies promises new insights in understanding the neural basis of normal and disordered behavior.

Finally, we discussed the use of artificial neural networks for modeling the kinds of computations that are likely to underlie certain elementary cognitive processes. Such procedures, which are based upon connectionist or parallel distributed processing models, may also be used to simulate the effects of brain damage, thereby providing another source of data with which to constrain models based on clinical lesion studies. In conclusion, clinical neuropsychology has enjoyed an extraordinary period of growth over the last two decades; given the broad scope of this book, our coverage of the relevant literature is therefore of necessity selective. In the chapters to follow, we will often appeal to recent synthetic reviews, rather than referencing all the relevant original studies. We will, however, supplement each treatment with a wide range of relevant new work from the last 5 years. The result, we hope, will enable the reader to appreciate the most recent advances in clinical neuropsychology in the context of a broader historical perspective.

Further Reading

Cohen, M. S., & Bookheimer, S. Y. (1994). Localization of brain function using magnetic resonance imaging. *Trends in Neurosciences, 17,* 268–277.

Cummings, J. L., & Hegarty, A. (1994). Neurology, psychiatry, and neuropsychiatry. *Neurology, 44,* 209–213.

Farah, M. J. (1994). Neuropsychological inference with an interactive brain: A critique of the "locality" assumption. *Behavioral and Brain Sciences, 17,* 43–104.

Gardner, D. (Ed.) (1993). *The neurobiology of neural networks.* Cambridge, MA: MIT Press.

Hinton, G. E. (1992). How neural networks learn from experience. *Scientific American, 267,* 104–109.

McCloskey, M. (1993). Theory and evidence in cognitive neuropsychology: A "radical" response to Robertson, Knight, Rafal, & Shimamura. *Journal of Experimental Psychology: Learning, Memory, and Cognition, 19,* 718–734.

Posner, M. I. (1993). Seeing the mind. *Science, 262,* 673–674.

Robertson, L. C., Knight, R. T., Rafal, R., & Shimamura, A. P. (1993). Cognitive neuropsychology is more than single-case studies. *Journal of Experimental Psychology: Learning, Memory, and Cognition, 19,* 710–717.

Sergent, J. (1994). Brain-imaging studies of cognitive functions. *Trends in Neurosciences, 17,* 221–227.

Shallice, T. (1988). *From neuropsychology to mental structure.* Cambridge: Cambridge University Press.

Vallar, G. (1991). Current methodological issues in human neuropsychology. In F. Boller & J. Grafman (Eds.), *Handbook of neuropsychology* (Vol. 5, pp. 343–378). Amsterdam: Elsevier.

Chapter 2

Disorders of Speech

The Aphasias

IN THE BEGINNING WAS THE WORD.

—JOHN 1:1

The terms *communication, speech, language,* and *vocalization* should not be used interchangeably, though in practice it is often not easy to maintain certain otherwise convenient distinctions. Although communication may refer to the general ability to influence other conspecifics (usually) via signals, language is a spontaneously acquired form of symbolic communication. Via symbols transmitted with gestures and vocalizations, language can convey perceptions, intentions, impressions, and actions. Speech is an articulatory manifestation of language. The fact that so much of the human brain is somehow involved in language, far more than just the primary speech areas, suggests that language plays more than just a communicatory role; it may in fact be deeply involved in how we model reality (Jerison, 1986).

Language is traditionally a predominantly left-hemisphere (LH) phenomenon, though currently a right-hemisphere (RH) contribution is increasingly evident. There are perhaps six aspects to language (Cutting, 1990), which are differentially lateralized to the two sides of the brain. These include the following:

1. *Phonology,* a largely categorical LH aspect involving the speech-related sounds (phonemes, the smallest sound differences that distinguish meaningful utterances). This aspect emphasizes the relationships between the positions adopted by the articulators and the resulting speech sounds.

2. *Morphology,* the denotative significance of meaningful units (words or word fragments, such as morphemes, which are the smallest individually meaningful elements). Thus phonemes are combined by rule into morphemes, largely under LH control.

3. *Syntax,* the grammatical rules to combine morphemic strings into uniquely meaningful propositions, again largely under LH control.

4. *Semantics,* the formal meaning of a string of morphemes, to which the LH may contribute denotative meaning, and the RH may contribute some connotative or contextual meaning or associations.

5. *Prosodic or suprasegmental aspects,* involving phonological variations within categorical limits, which do not change denotative meaning, to provide extra, interpretative, connotative meaning. Thus via changes in pitch, rhythm, stress, intonation, posture, and expression, a largely RH contribution may be made to the broader aspects of meaning.

6. *Pragmatic aspects,* the practical use to which we may put language, as with jokes, irony, metaphor, sarcasm, or context (i.e., the point of the discourse), to which the RH largely contributes. Thus RH damage may lead to problems in getting the point or problems with humor or connotative comprehension.

Our capacity for language is largely innate, even though its actual realization is highly specific, and reflects the speech of caregivers and experiences at critical developmental milestones. It is often learned against considerable odds (e.g., sensory impairments and flawed or defective models). This combination of innateness and learning, together with hemispheric specialization, is reminiscent of bird song (Bradshaw & Rogers, 1993), where male offspring learn the song of their conspecifics, if exposed at appropriate critical periods, via largely LH mechanisms.

Morphological Asymmetries in the Brain

Language lateralization to the LH has anatomorphological correlates in gross asymmetries that are often visible to the naked eye, and are present even in the fetus, with preadaptations apparent even in apes (Bradshaw & Rogers, 1993). Thus the left lateral ventricle is generally the larger, especially in males, and the anterior portion of the RH and the posterior portion of the LH protrude farther, and are wider, than their counterparts, especially in dextrals (see Figure 1), giving the brain an overall counterclockwise torque (Bear, Schiff, Saver, Greenberg, & Freeman, 1986). Asymmetries in the peri-Sylvian speech-related cortex have been observed for a century (Galaburda, 1984). Thus the Sylvian fissure, separating the temporal from the parietal regions, is longer in the LH and continues further in a horizontal direction on that side, before tending upwards (see Figure 2). Thus the temporal and parietal opercula, respectively constituting the floor and roof of the posterior portion of the Sylvian fossa, are also larger on the left (Geschwind & Galaburda, 1985). Indeed that part of the temporal operculum known as the planum temporale, a triangular region caudal to the first transverse auditory gyrus of Heschl on the superior surface of the temporal

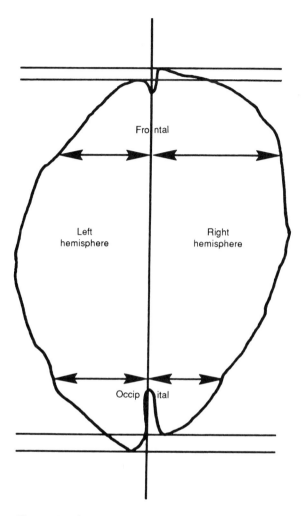

Figure 1 ASYMMETRIES OF FRONTAL AND OCCIPITAL REGIONS. THE WIDER RIGHT FRONTAL AND LEFT OCCIPITAL REGIONS AND THE MORE EXTENSIVE LEFT OCCIPITAL AND RIGHT FRONTAL POLES GIVE THE BRAIN AN OVERALL COUNTERCLOCKWISE TORQUE. (FROM BRADSHAW & NETTLETON, 1983.)

lobe, is larger on the left (see Figure 3); Heschl's gyrus itself, the primary auditory projection area, generally consists of two transverse gyri on the right and one (a longer one) on the left. Moreover, the central portion of Heschl's gyrus is more granular on the left, with a greater center-to-center distance on that side between cell clusters (Kertesz, 1991). Thus these posterior LH neurons may react more selectively to speech signals than those on the right.

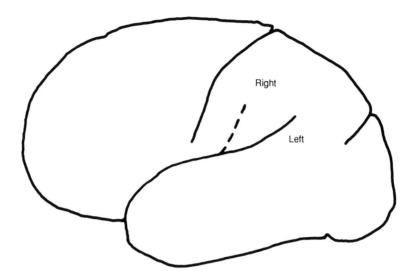

Figure 2   Left (solid) and right (broken line) Sylvian fissures, the right shown superimposed (and mirror-reflected) upon the appropriate part of the left hemisphere. The Sylvian fissure is generally longer on the left, but terminates at a lower level than the right. (From Bradshaw & Nettleton, 1983.)

Cytoarchitectonic differences in fact may be stronger correlates of *frontal*
asymmetry than gross morphological differences, although the frontal operculum may be more folded on the left in speech-related areas, with larger architectonic areas of the pars opercularis and pars triangularis on that side (Geschwind
& Galaburda, 1985).

Other Evidence for Evolutionary Adaptations for Language

Our vocal tracts, of course, are also adapted for the rapid production of a
wide range of easily discriminable speech sounds (Lieberman, 1989). This is
achieved by ultrarapid movement of the articulators, in parallel, toward target
positions. This adaptation of our vocal tracts for speech is at the cost of our being unable simultaneously to breathe and swallow, something that most mammals, and even human infants, before the larynx descends to its adult position,
can achieve. The enormously rapid rate of information transmission in speech is
both peripherally, in terms of the articulatory apparatus, and centrally, with respect to the requisite speech centers in the brain, beyond the ability of our near-
est relatives, the chimpanzees. The latter, however, especially the pygmy chimpanzee (*Pan paniscus*) or bonobo, is capable of understanding quite complex

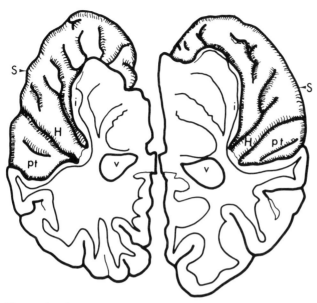

Figure 3 STANDARD ASYMMETRY OF THE PLANUM TEMPORALE (PT), THE ROUGHLY TRIANGU-LAR REGION LYING POSTERIOR (INFERIOR IN THE DIAGRAM) TO THE TRANSVERSE AUDITORY GYRUS OF HESCHL (H), AND BOUNDED LATERALLY BY THE SYLVIAN FISSURE (S). [NOTE: VENTRICLES (v) AND INSULA (i).]THIS PATTERN OF ASYMMETRY IS SEEN IN APPROXIMATELY TWO-THIRDS OF NORMAL HUMAN BRAINS. (FROM GESCHWIND & GALABURDA, 1987.)

messages by sign or even speech, so the species' limitation is more at the output end. Indeed, rapid reading suggests that we can process semantic information via the visual modality at an even faster rate. Thus our central nervous system (CNS) must contain spare capacity even beyond the demands of articulation, and we can only speculate about the evolutionary implications of such an observation.

Clearly, we are evolutionarily adapted for speech. Other evidence for such a proposition comes from finding that certain language disorders are genetically transmitted in families (e.g., delayed speech onset, articulatory difficulties, developmental dysphasia, reading difficulties, and problems with grammar). Often there may be associated anomalies of LH development; Galaburda (1994) provided just such an example in the context of developmental dyslexia. Williams syndrome (Bellugi, Bihrle, Jernigan, Trauner, & Doherty, 1990), on the other hand, exhibits a rare fractionation of higher cortical functioning, with selective preservation of complex syntax in the face of marked and severe cognitive deficits. There are critical periods for language acquisition, at infancy and possibly also puberty (Fromkin, Krashen, Curtiss, Rigler, & Rigler, 1974) with stages (babbling, one- and two-word utterances, the appearance of syntax) common to all races and even to congenitally deaf infants learning sign language (Stoel-Gammon & Otomo, 1986). The infant almost effortlessly acquires language, often from grossly flawed models and in very imperfect circumstances, whereas the young adult finds learning a second language a far harder task than the child, and one requiring quite different strategies. Moreover, infants at birth are sensitive to a broad range of speech sounds, and for the first few months can learn correctly to pronounce sounds in any language, a faculty that is later lost.

As Goodglass (1993) observed, every voluntary speech act normally initiates an interlocking pattern of neuromuscular activity. We are most aware of the finely adjusted movements of the tongue and lips as they rapidly move to their target positions. We are not normally aware of the uvula as it opens or blocks nasal passages to create distinctions between nasal and oral phonemes. Other muscle systems, of which we have partial voluntary control and relatively little awareness, control voicing and the respiratory patterns that modify the volume and phrasal patterns of speech. Yet others control the tension of the vocal cords to modify the melodic shape of our utterances. Given the degree of precision and timing required, and achieved partly through various systems (exteroceptive and proprioceptive) of sensory feedback, it is surprising, as Goodglass (1993) noted, how robust the system is in the face of possible damage.

Many acquired speech difficulties that do not involve disorders of planning (the true aphasias) come under the rubric of disorders of programming (speech apraxias) and of execution (dysarthrias). Oral or buccofacial dyspraxia encompasses problems with purposeful movements that involve the facial muscles in response to commands such as, "Purse your lips!" or "Pretend to blow out a candle!", while the patient may still be able spontaneously to do these things in an appropriate context (see Chapter 10). With speech apraxia there are problems in producing an orderly sequence of phonemes, even though individual phonemes may be correctly achieved, and overlearned speech automatisms or formulaic utterances may be preserved (see Chapter 4). Thus the damage cannot lie at a merely peripheral or articulatory level. As with all sequencing problems, oral or manual, the LH is predominantly involved. With dysarthria, anarthria, or motor speech difficulty, there are difficulties in clearly generating the articulatory gestures for individual phonemes, with repercussions upon phonation, articulation, and prosody. Unlike many of the specific aphasias (below), the conditions are little affected by the automaticity or familiarity of the utterance, or the presence or absence of a model to copy as in repetition. There is presumably damage to the motor speech pathways themselves, including the primary motor-sensory cortex, pyramidal, extrapyramidal, and spinal nerve tracts, cerebellum, and basal ganglia (BG). Depending on the locus, there may be varying degrees of weakness in the muscles for speech and facial movement. Both hemispheres may differentially contribute, with LH dysphonia resulting in weak, slurred, or imprecise articulation, and RH aprosodia resulting in a flat monotone without inflections (although see Ovellette & Baum, 1993, and below). Prosody of course conveys important "suprasegmental" information (i.e., involving units greater than one phoneme in length), via changes in stress patterns, word grouping, melodic contour, or voice volume imposed on words, phrases, or sentences.

Goodglass (1993) noted that the same errors of articulatory realization may result from dysarthria, apraxia, or aphasia, and that we can only make such distinctions on the basis of whether the disorder is sensitive to the parameters of linguistic automaticity, familiarity, or communicatory load. Dysarthria, apraxia, and aspects of Broca's aphasia (see below) can all result in changes in voicing (cf. the distinction between "ba" and "pa," or "da" and "ta") or in reductions in phonetic complexity. The latter may involve reduction of consonant blends (e.g., "pla" and "tra") or of combinations (e.g., "sta" or "spa") by omission of the continuant sound, with retention only of the stop consonant. Thus play → p'ay, train → t'ain, and spoon → 'poon. Vowel changes, less common than consonant changes, may cause the impression of speech with a foreign accent, via subtle changes in articulation and prosody. We might at this point note that the most important cue between voiced and unvoiced consonants (see above) is the timing (voice onset time, VOT) between release of closure of the vocal tract, and commencement of vocal-cord vibration for the following vowel. When VOT is less than 30 ms, we hear the speech signal as voiced, and as unvoiced when longer than 30 ms. Nor can we perceive the difference between two such tokens if both lie on the same side of this 30-ms boundary, only if they straddle it. This phenomenon of "categorical perception," however, may not be unique to humans or even primates (Bradshaw & Rogers, 1993).

Aphasia: General Considerations

The borderlines between speech apraxia and dysarthria and true disorders of planning (the aphasias) are, as we have just observed, often blurred, though the nature of the speech disregulation invariably involves LH dysfunction. In the discussion that follows we shall largely follow the aphasia taxonomy that originated with Wernicke, and was adopted by Geschwind and the Boston school of aphasiology.

Aphasia is a disturbance in language formulation or comprehension caused by a breakdown in the two-way process mediating between thought and language (Mesulam, 1990). It may relate to auditory or vocal, written, or signed material at the levels of morphology (phoneme sequence), lexicality (meaning), or syntax (grammar). It is not a sensory disorder (deafness, low vision), a perceptual problem (agnosia), a disorder of movement (dysarthria, apraxia), or of thought (schizophrenia, autism). Language and speech depend not just on a few discretely localized regions, but on the dynamic interplay between extended neuronal networks involving various cortical and subcortical regions. Thus we must be wary of trying to localize a particular linguistic symptom to damage to a single structure. Indeed, brain damage rarely respects gyral, sulcal, cytoarchitectonic, or anatomical boundaries; its effects may be more or less transient, and more or less severe as a function of size of lesion, age of onset, whether it was abrupt or slowly progressive, sex and handedness of the patient, and even educational background.

The aphasias are frequently, but by no means universally, cortical, largely but not exclusively LH, and consequent upon vascular damage, tumor, or trauma. Other cognitive and intellectual skills are typically involved (e.g., praxis and gnosis). The relationship of acquired speech and language difficulties to brain damage has been known from the days of ancient Egypt, and the Edwin Smith papyrus of the sixteenth century B.C.:

> . . . one having a smash in the temple; he discharges blood from his nostrils and his ear; he is speechless and has a stiff neck. (Lovell, 1993, p. 52)

There have been hints since then that the LH might be involved, as Lovell (1993) documented:

> . . . right side was paralysed and he had difficulty in speaking: he substituted one word for another so that his attendants had difficulty in determining what he wanted. (J. Schmidt, 1676, cited by Lovell, 1993, p. 52)

However it was not until the 1860s that Paul Broca enunciated his famous dictum:

> Nous parlons avec l'hémisphère gauche. (P. Broca, 1865, cited by Lovell, 1993, p. 55)

We should however still heed the caution of John Hughlings Jackson a few years later:

> To locate the damage which destroys speech, and to localize speech, are two different things. (Hughlings Jackson, 1871, cited by Lovell, 1993, p. 55)

Broca initially described the eponymously named Broca's aphasia, associated with left frontal damage, from the disabilities and subsequently determined lesions in the brain of his patient Tan, who had been able to utter little more than "Tan" after illness onset (Broca, 1861). Broca was revolutionary in proposing that different cortical regions and hemispheres may have different functions, an idea that Henderson (1990) noted was earlier associated with the phrenology of Franz Gall, and later discredited by the elegant animal studies of Pierre Flourens. However, we should note that Gustave Dax in 1863, 26 years after the death of his father, produced a manuscript purportedly prepared by Marc for presentation before a Montpellier medical congress in 1836, entitled "Lesions of the left side of the brain coinciding with the loss of signs for thought." There is, however, no proof that it was actually delivered (Henderson, 1990; though see Joanette, Goulet, & Hannequin, 1990). Although Broca favored the term *aphemia,* others proposed such synonyms as *alalia* and *aphasia,* which eventually triumphed. Aphemia is now reserved for *articulatory* dysfunction, with slow effortful speech that is nevertheless grammatic, unlike true Broca's aphasia (below). Lalia survives in *palilalia,* a pathological repetition of heard questions or statements.

After Wernicke (1874) described another acquired speech disability, this time consequent upon posterior damage, Lichtheim (1885) attempted the first connectionist taxonomy of what came to be known as aphasia. As an early diagram maker, effectively mapping the topography of hypothetical language functions on to the brain (see Figure 4), he anticipated many of the conclusions of modern connectionists and localizationists, such as Geschwind (1965) nearly 100 years later.

It is now apparent that an irreduceable core for spoken language consists of two closely adjacent and interconnected structures straddling the Sylvian fissure (see Figure 5), Wernicke's and Broca's areas (Mesulam, 1990). The former, a posterior "sensory" system is centered on the posterior part of the superior temporal gyrus adjacent to Heschl's gyrus (the primary auditory area), together with the planum temporale, which is effectively a horizontal continuation of the supe-

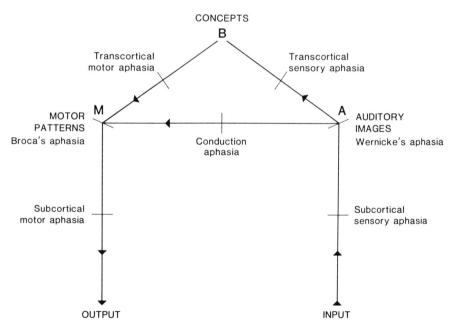

Figure 4 LICHTHEIM'S "HOUSE." THE PIONEERING NEUROLOGIST OVER 100 YEARS AGO THUS CONCEPTUALIZED THE RELATIONSHIP BETWEEN THE VARIOUS TYPES OF OBSERVED APHASIA (LOWER CASE) IN TERMS OF BREAKS (SLASHES) IN INFORMATION FLOW BETWEEN OR WITHIN PROCESSING REGIONS (UPPER CASE). VARIATIONS UPON THIS CONNECTIONIST APPROACH ARE STILL PROPOSED.

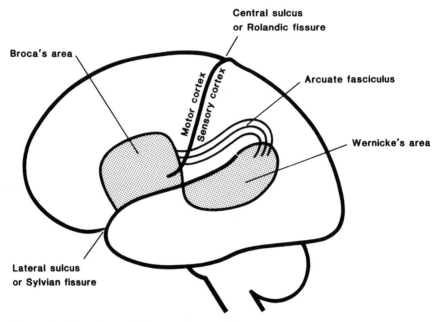

Figure 5   BROCA'S AND WERNICKE'S AREAS. THEIR EXACT BOUNDARIES ARE SUBJECT TO DISPUTE.

rior temporal gyrus within the Sylvian fissure. Adjacent parts of the middle temporal gyrus, inferior parietal structures, and the supramarginal and angular gyri, if damaged, add to the clinical picture. Broca's "motor" speech area is centered on the posterior portion of the third frontal convolution (the inferior frontal gyrus) next to the motor face area (see Figure 6). It is essentially a premotor region and therefore involved in speech organization and initiation. Neither area possesses strict cytoarchitectonic, physiological, or topographic limits; damage usually has to extend beyond these regions alone for lasting Wernicke's or Broca's aphasia to occur. The principle of mass action, moreover, typically applies in that the extent of deficit tends to parallel the size of the lesion.

Wernicke's Aphasia

In the accounts that follow of the classical aphasias, we have observed traditional nomenclature and taxonomy, and have largely adopted the descriptions given in such reviews as those of Benson (1993a), A. Damasio (1991), H. Damasio (1991), and Goodglass (1993). However we note with Bartlett and Pashek (1994) that major theoretical and practical issues surround the use of aphasia classification systems, which remain to be resolved.

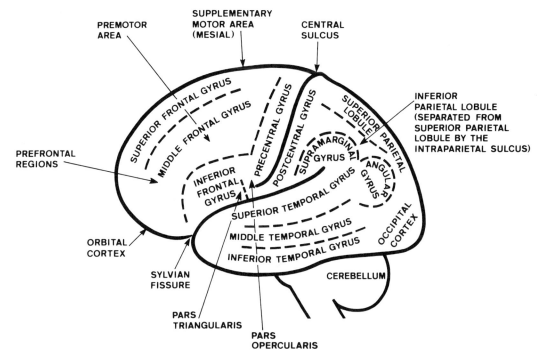

Figure 6 Major surface landmarks on the left hemisphere convexity, schematically shown.

Wernicke's aphasia is primarily a comprehension deficit, without necessarily any loss of ability to conceptualize. In practice, however, both comprehension and production of spoken and written language are normally affected. Thus Wernicke's area is not an auditory comprehension center per se; comprehension involves acoustic, phonological, syntactic, and lexical processing, to a greater or lesser extent in parallel, via a distributed network of processors in many cortical regions. Instead, Wernicke's area processes speech sounds so as to map conceptual meanings onto words, and vice versa, with full auditory comprehension occurring later, as a consequence of the operation of the entire system (Mesulam, 1990). Unlike Broca's aphasia, there are few accompanying focal neurological signs in Wernicke's aphasia, such as right hemiparesis or visual field defects. Thus the diagnosis of Wernicke's aphasia depends purely on the characteristic language problems.

Speech in Wernicke's aphasia is typically fluent, well articulated, syntactically and prosodically correct, with few primary deficits in oral movement and little speech apraxia or dysarthria. Word finding problems (anomia) on confrontation and in conversation, however, and neologisms or literal paraphasias (e.g., "loliphant" for "elephant") are common, as are semantic paraphasias (e.g., "binoculars" for "spectacles"). There may be substitution of a word at a more general taxonomic level for the target word (e.g., "people" for "women"). There is a general tendency to add extra syllables to words and words to phrases, resulting in an excessive pressure of speech (logorrhoea), with little underlying meaning. Thus an experimenter (HC) may ask a patient (RH) about coming to the speech therapy department after his stroke (Tyler, 1992, p. 165):

HC: Yeah, and then you came to this place. . . .

RH: Yes . . . well of course when they came there, I . . . em . . . he came there. I didn't know . . . there and I didn't know anything for it, any . . . I suppose we were there, when I went 1 2 3 4 5 and looked there and said well so and so and so and so. Now if you look here and I see, and then he said right, the next one I went there, so this is right, then the next one I'm going for him.

HC: What's this then? Is he teaching you to count or . . .

RH: Yeah . . . 1 2 3 4 5 6. Before. But the first, when I first went there, went there, and I didn't know anything. I didn't know anything and then he started—he's trying something, and was trying that . . .

HC: Trying to get you to say . . .

RH: And I was OK and try. And then I . . . you know, bad. (Laughs)

Because of impaired comprehension of speech, even one's own, the patient, unlike a Broca's aphasic, is often unaware of (anosognosic for) these problems, and is also impaired in relating words or phrases presented in the auditory mode. Indeed, reading and writing is also impaired as to content, although the letters may be well formed. Nevertheless, there is usually preservation of such social conventions as awaiting one's turn to speak, nodding the head at the right moment, and so forth. In due course adjacent structures in the inferior parietal lobule, such as the angular and supramarginal gyri, may substitute for Wernicke's area, if they are themselves undamaged. Wernicke's area is especially active, as shown by positron emission tomography (PET) imaging, in the normal brain during listening to speech, whereas if stimulated there may be interference in naming or comprehension. The superior temporal cortex thus recognizes the sounds of words, just as the inferior temporal cortex recognizes the sight of objects, and, if damaged, is associated with agnosia. Curiously, unlike Wernicke's aphasics, such agnosics are not usually anosognosic for their condition, and are therefore aware of their problems.

Wernicke's area occupies the semantic-lexical extremity of a language network, whereas Broca's area lies at the other, syntactic-articulatory, end (Mesulam, 1990). It accesses the lexicon, a distributed, multidimensional network of sound–word–meaning associations, and acts as a bidirectional interface between speech on the one hand, and meaning and thought on the other. A series of simultaneous approximations (initial sound, number of syllables, vowels, consonants) may serve to generate a word's initial representation; benign failure of such a system in everyday life may manifest in the tip-of-the-tongue phenomenon. In this situation, one is temporarily unable to generate a word or name (anomia—common indeed in Wernicke's aphasia), despite being able to recognize it, in a forced-choice situation, and being able to say a lot about what it probably sounds like (e.g., its initial or final letter or number of syllables).

Broca's Aphasia

Damage to Broca's classic area, the posterior portion of the third frontal convolution, leads at most to mild or transient "aphemia" (i.e., a speech rather than a language disturbance). Damage to motor and premotor areas also are required for true Broca's aphasia to manifest, together perhaps with the underlying white matter, the BG, and the insula. The latter structure constitutes the cortex hidden deep within the Sylvian fissure. Broca's aphasia typically only persists, in fact, with additional, deeper damage that usually includes the BG. The condition normally improves when damage is restricted to the frontal operculum. True Broca's aphasia ranges from complete muteness to slowed, deliberate speech constructed from very simple grammatical forms. It is nonfluent, effortful, dysarthric, telegraphic, and agrammatic, with an abnormal prosody. Word order is aberrant, with omission of articles (e.g., *the, a*) and prepositions. Content words, appropriate for the intended meaning, are favored, compared to grammatical functors like *to, and, if, when, for*. Thus in many ways Broca's speech is the converse of that in Wernicke's aphasia, with extreme verbal economy nevertheless generally manifesting the basic intended meaning (e.g., "Boy hit dog stick naughty smack cry"). Repetition, reading, and writing are also disordered, like the spontaneous verbal output. The gross simplification of verbal output is not merely an attempt to reduce articulatory effort, as comprehensional problems may also manifest, especially with difficult syntax. Thus nonreversible passives such as, "The pie was eaten by the girl," provide few problems; "The girl was eaten by the pie," is an unlikely statement. In contrast, reversible passives such as, "The boy was kissed by the girl," are more difficult, as either person could conceivably be agent, or object, and only the syntax disambiguates. Familiar or practiced words or phrases and such conversational stereotypies as greetings may be fairly well preserved in speech. Unlike Wernicke's aphasia, insight and awareness are usually retained. Stimulation of Broca's area during speech leads to arrest and sequencing problems, as when responding to such commands as, "Pick up the small blue triangle and place it on the large yellow circle." Clear focal signs typically accompany Broca's aphasia, such as right side motor deficits (maybe hemiparesis of the right upper limb or face), ideomotor apraxia (see Chapter 10), depression, and a "catastrophic reaction," which might extend to refusal to be tested; however there is usually no visual field loss. Diagnosis is therefore easy. (Note that the catastrophic reaction is a dysphoric state characterized by anxiety, loud crying, extreme negativity, withdrawal, and hostility, and is perhaps commonest with nonfluent Broca's aphasia after left-frontal lesions, see e.g., Berthier & Starkstein, 1994.)

Broca's area occupies the articulatory–syntactic extremity of the language network; neural word representations originating perhaps in Wernicke's area are transformed into the corresponding articulations, via sequenced phonemes and morphemes and syntactically arranged words (Mesulam, 1990). Neither center is exclusively motor or sensory. Thus recordings from the two areas during speech and listening to speech show that both are simultaneously active, in parallel, rather than successively. Word selection therefore occurs in parallel with the anticipatory programming of syntax and articulation; grammatical structure influences word choice, and word choice constrains syntax. During speech, Wernicke's area monitors the phonology, semantics, and syntax of the intended output, prior to its release for articulation; this may represent the operation of a prearticulatory editor whereby, for example, spoonerisms are normally detected before utterance, and are eliminated from a prearticulatory buffer. If the upcoming utterance is checked as satisfactory, Broca's area will then be released via the BG and thalamus (Crosson, Novack, & Trenerry, 1988). According to this model, which we shall shortly discuss in detail, in conversation we simultaneously decode our interlocutor's speech, generate our own from our mentation, monitor it syntactically, semantically, and phonologically, and then release it, in a continuous parallel fashion, via Broca's and Wernicke's areas, the arcuate fasciculus, BG, and thalamus. We are of course exposed to speech from birth, we have evolved special adaptations automatically to acquire it at a very early age (Bradshaw & Rogers, 1993), and we are under strong social pressures to do so. Occasionally, circumstances conspire to prevent our exposure to an appropriate model before the expiry of a critical period in the early years of life (Fromkin, Krashen, Curtiss, Rigler, & Rigler, 1974). When that happens, as with the girl Genie, both grammar and LH lateralization seem irretrievably lost, although like a Broca's aphasic the unfortunate person can learn to string words ungrammatically together.

Other Types of Aphasia

Wernicke (1874) predicted that difficulty in repetition would result from disconnection of Broca's and Wernicke's areas. The main interconnection is the arcuate fasciculus, a fiber bundle originating in the auditory association cortex, which passes posteriorly and superiorly before making a U-turn at the rear of the Sylvian fissure, to travel beneath the supramarginal gyrus and somatosensory areas in the parietal operculum to reach Broca's area. Another more direct route passes beneath the insular cortex beneath the Sylvian fissure. Lesions to either pathway may result in *conduction aphasia,* though unless subcortical routes are also interrupted, the degree of disruption may be less than expected, often being restricted to deficits in short-term auditory memory. Nevertheless, lesions to the white matter immediately beneath the supramarginal gyrus, or at least to the posterior peri-Sylvian region, insula, or arcuate fasciculus, may in practice suffice to disconnect Broca's and Wernicke's areas. In addition to Wernicke-type impaired naming and impaired reading aloud, with relatively intact reading and speech comprehension and conversation, there is an inability to repeat verbal material spoken by the examiner. Although this is most apparent for nonwords and word strings, nevertheless colloquialisms, stock phrases, and verbal automatisms may be preserved, though generally performance is poorer when trying to repeat longer material (Goodglass, 1993). The patient is typically aware of the problem (unlike the situation in Wernicke's aphasia), and may make repeated self-corrective attempts at repair, perhaps fixing one portion of the target word while introducing a new error elsewhere. Thus Goodglass noted the stability of this phonological skeleton around which the various attempts at correction revolve. When offered alternative pronunciations, patients are invariably accurate in rejecting those that are even minimally inaccurate, and in endorsing those that are correct, suggesting a retrieval deficit. Indeed, the problem may not even be primarily a disorder of repetition, nor just a disconnection of an auditory language center from one involved in motor speech planning.

Deep dysphasia, according to Goodglass (1993), is a rare form of repetition disorder, the patient typically substituting semantically related words (semantic paraphasia) for the target given by the examiner. Thus balloon → kite, beggar →tramp, kernel → shell, independence → elections, a phenomenon similar to the deep-dyslexic semantic paralexias that we shall encounter in the next chapter. Repetition is better for concrete than for abstract words, for nouns than for verbs, and grammatical functors or nonwords often cannot be repeated at all. The patient may complain of almost instantly having lost trace of the sound of a presented target word, although its meaning may be retained. Semantic paraphasias may occur also to picture naming. The problem may possibly be loss of phonological short-term memory due to temporal-lobe lesions, with semantic-paraphasic substitutions instead of the conduction aphasic's repeated attempts at phonologically guided self-correction.

With the transcortical aphasias (see Figure 7), lesions are outside of the two major language centers and their interconnections, so repetition is undisturbed. With *transcortical sensory aphasia*, Wernicke's area is disconnected from the rest of the parietal association area by damage at the parietotemporal junction, often with deep-seated lesions in the zone inferior to the angular gyrus, and between the posterior end of the Sylvian fissure and the parietotemporal junction. Comprehension, naming, reading, and writing are impaired; although repetition is generally fairly unimpaired, there may be echolalia and fluent, free-flowing, irrelevant verbal output similar to Wernicke's aphasia, though semantic substitutions may be more common than the paragrammatism and phonemic parapha-

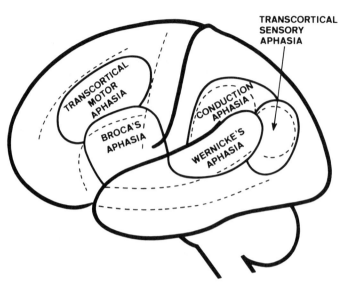

Figure 7   Cortical regions likely to be involved in aphasiogenic lesions. Dotted lines correspond to sulci as in Fig. 6.

sia, which may be more representative of Wernicke's aphasia (Goodglass, 1993). There is typically an anosognosic unawareness. The repetition, so characteristic of the disorder, is not, however, mere automatic echoing where it might be socially inappropriate. Indeed, patients may correct grammatical errors deliberately embedded in the model provided. The fact that they are good at repeating nonsense and foreign words suggests that repetition is based on phonological recoding without lexical support. Patients can recite previously memorized passages (e.g., prayers). Not unexpectedly, given the locus of the lesion, right-side visual defects, including right superior quadrantanopia, often accompany transcortical sensory aphasia as a focal sign.

With *transcortical motor aphasia*, or "dynamic" aphasia, the lesion (often deep within frontal white matter) now disconnects *Broca's* area from the adjacent (and mesial frontal) supplementary motor area (SMA), which is important in initiating voluntary speech. Verbal output is (not unexpectedly) nonfluent, effortful, short, and dysprosodic, with loss of grammatical functors and with impaired naming, whereas repetition and comprehension are relatively unimpaired. Again, not surprisingly in view of the locus of lesion, accompanying focal signs include right-side motor defects, typically of the lower extremity (Benson, 1993a).

With *mixed transcortical aphasia*, a combination of the above two syndromes isolates the speech areas and their interconnections from the rest of the brain.

The patient initiates little spontaneous output unless spoken to, and then may respond with a short, fluent, echolalic repetition. Though unable to comprehend it, the patient may correct a heard sentence if grammatically incorrect. All other language functions are typically lost, and accompanying focal signs include sensory and motor deficits on the right side.

With *anomic aphasia,* the only deficit is one of word finding during confrontation naming. It may be associated with a category-specific deficit (see Chapter 5), when the patient is unable to name items belonging to a particular category, such as furniture, vegetables, small animals, and so forth. Damasio and Tranel (1993) report two patients with, apparently, separate lexical systems for concrete nouns and verbs. One, with damage in the left anterior and middle temporal lobe, could not produce nouns, whereas the other, with a lesion in the left premotor cortex, could not produce verbs. Generally, nouns and proper nouns may be more affected than verbs, consequent upon lesions to the left anterior temporal cortex in regions associated with visual object recognition. However, the deficit occurs in most of the aphasias, and is not well defined either functionally or anatomically. Indeed it has been localized to the angular gyrus, the inferior parietal lobule, and the temporal pole. Lovell (1993) cited an early reference to the condition by van Swieden in 1742:

> They recovered well from apoplexy, except for this one deficit: in designating objects they could not find the correct names for them. (p. 54)

With *global aphasia,* there are deficits in comprehension, repetition, output, and naming, consequent upon a large lesion to much of the peri-Sylvian LH, cortex, and white matter, including also the BG, thalamus, and internal capsule. There may be preservation, with normal prosody and inflections, of nondeliberate automatisms, greetings, emotional expletives, verbal stereotypies, serial utterances (counting, days of the week, months of the year, etc.), together with associated comprehension of idiomatic expressions, and maybe a small number of discrete substantives. Associated signs include gross right-side sensory and motor loss.

The *foreign accent syndrome* (FAS) is not of course strictly an aphasia per se; nor is it a uniform syndrome, but rather the auditory impression of the emergence of a foreign accent in speech subsequent to a brain lesion. The phenomenon, though fairly uncommon, repeatedly reemerges in the aphasia literature typically after fluency, phrase length, and functional communication has been regained (see e.g., Berthier, 1994). A particular blend of segmental and prosodic speech-production deficits may be essential for the manifestation of FAS, and these may be associated with lesions involving the dominant precentral gyrus or the premotor region.

The primary auditory cortex (Heschl's gyrus) on the left is important in the fine temporal resolution of speech stimuli. Damage here may produce *pure word deafness* or auditory–verbal agnosia (Bauer, 1993; Goodglass, 1993). Only speech perception is affected, not the recognition of nonverbal sounds, speech production, reading, or writing. Affected individuals can still distinguish accents, speaker identity, and vowels, but not consonants, and the discrimination of place of articulation (e.g., *ba* vs. *ga*, *ka* vs. *ta*) may be worse affected than discrimination by voicing (e.g., *da* vs. *ta*, *ka* vs. *ga*), (see e.g., Goodglass, 1993). Speech may be heard as a rapid jumble of unintelligible speech sounds.

It should be emphasized that in all the conditions described above, lesions are rarely restricted, syndromes are rarely pure, and individual differences often make meaningful generalizations difficult and dangerous.

Subcortical Aphasias

With left-sided damage to the thalamus, BG, and internal capsule, a heterogeneous collection of atypical language disturbances may occur, perhaps after early mutism, which sometimes correlate poorly with the classical syndromes of cortical aphasia. There may be semantic paraphasias, perseverations, and reductions in spontaneous speech, mildly impaired auditory comprehension, and impairments at a phonological, lexical-semantic, syntactic, and even pragmatic level, although with a good prognosis for recovery. Hypophonia and dysarthria are common. With right-sided damage, however, aprosodia and dysphonia are commonly said to occur; however, see Ovellette and Baum (1993), who found little evidence of an RH involvement in linguistic prosody per se, but rather that such damage may instead affect responsivity to emotionality, with problems in the production of correct prosody being a function of left-side damage. Moreover, in a recent study Kirk and Kertesz (1994) compared the frequency, severity, and type of aphasia consequent upon cortical or subcortical lesions. They found that the two groups did not differ in severity, or in any of the Western Aphasia Battery subtest scores, with subcortical lesions leading to a variety of aphasia subtypes, the severity of which resembled that seen after cortical lesions of similar size. There was no clear correlation with lesion locus, and symptoms often depended strongly on the time elapsed between lesion and testing.

The BG and thalamic nuclei on the left, respectively, play a role in motor planning, programming, and execution on the one hand, and attention, activation, and alerting on the other. Crosson, Novack, and Trenerry (1988) provided a comprehensive account of the integrated functioning of the cortical centers reviewed above, their interconnections, and the BG and thalamus (pulvinar and ventral anterior nucleus), none of which operates in isolation from the others (see Figure 8). The arcuate fasciculus bidirectionally interconnects Broca's and Wernicke's areas directly, which also have reciprocal connections with the thalamic nuclei. The latter receive input from both the BG and Broca's and Wernicke's areas. The BG receive input from both of these cortical speech centers, but can only output to them via the thalamus. Crosson et al. saw the thala-

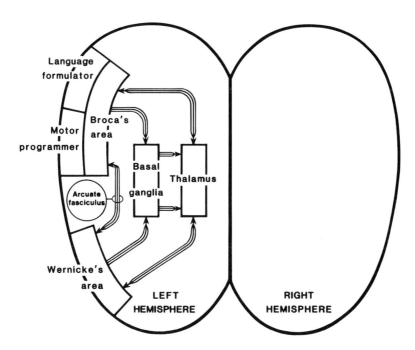

Figure 8 Possible cortical and subcortical interconnections in speech control. (From Bradshaw & Rogers, 1993.)

mus as activating two separate systems in Broca's area: a language formulator and a motor programmer. Thalamic pathways in conjunction with Wernicke's area perform *semantic* monitoring before articulatory release, whereas the arcuate fasciculus, also in conjunction with Wernicke's area, performs *phonological* monitoring before release. The BG in conjunction with the thalamus hold up (for monitoring) and release (after monitoring) the formulated information. According to this scheme, therefore, formulated language passes from Broca's to Wernicke's area via the thalamus for *semantic* monitoring. If it is correct, the BG and thalamus release it back to Broca's area for programming; otherwise it is reformulated. After phonological formulation in Broca's area, the information is sent via the arcuate fasciculus to Wernicke's area for *phonological* monitoring. Again, if correct, it is released back to the BG and thalamus to Broca's area for articulation. All systems must simultaneously operate for smoothly flowing conversation; while one segment is being executed in speech, the next is being monitored for phonological accuracy, the next is being motor programmed, the next is being verified for semantic accuracy, and the next is being initially formulated. The basic concept, therefore, is one of involvement of subcortical nuclei in preverbal semantic and phonological monitoring, *by* the posterior cortical speech centers and *of* the anterior regions. The cortical-striatal-pallidal-thalamic-cortical reentrant loop, with its internal chains of excitatory and inhibitory pathways, is essentially similar to those reviewed in Chapter 11, where we consider the functional operation of the BG in detail. Crosson (1992) extended these ideas in the context of BG regulation of information flow during the above stages of language formulation. He sees the BG–thalamus system playing two roles in relation to language (and movement generally):

1. A slow, tonic *neuromodulatory* role, as with (reduced) readiness to respond in Parkinson's disease (PD)
2. A faster, phasic *neuroregulatory* role, for the initiation, maintenance, and completion of a given speech element or movement.

The BG are involved in both of the above roles, while the thalamus plays an essentially activating or neuromodulatory role. Thus, careful testing of PD patients reveals subtle deficits such as problems in naming, and reductions in phrase length, grammatical complexity, and verbal fluency. Subcortical aphasias generally, therefore, may stem from (necessarily permanent) damage to cortico-cortical connections via the BG, or instead and perhaps more commonly, to loss of neuroregulatory and neuromodulatory functions that may recover. Metter (1992) came to similar conclusions; he found that lesions in the subcortical system lead to glucose hypometabolism (as imaged by PET) in the corresponding cortical areas to which the relevant parts of these regions project:

1. Posterior putamen and posterior internal capsule to ipsilateral temporoparietal cortex
2. Anterior putamen and anterior internal capsule to ipsilateral prefrontal cortex

Thalamic damage likewise on the left results in reduced activity over much of the language cortex, resulting in subtle "functional cortical lesioning," with problems in verbal readiness, planning, fluency, and linguistic set. Again as we shall see in Chapter 12, prefrontal damage can result in similar deficits, as the BG project to prefrontal structures.

Language and language deficits perhaps typify, more than in any other functional system, the tension between two historically contrasting approaches to neuropsychology and behavioral neurology:

1. The classical, cortical, localizationist approach, as realized, for example, by Wernicke, Lichtheim, and Geschwind
2. The neural network approach of cognitive architecture as conceived by cognitive scientists who emphasize information flow and function, before structure.

The latter and much more recent tradition does not appeal to specific neural circuits, and sees neuropsychological functions as not being localized in definite neural areas. Instead, such functions are thought to be subserved by a mosaic of often hypothetical systems, with their network of interconnections along which information flows and is transformed, between nodes. Each node performs a single operation, and all are orchestrated to perform different cognitive tasks. Thus each network can perform more than one function by altering the properties of the constituents. There is an emphasis on parallel more than on serial processing. As so often in science, a rapprochement is likely to occur, in time, between these two paradigmatically antithetical approaches. In practice, with formulations such as those of Crosson and his co-workers, we already see a shift from the traditional distinction between anterior-motor and posterior-receptive aphasia of the classical localizations; there is now an acceptance of the functionally important distinctions between phonological, lexical, semantic, and syntactic processing, though with an attempt firmly to ground them in anatomically identified patterns of connectivity.

Illustrative Case History of Paroxysmal Aphasia

Epileptic discharges over temporal and temporoparietal regions of the LH may result in paroxysmal aphasia, without major modification of consciousness and without any other obvious neuropsychological impairment. Lecours and Joanette (1980) reported just such a patient, whose thought and mentation remained normal during his aphasic paroxysms, an observation, incidentally, which indicates that thought does *not* depend upon language functions.

> Brother John is a right-handed 50-year-old who has completed 10 years of formal schooling. Being intellectually inquisitive, he has kept learning since then in the manner of an autodidact. Reading and writing have been part of his daily adult life. For the past decade or so, he has worked on the editorial staff of his order's pious periodical, his main job being to read and answer letters to the editor (up to 55 a day). He considers himself utterly untalented with regard to apprenticeship of foreign languages: indeed he remains very much of a French unilingual although he has been intensively exposed to both Italian and English for long periods. Brother John has now presented epileptic spells for more than 25 years. . . . He remains conscious throughout the spell provided he or an observer fights initial drowsiness; it has been observed that, although anosognosic for particular linguistic deviations (cf. below), he remains aware of severe language perturbation, is sad about it, and, on occasion, uses tricks deliberately aimed at hiding the fact of a spell actually taking place from those who need not know; the longitudinal evolution of his isolated linguistic impairment has been closely watched and described; he has proven capable of adequate manipulation of relatively complex tools, and of adequate recognition of things and events on the basis of nonlinguistic auditory, visual, or tactile informations; he has also proven capable of carrying on with previously given instructions. . . .
>
> *Praxis*: He remains capable, throughout his spells, of spontaneous actions

such as operating an elevator, tuning a radio, using a tape recorder (i.e., inserting a cassette, plugging the microphone, pressing buttons, rotating knobs in order to listen or record, etc.). Although his copy of a cube was not perfect, he is also capable of—indeed, prone to—tool manipulation, even at times when his comprehension difficulties are very severe: for instance, as a pastime during long spells, he fabricates elaborate montages with elastic bands and cigarette boxes he (carefully and with great precision) cuts up with scissors.

Gnosis: Routine agnosia testing was never carried out at peak of spell. From then until the end of the episode such tests (none pertaining to verbal deafness or blindness) did not evidence agnosia in the tactual and visual spheres: nonlinguistic tactile and visual stimuli were properly recognized whenever they could be paired with images or associated to gestures, and some were simply named; although a few nonlinguistic auditory stimuli apparently puzzled Brother John (including the noise of a typewriter, with which he is familiar, the neighing of a horse, and the grunting of pigs), most auditory stimuli were also unambiguously recognized. Furthermore the patient recognizes—but as a rule cannot name—known people he sees when dysphasic, as well as he recognizes, as witnessed by proper use, daily life objects such as comb and shaver, soap and towels, utensils and crockery, and so forth. He has been reported to listen to (instrumental) music during his long spells and says the latter do not diminish his (uneducated) capacity to appreciate music and recognize—but for their names as a rule—known melodies. He rightfully states himself capable of recognizing unseen familiar persons by listening to their voices (a nurse talking in the corridor, for instance) or even steps (a colleague who walks with a cane, for instance) although he can be utterly incapable of evoking their names, either for private or public expression.

Spatial orientation: He once had difficulty finding his way in Montreal and later explained: "I knew where to look for the names of the streets, I could see them but could not read them" (free translation).

Memory: Brother John remains capable, throughout long spells, to remember events—linguistic or otherwise—that have occurred previously. He will therefore carry out, in the course of a dysphasic episode, verbal instructions given days or weeks before; without being capable of any linguistic recapitulation, inner or overt, he has thus carried on with instructions such as sampling his own deviant speech on magnetic tapes, attempting to write a few words at peak of spell, alerting a third party in order to have one of us informed of the occurrence of a severe dysphasic episode. . . .

Thought and coping subterfuges: When he had to answer the telephone during a spell, he kept repeating "Allô! Allô! Allô!" louder and louder [with] the deliberate intention to deceive whoever was calling, that is, to lead him to believe that his interlocutor could not hear because of technical failure in the voice transmission apparatus. When he had visitors during a dysphasic episode, he saw to it that one of his colleagues, already informed of the expected visit, understood he was to replace him beside his friends, guide them to the convent's cafeteria, and entertain them through dinner time. In other words, the occurrence of a dysphasic episode does not prevent Brother John from thinking clearly in relation to co-occurring concrete situations, neither does it prevent him from confronting present to past events, planning accordingly, and acting at best. . . .

While he was traveling by train from Italy to Switzerland, Brother John once found himself at the height of paroxysmal dysphasia soon upon reaching the small town of his destination. He had never been in this town before but he probably had considered in his mind, before the spell began (or became severe), the fact that he was to disembark at the next stop of the train. At all events, he recognized the fact he had arrived when the time came. He consequently gathered his suitcases and got off the train and out of the railway station, the latter after properly presenting his transportation titles to an attending agent. He then looked for and identified a hotel, mostly or entirely on nonlinguistic clues since alexia was still severe, entered and recognized the registration desk, showed the attendant his *medic-alert* bracelet only to be dismayed and dismissed by a gesture meaning "no-room" and a facial mimic that perhaps meant "I-do-not-want-trouble-in-my-establishment." Brother John repeated the operation in search of a second hotel, found one and its registration desk, showed his bracelet again, and, relieved at recognizing through nods and gestures that there were both room and sympathy this time, he gave the receptionist (a "fat lady") his passport, indicating the page where she was to find the information necessary for completing his entry file. He then reacted affirmatively to her "do-you-want-to-rest-in-bed-now" mimical question. He was led to his room and given his key; he probably tipped as expected and went to bed. He did not rest long, however: feeling miserable ["it helps to sleep but sometimes I cannot because I am too nervous and jittery"] (free translation), then hungry, he went to the hotel's lobby and found the restaurant by himself. He sat at a table and, when presented with the menu, he pointed at a line he could not read but expected to be out of the hors-d'oeuvres and desserts section. He hoped he had chosen something he liked and felt sorry when the waiter came back with a dish of fish, that is, something he particularly dislikes. He nonetheless ate a bit ("potatoes and other vegetables"), drank a bottle of "mineral water," then went back by himself to his room, properly used his key to unlock his bedroom door, lay down, and slept his aphasia away. He woke up hours later, o-kay speechwise but feeling "foolish" and apologetic. He went to see the fat lady and explained in detail; apparently she was compassionate. (Lecours & Joannette, 1980, pp. 3–15)

Summary and Conclusions

Language is a form of symbolic communication employing gestures and vocalizations, which are readily acquired from local models within a certain critical developmental period, with speech its articulatory manifestation. However, it may go beyond mere communication, perhaps helping mediate our modeling of reality. With its six identified aspects, phonology, morphology, syntax, semantics, prosodic or suprasegmental aspects, and pragmatics, there is a clear shift from LH to RH involvement. Morphological asymmetries at a macrolevel in speech-related areas of the human brain, however, clearly mark out important LH structures, such as the planum temporale within the Sylvian fissure. Although the former is larger on the left, the latter is longer on that side, and terminates lower than its counterpart on the right. At a micro-, architectonic level, these speech-related asymmetries in both posterior and anterior regions may be even more marked. Indeed, language and speech depend not just on a few discretely localized regions, but on the dynamic interplay between extended neuronal networks involving various cortical and subcortical regions; we must not seek to localize a particular linguistic symptom to damage to a single structure. Adaptations of the supralaryngeal tract, to the probable detriment of its other functions, also indicate the evolutionary significance, if not uniqueness, of speech and language.

The true aphasias, involving disorders of speech planning (or reception), strictly exclude such other deficits as programming (speech apraxia) and execution (dysarthria). Similarly, aphasia is not a sensory, perceptual, motor, or thought disorder. It was not until the 1860s that structures in the LH were formally recognized as playing a major role in the realization of speech. We now know that its irreducible anatomical core consists of two closely adjacent and interconnected structures straddling the Sylvian fissure on the left, Broca's ("motor") and Wernicke's ("sensory") areas, respectively, in the posterior superior temporal gyrus and the posterior inferior frontal gyrus. However, neither is indeed purely "motor" or "sensory"; other areas are also involved, and neither area possesses strict limits. Indeed, for Broca's or Wernicke's aphasia to occur, damage to structures beyond the limits of the classical areas is usually necessary.

With *Wernicke's aphasia* a comprehension deficit is a primary manifestation, speech being typically fluent, excessively so in fact, well-articulated, and syntactically correct. There are, however, word finding problems and abundant neologisms and paraphasias. With *Broca's aphasia,* by contrast, speech is (often greatly) slowed, nonfluent, deliberate, and syntactically simplified to the point that grammatical (functor) words may be totally excluded. It is in many ways the diametric opposite of Wernicke's aphasia. Comprehension, especially where the syntax is at all complex, is probably also affected. Hemiparesis of the right upper limb or face, or ideomotor apraxia, may also occur. It must be emphasized that during conversation both regions are simultaneously active, in parallel rather than successively. If the regions or interconnections between Broca's and Wernicke's areas are selectively damaged, *conduction aphasia* may result. The classical deficit in this case is an inability to repeat verbal material, especially functors, spoken by the examiner. With *transcortical sensory aphasia,* Wernicke's area is disconnected from the rest of the parietal association area by damage at the temporoparietal junction. There is impairment of comprehension, naming, reading, and writing, with relative preservation of repetition. Right superior quadrantanopia frequently accompanies the disorder. With *transcortical motor (or dynamic) aphasia,* Broca's area is now disconnected from the adjacent SMA, which is important in initiating voluntary speech. A syndrome basically similar to

Broca's aphasia typically results, though repetition and comprehension are relatively intact. Right-side motor defects of the lower extremity often accompany the condition. With *mixed transcortical aphasia,* a combination of the above two syndromes isolates the speech areas and their interconnections from the rest of the brain. There is little spontaneous speech, and the patient may in fact be able to do little other than merely parrot what is said in an echolalic fashion, perhaps correcting the speaker's incorrect grammar. Finally, with *anomic aphasia* the only deficit is a word-finding problem, a deficit that indeed may occur in most of the aphasias. Nor is the syndrome well defined, functionally or anatomically.

With left-sided damage to the thalamus, BG, and internal capsule, a heterogeneous collection of atypical *subcortical aphasias* may occur, which may subsequently resolve. There may be semantic paraphasias, perseverations, and reductions in spontaneous speech, mild comprehensional problems, and impairments at a phonological, lexical-semantic, and even pragmatic level. The BG and thalamus on the left play a role respectively in motor programming and execution on the one hand, and attention, activation and alerting on the other. These structures almost certainly coordinate the activities of Broca's and Wernicke's areas during the give and take of conversation, as well as monitoring speech, before it is released, for phonological and semantic accuracy.

Further Reading

Benson, D. F. (1993). Aphasia. In K. M. Heilman & E. Valenstein (Eds.), *Clinical neuropsychology* (3rd ed.) (pp. 17–36). Oxford: Oxford University Press.

Crosson, B. (1992). *Subcortical functions in language and aphasia.* New York: Guilford.

Damasio, A. R. (1991). Signs of aphasia. In M. T. Sarno (Ed.), *Acquired aphasia* (2nd ed.) (pp. 27–44). San Diego: Academic Press.

Damasio, H. (1991). Neuroanatomical correlates of the aphasias. In M. T. Sarno (Ed.), *Acquired aphasia* (2nd ed.) (pp. 45–72). San Diego: Academic Press.

Goodglass, H. (1993). *Understanding aphasia.* San Diego: Academic Press.

Mesulam, M.-M. (1990). Large-scale neurocognitive networks and distributed processing for attention, language, and memory. *Annals of Neurology, 28,* 597–613.

Murdoch, B. (1990). *Acquired speech and language disorders.* London: Chapman & Hall.

Chapter 3

Disorders of Reading and Writing

The Alexias and the Agraphias

*A*lthough both aural and visual language clearly involve common mechanisms, at least at certain processing levels, speech differs from reading and writing in one important respect. Speech is more or less automatically acquired, at an early age, often even against great odds, and frequently from a considerably flawed model, as long as there is exposure to a local tongue before the passing of critical developmental milestones. Thus, unless the infant is grossly hearing-impaired, exposure to the frequently nonstandard, inconsistent, or ungrammatical models of older children or adults permits apparently effortless acquisition of syntactic rules of which speakers themselves are usually ignorant, at a conscious, introspective level. There seems to be a biologically inherited predisposition to learn, the product of a long evolutionary trajectory (Bradshaw & Rogers, 1993). Reading and writing, however, are very recent inventions, which may have occurred only two or three times in human history, in the Fertile Crescent running between Egypt and Mesopotamia, in China, and in central America. They have to be consciously and deliberately learned, and during this acquisition phase, at least, they are more or less parasitic upon the oral–aural channel.

Writing first appeared only a little more than 200 generations ago, such that it is unlikely to have led to noticeable evolutionary changes in neurological organization, particularly when for most of its history its use was confined to a small elite. Nevertheless, an enormous number and range of systems have appeared. These range from pictograms (essentially drawings), ideograms (conventionalized signs such as indeed are still employed for advisory and directive purposes along our roads), to syllabaries and the alphabet. The syllable is perhaps the most natural speech unit, as compared to phonemes and articulatory or acoustic distinctive features, and one to which we are all most immediately sensitive. There have been recurring suggestions (see e.g., Aaron & Joshi, 1989) that developmental dyslexia is less common among children taught a syllabary, where symbols correspond to syllables, compared to those learning an alphabet, which more or less roughly maps on to the phonemic level. The main disadvantage of a syllabary is that a relatively huge number of symbols is required, and even then there may be unintended artificialities of representation in the face of a run of consonant sounds in the natural language; this was apparent when an alien syllabary was imposed upon an early version of Greek, in the form of Linear B (Ventris & Chadwick, 1956).

The alphabet is certainly the most economical system, especially if, in the face of gradually changing and evolving pronunciations, and the adoption of loan words from other languages, an attempt is made to maintain a one-to-one correspondence with the 40 or so phonemes that are variously employed in most tongues. However, it remains a somewhat artificial system as the human "ear" is not always immediately sensitive to phonemes, and the invention of the alphabet in fact ranks as a great human achievement and insight.

It is therefore hardly surprising that the effect of a given language-area lesion on reading may depend to a considerable extent upon the nature of the script (e.g., phonological vs. ideographic). Some scripts (e.g., Japanese *Kana* and *Kanji*) combine, respectively, ideographic and phonological-syllabic characteristics, and while generalization may be premature, numerous lesion studies (Chen & Tzeng, 1992) suggest that during reading the angular gyrus is particularly involved with *Kana,* while inferior occipitotemporal structures may operate in *Kanji.* Generally, subcortical damage is more likely to result in graphomotor problems such as apraxic agraphia, whereas left parietal damage is necessary for true agraphia.

In visual word recognition, at least four processes may concurrently operate:

1. Apprehension of individual letters, independent of, for example, style
2. Application of graphophonemic (letter to sound) conversion rules
3. Association of the written letter sequence as a whole to the total phonological representation of the word
4. Direct association of the written word to a concept, *without* phonological mediation

This suggests that we can read via at least two routes (Marshall & Newcombe, 1973). These involve, first, graphophonemic conversion rules, which in fact alone permit us to read pronounceable nonsense strings (e.g., *gife, boaz*) which we may never previously have encountered, and, secondly, a system of direct access, which permits processing of irregular configurations, such as kg, Mrs., reign, Worcester, and so forth. To these two routes we can possibly add whole-word phonological activation (rather than graphophonemic recoding at the letter or letter-cluster level), and reading unfamiliar words (e.g., *fain*) by analogy with familiar ones (e.g., *rain*). Which route we employ may depend upon our reading experience (with native and nonnative languages), item frequency, and the possible occurrence (and locus) of brain injury. According to traditional connectionist theorists (e.g., Wernicke, 1874, and Geschwind, 1965), to understand written words it is always necessary first to transit the *spoken* word lexicon in Wernicke's area, after initial processing in the primary visual cortex and the angular gyrus. However direct, rather then mediated or phonological access, bypassing Wernicke's area, may be an option for skilled readers. Thus we normally have no interpretative problems when reading homophones like pair/pear (Coltheart, 1980a,b), and, as we shall shortly see, the semantic paralexias of deep dyslexics demonstrate the unavailability of phonological rather than semantic information. We shall also see that recent positron emission tomography (PET) imaging studies (Howard et al., 1992) suggest that when skilled readers read silently, Wernicke's area may not be active.

Aphasia for spoken language is usually associated with similar disorders of praxis, reading, and writing. Although reading disorders parallel speech disorders, they may be less systematic. Reading disorders may occur rarely in isolation (e.g., pure word blindness) (see below), and likewise speech disorders may very occasionally occur on their own (e.g., pure word deafness). Agnosic syndromes, with impaired object recognition, must of course be separated from reading disorders *per se*. Occasionally, capacities for oral and silent reading may dissociate, or there may be a dissociation between loss of ability to read grammatical functors (as in deep dyslexia, see below) and substantives, indicating that different systems process semantic and syntactic words. There may even be category-specific deficits in the ability to process nouns belonging only to certain taxonomic categories. Of course it only makes sense to talk about deficits where, after injury, someone who previously could read or write can no longer do so as before. We should, however, again note with Friedman, Ween, and Albert (1993) that reading and writing differ from spoken language in being often arduously learned rather than naturalistically and effortlessly acquired, and in involving a wider range of senses and centers. Moreover, more than is the case in any other syndrome, reading and writing disorders involve the widespread interplay of many areas and a wide range of idiosyncratic personal differences. Useful reviews of acquired alexias and agraphias are available in Friedman et al. (1993) and E. Kaplan (1991).

Alexia with Agraphia

In this "commonsense" deficit, both faculties, reading and writing, are lost. Damage to the angular gyrus at the posterior-superior temporal and inferior parietal junction, at the rear of the Sylvian fissure, has long been thought to be both necessary and sufficient (Déjerine, 1891, cited by Friedman et al., 1993). This polysensory region receives visual, auditory, and somatosensory input, as might be expected, and the patient can no longer spell aloud, read, write, or comprehend spelt words, though letter naming may be preserved. A right inferior quadrantanopia commonly co-occurs as a localizing sign.

Alexia without Agraphia, or Pure Alexia

In this apparently paradoxical syndrome of pure word blindness, the patient can still write but cannot read even his or her own writing. As predicted by early connectionist theorists (e.g., Déjerine, 1892, cited by Zaidel, 1990a), and revived by Geschwind (1965), this disconnection syndrome deprives the language zone in the left hemisphere (LH) of all visual input, as a result typically of occlusion of the posterior cerebral artery. Preserved comprehension of orally spelt words (e.g., c-a-t), and of letters traced on the patient's palm, indicates the overall intactness of orthographic codes, despite their visual inaccessibility. Patients may eventually be able to read short words via letter-by-letter assembly; Zaidel (1990a) suggested that letters (and objects) may evoke nonvisual associations in the right hemisphere (RH) which may cross to the LH via intact anterior fibers in the corpus callosum. In any case, patients have also been reported (Coslett & Saffran, 1989) to be able to perform rapid word–nonword lexical decisions, or semantic categorizations, by a form of preserved tacit or implicit processing, which we shall repeatedly encounter in many other areas of acquired deficit. Lovell (1993) quoted an early observation of this syndrome by G. Mercuriale in 1588:

> Following seizures with a partial loss of memory, this man could write but could not read what he had written. (p. 54)

Damage usually occurs to LH occipital structures (visual cortex and lingual gyrus, by occlusion of the posterior cerebral artery), leading to a right-visual-field defect (right superior quadrantanopia) and, frequently, right hemiachromatopsia (color blindness) and color anomia, together with disconnection of the splenial (posterior) regions of the corpus callosum (see Figure 1). Patients can therefore still see, via intact occipital structures on the right, and can still write, as the angular gyrus anterior to the damaged occipital structures on the left is intact. However, they cannot read even their own writing, as the posterior LH damage eliminates visual input to language regions, and the splenial damage prevents *verbal* input from the intact RH projecting across to the angular gyrus on the left. However, intact anterior regions of the corpus callosum (rostrum and genu) still permit object recognition information from the right inferior temporal lobe to signal the rest of the LH. Thus preserved language mechanisms in at least some individuals may still permit naming of objects seen via the RH.

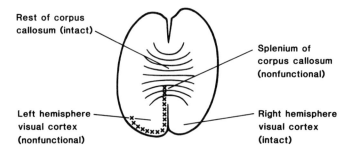

Figure 1 THE PATHOPHYSIOLOGY OF ALEXIA WITHOUT AGRAPHIA; SEE TEXT FOR EXPLANA-
TION.

Surface Dyslexia

This relatively common acquired reading disorder, involving temporal damage (Patterson, Marshall, & Coltheart, 1985), results in reading without comprehension, and may be of particular theoretical interest largely by comparison with
the much rarer, and more intriguing, deep dyslexia described below. Regularly
spelt words are "read" better than irregular ones (e.g., *sword* and *yacht*, which
may be wrongly regularized). Thus *pint* may be pronounced as if rhyming with
hint, and *bear* may be pronounced as *beer*, probably by analogy with *hear*. Legal
nonwords like *gife* and *boaz* may be read satisfactorily. Thus whole-word recognition may be lost, the patient having to fall back upon letter-by-letter or letter-
cluster processing. Thus the longer the word, the longer the reading time. In a
sense, patients are simultanagnosic (see Chapter 5), unable to see more than
one element at a time. No clear localization or pathology has been reported.

Spelling Dyslexia

As noted by Warrington and Langdon (1994), this distinctive clinical syndrome has been the subject of numerous, detailed single-case studies, though as
yet there is no consensus as to which subcomponent of the reading process is
impaired, or even whether it represents a reading-specific deficit. It resembles
alexia without agraphia in that writing is preserved, if labored, and also resembles surface dyslexia in that the patient can adopt a letter-by-letter reading strategy—its major characteristic. Thus letter naming is preserved, and the time tak-
en to read a word is proportional to word length. There may be further
disruption of already labored reading if script writing is provided or presentations
are tachistoscopic. Warrington and Langdon (1994) described a patient with a
left occipitoparietal infarction who, in addition to letter-by-letter reading with in-

tact spelling and writing, also had great difficulty in detecting a "rogue" letter attached to the beginning or end of a word (e.g., "ksong," "glassg") and in parsing two unspaced words (e.g., "applepeach"). However, she was immune to the effects of extraneous, interpolated colored letters in a word, which can make *normal* readers perform like spelling dyslexics and show word-length effects. She also had no tacit or implicit comprehension (unlike the pure alexics, see above, described by Coslett & Saffran, 1989); thus she could not categorize words semantically, had no partial knowledge of tachistoscopically presented words, and took just as long to make lexical (word–nonword) decisions as to read out the material letter by letter. The authors considered the possibility of a visual impairment not specific to reading (e.g., simultanagnosia, see Chapter 5), but concluded in favor of a prelexical deficit, with damage to an early stage in the reading process involving the visual word-form system.

Deep Dyslexia

Error analysis clearly differentiates this relatively rare disorder from other acquired dyslexias. However, there is no clear correlation with anatomy except for a large peri-Sylvian involvement, and the continued presence of the disorder after resolution of a major aphasia. Indeed, among individuals with otherwise similar and coextensive cortical damage, only a percentage are likely to evidence deep dyslexia. Numerous kinds of error may occur. They may be derivational (e.g., running → runner, length → long) or visual (e.g., error → terror, shallow → sparrow) or semantically paralexic. Except for the last type, the patient is typically anosognosic, although there may be some awareness of the semantic paralexias that so characterize the disorder. Indeed it is this feature that distinguishes *deep* dyslexia from the otherwise very similar *phonological* dyslexia (see below). Thus crocus → tulip, rock → stone, house → home, listen → quiet, air → fly, and sword → dagger. It is as if there occurs whole-word recognition for meaning, but not for form or sound. Indeed, patients are typically unable to read such intrinsically meaningless grammatical functors or particles as *be,* while having no problem with the slightly longer substantive *bee.* Shakespeare's "To be or not to be, that is the question," with its solitary (abstract) substantive should pose insuperable problems. Patients have great difficulty with nonwords (*gife, boaz*), because of their loss of phonology and of their capacity to sound out letter by letter; they have similar difficulty with the detection of homonyms (pail and pale), and rhymes with different spelling (light and kite). They typically have more difficulty with abstract nonimageable words (thought) as compared to concrete imageable ones (loaf). However, they can read, presumably via whole-word strategies, irregularly spelled concrete words (yacht, sword), though perhaps generating semantic paralexias (boat, dagger). However, paradoxically, and as yet inexplicably, they can also read case-alternating words (CaTtLe), or vertically written words,

d

o

g

or words whose overall shape is broken up by interposed asterisks (d*o*g). Patients can also successfully sort low-imageable substantives, and grammatical functors, from nonwords; this suggests intact access to the *orthographic* representations of all words, with instead a problem of accurate *semantic* access, even though they nevertheless seem to be able to go straight to semantic representations from the printed input. One explanation (Coltheart, Patterson, & Marshall, 1980) is that despite loss of LH phonology, residual RH (connotative) semantic information may be preserved and accessed prior to (LH) verbal output (see Figure 2), though as we shall see in Chapter 7 the RH reading of commissuroto-

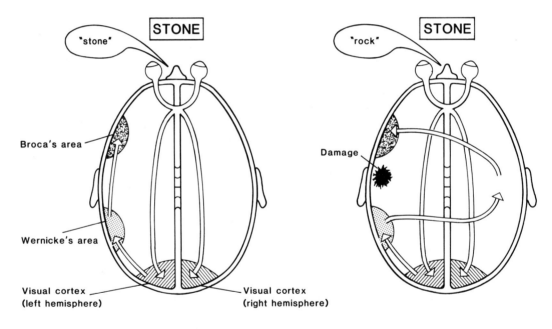

NORMAL BRAIN   BRAIN OF DEEP DYSLEXIC

Figure 2   When a normal reader sees the written word *stone*, information, initially reaching the occipital visual cortex, is transmitted to Wernicke's area in the LH, probably via the angular gyrus, and on to Broca's frontal speech area, when *stone* may be spoken. In the case of a deep dyslexic, a break between Wernicke's and Broca's areas may require rerouting via connotative language areas in the RH, with the result that the LH eventually utters the semantic-paralexic associate *rock*. (From Bradshaw, 1989, p. 127.)

my patients does not generally resemble that of deep dyslexics. Covert or automatic activation of phonology has been demonstrated (Katz & Lanzoni, 1992), the converse of the automatic activation of *meaning* in *surface* dyslexia. Moreover, some patients may be able to read legal nonwords by analogy with real, readable, words. Semantic paralexic errors may even be made to pictures, and if patients are given a single letter to name, a word response may be made where the letter occupies the initial position in the word. This all suggests an effort after meaning, a search for associations, with much covert, inaccessible knowledge. Patients seem to know a lot about what they cannot name; for example, if unable to say "lemon" to its black-and-white drawing, they may reply "it's yellow . . . a fruit . . . an orange."

Phonological Dyslexia

This disorder closely resembles deep dyslexia, with problems in reading legal nonwords (*gife*), functors, and low-frequency abstract words, except that patients characteristically fail to make semantic paralexic errors; it may merely be that they are more successful at "editing them out." High-frequency, concrete, imageable material is read best, and low-frequency, abstract, nonimageable material worst, with an intermediate level of difficulty for other combinations of material. Again localization is unclear. According to Goodglass (1993), the difficulty in reading functors may not be as marked as in deep dyslexia, and the condition may in fact occur during recovery from the latter. Despite the very considerable difficulty or inability to apply graphophonemic correspondence rules, once again the patient may be able to comprehend the material that cannot be "read."

Literal Alexia

Here, once again, whole-word reading (via the RH?) is preserved, especially when the words possess a unique or characteristic profile, contour, or shape by virtue of their (lower case) sequence of ascender or descender letters (e.g., *golf*, *apple*, or *play* vs. *rice*, *camera*, or *ear*). Patients cannot however read letters of the alphabet. The syndrome is perhaps most common among Broca's aphasics and was first described by Hinshelwood (1900, cited in Kaplan, 1991).

The Agraphias

With aphasia, especially Broca's, there is usually some impairment of writing. Block printing, rather than cursive writing is more likely to be preserved, with oversized letters awkwardly formed or reversed (e.g., И Ƨ). However, with Wernicke's aphasia, cursive letters may continue to be well formed, though the usual semantic paraphasias, neologistic jargon, and paragrammatisms may occur, this time in written form, with repetitive use of low information verbs and vague nouns. Misspellings occur through both omission and substitution (i.e., literal paragraphias). *Pure agraphia* is rare, and may or may not be associated with the disputed Exner's center at the foot of the second frontal gyrus beyond the hand area. Kaplan (1991) suggested that there may be two forms of the disorder, a frontal form involving a disorder of grapheme selection, and a parietal form involving spatiotemporal disorganization somehow specific to writing. With *parietal* agraphia, there is apraxic formation of letters, whereas with *surface* or *lexical* agraphia (cf. surface dyslexia) the patient may write regularly spelled words fairly well, but misapply rules for writing irregular words. Thus bought → bort, and cough → coff. As with surface alexia, there may be intact spelling of dictated pseudowords and of high-frequency words, but not so for homophones and irregularly spelled lower frequency words. With *deep agraphia,* reading may be intact, but writing exhibits semantic paragraphias, analogous to the paralexias of deep dyslexia. Thus yacht → boat, and laugh → smile. With *phonological agraphia* (cf. phonological dyslexia), inordinate difficulty may be experienced in spelling or writing nonsense strings to dictation, although real words may be written fairly well, especially if regular in spelling, common, and imageable or picturable; if the word can be written, it can then often be said, thereby demonstrating dissociation between written and spoken naming. Kaplan (1991) reviewed these disorders of writing in greater detail.

Preservation of Implicit Understanding in Language Processing

As Tyler (1992) observed, most aphasia research bears primarily upon processes and representations about which the listener can become aware, but generally fails to address aspects of the comprehension process that remain covert, aspects that may be crucial to language comprehension. As she noted, there may be two types of awareness that arise via quite different processes, and that may be selectively impaired in brain damage. One process is automatic, and provides "awareness" from a procedure that through injury cannot run through its normal operations; the other is at least partially under voluntary control, and involves the individual reflecting upon properties of the final representation. Thus patients may show relatively normal cognitive function when assessed via an *implicit* task, but not so when assessed by means of an *explicit* task. The former attempts to access internally represented knowledge without requiring the subject to become consciously aware of it, whereas with the latter the subject must consciously access certain information while carrying out the task.

Conventional explicit procedures, such as the latter, typically involve, for example, sentence–picture matching, acting out commands or statements, and grammaticality judgments where the patient has time to reflect on what was presented and what is required.

Tyler (1992) reviewed the few studies relevant to the implicit or explicit distinction in language comprehension by aphasics. Milberg and Blumstein (1981) reported Wernicke patients with severe comprehension deficits who nevertheless showed significant semantic priming effects in a lexical decision task. Thus, like normal controls, they were faster responding to target words that were preceded by semantically related compared to unrelated words, even though they could not overtly judge whether or not the words were semantically related. Tyler noted that a number of procedures may be employed to implicitly tap intermediate mental operations (e.g., associative priming and word monitoring). The latter becomes an implicit measure of language function when the target word for monitoring occurs after the relevant cue or priming stimulus, so that the listener's awareness of the properties of the target is not what is at stake. Rather, the experiment assesses how monitoring latencies to the target word are affected by what has been heard (as a cue) prior to that word. The listener in such "on-line" tasks, which tap into intermediate representations, must produce a response that is closely synchronous with relevant stretches of the speech input. Thus a response may have to be made immediately after the critical part of the stimulus has been heard, as in a word-monitoring task, or by stopping the input at a specified point and requiring the listener to respond on the basis of partial information, as in a gating task.

In her word-monitoring task, Tyler (1992) asked her Wernicke patient to press a response key as soon as he heard a target word such as *guitar,* which occurred in a series of matched sentences:

1. The crowd was waiting eagerly.
   John carried the guitar.
2. The crowd was waiting eagerly.
   John buried the guitar.
3. The crowd was waiting eagerly.
   John drank the guitar.
4. The crowd was waiting eagerly.
   John slept the guitar.

It was assumed that latencies to the target word reflect the successful processing of the prior material. Such latencies should be slowed by pragmatic (2), semantic (3), or syntactic (4) anomalies. Tyler (1992) found this pattern for both controls and her Wernicke patient, indicating that he accessed the content of the verb as soon as he heard it, and could use this information to constrain permissible verb agreement structures. She concluded that there may be a clear dissociation between relatively normal performance on tests of "on-line" language processing, with impairment on "off-line" tasks, such as grammaticality judgments and sentence-picture matching. Intermediate representations and the processes involved in their construction may therefore remain unimpaired, whereas conscious access to the final constructed representation is defective, so that explicit decisions cannot be made.

Although this discussion derives from the context of aphasia rather than alexia, the concept of preserved implicit understanding clearly bears upon performance in many of the alexias, where such performance in the laboratory typically tends to underestimate the patient's capacities in real life.

The Localization of Language Sites by Positron Emission Tomography

There has recently been an explosive increase in the use of positron emission tomography (PET) to study brain function. Brain areas actively engaged in current information processing can be visualized or imaged by the resultant reflex increases in regional cerebral blood flow (rCBF) or metabolism, as indexed by radioactive tracers. As Petersen and Fiez (1993) cautioned in their review, a functional area is not necessarily a task-mediating area, anymore than there is, for example, a "tennis forehand area"; complex functions like language cannot be localized to discrete unitary regions. We have already emphasized that any given task or function uses a complex and distributed set of brain areas, each of which makes a specific contribution to the performance of the task as a whole. Moreover, factors intrinsic to the design of experimental tasks, such as rate of stimulus presentation, practice, task difficulty, and attentional demands, may all affect the distribution of activated regions.

Petersen, Fox, Posner, Mintun, and Raichle (1989) pioneered the subtractive or control paradigm in the application of PET imaging in cognitive processes, a paradigm that derives from Posner's (1978) subtractive chronometric approach with reaction times. The active PET state at one level acts as the control condition for the next, higher, processing level; subtraction of the PET image of the former from that of the latter permits visualization of those areas that are uniquely active during processing in the higher level alone (see Figure 3). Thus the authors compared simple (visual or auditory) presentation of nouns versus no such presentation; repetition of nouns versus simple presentation; generation aloud of a verb appropriate to the presented noun versus repetition. They found that auditory presentations resulted in the bilateral activation of the superior temporal gyrus, and LH activation of temporoparietal cortex; visual presentations were associated with left-sided activation of medial extrastriate visual cortex. Speech output led to supplementary motor area (SMA) activation and that of the primary sensorimotor mouth cortex, and of LH premotor regions surrounding classical Broca's area. When verbs were generated, there was activation of the anterior cingulate, and of anterior inferior prefrontal regions on the left. They noted that prefrontal activation does not occur with practiced, automatic

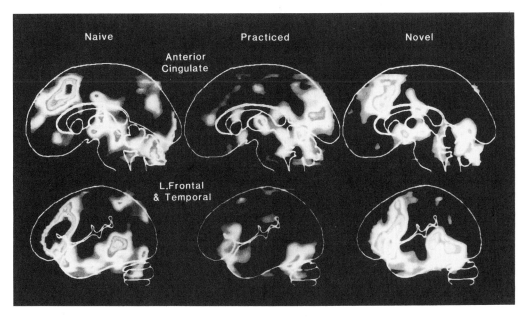

Figure 3   Thinking. A PET scan of the left hemisphere while the subject thinks of a use for each visually presented noun. (The neural activation for reading the nouns aloud is subtracted from the image.) The left column illustrates the initial performance of the task, the middle column is the activity after practice with the same list of words, and the right column is after switching to a new list. The top row (medial surface) shows activation of the anterior cingulate, thought to be involved in focal attention. The bottom row (lateral surface) shows language-specific activation in the frontal lobe and Wernicke's area. (After M. I. Posner (1993). Seeing the mind. *Science, 262,* 673–674, and from a figure kindly provided by M. Raichle. Copyright 1993 by the AAS.)

responding, only with difficult, nonhabitual responses involving higher order transformations of information, such as might be involved with working memory (Goldman-Rakic, 1988a). They placed the reading lexicon in mesial extrastriate cortex on the left, and the phonological lexicon in left temporoparietal cortex.

Howard et al. (1992) extended the subtractive control PET technique for rCBF changes during word reading and spoken repetition; rCBF was compared with control conditions of stimuli of comparable auditory and visual complexity. The control for *spoken* word repetition was hearing spoken words presented backwards, and this control led to bilateral activation of the superior temporal gyrus. Spoken word repetition relative to this control led to activation on the left in the superior and middle temporal gyri (i.e., Wernicke's area). The control of word *reading* was seeing stimuli "written" in "letterlike" shapes, and this control led to bilateral rCBF changes in striate and extrastriate cortex. Word reading relative to this control led to rCBF increases in the posterior part of the middle temporal gyrus on the left. They therefore concluded that the lexicon for *spoken* word recognition is localized in the middle part of the superior and middle temporal gyri on the left, whereas the lexicon for *written* word recognition is to be found in the posterior portion of the left middle temporal gyrus. Thus their findings, especially their auditory studies, agree fairly well with classical lesion studies; for their reading task, the posterior middle temporal gyrus lies on the edge of the classical angular gyrus. However, the medial extrastriate location for the visual lexicon proposed by Petersen, Fox, Snyder, and Raichle (1990) fits less well. Howard et al. (1992) also noted that absence of Wernicke's area activation in their written (unlike their auditory) condition goes against earlier claims that readers must first recode written words into auditory form in Wernicke's area.

Démonet et al. (1992) examined the auditory comprehension of phonological and lexico-semantic processing via PET. The control (reference) condition was the presentation of pitch, the phonological task required monitoring the sequential phonemic organization of nonwords, whereas in the lexico-semantic task subjects monitored concrete nouns according to semantic criteria. They found that phonological processing led to LH activation of the superior temporal gyrus and the anterior part of Wernicke's area, with some Broca's area activity. Lexico-semantic processing was associated with left-sided activation of the middle and inferior temporal gyri, inferior parietal and superior prefrontal regions, in addition to the (above) superior temporal activity during *phonological* processing. With respect to Broca's area involvement in the phonological task, Démonet et al. noted the extensive (arcuate fasciculus) connections to Wernicke's area, and speculated upon the possible role of Broca's area in the *motor* theory of speech perception (Liberman & Mattingly, 1985) or its involvement in an articulatory loop of working memory or in both. They also reviewed other findings of Broca's area and SMA activation during silent word generation or rhyming tasks.

Other PET or rCBF imaging studies are those of Paulesu, Frith, and Frackowiak (1993) and Mazoyer et al. (1993). The former group concluded that the articulatory loop of working memory (see Chapter 8) contains two components with visual presentations:

1. A phonological store localized to the supramarginal gyrus on the left
2. A subvocal rehearsal system associated with Broca's area on the left.

Mazoyer et al. (1993) similarly concluded that the large network of areas on the left, which is active when listening to fully meaningful material, corresponds closely to those identified in classical lesion studies. This network, however, is reduced to a sparse and weakly lateralized set when syntactically correct yet meaningless "jabberwocky statements" are given. They find

1. activation of the left inferior frontal gyrus (Broca's area) for lexical processing of single words,

2. activation of the left middle temporal gyrus for phonological processing of single words, and

3. activation of the left superior temporal gyrus (Wernicke's area) for conceptual or semantic processing.

The bilateral activation of the temporal poles alone, after subtraction of the above components with continuous speech, corresponds to activity of the left temporal pole for memory of linguistic content, and to activity of the right temporal pole for the prosodic and pragmatic aspects. They reject the idea of a neural *hierarchy* of areas that successively and automatically come into play whenever they receive an adequately structured stimulus, for processing phonology, lexicality, semantics, and so on; rather, there is coordination of a network of areas, in *parallel,* each of which may be specialized in one aspect of speech processing.

Generally, we can conclude that PET studies confirm and extend, rather than contradict, the findings from the classical lesion approach. Frackowiack (1994) noted in his summary a segregation into left inferior frontal and superior temporal activation for phonological processing, with a set of distributed foci in the inferior temporal, inferior parietal, and superior frontal regions for semantic aspects. Activation of the left posterior superior temporal cortex generally occurs in auditory comprehension tasks, with activation of the inferior frontal cortex (Broca's area) during speech output. He concluded that there is preliminary evidence for the involvement of posterior and middle temporal regions of the LH in access to the input lexicon, with the inferior left temporal cortex and the parieto-occipital junction (angular gyrus) being engaged during access to semantic representations of objects and lexical monitoring.

The Localization of Language Sites by Stimulation Mapping

The traditional model of language localization has been based on stroke and lesion studies. The investigation of language localization with electrical stimulation mapping during neurological operations under local anesthesia provides a different perspective on this model (Ojemann, Ojemann, Lettich, & Berger, 1989). A current applied to some cortical sites blocks ongoing object naming or results in errors. The inhibitory mechanism probably involves neuronal inactivation by depolarizing blockade. Such sites include but extend beyond the rolandic cortex of both hemispheres, and the SMA and classical Broca and Wernicke areas of the left. Lesion studies typically provide information about a more limited area for a given patient than is potentially available with stimulation mapping, and such lesion-based information may be contaminated by any functional reorganization that may have occurred between time of injury and time of testing. (Ojemann et al., 1989, however, noted few differences between patients with lesions of early or adult onset in their stimulation studies.) Studies that employ neuronal activity recording blood flow or metabolic measurements (e.g., PET) may indicate regions *participating* in language, but not whether they are *essential* for it, unlike the lesion and stimulation paradigms. Ojemann et al. (1989) assessed the localization of such cortical sites essential for language by stimulation mapping in the left, dominant hemispheres of 117 patients. The language center was often highly localized, forming mosaics 1 or 2 cm², usually one in the frontal and one or more in the temporoparietal lobe. At an individual level, despite considerable variability, areas were generally much smaller than the traditional Broca–Wernicke areas. Sex and verbal intelligence correlated with such variability, especially in the parietal lobe. However, females seemed to be overrepresented in the small subgroup of patients with only frontal language sites (and see also Kimura, 1983, for similar findings from lesion studies); moreover, differences in language organization (reading vs. naming errors), based on verbal IQ alone, seemed to involve primarily the middle and superior gyri of the temporal lobe. The biological substrate for language may therefore differ in patients with differing verbal abilities, with no anatomical landmark or differentiated cytoarchitectonic areas of cortex reliably indicating its presence or absence outside the posterior inferior frontal lobe. Indeed Ojemann et al. (1989) concluded that the classical Wernicke and Broca areas are an artifact of combining the locations of essential areas across different patients, with, occasionally, Broca's area not even being involved in language at all. Moreover, sometimes language sites have manifested in such unexpected locations as midfrontal, midparietal, or posterior temporoparietal locations, according to previous studies reviewed by Ojemann et al., but as they noted we must distinguish between arrest of speech activity during naming *per se,* and arrest of *general* motor activity *including* speech. At frontal premotor sites such a distinction may in any case even be inappropriate.

Ojemann et al. (1989) concluded that stimulation studies indicate a modular or mosaic organization of language cortex. Each patient will usually have several such units, generally at least one frontally and temporoparietally. During language these mosaics will probably be selected in parallel by a thalamocortical activating system producing intense local selective activity.

The Localization of Language Processes in Congenitally Deaf Users of Visual Signing

Poizner, Bellugi, and Klima (1990) described cases where, in the congenital absence of hearing, language and communication developed via the visual–spatial channel of signing, rather than the usual auditory–temporal modality, and where subsequent strokes rendered the patients aphasic for signs. Such questions could now be addressed as whether language is committed to the LH irrespective of the modality whereby it is made manifest, its relation to gesture, pantomime, and praxis, and whether hearing and speech are necessary prerequisites for the development of normal patterns of language localization. The medium employed was American Sign Language (ASL), a true living language rather than a manual pantomime or a simple derivation from English. It has a grammar and a sequential and segmental structure analogous to the phonemes and syllables of spoken English.

Gail D., previously effortless and fluent in her signing, resembled a Broca's aphasic after her anterior LH stroke. Her signing was effortful, agrammatic, and limited to single uninflected signs. Its content was almost exclusively restricted to nouns and verbs, without grammatical inflections. Her comprehension was, however, comparatively normal. From this and similar studies, some with posterior damage, the investigators noted the strong parallels between the different forms of aphasic impairment in signed and spoken language. Both systems, despite their striking spatiotemporal and modality differences, depend on LH structures, and the basic contrast between anterior (Broca's) and posterior (Wernicke's) aphasia seems also to hold up in ASL. It even proved possible to differentiate between the linguistic, symbolic, and motoric aspects of hand movement, with a dissociation between the capacities for initiating the linguistic gestures of sign language, and for producing and initiating communicative but nonlinguistic gestures such as miming.

Summary and Conclusions

Speech is more or less automatically, and effortlessly, acquired at an early age, against great odds and frequently from a very flawed model. Reading and writing however are recent inventions, and have to be painfully learned. Although there are many different forms of writing, including ideograms, syllabaries, and alphabets, syllabaries are perhaps the most naturalistic, and alphabets the most economical. Some scripts (e.g., Japanese *Kana* and *Kanji*) combine ideographic and phonological-syllabic characteristics, and may dissociate with respect to preferred locus of lesion in the event of acquired reading loss. Even in an alphabetic script we may read via two routes, one involving the application of graphophonemic conversion rules, and the other a system of direct access that may be available to skilled readers. Although reading disorders may parallel speech disorders, they are typically less systematic and may involve the interplay of a very wide range of centers.

In *alexia with agraphia,* both faculties are lost, reading and writing, although the patient may still be able to name letters. The left angular gyrus has long been thought both necessary and sufficient, as it receives the requisite polysensory input; a right inferior quadrantanopia may co-occur. With *pure alexia (alexia without agraphia)*, the patient can still write but cannot read even his or her own writing. The language zone of the LH is disconnected from all visual input. Surprisingly, some aspects of tacit or implicit understanding may be preserved. A right superior quadrantanopia may often co-occur. In *surface dyslexia* regularly spelt words are "read" better than irregular ones, which may be wrongly regularized, leading to reading without comprehension. Whole-word recognition is probably lost, the patient having to fall back upon letter-by-letter or letter-cluster processing. Letter-by-letter reading is also apparent with *spelling dyslexia,* the time taken to read a word being proportional to its length. However, it is not clear whether the phenomenon represents a deficit that is specific to reading. Error analysis differentiates *deep dyslexia* from the other acquired reading difficulties. A particularly characteristic error is semantic paralexia, where a word of related or associated meaning may be given to a visually presented word. Patients are typically unable to read grammatical functor words (though some tacit or implicit access to their orthographic representations may be preserved), and have more difficulty with abstract than concrete items. They can read, however, presumably via whole-word strategies, irregularly spelt concrete words. There is no clear correlation with anatomy, except for a large left peri-Sylvian involvement. Despite loss of LH phonology, it is possible that RH connotative or semantic information may be preserved and accessed. *Phonological dyslexia* closely resembles deep dyslexia, except that semantic paralexic errors are absent, whereas, with *literal alexia,* whole-word reading, again perhaps via the RH, is preserved, especially when words have a characteristic profile.

The agraphias largely parallel the alexias, for example, pure agraphia (though the condition is disputed, and any mechanism is likely to differ from that of pure alexia), deep agraphia, phonological agraphia, and surface or lexical agraphia.

A recurring phenomenon in neuropsychology that is not restricted to reading disabilities is the preservation of tacit or implicit processing or understanding, when the requisite function is apparently lost at an explicit, conscious level.

New imaging techniques permit us to visualize the brain regions metabolically active during a given reading (or speech) task. Although some studies have employed patients with brain injuries, many have used normal healthy individuals. It is reassuring that such techniques have generally confirmed and extended traditional findings made with conventional clinical populations. It is noteworthy however that practice and strategy can partly determine the regions currently active when inspecting words, reading them silently or aloud, or producing phonological or semantic associates. Stimulation mapping during reading (or speech) tasks provides yet another perspective on the mechanisms of visible and spoken language. A current applied to certain brain regions can block their function during an ongoing task. However, clinical patients prior to surgery are the usual candidates for such invasive procedures, thus accounting for possible discrepancies from findings with normal subjects studied by neuroimaging methods. The latter technique, moreover, can indicate regions *participating* in language, whereas electrostimulation and lesion mapping may show whether or not these regions are *essential* for it. Interesting discrepancies are emerging between the findings from electrostimulation and lesion-mapping techniques, the former tending to indicate a modular or mosaic organization of language cortex. On the other hand, studies with congenitally deaf individuals proficient in sign language who have experienced stroke indicate that sign language is represented in the brain in exactly similar ways to spoken language.

Further Reading

Friedman, R., Ween, J. E., & Albert, M. L. (1993). Alexia. In K. M. Heilman & E. Valenstein (Eds.), *Clinical neuropsychology* (3rd ed.) (pp. 37–62). Oxford: Oxford University Press.

Howard, D., Patterson, K., Wise, R., Brown, W. D., Friston, K., Weiller, C., & Frackowiak, R. (1992). The cortical localization of the lexicons. *Brain, 115,* 1769–1782.

Kaplan, E. (1991). Aphasia-related disorders. In M. T. Sarno (Ed.), *Acquired aphasia* (2nd ed.) (pp. 313–338). San Diego: Academic Press.

Ojemann, G., Ojemann, J., Lettich, E., & Berger, M. (1989). Cortical language localization in left, dominant hemisphere. *Journal of Neurosurgery, 71,* 316–326.

Petersen, S. E., & Fiez, J. A. (1993). The processing of single words studied with positron emission tomography. *Annual Review of Neuroscience, 16,* 509–530.

Poizner, H., Bellugi, U., & Klima, E. S. (1990). Biological foundations of language: Clues from sign language. *Annual Review of Neuroscience, 13,* 283–307.

Roeltgen, D. P. (1993). Agraphia. In K. M. Heilman & E. Valenstein (Eds.), *Clinical neuropsychology* (3rd ed.) (pp. 63–90). Oxford: Oxford University Press.

Chapter 4

A Right-Hemisphere Contribution to Language and a Calculation Deficit

A major if not exclusive role for the left hemisphere (LH) in language mediation is probably most secure for adult right-handed males, and for the phonological and articulatory aspects of speech, as shown by experiments with brain-lesioned individuals, split-brain patients, and normal healthy adults (Bradshaw, 1989). However, even with adult male dextrals, the right hemisphere (RH) almost certainly plays an important supportive role, an idea, as Baynes (1990) observed, which goes back to Hughlings Jackson in 1874. The latter noted that the RH might be the source of automatic, nonpropositional speech in global aphasia; more recently many have suggested that it might play a role in language recovery after LH damage, and that it might underlie the reading of deep dyslexics (Coltheart, 1980a, 1980b, 1985, though see Patterson & Besner, 1984). Zaidel (1990a) claimed that it may play a role in normal reading; we shall review evidence shortly that patients with RH damage may be deficient in the appreciation of humor and metaphor, and in maintaining the contextual flow of discourse or a narrative (Molloy, Brownell, & Gardner, 1990). On the other hand, RH linguistic skills of whatever sort may only develop in a subset of individuals, whether normal or brain-damaged. Indeed any RH contribution may depend partly on a person's age, sex, and handedness, and on whether receptive or expressive aspects of language are being addressed; thus most studies agree that the RH has little articulatory capacity. More importantly, any RH contribution may also depend upon whether it is operating in a fully normal brain, or in isolation following hemispherectomy or disconnection, or in the presence of a partially or an extensively damaged LH partner.

Injury can of course occur at any age, though the younger the patient, the more likely it is that plastic reorganization may occur within or between the hemispheres, with a good prognosis for recovery of function (Orsini & Satz,

1986). Thus infantile damage, especially the extreme case of hemispherectomy (see below), can inform us about the limits of plastic reorganization, whereas adult damage can tell us more about major and minor hemisphere involvement in a given language function in the normal mature brain. Of course the LH in many if not most clinical cases may continue to be the source of "recovered" aphasic speech. However, an apparently recovered aphasic with earlier LH trauma, and who again experienced speech loss after barbiturate (Wada test) suppression of the RH, would provide good evidence for RH involvement in language recovery. (In the Wada test, to determine speech lateralization prior to surgery, see e.g., Wada & Rasmussen, 1960, sodium amytal is directly injected into the internal carotid artery that supplies the ipsilateral hemisphere; in the case of the LH this typically leads to the temporary abolition of speech for about 5 min, apart from some clichés and automatisms, and the capacity sometimes to sing along. Comprehension is more or less preserved.)

A consideration of possible RH contributions to language must address the following issues:

1. Do surviving language capacities of LH patients reflect the suboptimal operation of damaged LH structures, or instead suboptimal RH takeover with or without LH interference?

2. If such LH interference does occur, after a putative takeover by the RH, might it be advisable surgically to remove residual LH structures?

3. Do any language difficulties, after RH injury, reflect deficits in that hemisphere's *normal* contribution to language (if any), or interference or noise from the right reaching the LH? The Wada test can of course address some of the above questions.

4. Do any verbal or indeed *spatial* deficits after unilateral injury reflect "cognitive crowding"? Thus infantile left-hemispherectomy studies suggest that language is highly valued biologically, and may preferentially take over RH processing space to the detriment of spatial skills, being maintained in whatever neurological substrate is available (Baynes, 1990).

5. What, if any, are the normal contributions of the RH to language? Indeed, performance after left hemispherectomy may reflect functional reorganization rather than the normal or initial capacities of the RH.

Major Left-Hemisphere Damage in Infancy

Such damage may require massive hemidecortication ("left hemispherectomy") because of intractable seizures. Near-normal language may be attained in later tests in adulthood (Dennis, 1988). Exceptions may occur where complex syntax has to be handled, such as passive negatives; in such instances the nouns themselves provide no immediate cue to the subject–object relationship (e.g.,

"The cow was not kicked by the horse"). Long, embedded sentences are also thought to be difficult for the surviving RH (e.g., "The thought that my admiration for your tan results from the fact that I have myself been indoors all winter really never crossed my mind"). However not everyone accepts the existence of such residual syntactic or phonological deficits in infantile left hemidecorticates tested as adults, and the studies themselves have been criticized on statistical and other grounds (Bishop, 1983).

Subtotal Left-Hemisphere Damage in Infancy

Subtotal damage is typically associated with a better prognosis the earlier the damage occurs. With early left-sided injury, an RH takeover of language functions, together with sinistrality, are more likely to take place, which Orsini and Satz (1986) believe are themselves good prognostic indicators. On the other hand, early lesions that do not result in a shift of hand preference tend not to shift language; the latter may not therefore develop or recover so well with retained, impaired LH dominance. Thus age of injury, rather than severity, may be the more important factor in determining both subsequent handedness and language functions in the young brain-injured. Indeed, account must be taken of the absolute age of occurrence of initial injury, when any subsequent surgical intervention may have taken place, and when testing occurred, together with the duration of associated intervals, in the context of brain reorganization.

Major Left-Hemisphere Damage in Adults

Left hemispherectomy theoretically provides the best evidence for any RH contribution to language. However, such an operation is naturally rare, with few survivors over any sufficiently extended period to permit adequate testing. Very often, moreover, initial damage or malfunction will have been sufficiently early to permit at least some brain reorganization. What evidence there is suggests severe expressive deficits with very little capacity for connected propositional speech, other than some expletives and automatisms, though a considerable capacity for comprehension may be retained. Thus the RH under such circumstances may behave like an intelligent nominal aphasic.

Subtotal Left-Hemisphere Damage in Adults

Damage of this sort typically results in one or other of the classical aphasic syndromes reviewed above. Such patients may evidence left ear, and left visual field, advantages for processing verbal material, suggesting RH takeover of function. While evoked potential (EP), transcranial magnetic stimulation, and positron emission tomography (PET) imaging studies may tend to support such conclusions (see e.g., Cardebat et al., 1994; Coslett & Monsul, 1994), it should be noted that the above ear and field asymmetries could merely reflect loss of the usual crossed route to the LH, leaving only the *ipsilateral* left ear and visual field pathway to the LH, with continuing language mediation by that structure. Thus a better indication of a gradual takeover of language function by the RH would be *fixed* right ear and visual field scores, with gradually improving performance on the left, especially perhaps for semantic rather than phonological aspects (Bradshaw, 1989). There is indeed some evidence for this effect occurring, and also for claims that nondextrals are likely to have a better prognosis after LH injury (and worse after RH injury) than dextrals (Bradshaw, 1989).

Evidence from Deep Dyslexia

Deep dyslexia is thought by some (e.g., Coltheart, 1980a, 1980b, 1985) to provide firm evidence for an RH involvement in reading by clinical patients. It is of course relatively rare, occurring usually in the recovery phase from severe aphasia, though it exhibits certain similarities with developmental dyslexia in children. It will be remembered (Chapter 3) that deep dyslexia is characterized by a preserved ability to read (by direct access) irregular words (*yacht, sword*) but not legal nonword letter strings (*grife, crobe*) because of a presumably impaired grapheme-to-phoneme conversion system. Thus patients typically cannot say whether *sail* and *sale* rhyme, but claim that *few* and *sew* do. They show frequent semantic paralexias (rock → stone, dagger → sword) with an inability to read grammatical functor words (*be,* but not *bee*). It is as if a read word has to be part-processed in the RH, where it acquires semantic associations that instead are ultimately articulated, presumably via residual LH systems. Interestingly, both deep dyslexics and normal readers may give more semantic paralexic responses to stimuli flashed in the left visual field, although the very sophistication of some semantic paralexic responses may suggest more than just RH mediation. Thus we may find hermit → recluse, oblivion → infinity, lunatic → moonman, muddy → quagmire, responses that are rare, low-frequency words. Indeed although it is dangerous to argue from a single case, there is a report (Roeltgen, 1987) of a deep dyslexic (consequent upon an initial LH stroke) who lost residual reading capacities after a second, LH stroke, thus arguing against the RH hypothesis. However, we shall argue that commissurotomy studies do show similar effects (to deep dyslexia) in RH reading (Zaidel & Schweiger, 1984). Thus both syndromes evince impaired grapheme–phoneme conversion, semantic paralexias, and better concrete than abstract word comprehension. The isolated RHs of commissurotomy patients tend to perform worse overall than deep dyslexics, who may therefore benefit from some residual LH assistance. Maybe the RH normally has very little reading ability, in health or after LH injury, except in a small minority of individuals, and it is *they* who become deep dyslexic if LH injured; deep dyslexia is certainly a rare syndrome. Zaidel and Schweiger (1984) however believed that the RH normally supports the recovery of LH aphasics, and that although the LH has both direct (lexical, whole-word) access and grapheme-to-phoneme conversion rules, the RH has only the former route to reading. They suggested that in deep dyslexia subtotal damage occurs to both LH routes, and that while the RH direct-access route largely takes over with its more diffuse semantic system, it then must still "report back" to residual LH mechanisms for output. We may then ask, with Baynes (1990), why *pure* alexics read so poorly. However, as noted above, with very *brief* presentations they *are* capable of performing "covert" lexical decisions, and semantic categorizations, just like deep dyslexics, while nevertheless *overtly* being only able to perform labored letter-by-letter reading (Coslett & Saffran, 1989).

Commissurotomy Studies

In Chapter 7 we describe in detail the studies with commissurotomy patients whose hemispheres have been surgically disconnected by section of the corpus callosum for the relief of otherwise intractable epilepsy. Zaidel (1990a) reviewed his findings with the Z-lens technique, which permits prolonged visual interrogation of the disconnected RH. He concluded that comprehension by the RH does not correspond to any one stage in a child's normal language acquisition; thus the RH has a visual vocabulary far smaller than its considerable auditory vocabulary, and a virtually complete expressive aphasia. It has a rich lexical, semantic, and conceptual system, a shorter verbal short-term memory than that of the LH, a limited syntactic system, an impoverished phonological system, and virtually no articulatory competence at all.

The pattern, moreover, corresponds to no single aphasia or acquired alexia type, though as we shall see its "reading" (as indexed by error analysis in multiple-choice responding) may sometimes resemble that of deep dyslexics by virtue of frequent semantic paraphasias. Thus words of similar or associated meaning may be chosen, suggesting that the RH has a looser, more associative lexico-semantic or conceptual organization than that of the LH (Zaidel, 1990a). Unlike a child who can read, recognize objects, and understand speech, the disconnected RH has no grapheme–phoneme conversion rules; thus it cannot match two printed rhyming words that are spelled differently (e.g., *rough* and *cuff*). It accesses meaning directly from orthography without intervening phonetic coding. It can therefore match the written word *sail* to a picture of a sail; it can even match such a picture to the spoken word sound. However, it cannot go directly from orthography to sound, matching the *written* word *sale* (homonym of *sail*) to the picture of a sail, even though it clearly must have some limited ability to evoke the phonological or acoustic representation of the name of a concept. In sum, it has a unique reading profile: it can read simple material (e.g., single words), but not via grapheme–phoneme conversion; it reads concrete nouns better than function words, and it often cannot tell apart close associates from the actual meanings of real words.

The fact that the disconnected RH can read and comprehend single words does not necessarily mean that it similarly participates in the normal undivided brain. Indeed, Gazzaniga (see e.g., Gazzaniga & Smylie, 1984) would deny that the RH normally plays any role in language, and that it is only in some few commissurotomy patients, with very long-standing epilepsy and cortical anomalies resulting in cerebral reorganization, that the RH has taken on any such functions. Zaidel (1990a) noted that seizure activity in such patients was unlikely to have been sufficiently severe to cause such reorganization. Much of the long-running Zaidel–Gazzaniga dispute stems from differences (real or interpreted) in the commissurotomy operations (e.g., total, partial, staged, or all-at-once), and differences in ages at time of initial pathology, operation, and testing.

A question nevertheless remains as to why LH aphasics do not normally show residual reading capacities (let alone auditory comprehension levels) compatible with the reading competence of the disconnected RH. Zaidel (1990a) suggested that the relatively few patients with pure word blindness (alexia without agraphia, see Chapter 3) lie at one extreme of a continuum, possessing no RH reading capacity; thus with so recent a historical development as reading, there is likely to be a wide range of individual differences, reflecting variations in brain organization, skill, and strategies. Alternatively, RH reading may only be released under certain circumstances from inhibition by the left (Regard, Cook, Wieser, & Landis, 1994), for example, after commissurotomy, or depending upon etiology, size, location, and nature of the aphasiogenic lesion in the LH (i.e., perhaps only after massive damage and/or with very "difficult" material; with "easy" material or subtotal damage the patient may be "stuck" with an incompetent LH). Similarly, the RH of the deep dyslexic may only be used for lexico-semantic access when the impaired reading mechanism of the damaged LH cannot cope, or is slower than the direct "ideographic" reader in the right (Zaidel, 1990a). Even then, of course, the phonological and articulatory mechanisms of the LH will be needed to *name* the words. It is also probable that, with pure alexics, the mechanism that might normally allow shift of control from LH to RH is also lost in the damage, again leaving the patient with only an incompetent LH. Zaidel therefore sees no real conflict between reading competence in the disconnected RH, and several of the acquired dyslexias or alexias. Transfer of responsibility to the RH is perhaps more likely in deep dyslexia than in pure alexia; reading in the disconnected RH does not provide an accurate estimate of the reading competence of the normal RH; the deep dyslexic has more reading competence than the disconnected RH, whereas the pure alexic has less.

Support for these conclusions comes also from the findings with the Z-lens technique of selective LH deficits, when *that* hemisphere is tested in isolation from the right. Although that hemisphere may show no gross impairment in the comprehension of spoken or written words, subtle deficits appear in reading discourse at a "parapragmatic" level. Thus there may be problems in interpreting affect as expressed by spoken prosody, and a tendency to associate spoken expressions with pictorial representations of *literal* rather than *metaphorical* meaning (cf. the studies, reviewed below, of the effects of RH lesions by Molloy, Brownell, & Gardner, 1990).

Right-Hemisphere Damage in Adults

Few or only minor language deficits have been traditionally associated with RH damage (e.g., perseverations, hesitations, mild anomia, reduction in verbal creativity and in the ability to use stock phrases, and verbal stereotypies). Prosodic alterations have also been reported after RH damage (for review, see Joanette, Goulet, & Hannequin, 1990), both at a linguistic and an emotional level. In the former case, intonational contour is used to stress the interrogative or declarative modality for questions, statements, or imperatives. In the case of emotional prosody (see e.g., Ross, 1981), anterior RH damage is associated with the expression, and posterior RH damage with the comprehension, of emotionality. Joanette et al. (1990), however, doubted whether the RH's contribution with respect to emotion exactly parallels that of the LH with respect to propositionality, and rightly caution against any absolute left–right dichotomy (and see also Bradshaw, 1989).

Recently, however, studies (see e.g., Molloy et al., 1990; Joanette et al., 1990) have reported RH-lesioned adults as missing the point in humor and jokes; they may adopt inappropriately literal or concrete interpretations with metaphor (whether "frozen" and conventional as in "to have a heavy heart," or newly coined as in "to stir the pot of popular resentment") and aphoristic sayings like "make hay while the sun shines," or "strike while the iron is hot." Such patients may have difficulty in detecting absurdities and *nonsequiturs*, and in understanding figures of speech. They may have problems grasping the connotation of the entire utterance and appreciating oxymorons ("intense apathy," "loyal opposition"), sarcasm ("you *are* a fine fellow!"), hyperbole ("tons of love"), and bathos ("one nuclear bomb can ruin your whole evening"). A general loss of context may result—a form of verbal hemineglect. Such deficits may affect our highest human capacities, such as creativity and sensitivity to contextual and social nuances. Meaning does not therefore proceed just from a string of isolated words or ideas, but from the contextual implications and inferences made by the listener, and the common assumptions and knowledge shared between the speaker and the listener. Thus an indirect request like "Can you bring the beer?"

may mean "Are you physically able to, whether it is needed or not," or "Please bring me some beer." RH patients—and computer-translator systems—both have problems with contextual differences and intended meanings such as these. With a joke, the critical utterance is not ambiguous, like the above indirect requests, or instances of irony or sarcasm ("you *are* a fine fellow!"), but instead serves to alter the interpretation of what (ambiguously) *preceded*. Consequently, the joke's punch line revises initial assumptions by providing new context, whereas irony or sarcasm modifies the interpretation of what follows in the light of prior information. RH patients are poor at choosing correct alternative punch lines, and generally at updating old assumptions. They have problems with *deductive* inferences, where speech or text must be interpreted on the basis of *preceding* information, and with *inductive* inferences, where the listener must anticipate the various implications of this interpretation in a top-down manner for the rest of the material to come. They also have general problems with the organization or comprehension of narrative speech, in recalling stories or organizing sentences into a coherent and logically sequential account, in evaluating the plausibility of an event in a given context, in extracting the implicit meaning or significance, and in interpreting the moral of fables (Joanette et al., 1990).

The RH may therefore function as a buffer to hold in store alternative meanings, while the LH selects and operates upon the "correct" one. Studies such as these support the notion that the RH is concerned with the social, situational, and *pragmatic* aspects of speech, with damage leading to tangentiality, unanticipated or irrelevant changes of topic, confabulatory or at times apparently even delusional changes of content, and inappropriate, incomplete, or excessively concrete interpretation of metaphor and humor. At an output or production level, the old idea that the RH might subserve the more automatic aspects of speech recently also received confirmation (Speedie, Wertmann, Ta'ir, & Heilman, 1993). The authors described a dextral with a lesion centered on the basal ganglia (BG) in the RH. Despite preservation of propositional speech, he could no longer recite familiar verses, and he experienced gross impairment of serial automatic speech, signing, rhyme recitation, and swearing, with preservation only of idioms and social greetings. His speech no longer contained overused phrases, though he could comprehend automatic speech. Thus the RH at least perhaps at the level of the BG (and see Chapter 11), may mediate the automatic aspects that come ready-made, more holistically, without effort, and in an invariant word sequence. This contrasts with the propositional constructions of the LH, where linguistic segments are less automatically assembled to express semantic information. Lovell (1993) documented an early observation by P. Rommel (1683) of exactly this dissociation:

> Following apoplexy with right hemiplegia, she lost all speech with the exception of the words 'yes' and 'no'. However she was able to recite some prayers. (p. 54)

Acalculia

Calculation resembles language, music, chess, and representation by maps as a semiotic system (Boller & Grafman, 1985), though its rules may be stricter. Thus numbers, like sentences, are sequences of symbols generated and interpreted in accordance with syntactic and semantic rules (McCloskey, Sokol, Caramazza, & Goodman-Schulman, 1990). Many of the questions arising in the study of sentence processing have their counterparts in the realm of number comprehension and production. The number systems are symbolic systems with circumscribed sets of basic symbols and syntax. Analysis of the mechanisms involved in number processing may tell us something about language mechanisms in general, and the process of lexical representation and retrieval in particular. Thus there are strong commonalities between, on the one hand, acalculia and aphasia, and on the other hand, acalculia and amusia. There are clearly sets of rules for addition, subtraction, multiplication, and division that resemble linguistic rules but are nevertheless largely independent of language (Kahn, 1991). Indeed, although arithmetic may initially be parasitic upon language as tables, facts, rules, and procedures are acquired, it must subsequently become semi-independent, except insofar as its algorithmic operations are verbally mediated (Geary, 1993). A commonality shared of course between arithmetic and reading is the fact that both involve invented processes that require active learning for their acquisition, unlike the rapid, automatic acquisition of the first spoken language (Levin, Goldstein, & Spiers, 1993).

Calculation heavily depends upon not only verbal processing mechanisms, as outlined above, but also on a number of functionally autonomous components, such as the perception and recognition of numerical orthography, number-symbol representation, visuospatial aspects (spacing and alignment), short-term or working memory (as in mental arithmetic), semantic long-term memory, syntactic or algorithmic reasoning (corresponding to syntax and grammar in language), and of course sustained attention (Boller & Grafman, 1985; McCloskey, Aliminosa, & Macaruso, 1991). Much of the processing is automatized (e.g., in the automatic application of addition and especially multiplication "tables"), but where such information is unavailable, controlled processing mechanisms (Schneider & Shiffrin, 1977) must be invoked.

Acalculia is of course an acquired calculation disturbance, in someone who previously possessed adequate calculatory capacity. It may occur as a relatively isolated deficit, or as a result of impairment in other cognitive abilities. As with amusia, the prior capacity is generally less widespread or available than is the case with language, which is nearly universal. Indeed, with the advent of calculators and machine computation to the populace at large, many more people in the future may never be initially competent. The term *acalculia* was introduced by Henschen (1925), and Berger (1926, cited in Kaplan, 1991) distinguished between primary acalculia, where the loss was not due simply to a secondary cause, and secondary acalculia, which was due to impairment of attention, memory, language, visuospatial functions, or the capacity to attend. Henschen invoked the role of the third frontal convolution for pronouncing numbers, and the angular gyrus for reading and writing them; he also rightly drew attention to the importance of education and premorbid intelligence and calculatory ability.

Gerstmann (1924, cited in Kahn & Whitaker, 1991) proposed a syndrome, centered in the angular gyrus of the LH, involving finger agnosia, agraphia, impaired calculation, and right–left disorientation, all a consequence of a defect in the body schema. Finger agnosia involves an inability to recognize or otherwise identify fingers on one's own or the examiner's hands. The syndrome has however rarely been confirmed as a co-occurring tetrad; some claim that other deficits (e.g., constructional apraxia) must also be present, and others that lesions bigger or other than the angular gyrus are involved (Kahn & Whitaker, 1991; Kaplan, 1991). One can of course conceptualize a relationship between spatial concepts, knowledge of finger position for counting, and language in counting (Ardila, 1993).

There have been very few systematic studies of acalculia per se, though many more normative studies have sought to determine the relative importance of the left and right cerebral hemispheres via visual field and ear asymmetries. Such studies cannot of course tell us much about the localization of function in acalculia, though just as with language and music we must remember that we are likely to be dealing with a distributed network of locally specialized subsystems. Indeed, as with amusia, most of the few published studies have relied upon detailed analysis of single cases, to offer insight into the general architecture of a specialized cognitive system. As with amusia, acalculia may occur together with or independently of aphasia. Brain damage may impair the production of Arabic numbers while leaving intact their comprehension, or vice versa, and there may again be a dissociation between Arabic and verbal number processing. Thus McNeil and Warrington (1994) described a patient with mild dyslexia for speaking Arabic numbers and a marked calculation deficit that was specific to written (as opposed to spoken) input and where additions rather than subtractions were involved; there was no problem writing Arabic numbers. Hécaen (1962) proposed the following taxonomy (and see E. Kaplan, 1991):

1. *Alexia and agraphia* for digits and numbers, with or without corresponding deficits for letters and words, and with preserved computational ability as demonstrated by multiple-choice solutions. Thus arithmetic operations were affected due to impaired reading and writing of digits and numbers. Hécaen believed that temporoparietal regions on the left were involved.

2. *Spatial acalculia*, an RH parietal disorder of spatial organization with respect to the rules for setting out written digits in their proper order, column alignment, directional sequencing, and so forth.

3. *Anarithmetia,* a (bilateral?) disorder of the conduct of the fundamental arithmetic operations involving mathematical thinking and understanding the concepts of mathematical operations, without having as its source number alexia or agraphia, or spatial disorganization.

To the above we would perhaps currently add the following:

1. Disordered ability to remember and retrieve mathematical facts (semantic memory) and use them in a proper sequence
2. Disordered ability (lexical?) to appreciate the meanings and names of numbers.

Boller and Grafman (1985), in one of the few multisubject studies of acalculia, found that LH posterior lesions were in fact important in both anarithmetia per se, and in deficits corresponding essentially to spatial acalculia. These authors concluded that acalculia could stem from any one or more of the following:

1. An inability to appreciate the meaning of number names, a condition often associated with anterior or Broca's aphasia
2. Visuospatial deficits interfering with the spatial arrangement of numbers and the mechanical aspects of mathematics, a condition associated with posterior right damage
3. An inability to recall mathematical facts and to appropriately use them
4. Deficits in mathematical thinking and understanding underlying operations.

One might be tempted to add the following two factors:

1. An inability to conceptualize quantities (numerosity), and to perform adding and subtracting operations (Ardila, 1993)
2. Problems with working or scratch-pad memory for calculation.

McCloskey, Aliminosa, and Macaruso (1991) distinguished between the following two factors:

1. A number-processing system, encompassing mechanisms for number comprehension and production (spoken or written, Arabic or word, etc.)
2. A calculation system for such operations as facts (e.g., multiplication tables), rules (e.g., $N \times 0 = 0$), and procedures (e.g., how to divide).

Consequently, calculation errors in both normal and brain-injured individuals can stem from inappropriate fact retrieval, misuse of arithmetic rules, or procedural errors. They therefore follow Hécaen (1962) and link together developmental and acquired dyslexia into three categories (McCloskey, Aliminosa, & Sokol, 1991):

1. *Alexia and agraphia* for numbers, an LH syndrome often associated with aphasia

2. *Spatial acalculia,* a posterior RH deficit involving misalignment or omission of numbers, problems with place and value, number rotation, etc.
3. *Anarithmetia,* a posterior LH deficit involving problems in retrieving basic arithmetic facts from long-term memory.

At this point we should note that two views may be distinguished concerning the cognitive architecture underlying numerical skill, their major differences relating to the abstractness of the relevant number codes, and the modularity of number processing. The approach considered above of McCloskey and colleagues invokes a fairly abstract and modular architecture, with calculation processes taking place in a manner that is functionally independent from and not interacting with other mental processes. Clark and Campbell (1991), on the other hand, considered that more general cognitive mechanisms are involved that are integrated with other operations and functions. Their nonmodular architecture involves a complex network of multiple brain regions all differentially contributing to numerical processing, with injury to different areas producing a diverse range of dysfunctions. This latter approach clearly may better reflect the interactive and distributed properties of anatomical structures, and the physiological processes likely to underlie complex cognitive functions.

Given the complexity of the neural computational processes involved in arithmetic processing, the range of premorbid individual differences in skill and strategy, and the probably distributed nature of the relevant cognitive architecture, we shall probably not learn much more from lesion studies. Although the same brain areas may be affected across a number of acalculics, nevertheless the disorder has inevitably been reported after a range of cortical and subcortical, anterior and posterior, LH and RH sites (Kahn, 1991). Computational ability clearly involves both language and spatial processing, and both perception and strategic action (Ardila, 1993); however, the LH may be somewhat more important than the RH, and posterior regions than anterior (Kaplan, 1991). Nevertheless, the number of processing components virtually guarantees that a focal brain lesion will rarely produce a circumscribed mathematical deficit (Geary, 1993). For definitive information on localization we shall have to await subtractive PET studies with normal subjects, in a graded series of conditions involving, for example, inspecting numbers, subvocally reading them, and performing calculations, with spoken, written, and remembered output, and subjects of varying abilities and strategies. A recent lesion study (Takayama, Sugishita, Akiguchi, & Kimura, 1994), however, reported three patients with isolated calculation disturbances due to stroke with lesions in the left intraparietal sulcus, which separates the inferior and superior parietal lobules. The patients made calculation errors in the process where a number of steps had to be performed simultaneously, suggesting a disruption of the working memory for calculation. They displayed no aphasia or visuospatial disturbances in number operations, fully understood the basic calculatory process; and had no difficulty in retrieving table values. Their problems apparently involved the temporary storage of information along with the mobilization of long-term knowledge in a working memory for calculation. This study is perhaps the first to propose a specific locus for such operations, a locus that, it is interesting to note, is not very distant from the angular gyrus as proposed 60 years ago by Gerstmann.

Summary and Conclusions

The RH may have little articulatory capacity, but it is increasingly being found to play an important role in the nonphonological, nonsyntactic aspects of language. There is still some debate, however, as to whether this situation applies in the normal brain (and, if so, whether equally for everyone), or largely or only as a consequence of LH injury. In the latter instance, age of injury and handedness may be important variables, just as they are generally for language recovery after brain injury.

Massive LH damage in infancy such as hemispherectomy permits a surprisingly sophisticated level of linguistic functioning to be achieved in due course, although claims about residual syntactic deficits in adulthood have been disputed. Subtotal damage is generally associated with a better prognosis for language development the earlier the damage occurs. In fact, age of injury may be a more important factor than severity. With adults, left hemispherectomy, in the comparatively few cases that have been investigated, results in severe expressive deficits with very little capacity for connected propositional speech other than some expletives and automatisms, though some comprehension may be retained. With subtotal adult damage to the LH, numerous experimental paradigms converge in indicating an RH contribution to recovered capacities, especially in nondextrals.

Deep dyslexia, with its semantic paralexias, seems to provide good evidence for a connotative semantic contribution by the RH, probably involving direct access rather than graphophonemic recoding, though the condition may be rarer than it should be if the phenomenon of an RH involvement were general; moreover, the evidence itself is subject to some dispute. Also disputed is whether the isolated RHs of commissurotomy patients exhibit similar deep-dyslexic and semantic-paralexic phenomena.

Commissurotomy studies, where the isolated RH is interrogated, certainly suggest considerable capacities for auditory and visual comprehension, despite a virtually complete expressive aphasia. However, the pattern fails to correspond to any single clinical or developmental pattern, and there is considerable intersubject variation as a function of age of original injury, age at surgery, and age at testing, together with the time intervals between these events. Thus considerable reorganization may have occurred in at least some brains, and once again we are left wondering how relevant these findings are for the normal brain. There is also the problem of discrepancies in performance between clinical alexics and RH performance after commissurotomy.

The effects of RH damage in adults have recently been shown to include complex deficits over and above the traditional hesitations, reductions in verbal creativity, aprosodias, and reduced ability to employ verbal stereotypies. Thus RH-lesioned adults may miss the point in humor or sarcasm, interpret metaphor in an inappropriate or concrete fashion, and have problems with connotative or contextual implications, and with the organization or comprehension of lengthy connected discourse. Such deficits may affect our highest human capacities for intelligent communication. The RH may therefore function as a buffer to hold in store alternative meanings, whereas the left selects and operates upon a particular version. It may also mediate the more automatic aspects of speech that come ready-made for invariant, stereotypic utterance.

Calculation is another "recently" acquired capacity, in evolutionary terms, like visible speech. Numbers, like sentences, are sequences of symbols employed according to syntactic and semantic rules. Although arithmetic may initially be parasitic upon language as tables, facts, rules, and procedures are acquired, it subsequently may become semi-independent. Calculation of course depends also on visuospatial and memory aspects at various levels. Acalculia is an acquired calculation disturbance in someone who previously possessed adequate arithmetic skills. Although it has been linked with other disorders (finger agnosia, agraphia, and left–right disorientation) as Gerstmann's syndrome, the reality of the latter is disputed, as are the regions likely to be involved in arithmetic processing or its disturbance. Various taxonomies have been proposed for acalculia; what is clear is that arithmetic involves a number of complex and differentially localized operations, some on the left and some on the right, and that individuals differ greatly in their premorbid experience and skills. It is probably pointless to seek any uniquely important region or regions, although a recent report identifies the intraparietal sulcus in an isolated calculation deficit.

Further Reading

Baynes, K. (1990). Language and reading in the right hemisphere: Highways or byways of the brain. *Journal of Cognitive Neuroscience, 2,* 159–179.

Joanette, Y., Goulet, P., & Hannequin, D. (1990). *Right hemisphere and verbal communication.* New York: Springer.

Kahn, H. J., & Whitaker, H. A. (1991). Acalculia: An historical review of localization. *Brain and Cognition, 17,* 102–115.

Levin, H. S., Goldstein, F. C., & Spiers, P. A. (1993). Acalculia. In K. M. Heilman & E. Valenstein (Eds.), *Clinical neuropsychology* (3rd ed.) (pp. 91–122). Oxford: Oxford University Press.

McCloskey, M., Aliminosa, D., & Sokol, S. M. (1991). Facts, rules, and procedures in normal calculation: Evidence from multiple single-patient studies of impaired arithmetic fact retrieval. *Brain and Cognition, 17,* 154–203.

Molloy, R., Brownell, H. H., & Gardner, H. (1990). Discourse comprehension by right hemisphere stroke patients: Deficits of prediction and revision. In Y. Joanette & H. H. Brownell (Eds.), *Discourse ability and brain damage: Theoretical and empirical perspectives* (pp. 113–130). New York: Springer.

Zaidel, E. (1990). The saga of right-hemisphere reading. In C. Trevarthen (Ed.), *Brain circuits and functions of the mind. Essays in honor of Roger W. Sperry* (pp. 304–319). Cambridge: Cambridge University Press.

Chapter 5

Disorders of
Object Recognition

The Agnosias and Related Phenomena

Our ability to recognize visual objects so rapidly, effortlessly, and accurately far transcends the capacities of even the most powerful modern computers. When we add to that our capacity to cope with an enormously noisy and variable input under all sorts of different perceptual conditions, we realize just how sophisticated is our visual object recognition system. Moreover, it can operate at such levels as correct overt naming or verbal identification, nonverbal categorization, sorting or grouping of objects, and their appropriate use in context, levels that may dissociate under conditions of partial breakdown or agnosia (Bauer, 1993; Humphreys & Riddoch, 1993). It is of course difficult to assess agnosia without some form of verbal or gestural response, and it is not always clear that deficits are occurring at a purely visuoperceptual level, uncontaminated by problems at a response level (Beaumont & Davidoff, 1992). Recognition deficits are also often modality-specific, and it is no accident that, as essentially visual animals, we know more about visual than auditory or tactile agnosia. Even then, however, it is often unclear whether a given deficit is essentially perceptual or a matter of interpretation or meaning. The dividing line of course between perceptual and memorial processes can never be sharply drawn, as Munk (1881, cited in Bauer, 1993) long ago recognized when he described "mind-blindness"—*Seelenblindheit*—in dogs with bilateral occipital lesions; they avoided obstacles but in a typically agnosic way failed to react appropriately to them.

Until recently, research into spatial functions and object recognition has been largely descriptive in terms of localization of function. Perhaps more than in any other syndrome, except maybe the amnesias, the adoption of a more cognitive or information-processing approach has proved fruitful in determining the underlying processes and modes of disruption (Farah, 1990). This is because object recognition is so essentially a multistage process:

1. The coding of light intensity and the elementary perceptual processes of linearity, angularity, depth, and color
2. The integration of such elementary aspects in terms of spatiotemporal relationships, figure–ground relationships, overall form, and possible movement
3. The effect of previous memory, associations, and biological significance
4. Output aspects in terms of how appropriately an object is used or described or verbally identified

Such an account may seem to imply a serial, bottom-up sequence of information processing; however, as is so often recognized nowadays in such other cognitive areas as speech recognition and output, reading, and remembering, processes may interact in a more or less parallel or cascade fashion, updating each other continuously. Indeed, the very occurrence of impairments at a "local" level with intact apprehension of global shape, as may occur with left hemisphere (LH) compared to right hemisphere (RH) dysfunction, argues against seriality of processing from the first to the second level. Nevertheless, as we shall see, most attempts at categorizing the agnosias, and even at deriving a model of visual recognition processes, have tacitly adopted a largely serial account.

Two pathways may be identified in the primate visual system, according to Mishkin, Ungerleider, and Macko (1983):

1. A *ventral* pathway for object recognition, and directed towards the inferotemporal cortex; in humans there might be an LH bias.
2. A *dorsal* pathway for spatial localization, and directed towards the inferior parietal lobule; in humans there is a decided RH bias.

It is unclear whether elementary object-processing aspects are equally mediated by either hemisphere, or whether there is an RH bias for line orientation and elementary shape discrimination (occipital), stereopsis (prestriate), and color (mesial occipitotemporal). However, we do know that the anterior portion of the inferotemporal cortex (IT) plays a key role in object (and face) recognition (Miyashita, 1993; Tovée & Cohen-Tovée, 1993), and is the region where visual perception meets memory and imagery. It may store the "prototype" of a visual object, if indeed this is how pattern recognition takes place, and damage here affects size constancy (see below) and our ability to discriminate objects transformed by size, orientation, or shadow configuration. Adjacent regions within the inferomesial part of the occipitotemporal cortex, which includes the fusiform and lingual gyri and the posterior part of the parahippocampal gyrus, are all supplied by branches of the posterior cerebral arteries that originate from the basilar artery. And, as we shall see, such regions are involved in face, color, and topographical processing.

Cells in the primary visual cortex (V1) extract orientation, size, color, and simple textural information. They project such information to IT for object recognition, with subsequent onward projections to the amygdala, entorhinal cortex, and hippocampus, and onward again to the ventromedial prefrontal regions via the medial diencephalon, and, separately, to other prefrontal regions via the uncinate fascicle (Miyashita, 1993; Tovée & Cohen-Tovée, 1993) for the initiation of appropriate response cycles. Tanaka (1993) noted that cells in IT specify more complex features than those in V1, together with combinations of the simple features extracted there. Activation of numbers of IT cells can specify a particular natural object; the system learns through experience and is modified by practice. IT is organized in columnar modules, like V1, in which cells with overlapping but slightly different selectivity cluster together. The columnar organization indicates that object recognition requires the activity of many cells within a single modular column. Selectivity varies from cell to cell, and effective stimuli (i.e., those able to be recognized) largely overlap these cells. This arrangement permits robustness to all the subtle changes that continually occur in input images as a function of lighting, orientation, size, and adventitious irrelevancies—the problem of the perceptual constancies, see below—via a clustering of cells with slightly different selectivities. It also permits, we would add, robustness in the face of local damage to the network. According to Tanaka (1993), the selectivity of cells gradually changes along an axis parallel to the cortical surface within a single column, so that a variety of object features are systematically, and continuously, represented along the cortical surface, with discontinuous or step changes at column boundaries.

An object typically consists of an inextricable association of qualities and attributes (Treisman, 1986). How does the brain integrate fragmentary neural events at multiple locations to produce a unified perceptual experience? Bressler, Coppola, and Nakamura (1993) addressed this "binding problem" in terms of correlated activity at different cortical sites during perceptuomotor behavior, perhaps by synchronization of narrow-band oscillations in the gamma-frequency range (30–80 Hz). They described episodes of such increased broadband coherence among local field potentials from sensory, motor, and higher order sites in the macaque during visual object-discrimination tasks. Widely distributed sites became coherent without involving other sites. Thus, spatially selective multiregional cortical binding, in the form of broad-band synchronization, may underlie visual object recognition. All routes to the cortex pass through the thalamus, a collection of nuclei relaying incoming signals to cortical target areas. Each nucleus receives massive back projection from the cortical region to which it connects, thereby perhaps selecting those signals that are most appropriate for subsequent cortical processing, and maybe synchronizing the responses of cells projecting to the visual cortex (W. Singer, 1994). Thus the thalamus may play a key role in integrating and binding the fragmentary neural events that constitute the different aspects and attributes of a stimulus "object" or "event." Such a mechanism permits subsets of the same cells to be selected and bound with others within unchanged hardware, such that highly selective input configurations can be defined functionally, without the need for matching selectivity of anatomical connections. As W. Singer (1994) in his review observed, this procedure reduces the required number and precision of thalamo-cortical connections, and is highly economical on neuronal numbers, as different stimuli can be represented by varying relations among a limited set of active neurons. He concluded that thalamic synchronization may play an integral part in the selection and binding operations that occur simultaneously at different levels of visual processing, and that influence one another through reentry loops. The binding problem, of course, is itself an aspect of the more general problem of consciousness, which Crick (1994) saw as closely related both to attention (Chapter 6) and working memory (Chapter 8). Cortico-thalamo-cortical loops involving parietal and prefrontal structures are likely to be involved in consciousness, in a distributed network, along with the anterior cingulate and supplementary motor area (SMA) (see Chapter 10) for the contribution from will and volition.

With true visual object agnosia, there is impairment of higher visual processes necessary for the recognition of objects, their meaning, significance, or correct manipulation, with, typically, preserved auditory or tactual recognition. It is not a matter of apraxia, where *automatic* manipulative ability may be preserved despite loss of the ability *deliberately* to perform or mime appropriate actions; nor is it a matter of anomia, involving word- or name-finding difficulties. Thus agnosia manifests with nonverbal tests of recognition (e.g., sorting objects into category groups).

Lissauer (1890) distinguished between apperceptive and associative agnosia. It is still a useful distinction between impairment at a visuoperceptual (not sensory) level, and a "higher order" associative or semantic deficit leading to lack of recognition, a "normal percept stripped of its meaning" (Milner & Teuber, 1968). It is of course far from clear, as we saw, whether we *can* separate perception from memory, whether processing proceeds serially in this way, whether perception *is* normal in associative agnosia, and whether in apperceptive agnosia there is a major RH involvement, with a corresponding bias in associative agnosia to the left (Warrington, 1982). However, as Bauer (1993) and Benton and Tranel (1993) observed, the distinction still has appeal and is heuristically useful.

Apperceptive Agnosia

Object recognition is not a template match between input image and stored representation, but flexibly accommodates considerable changes in images due, as we saw, to changes in illumination, viewing angle, orientation, and so on (Tanaka, 1993). System failure to accommodate the perceptual constancies (Rock, 1975) may partly underlie apperceptive agnosia. Patients, who often are in a recovery phase from cortical blindness (Bauer, 1993) typically have preserved such elementary visual functions as acuity, brightness discrimination, color vision, depth, and contour perception. They cannot however group elements into higher order configurations, and cannot perceptually recognize or match objects, although they often can trace these visually unrecognized shapes by hand or head movements (Farah, 1990), which may then provide the necessary substrate of recognition. The slavish quality of such tracings easily leads to derailing along extraneous, irrelevant, or superimposed additions or contours. The copying performance is in any case very poor. Anosognosia may manifest in complaints about poor spectacles, lighting, quality of target drawings, or the absence of previous experience in such tasks (Bauer, 1993).

Farah (1990) distinguished between the above conception of severe apperceptive agnosia (where the perception of local contour alone is preserved, due, she says, to damage at a low level of perceptual processing that affects the grouping of elements into larger contours, regions, and surfaces) from where the deficit is one of perceptual categorization. In this much milder disability, there may be few problems in recognizing or naming objects in normal daily life, difficulties only emerging in laboratory situations, or where objects appear in a nonstandard form. Such apperceptive agnosics typically have difficulty perceiving anomalies in sketchily drawn scenes, or in processing incomplete or overlapping figures, degraded or fragmented letters, or three-dimensional objects that have to be recognized from noncanonical perspectives, in silhouette, from a nonstandard viewpoint, or under uneven lighting (see Figures 1–5). However, an apperceptive agnosic, unable to reorganize objects by sight, nevertheless was reported as correctly and appropriately molding the hand for grip when reaching to pick up objects, implying some level of covert or implicit object recognition (Goodale, Milner, Jakobson, & Carey, 1991).

The concept of apperceptive agnosia has not been accepted without criticism, as not all patients show the same full complement of symptoms, and indeed there may be different underlying deficits (e.g., loss of stored material in long-term memory, inability to access it, or to use it when accessed so as to "drive" object constancy mechanisms). Indeed some apperceptive agnosics with classic perceptual problems may nevertheless draw quite well from memory.

Diffuse RH parieto-occipital damage is frequently reported in cases of apperceptive agnosia. It is no accident that studies with normal healthy subjects (Bradshaw, 1989) show that the left visual field (RH) is generally better when discriminating stimuli of low levels of brightness or contrast, of brief duration, small size, or when blurred or subjected to filtering of high spatial frequencies.

Dorsal Simultanagnosia

Closely related to apperceptive agnosia, but each progressively slightly less restrictive, are dorsal and ventral simultanagnosia. The continuum now also may

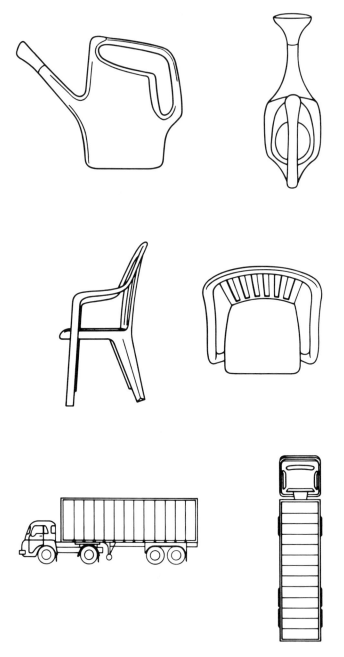

Figure 1   APPERCEPTIVE AGNOSICS MAY HAVE DIFFICULTY IN RECOGNIZING AN OBJECT (LEFT) SHOWN FROM AN UNUSUAL ANGLE (RIGHT).

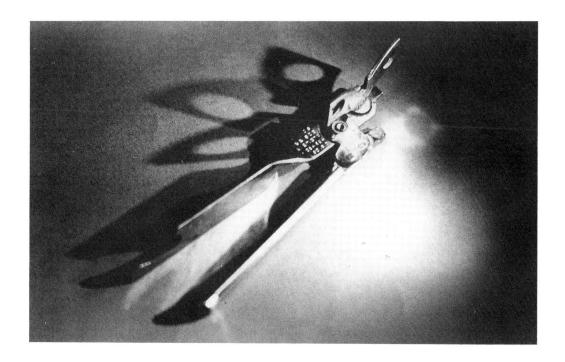

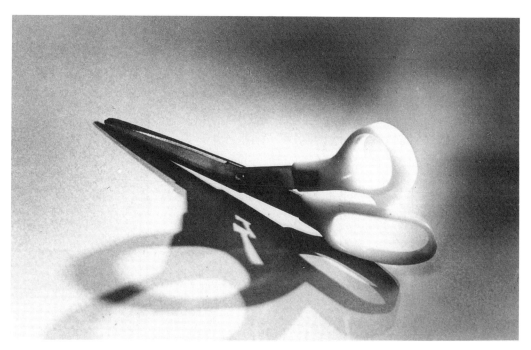

Figure 2 Apperceptive agnosics may have difficulty in recognizing an object portrayed under conditions of uneven lighting, or in shadow.

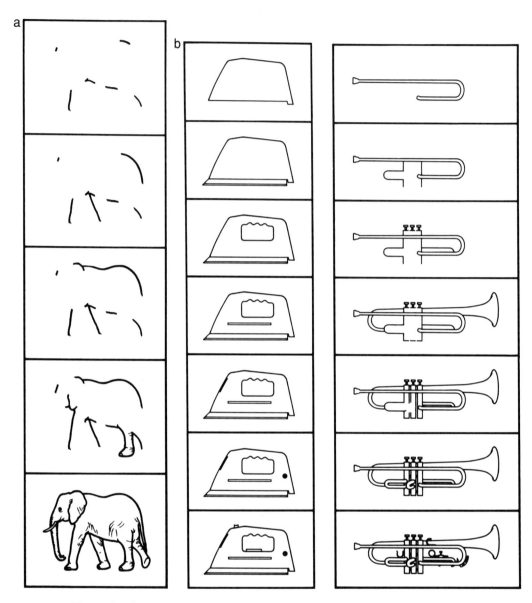

Figure 3 APPERCEPTIVE AGNOSICS MAY HAVE DIFFICULTY IN RECOGNIZING EVEN A COMPLETE DRAWING, WITH INORDINATE PROBLEMS WHEN VIEWING (a) MORE FRAGMENTED OR (b) MORE INCOMPLETE MATERIAL.

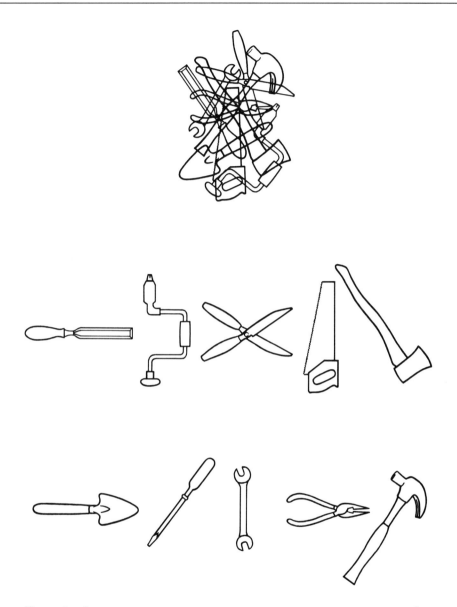

Figure 4 APPERCEPTIVE AGNOSICS MAY EXPERIENCE INORDINATE DIFFICULTY IN "DISEMBED-DING" OUTLINE DRAWINGS OF OBJECTS, HERE HAND-TOOLS, WHEN PRESENTED IN OVERLAPPING OR OVERLYING FORMAT.

Figure 5   APPERCEPTIVE AGNOSICS MAY HAVE DIFFICULTY IN RECOGNIZING LETTERS THAT CONTAIN EXTRA OR SUPERFLUOUS ELEMENTS OR CONTOURS

possibly extend from an RH bias (apperceptive agnosia) via bilateral superior parieto-occipital damage (dorsal simultanagnosia) to LH involvement (ventral simultanagnosia and associative agnosia). Dorsal simultanagnosics have reasonable perception of the individual elements or details of a complex picture, but cannot apprehend the whole scene. They often can only see one object at a time, or part thereof; they may see individual elements clearly, but not how they fit together, with the environment appearing fragmented. Thus when two contestants are viewed fighting in a video, one is seen to be inexplicably knocked around by an invisible opponent. Route finding, similarly, is more easily performed with the eyes closed. There may be problems in redirecting gaze, and in pointing to and reaching objects, with attentional deficits such that even the attended item may spontaneously slip from view and vanish. Patients grope around and fall over things if more than one object is present, leading to an extreme form of visual neglect or extinction (see Chapter 6) from competition between copresent items. It is noteworthy that, in both unilateral neglect and dorsal simultanagnosia, parietal damage is common. There may likewise be a "stickiness" or difficulty in disengaging attention or shifting it to something else. Performance therefore is better when objects are small or closely grouped. Even with a small outline square composed of eight dots, however, a patient may see a square (but no dots) or a single dot (and no square). Similarly, a face may be seen either as such, or as a single feature. Typically, local over global aspects may be preferred. A patient shown a complex picture may gradually build it up, item by item, but may eventually be misled by erroneous perception of one of the components, or may experience slipping away from attention, memory, or consciousness, of individually recognized components while working on others.

Ventral Simultanagnosia

This syndrome, typically involving inferior mesial temporo-occipital damage to the LH, resembles dorsal simultanagnosia except that multiple objects may now be seen, counted, and manipulated. Patients can walk around without bumping into things, read slowly, often spelling out the word letter-by-letter (see Chapter 3), and can describe pictures element-by-element, though now typically *without* being able to understand the scene as a whole.

All three syndromes considered to date seem therefore to lie along both a behavioral and an anatomical continuum. All involve piecemeal perception. However, in apperceptive agnosia, only very local contour is perceived, and the lesion tends to involve parieto-occipital structures on the right. In dorsal simultanagnosia, the whole object may be recognized, even if composed of discrete elements, but more than one such object cannot be perceived, with attentional processes determining what is seen. The damage tends to be superior parieto-occipital on both sides. With ventral simultanagnosia, multiple objects may be seen, though the meaning of the whole configuration may be unavailable. LH damage to mesial temporo-occipital structures is common. We shall see that ventral simultanagnosia is very closely related to Lissauer's (1890) traditional associative agnosia, with LH IT or temporo-occipital injury. We might however first observe that simultanagnosia features prominently in Balint's syndrome, a supposedly bilateral visuomotor disturbance, which in fact may or may not really exist as a functional entity; it is said to involve the co-occurrence of three elements, which indeed often seem to dissociate:

1. *Gaze apraxia* is an inability to shift gaze so as to bring peripheral stimuli into fixation. The gaze wanders until it happens by chance to hit a target; it may be hypometric, falling short of (stationary) targets, though able to track moving ones, and the subject may resort to artificial means to shift gaze (e.g., closing the eyes or bobbing the head).

2. *Restriction of attention.* The problem of shifting attention parallels that of shifting gaze, and leads to simultanagnosic difficulties in seeing more than one object at a time, with loss of salience of peripheral stimuli.

3. *Optic ataxia,* with impaired manual reaching to visual stimuli, without such problems when under, for example, auditory control.

Associative Agnosia

The perceived meaning or significance of objects is now compromised, rather than the ability to discriminate them from the background. Although the disorder may manifest as an inability to perform object naming, the problem is not one of anomia per se. LH temporal or temporo-occipital regions may be affected, with a dense right homonymous hemianopia commonly occurring as a localizing sign. Disconnection of visual from language areas results in loss of visuosemantic identification and intermodality associations. Confabulatory effort-after-meaning may result, with guessing (often helped by cues) and listing of distinctive features. Perceptual processes in themselves are relatively intact, and problems with nonverbal recognition tests, pantomime, and gesture distinguish it from optic aphasia (below). Line drawings are processed worse than photographs, which fare worse than real moving objects. However, the patient may be able to say whether the stimulus is indeed a real-life object rather than an imaginary or artificial composite or concoction (see Figure 6), again suggesting preserved access to implicit or tacit knowledge. As so often is the case with brain

Figure 6   Associative agnosics may still be able to judge the reality or otherwise of pictorial representations. This beast is clearly imaginary and may be recognized as such.

damage, performance in terms of superordinate categorization (e.g., an orange, seen as a "fruit") is better preserved than discrete identification. Stimuli coming from a large subset of physically similar or confusable members prove especially difficult (e.g., particular breeds of dog, types of car, etc.), a phenomenon that will reappear in the context of prosopagnosia, itself a form of associative agnosia. Thus the likelihood of an associative-agnosic confusion in a particular circumstance may depend upon class size, as constituted by members with physically similar appearance, how similar they are, how frequently they occur in life, their importance to the individual, and so forth.

Nevertheless, hitherto familiar objects may now be treated as unfamiliar or novel. Thus an affected physician "saw" his stethoscope as a "long cord with a disk at one end," and a glove as a "container with five outpouchings." While unable to recognize them visually for what they really were, he nevertheless could immediately recognize them by feel, and could painstakingly copy objects perfectly, line by line, without knowing what they were. Indeed, in associative agnosia, perceptual processes are probably never completely intact, as evidenced by slow feature-by-feature or slavish drawing or analysis, with reliance on local rather than global attributes (Bauer, 1993; Farah, 1990). Although we are normally slower at copying "impossible" figures compared to normal ones (see Figure 7), being misled by the internal contradictions, associative agnosics have no such trouble, slavishly copying both types of material, line by line.

Although apperceptive agnosics may typically have great difficulty in matching objects by shape or appearance, an associative agnosic's problem is more one of matching by use, function, or category. It may be that there are three systems in object recognition (see e.g., Warrington, 1985), which may operate in a parallel-interactive fashion, rather than serially:

1. Visual sensory analysis—mediated largely bilaterally
2. A fully organized structural description—largely RH
3. Assignation of meaning—a largely LH process except perhaps in the case of prosopagnosia.

This approach, to which we shall shortly return in more detail, parallels current thinking in imagery (Kosslyn, 1990) from both a clinical and a normative perspective. The LH may mediate the categorical aspects, the presence or absence of elements, and their absolute location in multielement and multipart imagery, whereas the RH may be responsible for the relative distances and positions.

Optic Aphasia

The chief difference between associative agnosia and optic aphasia is the preserved ability, in the latter, to pantomime an object's use, demonstrate knowledge of its meaning, or select a named item from among alternatives, while being

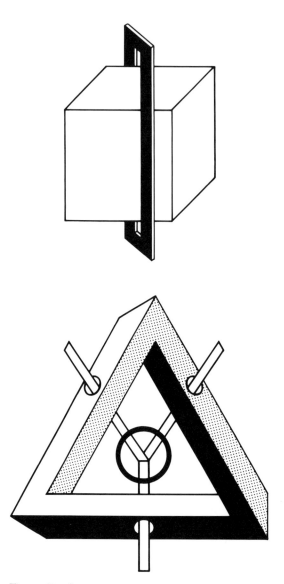

***Figure* 7**   Associative agnosics, unlike controls, may not experience undue diffi-
culty in copying "impossible" figures, as they are less likely to be misled by conflicting interpretations. (From Molins, J. V., 1991. Some thoughts on impossible
figures. *Perceptual and Motor Skills,* 73, 107–114. Copyright Perceptual and
Motor Skills, 1991.)

unable to name it when visually presented in isolation. Naming of objects presented or demonstrated via nonvisual modalities (auditory, or tactual) may be preserved, indicating the modality-specificity of naming systems. Nevertheless, as Bauer (1993) observed, there again may still be covert or implicit access via vision as patients often can correctly decide whether drawings represent real or imaginary objects. As with associative agnosia, to which the disorder is probably related, a posterior LH lesion is common, with now only partial disconnection of visual processes from speech and praxic regions. The frequent co-occurrence of color agnosia suggests inferomesial damage. Their own misnaming may confuse patients, so keeping them quiet and allowing them to pantomime an object's use may permit eventual self-cued recognition of an object. Similarly, correct naming of an object's parts may eventually lead to recognition of the object as a whole. Indeed such patients often give semantically related responses, or even successive approximations, *conduite d'approche,* (e.g., a picture of a goat → "looks like a horse, no it's got horns, must be a cow, no, tail's wrong, an antelope"). Drawings by optic aphasics are characteristically good when copying, and typically represent the presented object, not what the patient may continue to claim it is.

The taxonomy of the visual agnosias given so far clearly derives from that originally proposed 100 years ago by Lissauer (1890), and is largely followed, with reservations and modifications, by Bauer (1993) and Farah (1990). Humphreys and Riddoch (1993) noted that Warrington (1982) built upon this approach but applied finer distinctions, including breaking down apperception into three hierarchical stages:

1. *Shape coding,* where component orientations within shapes are integrated so as to provide a coherent preliminary description; capacities at this level may be assessed via shape-matching tasks.
2. *Figure–ground segmentation,* where there is primitive coding of one part of a complex display as foreground, segregating it from the rest as background; capacities at this level may be assessed in terms of performance with overlapping figures.
3. *Perceptual classification,* where objects can be judged as the same when viewed from a range of different viewpoints; capacities at this level may be assessed via, for example, unusual or contrasting views tests.

Her next stage in the processing hierarchy, semantic classification, corresponds of course to Lissauer's associative agnosia, the ability to classify objects on the basis of *function* rather than *perceptual* characteristics; this ability, in turn, is testable by the requirement to match two usually different objects on the basis of whether or not they are used for similar functions.

Humphreys and Riddoch (1993) themselves proposed a similar hierarchical analysis:

1. Feature coding at several spatial scales, to encode edges and parse the visual world into discrete spatial regions; failure due to occipital damage (probably bilateral) leads to problems in shape perception.

2. Feature integration and grouping, to perform figure–ground segregation. Component lines are linked by processes sensitive to such gestalt features as collinearity and parallelism, thereby determining underlying regularities, grouping, and segmentation. Occipitotemporal regions are involved, probably bilaterally.

3. Mapping of the above perceptual descriptions on to stored knowledge about the object, involving stored structural descriptions and perceptual classifications. This stage of the analysis must cope with unusual orientations, foreshortenings, and so on. RH parietal structures may be involved.

4. Stored structural descriptions and semantic knowledge about objects, including information on object function and associations. LH temporal structures are probably involved.

All three systems, Lissauer's, Warrington's, and that of Humphreys and Riddoch, place perceptual processes at one end, with a largely bilateral or RH involvement, and semantic processes at the other end, with a largely LH contribution. The three systems are also hierarchically serial, although they could be recast in a parallel-interactive form, and differ largely in their fineness of grain. The account of Humphreys and Riddoch explains well such discrete deficits as preserved discrimination of single-target-oriented lines, with breakdown when the same lines are embedded among many such nontargets. Other similar discrete deficits, which the account handles well, include problems with shape matching, grouping of local contour, judgment of item orientation, matching of objects across different viewpoints, or drawing such characteristically unique animals as giraffes or elephants. The account also copes well with preserved ability to distinguish familiar from unfamiliar objects, although not recognizing them as such or knowing how to use or group them. It was recently largely confirmed and extended by a posterior cerebral artery amobarbital test (R. F. Kaplan et al., 1994), which aimed to determine how each temporal lobe mediates memory for objects. The left temporal lobe was found to have an advantage in encoding the verbal representation of an object, whereas the right temporal lobe was found to be critical for memory of objects' specific visual attributes.

Category-Specific Deficits

Sometimes a patient may apparently be unable to process pictorial, written, or spoken instances of certain specific categories of item (see e.g., Etcoff, Freeman, & Cave, 1991; Sartori & Job, 1988), as if there is localized impairment within a taxonomically organized system (Bauer, 1993). Although clearly this should not necessarily be taken to mean that adjacent brain regions code closely related concepts, the observations may seem to suggest adjacency of access routes or network interconnectivities. Lesions in the posterior white matter are indeed commonly reported. Highly selective deficits of these sorts may manifest in the context of associative agnosia or alexia. Some patients may be unable selectively to name certain pictures, but have an intact reading vocabulary; in others the situation may reverse, the patient being unable to define test words, while in yet a third group the deficit may include both input channels (i.e., words and pictures). There may therefore be two separate representations of objects—visual and verbal. This may not be so surprising when the deficit involves rarely encountered items of low frequency or familiarity, or abstract concepts; however, the converse situation is not uncommon (McKenna & Warrington, 1993). The patient may be able to recognize pictures of a sphinx or abacus, but not of a dog or cat. McKenna and Warrington (1993) reviewed cases where there seemed to be selective loss of concrete concepts ("alligator," "barber") with preservation of abstract ideas ("hint," "opinion"); patients may be able to provide sophisticated definitions of a briefcase ("a small case used by students to hold papers") or a whistle ("a thing you blow to make a piping sound"), while answering "I don't know" to a picture of a parrot or kangaroo. The patient may be unable to distinguish between pictures of a squirrel and a cat ("Both must be mice, because both have whiskers"), but may have no trouble differentiating between an orchid and an iris. Selective deficits have been documented in different patients in such categories as foodstuffs, minerals, vegetables, things found indoors, and so on. One patient had a good understanding of human-made artifacts except for musical instruments, whereas another, with a loss of knowledge of living things, was quite competent with human-made objects except for vehicles. Large outdoor objects may be inaccessible, but not small indoor items. One patient could not retrieve the names of famous people, but had no problems with ordinary, family, or geographical names. Another patient could comprehend proper nouns (names of famous people and countries), but common nouns (names of objects) were poorly understood under auditory (spoken) presentation (McNeil, Cipolotti, & Warrington, 1994). Her ability to identify visual stimuli was fairly intact, and the pattern of her deficits suggested a difficulty in accessing representations within the semantic store, rather than a breakdown of the store itself. As McKenna and Warrington (1993) observed, these category-specific deficits are so counterintuitive that they have long been ignored. There may be deficits with visual or tactual presentations, but not when the name is spoken. The patient may be unable explicitly to recognize the picture of an object, like a carrot, but can still implicitly grasp its abstract, connotative, or metaphorical meaning. Thus a carrot shown dangling before a donkey passes unrecognized as such, while the composite idea of encouragement is readily accepted (see Figure 8).

Rarely, however, is the division between recognized and unrecognized items

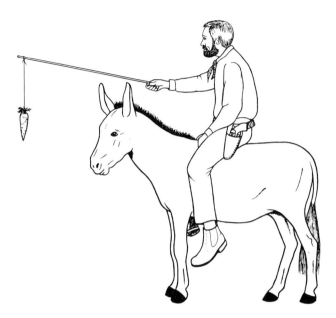

Figure 8   A PATIENT WITH A CATEGORY-SPECIFIC DEFICIT FOR OBJECT RECOGNITION MAY FAIL EXPLICITLY TO IDENTIFY THE CARROT, WHILE CLEARLY PROCESSING IT AT A TACIT LEVEL, AS EVIDENCED BY CORRECTLY PERCEIVING THE COMPOSITE CONCEPT OF ENCOURAGEMENT.

totally black and white; rather there is a continuum, with residual implicit knowledge elicitable by cues or prompts, and as so often is the case, superordinate category knowledge is less vulnerable than specific details. Thus a patient may still know that canaries are living, just not that they are yellow. The patterns of errors for the same objects may also fluctuate over time. There may be selective loss of the ability to recall facts about an object (e.g., its color, size, or locus), with preservation of information about its potential usage, or vice versa.

     With a category-specific deficit for animate objects, it might be argued that the problem is merely peripheral, in perceptual processing. However, category-specific deficits have, as we saw, also been reported for *inanimate* objects (Humphreys & Riddoch, 1993; Warrington & McCarthy, 1987), with preservation for living things. However, Damasio, Tranel, and Damasio (1993) argued that the general phenomenon of category-specific deficit could be an artifact of similarities between chosen instances in terms of physical structure, value, or familiarity to the patient, or the type of sensory and motor interaction required to map the exemplars, and how the exemplars operate or behave. Indeed Damasio et al. noted that careful analysis generally reveals exceptions to the "rule" in both the "normal" and the "impaired" categories. Funnell and Sheridan (1992) similarly claimed that, at least in their patient, an apparent deficit in processing living things, either by naming pictures or defining words, was due merely to the generally lower familiarity of instances of living compared to inanimate things. They argued that there is no convincing evidence that semantic memory is organized into dissociable categories of living and nonliving things. They claimed that it is all a matter of differential frequency of handling, examining, or using objects or concepts, so that details of physical features are thoroughly processed. Thus in our artificial modern world some people are far more familiar with inanimate objects. Generally, they say, apparent category-specific deficit is likely to be an artifact of the actual test items employed.

McKenna and Warrington (1993) however disputed such a conclusion; if items are matched for frequency, the effects still hold. Moreover, animals may be impaired and objects spared in one patient, with the reverse in another. Nor, they said, is it simply the case that impaired objects are more visually similar and confusable (e.g., animate vs. inanimate objects), as again the effects persist with adequate controls, and may even reverse in some patients. It might, they said, be argued that animate objects are known by their sensory properties, whereas inanimate objects are known by their function; however, this explanation cannot extend to selective loss of small manipulable objects. Nevertheless De Renzi and Lucchelli (1994) further developed the argument that visuoperceptual attributes are more important for living things, and functional aspects for nonliving human artifacts that have been constructed for a specific purpose. Form, they said, matches function only for human-made objects, whereas the form of animate things bears no relation to the possible human use that might be made of them. Although we would agree with them that the deficit probably results from a block of access or retrieval, rather than storage degradation, we feel that the existence of reversed deficits in some patients, and extreme fractionation in others, continues to argue against any simple or all-encompassing explanation. It is possible however that individual developmental differences in time of acquisition of concepts involving predominantly perceptual ("looks") or motor ("usage") channels of apprehension partially account for the phenomenon. Moreover, an object may be conceived of in more than one way; thus a single fossil shell may be both an inanimate mineral and an animal, and whether or not it is recognized may depend upon its context—both in the patient exhibiting category-specific deficits, and in normal, everyday life.

Tactile Agnosia or Astereognosis

As Bauer (1993) observed, deficits in this modality, which is not a major source of our knowledge about the world, are less understood than those in vision or hearing. The phenomenon involves impaired perception of shape, density, and texture, including the three-dimensional aspects, and of identity. The distinction between apperceptive and associative agnosia, described earlier in the context of visual deficits, may be usefully employed here, though it may now be even more difficult to distinguish empirically between perception and identification or meaning. Medial lemniscal pathways send afferent touch signals to the parietal (sensorimotor) cortex in the postcentral gyrus, where motor, proprioceptive, and spatial functions are elaborated in terms of sensorimotor exploration by touch. Perception is of course an active process. The system is important, as Bauer (1993) noted, for intentional, spatially guided movements that bring the subject into contact with environmental objects. Although damage to hand areas clearly affects, contralaterally, tactile object recognition, the left-hand–RH system may, in primates generally, be especially proficient in tasks where spatial factors are important, and for complex, spatially patterned discriminations (Bradshaw & Rogers, 1993). Thus RH damage may be more detrimental for such tasks than corresponding lesions on the left, almost irrespective of hand. As with vision, hemispheric specialization may be less apparent for the more elementary (somatosensory) functions. Sometimes patients may be able to draw objects that they cannot recognize by feel, and then recognize them (Bauer, 1993). One clearly must distinguish true astereognosis from tactile *anomia* or *general* agnosia, which is not limited to the tactile modality.

Auditory Agnosia

Schnider, Benson, Alexander, and Schnider-Klaus (1994) described the effects of RH and LH lesions in 52 patients in the superior temporal gyrus (including the planum temporale), inferior parietal lobule, and parietal operculum. Left-sided lesions were associated with confusion over semantically related sound sources in multiple modalities, whereas right-sided lesions affected the patients' ability to match acoustically related nonverbal environmental sounds. Indeed, with auditory agnosia for nonspeech sounds, patients may be generally unable to recognize such noises as a toilet flushing, a dog barking, a cock crowing, or a bell ringing. In the case of pure word deafness or auditory verbal agnosia (Bauer, 1993), the patient can read, write, and speak but not comprehend *spoken* language. Left-sided lesions of the superior temporal gyrus (i.e., Heschl's gyrus, the primary auditory cortex) may suffice, with sparing of Wernicke's area; otherwise aphasia would also be present. The patient may complain of voices in the absence of words, or of foreign-language sounds. Auditory acuity in the pure form of the disorder is unimpaired. According to Goodglass (1993), in pure word deafness the discrimination of place of articulation (e.g., *ba* vs. *ga*, *ka* vs. *ta*) may be worse affected than discrimination by voicing (e.g., *da* vs. *ta*, *ka* vs. *ga*), and the perception of consonants may be more affected than the perception of vowels. There may be loss of the ability to integrate auditory speech patterns beyond the level of individual sounds.

Amusia

Some form of music is evident in all cultures (Gardner, Winner, & Rehak, 1991), and musical performance and rites date back tens of thousands of years (e.g., at least to an Aurignacian flute of 30,000 years ago, Bradshaw & Rogers, 1993). Why has such an apparently "nonadaptive" symbol system continued to survive? Is it to facilitate bonding, mating, or group coordination? Music, like language, is after all a system of communication between a producer (who, if not the original composer, plays an interpretive role) and the listener. Unlike language, it is nonreferential, nonpropositional, and ultimately emotional and subjective (Sergent, 1993), while following its own combinatorial rules and syntax. As Henson (1985, p. 489) observed, "There is an ultimate mystery of musical experience which is not susceptible to neurological study." As the medium rather than the message, and in that respect the antithesis of speech, music ultimately consists of organized sound whose effect upon the recipient is determined by the individual's uniquely subjective mix of interest, education, culture, experience, and current emotional state. Both music and speech, however, are used expressively and receptively, both involve fine, sequential motor activity, and both are constructed from perceptually discrete sounds that can also be graphically represented (Sergent, Zuck, Terriah, & MacDonald, 1992). In music, of course, the significant auditory elements are the acoustic properties of a note in terms of its pitch, duration, timbre, and intensity, its place in a pitch set or scale, and the consecutive (melodic) or simultaneous (harmonic) sounding of notes in a time series (rhythm) (see e.g., Henson, 1985). Moreover, although the written alphabets of speech are essentially categorical, musical notation is spatial and relative. Both media, however, employ similarly complex multimodal operations—visual (for reading score or text), auditory, motor, cognitive, and emotional. It is therefore hardly surprising that although amusia and aphasia frequently co-occur, they also often dissociate. Thus LH aphasics may preserve singing, whereas RH amytal suppression may suppress singing but not speech, even though the latter may become monotonous and aprosodic (Gordon & Bogen, 1974). (Speech aprosodia, incidentally, offers a commonality between aphasia and amusia in that an apparently musical component such as pitch contour is involved, which happens to be an integral part of language; a common system may therefore process pitch contours in both modalities, speech and music.) In any case, language is almost universal, although few of us are musically expert, and it is only such people who can provide us with information about amusia per se—an acquired impairment of musical function in someone with a preexisting musical aptitude (Henson, 1985). Consequently, the history of the neurological study of amusia has largely been one of a fascinating, if ultimately inconclusive succession of single-case studies, frequently of well-known figures, who have suffered musical loss through brain damage. Indeed the condition is inevitably underreported as only those who know they have lost a capacity are likely to present clinically. Moreover, these individuals are likely to exhibit far greater variability in premorbid ability than aphasics.

Gall (1825) was the first to assert the existence of a "musical organ" in the human brain (Basso, 1993). Henschen (1920) wrote the first monograph on amusia, summarizing the cases to that date; he proposed a parallel between music and language disturbances, and suggested a localization for different aspects of music. Last century there was generally the uncritical assumption that the LH was dominant for most functions, including both language and music, and the RH had no specific competence, and could only help in functional recovery after left-sided injury. Later, the pendulum swung to favor the RH, though nowadays the complexity of musical ability is increasingly recognized, with both hemispheres and many regions being involved. Thus Bever and Chiarello (1974) attempted to resolve the issue in terms of musical expertise; in normative dichotic listening studies, musicians were found to process melodies better by the right-ear–LH system, whereas nonmusicians were found to favor the left-ear–RH. Because most case studies had generally been proficient musicians, not surprisingly, the LH seemed to have been involved. This dichotomy in terms of musical expertise has since been refined and extended in terms of an LH mediation of "local" information processing in melodies, and an RH superiority at a global level. Thus Peretz (1990), in a study of melody processing in left and right brain-damaged patients, found evidence for one subsystem for processing pitch and another for temporal variations. Although the former tended to favor the RH, especially where melodic contour was important, the LH also played a role with respect to intervallic structure, especially where local cues were significant. With respect to temporal or rhythmic variations, the LH was generally favored, except where global cues were relevant. In this way the local–global distinction was seen to cut across the more traditional RH pitch–LH rhythm distinction (Gates & Bradshaw, 1977). The single case study of Mazzoni et al. (1993) reached an essentially similar conclusion; their patient with temporoparietal damage to the RH experienced problems in interpreting music as intellectually or emotionally meaningful as a whole, while retaining ability to process individual elements such as rhythm, melody, or harmony. However, Robin, Tranel, and Damasio (1990) compared two groups of patients with temporoparietal lesions of either the LH or the RH in the auditory perception of temporal or spectral information. The former task involved gap detection and the identification of those two tones (in a sequence of six) that were separated by the shortest gap; the latter task involved pitch matching and frequency discrimination. They found a double dissociation along traditional lines, the LH-damaged patients being poor at temporal processing, and the RH-damaged at spectral information processing, and concluded that problems in processing signal duration and intervals between sounds are likely to be associated with aphasia, but not so problems in processing pitch (fundamental frequencies) and harmonic structure. The latter difficulties may instead be associated with aprosodia (Ross, 1988).

At this point we would do well to heed the cautions of Basso (1993). Although a dichotomy in terms of LH analytic, local, sequential, and temporal, and RH holistic, global, simultaneous, and spectral has long been attractive (Bradshaw & Nettleton, 1983), there is a grave danger of circularity; if in a given musical task an LH superiority is found, it is *assumed, a posteriori,* that this effect reflects sequential, analytic processing. Moreover, hemisphere differences may not neatly reduce to a single principle. In any case, music probably does not closely match or parallel corresponding language functions. There is certainly no close parallel between reading music and text in terms of direct and mediated access, even though the skilled sight reader of music may chunk several notes together, while the tyro may operate note-by-note. Not only do we lack an adequate psychology of normal musical functioning, which is necessary before we can develop a neuropsychology of music, but it is uncertain whether there are brain structures dedicated to music any more than there are, for example, to chess. Thus there is no evidence of selective musical (or chess) impairments after localized lesions, even though the latter may affect functions necessary for the correct operation of such skills. Similar arguments of course apply in the case of acalculia (see Chapter 4).

As Henson (1985) observed, there are probably as many varieties of amusia as there are components of musical ability, and each variety may be contaminated or complicated by coincidental impairment or loss of functions associated with perception or performance. Benton's (1977) classification is as follows:

1. *Oral, expressive,* or *vocal amusia,* affecting humming, whistling, or singing a tune, often associated with anterior RH temporal or especially frontal damage

2. *Instrumental apraxia,* a rare disorder in its own right

3. *Musical agraphia,* involving problems in transcribing heard notes or copying musical notation, and often associated with other musical deficits or visuoperceptual problems

4. *Musical amnesia,* a (generally right temporal) inability to identify or produce tunes or melodies that should be known

5. *Musical alexia,* a problem in reading musical notation

6. *Disorders of the sense of rhythm,* often associated with left-sided damage

7. *Receptive amusia,* a failure to discriminate the basic characters of notes or a series of notes in terms of pitch, intensity, timbre, duration, or rhythm. It is often associated with auditory agnosia for natural sounds and word deafness and temporal damage.

Just as not all aspects of language are lateralized to a single hemisphere, music similarly consists of a number of discrete components all differentially lateralized to left and right, and differentially localized in an anterior and posterior direction. At a purely perceptual level, a right-temporal lobectomy is associated with deficits in timbre, tonal memory, loudness, and pitch discrimination, complex tone discrimination, and melody recognition (Zatorre, 1984, and see also

Kaplan & Gardner, 1989), whereas LH damage affects recognition of familiar tunes and lyrics.

Subtractive PET imaging techniques perhaps provide the best information on localization of function during music processing in the normal brain. Sergent, Zuck, Terriah, and MacDonald (1992) studied 10 professional musicians in the contexts of sight reading alone, sight reading and playing, and listening. They found that playing (with the right hand) involved, perhaps not unexpectedly, LH motor and premotor cortex activity and the right cerebellum. Listening to scales involved the secondary auditory cortex (area 42) of both hemispheres, together with the superior temporal gyrus of the LH (area 22). Listening to musical pieces additionally involved the superior temporal gyrus on the *right* (i.e., now there was *bilateral* temporal activity). Reading a score without listening or playing activated the extrastriate visual areas bilaterally, but the lingual and fusiform gyri normally responsive to words were now silent; instead, the occipitoparietal junction on the left responded; in other words, the LH dorsal visual system associated with spatial processing was apparently responsive to the relative heights of the written notes on the score. Reading the score and listening added activity from the superior and posterior part of the supramarginal gyrus in the inferior parietal lobule of both hemispheres, in a region just outside that involved in reading and writing words; it presumably mediated the mapping between musical notation and the corresponding sounds and melodies. Finally, when all three activities of playing, reading, and listening were combined, there was the additional activation, bilaterally, of the superior parietal lobule, a spatial region that the authors note is important for sensorimotor transformations for visually guided manual actions. Additionally, there was activity in the premotor regions of the LH and the inferior frontal gyrus above Broca's area, regions responsible for the sequencing and timing of finger movements. Thus, just as in language, we see the activity of an extensive neural network of distributed but locally active regions, with music even more bilaterally represented than language. These findings highlight the observation that clinical and lesion studies on their own cannot capture the intricate pattern of discretely localized regions involved in such complex activities as language and music.

Prosopagnosia

Faces provide us with information on gender, age, personality, emotional, attentional and intentional states, health, and personal identity (Young, Newcombe, de Haan, Small, & Hay, 1993), information that along with eye contact and gestures may be used to regulate social interactions. Similarly, observation of lip movements may aid speech comprehension. Some or all of these sources of information are mediated by different processes. Indeed, visual identification may itself involve independent processes for recognizing familiar faces and for discriminating or matching unfamiliar ones, as indicated by the phenomenon of double dissociation. Thus some patients have problems only with recognizing familiar faces, and others only in distinguishing between unfamiliar ones. As Young et al. (1993) observed, these two processes, along with the analysis of expressions, are likely to involve at least three separate processing systems. As yet we have no evidence of specific deficits in the recognition of age or gender, or of a preserved ability to name familiar faces while being unable to retrieve other aspects of identity-specific information, and as Bruyer (1993) noted we probably never will. Indeed prosopagnosia should not be associated with *person* agnosia, as by definition the patient can identify people via other cues (e.g., sound of voice, silhouette, gait, or car), though some patients are unable to recognize people by face, name, or voice. Bruyer (1993), conversely, described some patients whose only deficit is an inability to *name* familiar faces, but who could still retrieve identity-specific semantic information, and could recognize the name itself. This deficit seems to us to resemble an extreme form of the tip-of-the-tongue phenomenon (see Chapter 8).

RH damage may be associated with problems in recognizing people by voice, and is indeed known to be associated with problems in correctly perceiving facial emotions (Ekman, 1993), and we shall shortly see that similar damage may affect ability to recognize familiar faces. However, the latter deficit typically occurs quite independently of any problems in recognizing emotions, age, gender, and so forth. Indeed prosopagnosics are usually relatively unimpaired in categorization of faces by age, gender, or expression, though *some* prosopagnosics inevitably have concomitant impairments in *some* of these capacities.

Bruce and Young (1986) suggested that outputs from putative face recognition units, which determine whether or not a face is familiar, also access information on personal identity (occupation, association with other individuals, etc.), and, separately, initiate name retrieval. The latter can also presumably be separately accessed via written or spoken information. We shall see that prosopagnosics, with a specific disability in recognizing *familiar* faces, can usually recall information about individuals from their *names*. Such information is normally accessible from the face input on its own.

Bodamer (1947) described a patient of normal intelligence who, after brain injury, could no longer recognize his own face in a mirror, or the faces of family and friends; there was also a total loss of a sense of familiarity. Dogs were described as "like people with curiously hairy faces." He could however identify all the features in a face, and see faces, and later could recognize people by nonfacial aspects such as dress, posture, gait, voice, mannerisms, and so forth. (Indeed *prosopagnosics,* the term introduced by Bodamer for people with this deficit, often can recognize individuals by distinctive features such as a mole or beard; however, recognition by faces is normally far more efficient than any other channel or process.) Though the term was originally defined as a deficit in the recognition of *familiar* faces, and is still correctly used as such, there are currently increasing debates as to the extent to which there may also be concomitant deficits in matching *unfamiliar* faces, whether the syndrome of prosopagnosia is necessarily restricted to faces alone, and where it is localized.

We do of course have a quite amazing capacity to recognize thousands of faces over a lifetime, despite the presence or absence of distracting "paraphernalia" such as spectacles, makeup, beards, and so on, the aging of the face, and changing emotional expressions. In a very real sense normal face recognition depends upon two processes:

1. Whole-face *gestalt* processing via patterns of interrelationships between and largely independent of individual features. We often cannot verbalize this "*je ne sais pas*" about a face, an attribute which may be largely dependent upon the RH.
2. Piecemeal feature-by-feature analysis, largely the province of the LH.

A different mix of these two processes may be needed for recognizing familiar faces, compared to matching or discriminating between unfamiliar ones. Traditionally, prosopagnosia has been thought to require bilateral posterior damage (Damasio, Damasio, & Van Hoesen, 1982), though recently RH damage has often been thought both necessary and sufficient (Benton, 1990, and see also De Renzi, Perani, Carlesimo, Silveri, & Fazio, 1994, who concluded that the acquired inability to recognize familiar faces is a rare disorder, not manifested by the majority of patients with right occipitotemporal injury; they suggested that only in a minority of dextrals is RH specialization for face processing so marked that it cannot be compensated for by the healthy LH). Indeed, the frequent co-occurrence of a left superior quadrantanopia as a localizing sign itself suggests a right-sided inferomesial occipitotemporal lesion perhaps involving the fusiform gyrus. Such lesions also lead to achromatopsia (disturbed perception of color) and topographagnosia (problems in route finding), which also both involve disturbances in the sense of familiarity, as in prosopagnosia and the other disorders of misidentification, such as Capgras and Frégoli syndromes (to be described shortly). As Bauer (1993) noted, these regions form a functional interface between the visual association cortex and the temporal lobe that projects to the limbic system. Prosopagnosia is therefore a visual–limbic disconnection. Indeed, recent neuroimaging studies (Sergent & Signoret, 1992) indicate the involvement, in recognition of familiar faces by normal individuals, of the lingual, fusiform, and parahippocampal gyri and anterior temporal cortex of the RH. Sergent and Signoret (1992), however, commented on the absence of activation of the superior temporal sulcus or the lateral inferotemporal cortex, in which cells, selectively responsive to faces, had been found in the monkey. On the other hand Lu et al. (1991) reported activity in the middle temporal gyrus during face-processing tasks, and Ojemann, Ojemann, and Lettich (1992) reported single-cell activity in the middle and superior temporal gyri on the right during such tasks. We shall return to the question of localization when we consider whether there may be more than one form of prosopagnosia.

Recently it has been asked whether prosopagnosia is restricted to faces, or whether it may occur also with other generic stimuli requiring unique individuation, where category members are ambiguous or confusingly similar (see e.g., Levine & Calvanio, 1989). Such items may share the same global shape but differ quantitatively along continuous dimensions, and may require arbitrary discrimination on the basis of individual "episodic" memory or experience. Thus a farmer may now come to have problems recognizing his cows, a tailor her cloths, a dogfancier his breeds, an ornithologist her bird species, an architect different kinds of buildings, a veterinary surgeon different skin diseases. Deficits have even been reported for cars and chairs (Bauer, 1993). However, it is unclear whether this spread occurs with all patients, and to all such categories. Although we might think that such a problem is one of increasing categorical *specificity,* and it is certainly true that recognition of *superordinate* categories is always easier, there is little evidence that prosopagnosics have difficulty recognizing their *own* belongings, though see Damasio, Tranel, and Damasio (1990) and Etcoff et al. (1991). Faces of course, above all other object categories, are not only "many," "different," and "unique," but they are also "similar" (Damasio et al., 1990). Perhaps we should view prosopagnosia as a form of facial simultanagnosia, such that the patient is forced to concentrate at a fragmentary, featural level as mediated by the LH.

Milders and Perrett (1993) reviewed the evidence for the preservation in prosopagnosia of tacit or implicit recognition, via nonstandard or indirect tests, where face recognition is only an incidental aspect of a task that seems to demand other abilities (and see also Humphreys et al., 1992). Two broad categories of task may be distinguished:

1. *Physiological measures,* such as changes in pupillary dilation, skin conductance, or the P300 evoked potential to familiar faces that the patient denies knowing. Rizzo, Hurtig, and Damasio (1987) reported that, in prosopagnosics, eye scan paths differed depending upon whether the patient inspected "previously known" or "unfamiliar" faces, suggesting that previously acquired and still accessible schemata could drive the scan path for familiar faces. Such effects could reflect activity of intact face-processing mechanisms that are disconnected from systems normally controlling overt behavior. Alternatively, they could reflect very weak signals from a damaged recognition system unable to support overt behavioral choice, but still strong enough to bias other aspects of behavior. Normal individuals with subthreshold stimulation may also demonstrate tacit recognition of familiar faces, while denying that their choice is conscious, deliberate, or accurate. Prosopagnosics, if cued, often are able overtly to recognize faces with confidence.

2. *Behavioral measures,* such as faster learning and naming of correct

name–face pairings than of incorrect pairings, interference, and priming tasks where an "unknown" face precedes a name about which a judgment is to be made, and matching pairs of faces as being of the same or of different ("familiar" or "unfamiliar") people. Other behavioral measures indicative of tacit or implicit recognition in the absence of overt recognition include preference measures, and forced-choice decisions (Bruyer, 1993).

It is noteworthy that these effects of covert or implicit face processing occur via indirect testing, even when the patients fail to perform above chance on forced-choice discriminations of face familiarity; with the classic blind-sight experiments (Cowey & Stoerig, 1992), cortically blind individuals demonstrate above-chance performance in judgments of location, movement direction, velocity, size, and orientation, with forced-choice *direct* testing, despite absence of conscious stimulus experience (Humphreys et al., 1992). However, not all prosopagnosics manifest covert face recognition. Indeed, Etcoff et al. (1991) described a patient with no obvious perceptual deficits (i.e., no "perceptual" prosopagnosia, see below), a total inability to recognize familiar faces and to answer related questions from memory (i.e., a profound "associative" prosopagnosia, see below), a preserved ability to match faces, and to recognize age and sex, and personal identity via nonfacial channels, and yet a complete absence of covert or implicit face recognition as indexed by skin responses, pupillary dilation, and so on. In this case, at least, the authors argued that brain injury seems to have *eliminated* the storage of information about familiar faces, rather than merely disconnecting it from consciousness.

When covert, tacit, or implicit face recognition does occur, Humphreys et al. (1992) suggested that it does so only when there is a mnestic problem affecting the stored recognition units or their output, not when prosopagnosia results merely from some more superficial perceptual impairment. Such a distinction, which is also proposed by Milders and Perrett (1993, and see also De Renzi, Faglioni, Grossi, & Nichelli, 1991), is of course analogous to the associative–apperceptive distinction of agnosia, of which prosopagnosia may seem to be a special instance. Thus, according to this view, prosopagnosia might merely be a mild form of object agnosia, affecting only the most difficult stimuli, like faces. However, as Etcoff et al. (1991) observed, if this were the case there should not exist—as there do—patients with object agnosia and intact face recognition. Damasio et al. (1990) suggested that the converse—profound prosopagnosia, with intact abilities to discriminate patterns, copy, or construct patterns, etc.— represents an amnesic or associative form of prosopagnosia, involving bilateral temporal and hippocampal damage, including also the entorhinal cortex and amygdala. Conversely, prosopagnosics with difficulties at a more *perceptual* level, according to Damasio et al., may have lesions in the *RH* visual association cortex (i.e., occipitoparietal and mesial occipitotemporal). Alternatively, as we saw, covert, implicit, or tacit recognition of faces may merely reflect disconnection, rather than destruction, of the face-recognition system. In the former case, McNeil and Warrington (1993) reported cases where signs of covert recognition occur even for faces encountered after the occurrence of injury, suggesting that new, inaccessible records can be laid down.

Another model that attempts also to accommodate the implicit–tacit–covert issue, is that of Young et al. (1993). These authors proposed two routes in face processing, a ventral route to the temporal lobes for overt recognition, and a dorsal route via the inferior parietal lobule to the limbic system, which mediates emotional salience; prosopagnosia is then due to damage to the ventral route, and recognition without awareness is still possible if the dorsal route is intact. The dorsal route would permit prosopagnosic patients (and others) to interpret emotional expressions. If damaged it may lead to disorders of misidentification (see below). However, the authors noted that covert recognition still occurs with faces devoid of emotional significance for the patient. They suggested that the fact that provision of cues (e.g., about occupation) can convert covert-only recognition to confident overt recognition indicates the invalidity of an overt–covert distinction; it all may be a matter of relative confidence levels. As we saw, normals in threshold experiments often claim to be guessing even when consistently correct. However, we would ask how this conclusion relates to all the *other* instances of apparent tacit or implicit processing, with indirect testing, in such other neuropsychological domains as amnesia, unilateral neglect, alexia, and agnosia.

Other unresolved issues include the following:

1. Is there a unique, specialized, dedicated processor for faces in the RH complementary to an LH-language mechanism? The recognition of emotions, personal identity, age, sex, and language itself all involve communication. Such a faces-specific processor would presumably operate as one of Fodor's (1983) modules, even though, as we have seen, an extensive neural network subserves face processing, encompassing most of the ventromedial regions of the RH from the occipital to the temporal poles (Sergent & Signoret, 1992). According to Fodor we can invoke modularity of operation under conditions of cognitive impenetrability (with little conscious awareness of the underlying operations) when there is (1) informational encapsulation (with photographic, caricature, or schematic representation being perceived as real faces even though we know they are not); (2) when there is a domain specificity (requiring operations unlike those involved in processing other types of visual objects); (3) when processing is innately specified (e.g., the remarkable abilities of neonates in face processing); and (4) when there is neural specificity. With respect to the last criterion, focal damage can indeed selectively impair face processing, although we now know many areas participate, and it is not clear whether all are specifically or uniquely devoted to face processing. We asked, at the beginning of this subsection, whether there is a unique, specialized, dedicated processor for faces in the

RH, complementary to an LH-language mechanism. Rentschler, Treutwein, and Landis (1994) reported two patients with inferomesial occipitotemporal strokes. The patient with the stroke on the *right* became prosopagnosic, could still read but could not recognize whose handwriting it was. She also had a selective loss of global visual perception, with deficits in the recognition of facial expressive gestures, but could lip-read speech. The patient with the exactly similar damage on the *left* became alexic for text, but could still recognize who wrote it, and while she had no deficit in face recognition nor in the classification of facial expressions, she was impaired at lipreading and the integration of stimulus features. The authors reviewed similar double dissociations reported in the earlier literature. We believe that the case for a left–right complementarity of function is strong, and, with Rentschler et al. (1994), argue for the conclusion that prosopagnosia may be one aspect of a more *general* alteration of perception.

2. Is face processing displaced to the RH because language has already preempted processing space in the LH, or is language in the LH because of an evolutionarily older system for mediating spatial and emotional processing (Bradshaw & Rogers, 1993)?

3. Is any such RH specialization for faces separate from a general mechanism for spatial and/or object processing? As we have seen, many parts of the brain are involved, in parallel, in different aspects of face processing; the same applies with respect to other complex objects, although the processing power devoted to faces may be substantially greater because of their social importance to the individual.

Damasio et al. (1990) provided a good, illustrative example of a case of prosopagnosia:

A 65-year-old woman suddenly developed an inability to recognize the faces of her husband and daughter. She could not even recognize her own face in the mirror—although she knew the face she was observing must belong to her, she did not experience a sense of familiarity when viewing it. The faces of other relatives and friends became equally meaningless. Yet she remained fully capable of identifying all those persons from the voice. About two months prior to the onset of the prosopagnosia, she had noticed some difficulty in perception of colors. Specifically, colors appeared "washed out" and "dirty," although she continued seeing details of shape, depth and movement without any problem. She had no other history of neurological problems and no history of psychiatric disease. . . . She was entirely unable to recognize a single face of any relative or friend, either in person or from photographs. Furthermore, she did not even have a hint that the faces belonged to persons with whom she was well acquainted. She was equally impaired in the recognition of celebrities from photographs or movies. . . . The profound face recognition defect of this patient occurred against a background of otherwise virtually normal neuropsychological capacities. She had normal intellect, intact speech and language, and normal memory and learning for both verbal and nonverbal material (outside the visual realm).

Recognition and naming of objects were normal in the auditory and tactile modalities. . . . She could read normally and she could write, with normal spelling. Of particular importance, visual acuity was 20/20 in both eyes, and she performed normally on a wide range of visuoperceptual, visuospatial, and visuo-construction tasks.

The exceptions to the intactness of basic visual perception were: (a) a defect in color perception and (b) small form vision scotomata in the right superior and left superior quadrants. The patient could not perceive colors in any part of the visual field and saw form always in shades of black and white. . . .

The defects in this patient were caused by infarctions in ventral visual association cortices, bilaterally, in the inferior occipital/posterior temporal region. The infarction on the right side occurred first and caused left hemiachromatopsia, as suggested by her complaints of color perception loss. The second infarct, on the left side, further impaired her ability to process color and, in combination with the first lesion, created the neuroanatomical setting for prosopagnosia. (pp. 89–91)

Color-Related Disorders

As Davidoff (1991) observed, central achromatopsia (loss of all color vision after cortical lesions) provides the best evidence for a specific color module. The proposal for modular processing (Fodor, 1983) requires that brain damage can completely remove a given function (e.g., color vision) and leave all other (visual) functions intact. Davidoff (1991) noted that acquired achromatopsia of central origin was first described by Samelsohn in 1881. Verrey in 1888 reported achromatopsia confined to a single visual field after an infracalcarine lesion involving the lingual gyrus and underlying white matter (see below).

Three types of color disturbance may be identified, in descending order of severity (Grüsser & Landis, 1991): achromatopsia, color agnosia, and color anomia. With *achromatopsia,* there is typically loss of color vision, with preserved form vision, affecting the whole or part of a visual field (e.g., hemianopia or upper—often left—quadrantanopia for color, or hemiachromatopsia and upper quadrantachromatopsia, respectively). The inferomesial (i.e., ventromedial) occipitotemporal cortex of either hemisphere (though often of the right alone, in the case of pure achromatopsia, in the absence of language deficits) is likely to be affected. Critical regions are the fusiform and lingual gyri, (Benton & Tranel, 1993, reviewing the work of Damasio and colleagues, concluded that achromatopsia is associated with infracalcarine lesions affecting the middle third of the lingual gyrus, and the white matter immediately behind the posterior tip of the lateral ventricle; these conclusions are supported by the neuroimaging studies of Lueck et al., 1989, and of Corbetta, Miezin, Dobmeyer, Shulman, & Peterson, 1990.) Achromatopsia is often associated with prosopagnosia and topographagnosia (route finding difficulties—see below), especially with RH injury, and with associative agnosia if the left side instead is damaged.

With *color agnosia,* a complete right homonymous hemianopia is not uncommon; the angular gyrus on the left may be affected. The patient cannot name colors on presentation, or point to a given color if named, or correctly associate colors with objects. Thus green elephants seem acceptable. Although colors cannot be conceptually grouped, color vision itself is intact, as demonstrated by the preserved ability to match color chips. Similarly, there is no language impairment, as the patient can, for example, *verbally* associate a fire engine with red. An accompanying parieto-occipital alexia is not uncommon.

With *color anomia,* seen colors cannot be named, but the correct color can be matched or indicated when the name is given, unlike color agnosia. However, patients now cannot answer what is the color of a carrot or a fire engine, although they may be able to choose the correct color chip for it, and can color match. The disorder may accompany alexia without agraphia. As Davidoff (1991) observed, LH occipitotemporal damage (with a right homonymous hemianopia) frequently accompanies it, and these regions have been associated with imagery.

For completeness we should briefly mention other color disorders where language is also involved. Thus Benton and Tranel (1993) described patients who can identify and name objects but not colors—a category-specific deficit suggestive of different loci for object naming (LH anterolateral temporal cortex) and color naming and identification. Similarly, with *specific color aphasia* the patient cannot match similar color names like *rosy* and *red.*

Disorders of Route Finding: Topographical Agnosia or Topographagnosia

As Grüsser and Landis (1991) observed, to interact appropriately with the environment, to correctly reach for and grasp objects, and to successfully navigate, one must formulate an integrated viewpoint that appropriately represents objects in the real world in terms of their spatial positions relative to oneself and to each other. We must therefore be adequately able to process visual inputs at an early perceptual level, and to associate the result with information already stored in long-term memory. Attentional processes must also be sufficient to select out the currently relevant regions. Disorders at any of these three levels may affect our route-finding ability. Thus the apperceptive and the associative agnosic are both likely to experience deficits in this area, as indeed also is the patient suffering from unilateral neglect or hemi-inattention (see Chapter 6). True spatial deficits involving route finding should therefore be independent of any more primary perceptual or agnosic deficit. We normally learn our way around unconsciously and nonverbally, in a manner suggestive of RH mediation. It is therefore probably no accident that prosopagnosia and achromatopsia often accompany associative topographagnosia (see below), in inferomesial occipitotemporal lesion sites, especially on the right.

Grüsser and Landis (1991) distinguished between perceptive, apperceptive, and associative topographagnosia. With *perceptive topographagnosia,* an incomplete perceptual representation of a habitat is mismatched to otherwise adequate stored spatial memories, leading to disorientation. Thus drawing of remembered routes, and real route-finding abilities themselves, require testing. Unilateral neglect often accompanies this frequently right-parietal damage.

With *apperceptive topographagnosia,* lateral temporoparietal-occipital damage may lead to a dissociation: left-sided lesions may be associated with problems with "local" information, whereas right-sided lesions may be accompanied by problems with "global" or overall spatial integration or interrelationships. In the former case the patient may rely upon salient local cues (e.g., traffic lights, roundabouts, yellow buildings) much as a prosopagnosic comes to rely upon moles, mustaches, and so forth.

With *associative topographagnosia,* mesial inferior occipitotemporal damage, especially of the RH, may result in the patient getting lost in familiar environments and recognizing this. He or she may develop complicated compensatory strategies to cue or guide, paying abnormal (and successful) attention to local cues, while experiencing disturbing feelings of strangeness and unfamiliarity, as with prosopagnosia (above) and delusions of misidentification (below). There may be inability to locate countries, cities, or buildings on a map.

It may be necessary to test the patient both in real-life situations and with artificial maze-type tests, to discriminate between the above syndrome variants, and between problems in accessing otherwise stored information, and actual loss of stored information.

Horologagnosia: A Recently Described Impairment of the Ability to Tell the Time

Kartsounis and Crewes (1994) reported a case of profound difficulty in telling the time from a clock or watch after a right posterior stroke, which also resulted in decreased sensation and muscle weakness on the left side of the body, left homonymous hemianopia, and left visuospatial neglect. Problems were evident in telling the time, or clock setting, even for numbers on the right side of the clock. The patient could not even identify the time from a clock face on which he himself had only just correctly inserted the hands to depict a specified time (cf. alexia without agraphia, Chapter 3). The authors propose the term horologagnosia from ὡρολογίον, clock.

Disorders of Distortion and Perseveration

Vascular disturbances of the occipital cortex lead to the commonly experienced glitter, scintillations, and scotomata of migraine, with its simple, perceptual effects. With temporal cortex dysfunction, more frequently of an epileptiform nature, there may be evocation of more complex scenes with affective connotations of a frightening, pleasant, or bizarre nature.

With *metamorphopsia* (see e.g., Beaumont & Davidoff, 1992), alterations in an object's customary perceptual quality lead to apparent changes in size (micropsia or macropsia, see e.g., Cohen, Gray, Meyrignac, Dehaene, & Degos, 1994, where lesions of the ventral occipitotemporal visual pathway were implicated), orientation, distance (*teleopsia*), or three-dimensional quality. It may be associated with prosopagnosia, and may occur only on the left side, again implying RH pathology. Frank hallucinations may be involved; Grüsser and Landis (1991) cited the example of a lady who sees a poodle on a bus and subsequently all the passengers seem to acquire poodle faces. With *polyopsia,* an object may be multiplied repeatedly.

With *palinopsia,* an image persists (maybe for several hours) or recurs after the stimulus is removed, perhaps appearing in the blind half field of a homonymous hemianopia (see optic allesthesia, below). This typical acute, ictal condition is curiously reminiscent of the debated and disputed phenomenon of eidetic imagery (Haber, 1969), whereby certain individuals, typically nonliterate children, are said to be able to recall in vivid photographic detail a viewed scene or picture. Patients afflicted with palinopsia do not confuse the visual hallucination with reality. The commonly co-occurring left-sided field disturbances suggest posterior right lesions (Bender, Feldman, & Sobin, 1968; Cummings, Syndalko, Goldberg, & Treiman, 1982; Michel & Troost, 1980).

The "Dreamy State," Déjà Vu, and Sensory Illusions

At the turn of the last century John Hughlings Jackson observed that medial temporal lobe seizures may result in a "dreamy state" of vivid memorylike hallucinations, an apparent doubling of consciousness ("mental diplopia," with images apparently from the past added to the state of present awareness), and/or a sense of previous experience (déjà vu, déjà vécu). Penfield and Perot (1963) showed that such a state can be elicited with electrical stimulation of the lateral temporal neocortex, especially the superior temporal gyrus. Later studies (reviewed by Bancaud, Brunet-Bourgin, Chauvel, & Halgren, 1994) showed that the phenomenon may appear with hippocampal and amygdala stimulation; lateral stimulation may evoke it perhaps only when the resulting after-discharge spreads medially. Bancaud et al. found that the dreamy state is often associated with epigastric phenomena, oro-alimentary automatisms, simple gestural automatisms, fear and loss of contact—all characteristic of partial seizures within the medial temporal lobe—and, less frequently, sensory elusions within which visual forms predominate. They concluded that Hughlings Jackson's dreamy state probably depends upon a network engaging both medial and lateral aspects of the temporal lobe, and that the anterior hippocampus, amygdala, and superior temporal gyrus have a relatively privileged access to this circuit. (It is unclear why the superior, i.e., auditory, rather than the inferior, i.e., visual, temporal cortex is involved.)

Visual or Optic Allesthesia

A visual image is transposed from one visual field (or quadrant) to another, often after a delay or with distortion. It may be from a seeing to a "blind" visual field (see palinopsia, above), or from a neglected to a "good" visual field (see Chapter 6).

Delusional Misidentification Syndromes

Joseph (1986) brought together a number of misidentification syndromes. By including *reduplicative paramnesia*, he linked mistaken misidentification for place with delusional misidentification of person. First described by Pick (1903), reduplicative paramnesia refers to the belief that a place familiar to the patient exists simultaneously in more than one physical location. Thus a patient may insist that the hospital she or he is in has been duplicated and relocated from one site to another, so that the two hospitals coexist in different places at the same time (Anderson, 1988; Fuller, Marshall, Flint, Lewis, & Wise, 1993). Strikingly, the patient puzzles over apparently compelling counterevidence, without shifting belief. RH rather than LH damage seems prevalent (Benson, Gardner, & Meadows, 1976; Kapur, Turner, & King, 1988; see also Fuller et al., 1993; Levine & Grek, 1984; Price & Mesulam, 1985; Young, Flude, & Ellis, 1991), and there may be an association with prosopagnosia. Thus faces and probably objects generally receive elaborative processing in inferomesial temporal regions, while adjacent limbic regions may be involved with their associated emotional significance (and see Bruce & Young, 1986; Young et al., 1991). Damage to the former regions may lead to prosopagnosia and covert, implicit recognition, whereas damage to the latter regions may result in loss of emotional significance, and a change in the feeling of recognition for objects, places, and faces (see below). Indeed inappropriate, ictal activation of limbic structures may lead to a deranged view of a person's, object's, or place's identity, significance, appearance, relevance, or meaning. According to a related formulation (H. Ellis & Young, 1990, and see above under prosopagnosia), there are two routes to face recognition: a ventral route to the limbic system, which is the principal pathway for facial identification, and a dorsal route, which has the same termination but passes via the inferior parietal lobule. In prosopagnosics, the ventral route is damaged, leading to the loss of conscious recognition, though with preservation (via the intact dorsal route) at an unconscious level of emotional significance. With Capgras' syndrome (see below), the mirror reverse situation obtains, with preservation of the ventral route and conscious recognition, and damage to the dorsal route and resulting loss of emotional significance.

We speculate whether similar limbic involvement underlies the strange and rare phenomenon of synesthesia, where a stimulus in one sensory modality gives rise to an otherwise unexpected response in another; the subject may "taste" words, or "smell" the color of paint (Baron-Cohen, Harrison, Goldstein, & Wyke, 1993; Cytowic, 1989), or associate certain days of the week, or numerals, with certain colors. These associations, for any given synesthete, are usually entirely consistent over time. The phenomenon, though extremely rare, may be more common in females and nondextrals, and may be associated with mild topographagnosia, again indicating limbic and/or mesial-occipital involvement. The limbic system with its rich intermodal interconnections, is of course uniquely placed for such involvement. We must await the results of brain-imaging studies for a definitive answer.

Capgras' syndrome was first described by Capgras and Rebone-Lachaux in 1923. Someone familiar to the patient is believed to have been replaced by a double or imposter with identical or very similar appearance (Berson, 1983). Thus a woman may believe that her spouse and children have been replaced by doubles, together with the police and neighbors, typically with evil intent. The Germanic myth of the "doppelgänger" is intriguingly reminiscent (Grüsser & Landis, 1991). RH involvement again seems to predominate (Feinberg & Shapiro, 1989).

Frégoli syndrome was first described by Courbon and Fail (1927), named after the Italian actor and mimic Leopoldo Frégoli, famous for his stage impersonations. Characteristically, there is delusional misidentification of individuals as being familiar persons in disguise. Thus patients may believe that they are being persecuted by someone who has assumed the guise of a person normally encountered in everyday life. It is as if biographical information is automatically accessed for a certain person, regardless of who is actually there, and wrongly applied. Again, RH involvement seems to predominate, and may be coupled with a form of erotomania known as de Clérambault's syndrome (Wright, Young, & Hellawell, 1993), in which patients suddenly arrive at the delusional belief that someone (usually of higher social standing) is in love with them.

In the *syndrome of intermetamorphosis* (Courbon & Tusques, 1932), someone, familiar or not, is believed to have been changed into someone else entirely, with altered identity and physical appearance (de Pauw, Szulecka, & Poltock, 1987; Joseph, 1987; Staton, Brumback, & Wilson, 1982). Thus there is false recognition of both appearance and associated identity, whereas with the Frégoli syndrome patients are confused by the identity but not the appearance of a stranger. We can ascribe all such deficits to Bruce and Young's (1986) second stage (see Ellis & Young, 1990) where faces are judged as familiar or not. Although with Capgras' syndrome there is replacement by a double, and with Frégoli syndrome it is a matter of disguise rather than replacement, with inter metamorphosis there is interchange between two different people.

Finally, in the *syndrome of subjective doubles,* someone else is believed to have been transformed into oneself (Christodoulou, 1978), or another person is believed to have taken on the physical characteristics and identity of oneself.

Summary and Conclusions

Our object-recognition capacity matches language in terms of rapid, effortless, and accurate computational power, which in fact far exceeds the capacity of any artificial system. Indeed, perhaps more than in any other syndrome except maybe amnesia, the emphasis in the study of agnosia has been on information-processing aspects rather than purely structural mechanisms. We do know however that a ventral pathway, directed toward the IT cortex, is involved in object recognition, whereas a dorsal pathway, incorporating the inferior parietal lobule, mediates spatial localization. In fact, the anterior portion of the IT cortex may be where visual perception meets memory and imagery. Adjacent regions within the inferomesial part of the occipitotemporal cortex are involved in face, color, and topographical processing. However, an object typically consists of a multiplicity of qualities and attributes; a unified perceptual experience may depend on widely distributed sites becoming coherent via broad-band synchronization.

One hundred years ago Lissauer distinguished between apperceptive and associative agnosia, between deficits primarily at a perceptual level or at a level involving meaning and significance. That distinction, though modified and adapted, is still with us today in various guises. Thus diffuse RH occipitoparietal damage is frequently associated with problems in perceiving anomalies in sketchily drawn scenes, or in processing incomplete or overlapping figures, degraded or fragmented characters, or three-dimensional objects that have to be recognized from noncanonical perspectives. Nevertheless, the concept of apperceptive agnosia has not been universally accepted.

Bilateral superior occipitoparietal damage has been linked to dorsal simultanagnosia, where the patient is said to have reasonable perception of individual elements or details of a complex picture or scene, but an inability to apprehend the whole. Competition between copresent items seems to lead to a form of extinction, with a difficulty in disengaging and shifting attention.

Continuing along a conceptual if not an anatomical continuum, ventral simultanagnosia may occur with inferomesial temporo-occipital damage to the LH. Multiple objects may now be seen and recognized, even though the scene as a whole may not be comprehended. The condition in turn closely resembles the traditional concept of associative agnosia, with LH IT or temporo-occipital injury. Now the perceived significance of objects is compromised, rather than the ability to discriminate them from the background. Confabulatory interpretations may be attempted, though access is often available to tacit or implicit knowledge. Recognition in terms of superordinate categories is also often preserved. Although the apperceptive agnosic may have great difficulty in matching objects by shape or appearance, the associative agnosic's difficulty is more in terms of matching by use or function.

Although the associative agnosic cannot even demonstrate knowledge of the meaning of an object by pantomiming its use, an optic aphasic can do so, merely being unable to name it when visually presented in isolation; thus optic aphasia involves a modality-specific anomia, due probably to partial disconnection of visual processes from speech and praxis on the left. Inferomesial damage is suggested by the frequent co-occurrence of color agnosia.

Accounts such as the one above, which in various versions, past and present, dates back 100 years, place perceptual processes at one end, with a largely bilateral or RH involvement; semantic processes lie at the other end, with a largely LH contribution. Sometimes, however, a patient may be unable to process instances of certain specific categories of item. These highly selective, category-specific deficits have sometimes been invoked to claim adjacency of locus of storage or mediation of related concepts, though clearly other explanations are possible and even preferred. Were the deficits restricted largely to, for example, low-frequency or abstract items, or even to animate objects to which city dwellers may have had limited prior access, it could be argued that the problem is merely one of unfamiliarity. However, a bewildering range of specific deficits have been reported, to visual object or word presentations, with some being spared in one patient and impaired in another.

Agnosia may not be limited to the visual modality, as witnessed by the phenomenon of tactile agnosia (astereognosis) and auditory agnosia. The latter may include inability to recognize naturally occurring sounds, or to process music, when the condition is known as amusia. In some respects amusia parallels aphasia in its complexity and manifestations, although clear dissociations are the rule rather than the exception. Of course, many people who are naturally nonmusical may never know if they have become amusic. Nor is there clear lateralization of musical abilities, though familiarity and expertise may induce hemisphere-linked strategy differences. However, there are few group studies, and a process as complex as music is unlikely to involve only one or two critical processing loci. Thus melody, timbre, and rhythm are all likely to be processed in different regions of the brain. Indeed, there are likely to be as many varieties of amusia as there are components of musical ability, and ways of representing a piece, receptively or productively, visually, or via the auditory modality. In the future, we can expect to learn most about the localization of regions intimately involved in the various aspects of music from subtractive PET imaging techniques. From the data currently available, active regions have generally corresponded to those that would have been predicted from clinical evidence and an understanding of the likely stages involved in information processing.

Faces provide us with an enormous range of information about gender, age, personality, attentional and emotional stress, health, and personal identity. Some or all of these aspects are mediated by different processes. Prosopagnosia, or loss of the ability to recognize familiar individuals by the face alone, irrespective of preservation or loss of the ability to discriminate between unfamiliar faces, has been associated with posterior RH or bilateral damage. Opinion now tends to hold that a right-sided inferomesial occipitotemporal lesion, in the lingual or fusiform gyri, may be sufficient for the manifestation of prosopagnosia, perhaps via loss of whole-face gestalt processing; the preserved ability still to identify a person via distinguishing facial characteristics such as a mole may be mediated via feature-by-feature analysis, perhaps largely the province of the LH. There is debate, however, as to whether the recognition of other classes of generic stimuli, with category members that are ambiguous or confusingly similar, might also be affected. There is also debate about the extent to which there may be preservation, in prosopagnosia, of tacit or implicit recognition; if so, it is possible that there are two routes for face processing, a ventral route to the temporal lobes for overt recognition, and a dorsal route via the inferior parietal lobule to the limbic system for tacit or covert "recognition" or a feeling of knowing.

Three types of color disturbance have been identified: achromatopsia, color agnosia, and color anomia. With achromatopsia there is typically loss of color vision, with preserved form vision, affecting the whole or part of a visual field. The inferomesial occipitotemporal cortex is likely to be affected, especially the fusiform and lingual gyri. With color agnosia, the angular gyrus on the left may be affected, and the patient cannot name colors or correctly associate them with objects. With color anomia, again colors cannot be named, but the correct associations can now be made. LH occipitotemporal damage may be responsible.

Disorders of route finding, topographical agnosia or topographagnosia, can of course involve deficits at a number of processing levels, and in a number of different manifestations, though posterior-parietal spatial processing mechanisms are clearly important for perceptive topographagnosia. However, because prosopagnosia and achromatopsia often accompany associative topographagnosia, it is not surprising that right-sided inferomesial occipitotemporal lesions have been implicated in this manifestation.

A number of bizarre disorders of distortion, perseveration, and delusion have been reported, frequently after posterior RH damage. These have included metamorphopsia (i.e., alterations in an object's customary perceptual quality), palinopsia (persistence of an image), visual or optic allesthesia (apparent transposition of a visual image in space), and a number of misidentification syndromes. The latter involve, in various ways, belief that a place (e.g., home or hospital) known to the patient has been somehow duplicated and/or translocated elsewhere, or that a person known to the patient has been replaced by an imposter, or that individuals have somehow changed identities. Although faces may receive elaborative processing in inferomesial temporal regions, adjacent limbic structures may be involved with their associated emotional significance; damage to the former regions may lead to prosopagnosia, whereas if the latter structures instead are affected there may be a change in the feeling of recognition for objects, places, and faces.

Further Reading

Basso, A. (1993). Amusia. In F. Boller & J. Grafman (Eds.), *Handbook of neuropsychology* (Vol. 8, pp. 391–410). Amsterdam: Elsevier.

Bauer, R. M. (1993). Agnosia. In K. M. Heilman & E. Valenstein (Eds.), *Clinical neuropsychology* (3rd ed., pp. 215–278). Oxford: Oxford University Press.

Benton, A. L., & Tranel, D. (1993). Visuoperceptual, visuospatial, and visuoconstructive disorders. In K. M. Heilman & E. Valenstein (Eds.), *Clinical neuropsychology* (3rd ed., pp. 165–214). Oxford: Oxford University Press.

Bruyer, R. (1993). Failures of face processing in normal and brain-damaged subjects. In F. Boller & J. Grafman (Eds.), *Handbook of neuropsychology* (Vol. 8, pp. 411–436). Amsterdam: Elsevier.

Damasio, A. R., Tranel, D., & Damasio, H. (1990). Face agnosia and the neural substrates of memory. *Annual Review of Neuroscience, 13,* 89–109.

Damasio, A., Tranel, A., & Damasio, H. (1993). Similarity of structure and the profile of visual recognition defects: A comment on Gaffan and Heywood. *Journal of Cognitive Neuroscience, 5,* 371–372.

Davidoff, J. (1991). *Cognition through color.* Cambridge, MA: MIT Press.

Etcoff, N. L., Freeman, R., & Cave, K. R. (1991). Can we lose memories of faces? Content specificity and awareness in a prosopagnosic. *Journal of Cognitive Neuroscience, 3,* 25–41.

Farah, M. J. (1990). *Visual agnosia: Disorders of object recognition and what they tell us about normal vision.* Cambridge, MA: MIT Press.

Grüsser, O.-J., & Landis, T. (1991). *Visual agnosias and other disturbances of visual perception and cognition.* London: MacMillan Press.

Humphreys, G. W., & Riddoch, M. J. (1993). Object agnosias. *Baillière's Clinical Neurology, 2,* 339–359.

McKenna, P., & Warrington, E. K. (1993). The neuropsychology of semantic memory. In F. Boller & J. Grafman (Eds.), *Handbook of neuropsychology* (Vol. 8, pp. 193–214). Amsterdam: Elsevier.

Sergent, J. (1993). Music, the brain and Ravel. *Trends in the Neurosciences, 16,* 168–172.

Sergent, J., & Signoret, J.-L. (1992). Functional and anatomical decomposition of face processing: Evidence from prosopagnosia and PET study of normal subjects. *Philosophical Transactions of the Royal Society of London B, 335,* 55–62.

Young, A. W., Newcombe, F., de Haan, E. H. F., Small, M., & Hay, D. C. (1993). Face perception after brain injury: Selective impairments affecting identity and expression. *Brain, 116,* 941–959.

Disorders of Spatial Cognition

Unilateral Neglect

$\mathcal{U}$nilateral neglect is a lateralized disorder of spatial cognition and space-related behavior that may occur following damage to either cerebral hemisphere. Other space-related deficits, such as disorders of route finding and topographagnosia, we have chosen to discuss in the previous chapter, as they relate to color agnosia. The disorder of unilateral neglect is "lateralized" in the sense that patients typically fail to respond to stimuli occurring in the side of space opposite the lesioned hemisphere (i.e., in the contralesional side). Although the central anchor determining the inherently asymmetrical nature of unilateral neglect may ultimately be the body (or "trunk") midline (Bradshaw, Nettleton, Pierson, Wilson, & Nathan, 1987; Karnath, Schenkel, & Fischer, 1991), as we shall see, the system of spatial coordinates in which the disorder is manifest may change according to the size, contents, and location of the stimulus array, and the nature of the response required. Indeed, the difference between normal processing of ipsilesionally located stimuli (i.e., those arising on the same side as the lesioned hemisphere) and impaired processing of contralesionally located stimuli may be probabilistic, reflecting a continuum across the horizontal (or azimuthal) plane, from contralesional to ipsilesional space (Kinsbourne, 1993).

The earliest clinical descriptions of unilateral neglect date from the latter half of the 19th century (e.g., Hughlings Jackson, 1876; Anton, 1893). However, the existence of unilateral neglect as a discrete neurological syndrome was perhaps not generally recognized until the mid-20th century, when workers such as Brain (1941), Paterson and Zangwill (1944, 1945), and Critchley (1953) described the features of lateralized neglect and "inattention" as being dissociable from agnosia, spatial disorientation, and topographical memory disturbances (see Chapter 5). In all these studies, the descriptions were of patients with right hemisphere (RH) damage and left-sided unilateral neglect. It was therefore assumed that the RH played a predominant role in mediating space-related behavior, a view that continues to influence current models of RH functioning. The delay in recognizing the heterogeneity of disorders of spatial cognition—when contemporaneous descriptions of language disorders had already reached a level of complexity that was at least an order of magnitude more advanced (see Chapter 2)—is perhaps attributable to the fact that the psychological dimension of space was, and continues to be, an elusive construct. We shall see in this chapter that the absence of suitable heuristics with which to conceptualize the cognitive operations involved in mediating space-related behavior has proved to be the principal challenge facing those researching the disorder of unilateral neglect.

For this reason, in the absence of preexisting taxonomies, conceptual schemata or even recent synthetic reviews of disorders of spatial cognition, we shall present in this chapter a considerably more detailed analysis and treatment of the relevant phenomena than will be found elsewhere in this book. Indeed, we shall argue that disorders of attention underlie many aspects of disordered spatial cognition and unilateral neglect; as intact attentional processes are central to almost all other aspects of clinically normal functioning, we shall deal at some length with attentional mechanisms.

Clinical Descriptions of Unilateral Neglect

The early reports mentioned above provide some of the best descriptions of the clinical manifestations of unilateral neglect. For example, Brain (1941) described three patients, all with right parieto-occipital damage and left homonymous hemianopic field defects, who inappropriately made right- instead of left-hand turns in route finding. Of his Patient 5, he wrote:

> When she set out for the bathroom she arrived at the lavatory, which was a door on the right, and when she tried to go to the lavatory she made a similar mistake, took a turning to the right and got lost again. . . . When asked to describe how she would find her way from the tube station to her flat she described this in detail correctly and apparently visualizing the landmarks, but she consistently said right instead of left for the turnings except on one occasion. When offered objects she tended to pass point to the right with both hands. (p. 259)

Such biased behavior in route finding has also been reported in contemporary descriptions of unilateral neglect (Bisiach, Brouchon, Poncet, & Rusconi, 1993). Brain attributed unilateral neglect to a lateralized disorder of the body schema, and suggested that neglect of the contralesional half of the body and neglect of the corresponding region of external space could be ascribed to the same underlying impairment.

One of the patients described by Paterson and Zangwill (1944), who had suffered a penetrating wound of the right parieto-occipital region (including the angular and supramarginal gyri), serves to illustrate some further clinical symptoms of left unilateral neglect. In addition to a left homonymous hemianopia,

> there was a marked tendency to neglect the left half of visual space. The patient often collided with objects on his left which he had clearly perceived a few moments before. He was liable at table to knock over dishes on his left-hand side and occasionally missed food on the left of his plate. He commonly failed to attend to the left-hand page in turning the pages of a book and in reading lines of disconnected words commonly omitted the first word or two. (p. 339)

Another patient reported by Paterson and Zangwill (1945) also exhibited such lateralized anomalies of space-related behavior following a penetrating wound of the right parietal lobe:

> It was noticed early that the patient *totally neglected his left upper extremity* and made virtually no use of it despite good preservation of motor power. The left hand had literally "dropped out of the patient's spontaneity" . . . and its spontaneous use was first noted only a month after admission. . . . A corresponding neglect of the *left half of visual space* was also observed. The patient tended to deviate to the right in walking and consistently failed to appreciate doors and turnings on his left-hand side even when he was aware of their presence. Thus, he would automatically turn to the right whenever the opportunity afforded and consequently tended to wander in a circle. . . . It was also noticed that the patient neglected the left-hand side of a picture or the left-hand page of a book despite the fact that his attention was constantly being drawn to the oversight. . . . In playing draughts, it was observed that the patient neglected the left side of the board and when his attention was drawn to the pieces on this side he recognized them but immediately thereafter forgot them. (p. 195)

Paterson and Zangwill (1945) also noted that several elementary and apparently unrelated skills were disrupted by the disorder:

> In free-hand drawing, the patient always began on the right-hand side and often omitted the left half of the object he was attempting to draw. . . . On one occasion we called the patient's attention to an error of dress and he removed his trousers in order to replace them correctly. . . . He showed the greatest difficulty in orientation, at first trying to put on the trousers inside-out. Eventually, he got the right trouser-leg turned right way out and introduced his leg into it correctly but the left trouser-leg remained inside-out. He tried nevertheless to get his left leg into the trousers with understandable lack of success. At one point, he appeared to "lose" the left trouser-leg completely and endeavoured to introduce both legs into the right one. (p. 196).

There are several salient points to be noted from these descriptions of unilateral neglect. First, the disorder is characterized by a strong lateral bias. More specifically, patients fail to respond to stimulation occurring in the contralesional side of space, while at the same time being pathologically biased toward the ipsilesional side. Second, the appearance of (left) unilateral neglect is strongly associated with damage to the posterior (usually parietal) region of the RH. Third, the clinical manifestations of unilateral neglect are strikingly heterogeneous, with patients exhibiting anomalies in tasks such as eating, dressing, reading, writing, drawing, copying, route finding, and even goal-directed movement. This heterogeneity illustrates an important concept in neuropsychology, namely, that the deficits shown by patients on any given task may arise as a consequence of damage to one of a number of underlying functional mechanisms. In the case of unilateral neglect, the deficits shown by patients reflect a lateralized impairment in representing and attending to specific regions of the spatial environment. Thus, individuals with left unilateral neglect may fail to read the first few words in each new line of written text, yet they would not be considered to have a primary language disturbance. Similarly, such patients may incorrectly make ipsilesional turns in route finding, yet they would not be considered to have topographagnosia (see Chapter 5).

Performances by unilateral neglect patients on standard clinical tests provide a striking illustration of the nature of the disorder in its full elaboration. Patients may read words only from one side of a printed page, or make errors in reading individual words in a sentence. In the latter case the errors may be ones of *omission*, where the word is truncated at its beginning or end (depending upon which hemisphere is damaged), or of *substitution*, where individual letters on the contralesional side are changed to form a different word. Errors in reading are known collectively under the title of "neglect dyslexia" (Ellis, Flude, & Young, 1987). Patients may omit details on the contralesional side when asked to copy pictures or to draw from memory. When asked to cross out visual targets scattered over a page, they may perform adequately for ipsilesionally located targets, but fail to cross out those located more contralesionally. When given a horizontal line to bisect, they may place their transection mark away from the true midpoint toward the ipsilesional end point, as if some portion of the contralesional extent of the line had not been perceived (see Figure 1).

In describing familiar scenes from memory, patients may even omit contralesionally located landmarks (Bisiach & Luzzatti, 1978), thereby indicating that their deficit is not restricted to external sensory information. The fact that unilateral neglect occurs in such imaginal tasks implies that the disorder cannot be attributed to primary sensory loss, such as hemianopic visual field defects. Indeed, unilateral neglect is also manifested in other sensory modalities. Patients may fail to respond to auditory stimuli arising in the contralesional side of space, or they may incorrectly localize the source of such stimuli to a more ipsilesional position in space (Bisiach, Cornacchia, Sterzi, & Vallar, 1984). Similarly, left unilateral neglect patients may be slower to respond to tactile stimuli presented to the ipsilesional hand when it is located in the contralesional compared with the ipsilesional hemispace (Pierson-Savage, Bradshaw, Bradshaw, & Nettleton, 1988).

Figure 1   Performances of a patient with left unilateral neglect following right hemisphere damage. (a) Three different cancellation tasks, in
which the patient has omitted (from top to bottom) scattered line segments, circles, and small stars on the left. (b) Rightward deviation from
the true midpoint in line bisection. (c) Copy of a simple scene. (d)
Spontaneous drawing of a woman. (From Mattingley, Bradshaw, &
Bradshaw, in press.)

Although clinical observations provide useful vignettes of the behavioral anomalies shown by individual patients, quantitative analyses based on carefully controlled experiments must ultimately provide the critical data upon which a principled understanding of unilateral neglect may be achieved. In this context, it is again pertinent to refer to the pioneering studies conducted half a century ago by Paterson and Zangwill (1944, 1945; see also Mattingley, in press). These researchers were among the first to conduct detailed experimental studies of patients with unilateral neglect. Their battery of tasks included line bisection, figure copying, spontaneous drawing, letter completion, tachistoscopic presentation of visual stimuli, pointing to discrete targets, and a number of other visuoperceptual and visuoconstructional paradigms. As we shall see, many of these tasks, or variants thereof, continue to be used by contemporary researchers.

Terminology and Deficits Related to Unilateral Neglect

Although it has been suggested that the term *unilateral neglect*, like the terms *aphasia* and *agnosia*, is in fact no more than a convenient shorthand for describing a number of functionally distinct deficits (e.g., Halligan & Marshall, 1992), it has nevertheless been adopted here, partly as a matter of convenience, but principally because there is as yet no widely accepted taxonomy for classifying its (putative) subcomponents. However, with recent data suggesting the existence of discrete functional losses in the domain of spatial cognition, nascent taxonomies have emerged, and these will be described where appropriate.

The terminology associated with disorders of spatial cognition is further complicated by the fact that there are several other, apparently interchangeable, terms that have been used to describe the "syndrome" of unilateral neglect. Some of the more commonly used terms include *contralesional neglect, visuospatial neglect, hemineglect, hemi-inattention,* and *hemispatial inattention.* A more extensive, though by no means exhaustive, listing has been provided by Halligan and Marshall (1993a). Unless otherwise stated, the term *unilateral neglect* will be used throughout this chapter, even if the author(s) whose work is being described used a different (but equally arbitrary) term. In those rare instances in which an alternative (usually more specific) term has been used, it will be introduced and defined accordingly.

In order to provide an adequate explanation of the behavioral manifestations of unilateral neglect, it is first necessary to consider the context in which they occur. As previously hinted, the weight of evidence suggests that unilateral neglect is more frequent, severe, and persistent after RH damage than after left hemisphere (LH) damage (Colombo, De Renzi, & Faglioni, 1976; Stone et al., 1991; though see Ogden, 1985a, 1987). In fact, there are good reasons for suspecting that the manifestations of unilateral neglect arising from LH versus RH damage reflect impairments of functionally distinct cognitive mechanisms (Egly, Driver, & Rafal, 1994; Halligan & Marshall, 1994a; L. Robertson & Lamb, 1991), and that left-sided unilateral neglect after RH damage is easier to detect and more pervasive than the reverse situation (Humphreys & Riddoch, 1994). We shall consider several possible explanations for this hemispheric asymmetry in a later section of this chapter.

The phenomenon of *extinction* has been considered by some authors to be an integral part of unilateral neglect (Heilman, Watson, & Valenstein, 1993), rather than a co-occurring but functionally distinct impairment (Schwartz, Matchok, Kreinick, & Flynn, 1979; Vallar, Rusconi, Bignamini, Geminiani, & Perani, 1994). It involves a failure to respond to one of two horizontally aligned stimuli delivered simultaneously, although normal responses may be elicited by either stimulus presented in isolation (Heilman et al., 1993). Extinction is most severe when simultaneous stimuli are presented bilaterally (traditionally, on the left and right sides of the body midline), in which case the contralesional stimulus is typically not responded to (i.e., it is "extinguished"). However, it may also occur when both stimuli are given on the same side of the body midline (Rapcsak, Watson, & Heilman, 1987), or to the same limb (Behrmann & Moscovitch, 1994). Like unilateral neglect, extinction has been found to occur in the auditory (De Renzi, Gentilini, & Pattacini, 1984), olfactory (Bellas, Novelly, Eskenazi, & Wasserstein, 1988), visual (Rapcsak et al., 1987), and tactile (Behrmann & Moscovitch, 1994; Bisiach, Perani, Vallar, & Berti, 1986) modalities. Furthermore, in different patients extinction may be present in some but not all sensory modalities (De Renzi et al., 1984; Vallar et al., 1994). Extinction has also been observed in motor activities (*motor extinction*), where a patient may move either limb independently, but fails to move the contralesional limb when bilateral activity is required (Heilman et al., 1993). The phenomenon occurs after damage to either cerebral hemisphere (Ogden, 1985a), though again it may be more prevalent and longer lasting after RH damage.

Whether or not extinction and unilateral neglect are functionally related is a matter of some controversy. It may be noteworthy that extinction is amenable to manipulations that alter patients' expectations of the likely locations of upcoming stimuli (e.g., R. Kaplan, Verfaellie, De Witt, & Caplan, 1990). Such susceptibility to attentional cues is also apparent in patients with unilateral neglect and, as will be outlined later, several theorists have suggested that both disorders may be attributable to a failure of mechanisms underlying spatial attention. Moreover, it is now clear that specific attributes of visual stimuli play an important role in modulating extinction effects. For example, Baylis, Driver, and Rafal (1993) examined the responses of patients with clinically documented visual extinction to colored letter stimuli presented simultaneously in left and right visual fields. Extinction of the contralesional stimulus occurred more often when the two stimuli shared the same, task-relevant attribute (shape or color) than when the two stimuli differed on this attribute. For example, when patients were required to report the *color* of the two stimuli, they made more extinction errors if both stimuli were red than if one was red and the other was green. No effect of similarity was found on the incidence of contralesional extinction when the stimulus pair shared a *task-irrelevant* attribute. Thus, when required to report *shape*, the same number of extinction errors was made with two red (or green) stimuli as with pairs of stimuli with different colors.

Other studies have shown that manipulations of the relative timing of onset of each stimulus (Dick, Wood, Bradshaw, & Bradshaw, 1987), and the extent to which the stimuli are likely to be grouped as one object or two separate objects (Ward, Goodrich, & Driver, 1994), may also determine whether or not extinction occurs. In the latter study, the frequency of visual extinction exhibited by RH-damaged patients was found to be reduced when the left- and right-sided stimuli could be grouped, according to Gestalt principles, as a single object (e.g., when the stimuli were two inward-facing brackets, thus: [ ]). In cases where bilaterally presented stimuli could not be grouped (e.g., when the stimuli consisted of a bracket and a dot, thus [•), the frequency of extinction was significantly greater than in the grouped conditions. We shall return to consider disorders of object-based attentional mechanisms in a later section of this chapter.

These stimulus-specific effects clearly suggest that, as with unilateral neglect, basic perceptual mechanisms play a role in mediating the manifestations of visual extinction. It is important to note, however, that double dissociations have been demonstrated between the two phenomena. In a recent investigation of 159 RH-damaged stroke patients, Vallar, Rusconi, Bignamini, Geminiani, & Perani (1994) found that of 46 individuals showing visual or tactile extinction, only 13 also exhibited unilateral neglect. Similarly, of the 58 patients showing unilateral neglect, only 13 also had visual or tactile extinction. It therefore seems sensible to conclude that extinction and unilateral neglect may themselves be heterogeneous disorders, arising from dysfunction at different levels of the central nervous system (CNS) (Bisiach, 1991; Vallar, Rusconi, Bignamini, Geminiani, & Perani, 1994). In some patients, common impairments may underlie the behavioral manifestations of both disorders, whereas in others disruption of functionally distinct mechanisms may be responsible for each.

Allochiria and *allesthesia* are terms used to describe, respectively, mislocalizations of stimuli delivered to the contralesional side of the body or space to an approximately homologous location on the ipsilesional side, and displacements of stimuli to a different location on the same side of the body (Meador, Allen, Adams, & Loring, 1991). Unfortunately the terms *allochiria* and *allesthesia*, like those associated with unilateral neglect, have been used interchangeably with numerous others, a situation that has served to obfuscate clinical and experimental findings in this area (Meador et al., 1991). Allochiria, which is of primary interest here because it involves distinctly lateralized transpositions of stimuli, has been observed in the auditory (Diamond & Bender, 1965), tactile (Obersteiner, 1882; cited in Meador et al., 1991), and visual (Halligan, Marshall, & Wade, 1992a,b) modalities.

Some of the most striking examples of allochiria have been observed in patients' drawings from memory of an analog clock (Figure 2), in which numbers belonging on the contralesional half-face are instead drawn on the ipsilesional half-face (e.g., Bisiach, Capitani, Luzzatti, & Perani, 1981; Halligan & Marshall,

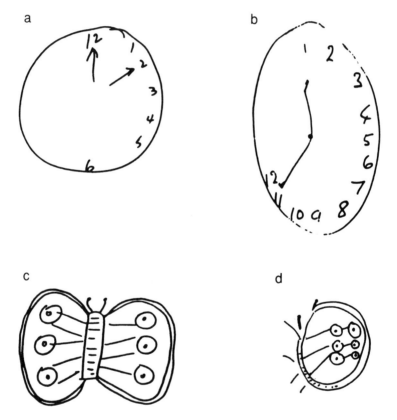

Figure 2   EXAMPLES OF ALLOCHIRIA IN PATIENTS WITH LEFT UNILATERAL NEGLECT. ABOVE, TWO ATTEMPTS BY A SINGLE PATIENT TO DRAW A CLOCK-FACE FROM MEMORY. IN (a), PRODUCED DURING THE ACUTE PHASE POSTINJURY, NUMBERS NORMALLY APPEARING ON THE LEFT HAVE BEEN OMITTED. IN (b), PRODUCED AFTER A PERIOD OF RECOVERY, ALL THE NUMBERS APPEAR BUT THESE HAVE BEEN CRAMPED INTO THE RIGHT HEMICIRCUMFERENCE. (FROM HALLIGAN & MARSHALL, 1994.) (c) EXAMINER'S MODEL OF A BUTTERFLY. (d) PATIENT'S COPY, WITH LEFT-SIDED DETAILS TRANSPOSED TO THE RIGHT. (FROM HALLIGAN, MARSHALL, & WADE, 1992.)

1993a; though see also Ishiai, Sugishita, Ichikawa, Gono, & Watabiki, 1993, for a cautionary note on interpreting such findings). Halligan et al. (1992a) also found that a patient with left unilateral neglect transposed left-sided details onto the right side of her copies of butterflies (Figure 2). Interestingly, the same patient, when asked to draw a plan view of an object-filled room, also transposed left-sided objects over to the right. Allochiria has also been found among patients with left unilateral neglect on tasks of mental imagery (Bisiach et al., 1981). When asked to mark the locations of cities on a map, patients may misplace landmarks from the contralesional to the ipsilesional side.

As with extinction, the phenomenon of allochiria may be functionally independent of unilateral neglect, although several authors have suggested that they are theoretically related (Bisiach & Berti, 1987; Halligan & Marshall, 1994b). Unlike unilateral neglect, which in severe cases involves a complete failure to respond to contralesionally located stimuli, allochiria implies adequate perception of such stimuli coupled with a tendency to mislocate them from the contralesional to the ipsilesional side. The possibility of a dissociation between intact perception of contralesional stimuli, and impaired assignment of their relative positions in space, is consistent with the concept of parallel ("what"—ventral vs. "where"—dorsal) processing streams in the primate visual system (Mishkin, Ungerleider, & Macko, 1983).

The term *anosognosia* was first used by Babinski (1914; cited in Levine, Calvanio, & Rinn, 1991) to describe unawareness or denial of hemiplegia, though it may also occur for blindness (Anton's syndrome) and hemianopia (Bisiach, Vallar, Perani, Papagno, & Berti, 1986). The condition, which ranges from obstinate denial to an inappropriate lack of concern about the impairment (*anosodiaphoria*; Critchley, 1953), is more frequently associated with RH damage than with LH damage (Cutting, 1978). Anosognosia is often associated with delusions and confabulations relating to body parts on the affected side (*somatoparaphrenia*), and with negative emotional reactions toward limbs on the affected side (*misoplegia*; Critchley, 1953).

The possibility of a theoretical relationship between anosognosia and unilateral neglect has been suggested (Bisiach & Berti, 1987), though there have been instances of double dissociation between the two disorders (Bisiach, Vallar, Perani, Papagno, & Berti, 1986). In a study by Levine et al. (1991) it was found that RH-damaged patients exhibiting persistent (i.e., lasting more than 1 month poststroke) anosognosia for hemiplegia tended to have more severe unilateral neglect than those who became aware of hemiplegia within a few days after onset. However, the presence of severe unilateral neglect was found to be neither a necessary nor a sufficient condition for the occurrence of persistent anosognosia.

Although the pathogenesis of anosognosia remains unclear, it may arise as a consequence of the patient's failure to discover, through self-observation and diagnosis, the presence of sensory loss (Levine, 1990; Levine et al., 1991).

Unilateral neglect may therefore exacerbate or perpetuate anosognosia by reducing the likelihood of discovery (Levine et al., 1991). Bisiach and Berti (1987) have provided an alternative explanation in which disorders such as anosognosia and unilateral neglect form part of a wider syndrome, which is itself traced back to an underlying deficit in representing the contralesional side of personal (i.e., involving the body) and extrapersonal (i.e., beyond the body) space. This latter explanation will be considered in more detail later, in the context of a discussion of the various models of unilateral neglect.

Etiology of Unilateral Neglect

The most common cause of unilateral neglect is acute cerebrovascular ischemia or hemorrhage (see Chapter 1), although the disorder also occurs in patients with such cerebral tumors as meningiomas and glioblastomas (Albert, 1973; Heilman et al., 1993; Ogden, 1985a). In the latter cases, rapidly growing malignant tumors are considered more likely to produce unilateral neglect than relatively slow-growing tumors (Heilman et al., 1993), implying that pathologies of sudden onset are the most likely precursors of unilateral neglect. The extent to which the disorder is associated with other pathologies remains unclear, though it has been reported in patients with penetrating missile wounds (Paterson & Zangwill, 1944, 1945), Parkinson's disease (Villardita, Smirni, & Zappalà, 1983), ventrolateral thalamotomy (Vilkki, 1984), and Marchiafava-Bignami disease (Kamaki, Kawamura, Moriya, & Hirayama, 1993); the latter involves demyelination of the corpus callosum consequent upon chronic alcohol abuse.

Temporary manifestations of unilateral neglect have also been reported following right intracarotid amobarbital testing in epileptic patients (Spiers et al., 1990), in the postictal period following right-sided complex partial seizures of temporo-parieto-occipital origin (Heilman & Howell, 1980), and immediately after unilateral right-sided and bilateral electroconvulsive therapy (Heilman et al., 1993; Sackeim et al., 1992). In normal healthy subjects, it has also been found that focal transcranial magnetic stimulation of either the left or right parietal lobe causes extinction of contralesional visual targets during double simultaneous stimulation (Pascual-Leone et al., 1994).

Neuroanatomical Correlates of Unilateral Neglect

As stated earlier, although unilateral neglect occurs in association with lesions of either hemisphere, it is more frequent, severe, and persistent after RH damage, at least insofar as the more obvious clinical manifestations are concerned (Vallar, 1993). In addition to this apparent asymmetry, Ogden (1985a) provided evidence to suggest that, although in RH-damaged patients, unilateral neglect is most often associated with lesions involving cortical structures posterior to the central sulcus, in LH-damaged patients it is most often associated with lesions involving cortical structures anterior to the central sulcus. This finding suggests that the mechanisms subserving space-related behavior are located in different regions of the two hemispheres. Ogden (1985a) speculated that the relative specialization of left posterior regions for language functions may have required that visuospatial functions be subserved by more anteriorly located structures. Indeed, as suggested earlier, it is likely that the two cerebral hemispheres participate in essentially different ways in the processing of spatial information; we shall return later to consider evidence in favor of this suggestion. At this point, we wish simply to highlight the fact that there is a striking paucity of empirical data on such deficits following LH damage (though see Ogden, 1985a,b, 1987). This situation will need to be rectified if we are to gain a comprehensive understanding of the mechanisms underlying space-related behavior.

By far the largest body of data on lesion sites in unilateral neglect has been obtained from patients with RH damage. The early descriptions of unilateral neglect by workers such as Brain (1941), Paterson and Zangwill (1944, 1945), and Critchley (1953) were of patients with RH lesions involving the parietal lobe or the parieto-occipital junction. These authors emphasized the critical role of such regions in mediating space-related behavior. More recently, in a large group study of patients with acute ischemic or hemorrhagic lesions, Vallar and Perani (1986) found that unilateral neglect was indeed more frequently associated with posterior damage (i.e., damage posterior to the central sulcus) than with anterior cortical damage. Furthermore, the incidence of unilateral neglect after cortical damage was highest in patients with lesions involving the inferior parietal lobule. Posterior damaged patients without unilateral neglect tended to have lesions involving the superior parietal lobule or occipital lobe. There is some evidence to suggest that damage to the superior parietal lobule is associated with disorders of reaching toward visual targets located in the contralesional hemispace (Ratcliff & Davies-Jones, 1972; Levine, Kaufman, & Mohr, 1978). Perenin and Vighetto (1983) reported two RH-damaged patients with optic ataxia, both with predominantly superior parietal damage. These patients were inaccurate in reaching toward discrete visual targets in the contralesional hemispace, regardless of whether they used their contralesional or ipsilesional upper limbs.

Vallar and Perani (1986) also found one severe unilateral neglect patient with damage involving the dorsolateral frontal cortex and underlying white matter, suggesting that the disorder may arise from damage to cortical structures other than the inferior parietal lobule. Unilateral neglect has also been reported after damage to the anterior cingulate gyrus and supplementary motor area (Heilman & Valenstein, 1972). However, unilateral neglect in humans is not frequently associated with discrete frontal cortical damage. By contrast, in monkeys, damage to the frontal eye fields and inferior Area 6 has been found consistently to produce behavior resembling unilateral neglect in humans (Rizzolatti & Gallese, 1988). In the monkey, these areas are connected, respectively, with the rostral inferior parietal lobule (area PF) and the caudal inferior parietal lobule (area PG), thereby implying the existence of (at least) two distinct circuits for mediating space-related behavior. As we shall see, these circuits may be involved in the control of attention and action in distal (i.e., extrapersonal) and more proximal (i.e., peripersonal) regions of space (Rizzolatti & Berti, 1993; Rizzolatti & Camarda, 1987).

There is also evidence from human studies that structural damage to any one of a number of *subcortical* structures may produce unilateral neglect, including the thalamus (Watson & Heilman, 1979), caudate nucleus (Caplan et al., 1990), putamen (Hier, Davis, Richardson, & Mohr, 1977), and internal capsule (Sakashita, 1991; Vallar & Perani, 1986). The frequency of unilateral neglect has been found to differ according to the specific locus of subcortical damage, with the thalamus being most often involved, followed by the basal ganglia (BG), and finally the white matter (Vallar & Perani, 1986). There is also some evidence to suggest that nuclei in the medial and posterior thalamus (e.g., the pulvinar), rather than those located more anteriorly, are more often damaged in patients with unilateral neglect (Vallar, 1993). The likely role of the thalamus, and particularly the pulvinar, in mediating space-related behavior is consistent with anatomical evidence showing that it receives afferents from several structures, including the occipital cortex and superior colliculus, and has strong projections to the posterior parietal cortex. On the basis of an extensive review of relevant studies, Cappa and Vallar (1992) concluded that subcortical lesions consistently produce unilateral neglect only when grey nuclei are involved.

The phenomenon of extinction following RH damage may primarily be associated with subcortical damage. In the case series reported by Vallar et al. (1994), 52% of patients with tactile or visual extinction had circumscribed subcortical lesions, with the lenticular nucleus being damaged in the highest proportion of cases, followed by the anterior periventricular white matter, and the posterior limb and genu of the internal capsule. Thalamic damage was documented in 17% of cases. In contrast, only about 25% of patients with unilateral neglect but no extinction had lesions confined to subcortical structures. In these individuals, the most frequently damaged structures were the posterior limb of the internal capsule and the thalamus. In keeping with the findings of previous studies, over 90% of patients with unilateral neglect had either anteroposterior or posterior cortical lesions. These anatomoclinical data confirm our earlier suggestion that extinction and unilateral neglect may reflect damage to different neural systems. Patients with unilateral neglect have a relatively high-level disturbance of spatial cognition consequent upon damage to posterior (mainly parietal) cortical structures. In contrast, extinction may arise either from subcortical damage (affecting the ascending sensory pathways), resulting in impaired processing of sensory information on the contralesional side, or from thalamic or posterior cortical damage, resulting in a higher level disturbance akin to, but perhaps less severe than, that shown by patients with unilateral neglect. This may account for double dissociations between extinction and unilateral neglect, while at the same time acknowledging that the two disorders may under certain circumstances share a common functional underpinning.

Collectively, the results of lesion analyses suggest that spatial cognition is subserved by a network of anatomically distinct but functionally interconnected structures. However, one important point that has not yet been considered in our discussion of lesion sites in unilateral neglect is the extent to which circumscribed structural lesions may disrupt the normal functions of anatomically distant regions (*diaschisis*; see Chapter 1). All of the anatomoclinical studies discussed so far have relied upon *structural* imaging techniques that only document primary tissue damage. It is therefore possible that the apparent association between damage to specific structures and manifestations of unilateral neglect is in fact attributable, at least in part, to a secondary (*functional*) impairment of one or more distant structures.

We have already seen how functional neuroimaging techniques such as positron emission tomography (PET) and single photon emission computed tomography (SPECT) can provide clues as to the roles of specific brain regions in mediating specific cognitive functions (see e.g., Chapters 1 and 3). These techniques have also been used in patients with unilateral neglect to document physiological abnormalities in brain regions removed from the site of primary tissue damage. Several recent studies have shown reduced regional cerebral blood flow (rCBF) in cortical regions throughout the ipsilesional hemisphere in patients with unilateral neglect following primary lesions of subcortical structures (Bogousslavsky et al., 1988; Perani, Vallar, Cappa, Messa, & Fazio, 1987). In contrast, patients with primary subcortical damage but no evidence of unilateral neglect have been shown to have only mild ipsilesional hypoperfusion (Perani et al., 1987). There is also evidence that left unilateral neglect patients with RH damage may exhibit severe hypometabolism in both the ipsilesional and *contralesional* hemispheres soon after subcortical stroke (Perani, Vallar, Paulesu, Aberoni, & Fazio, 1993). Moreover, subsequent recovery from unilateral neglect may be correlated with restitution of more normal levels of blood flow and metabolism in both ipsilesional and contralesional hemispheres (Perani et al., 1993; Vallar et al., 1988).

PET activation studies in normal subjects have confirmed the critical role played by the parietal cortex in mediating shifts of focal attention in extrapersonal space. In a study by Corbetta, Miezin, Shulman, and Petersen (1993), cerebral metabolic activity was measured while subjects performed a psychophysical task that involved shifting focal attention to visual stimuli presented at different eccentricities within either the left or right visual field. Regions of superior parietal and superior frontal cortex were found to be more active when subjects were engaged in attention shifting than when they maintained central fixation, although the frontal region was only active when the attended stimulus required a motor response. This latter point is of considerable interest in terms of the likely interactions between neural systems responsible for coordinating perception and action. We will return to discuss models of unilateral neglect that emphasize the reciprocal and interdependent nature of mechanisms underlying spatial attention and those involved in coordinating goal-directed actions.

One final point of interest is the apparent discrepancy between the finding by Corbetta et al. (1993) of *superior* parietal activation among normal subjects performing an attentional task, and the strong association between unilateral neglect and damage to the *inferior* parietal lobule. In the case of patients, it is likely that widespread dysfunction of regions distant from the site of *primary* damage may be partly responsible for this discrepancy.

Models of Unilateral Neglect

In the next section, we provide a brief description of the various models proposed to account for unilateral neglect. We will then consider a number of important conceptual issues arising from recent clinical and experimental observations, and determine which of the models (if any) deals most adequately with them.

Early attempts to explain unilateral neglect assumed that the disorder was attributable to a combination of primary sensory and motor dysfunction, together with a global reduction in the capacity for information processing (e.g., Battersby, Bender, & Pollack, 1956). This view would at first blush seem to be at least intuitively plausible, as primary sensory (e.g., hemianopia, hemianesthesia) and motor (e.g., hemiplegia) disturbances are frequent concomitants of unilateral cerebral damage. However, many patients with unilateral neglect of visual stimuli can be shown *not* to have visual field defects, and in those that do, the severity of impairment seems not to be exacerbated relative to those with full visual fields (Halligan, Marshall, & Wade, 1990). Moreover, patients with visual field defects alone typically cope with their constricted field of vision by making compensatory head and eye movements, whereas patients with unilateral neglect, with or without visual field loss, rarely engage in such strategies. In addition, over time many patients recover from unilateral neglect, but continue to exhibit homonymous field defects (Mattingley, Bradshaw, Bradshaw, & Nettleton, 1994a,b). Finally, unilateral neglect may also occur with auditory, tactile, and even olfactory stimuli, despite evidence of normal thresholds in the detection of basic sensory information (Bellas, Novelly, Eskenazi, & Wasserstein, 1988; Bisiach et al., 1984; Vallar, Sandroni, Rusconi, & Barbieri, 1991). Thus, current attempts to explain unilateral neglect assume that the disorder reflects a relatively high-level disturbance in spatial cognition.

Having ruled out a simple sensory explanation, the remaining models of unilateral neglect can be broadly conceptualized in terms of the level at which the underlying cognitive deficit is assumed to reside. One set of models posits a disruption of *attentional* mechanisms: mechanisms involved in maintaining arousal or vigilance and in controlling orientation toward, and selection of, discrete stimuli as targets to be elaborated for further processing. The second set of models posits disruption of mechanisms involved in the *representation* of space: mechanisms involved in reconstructing an internal "map" of space, based either on processing of external sensory data, or on internally derived information. Of course, mechanisms underlying attention on the one hand, and representation on the other, are not necessarily mutually exclusive. Indeed, some theorists have considered that the two are likely to be inextricably linked (e.g., Bisiach, 1993). Thus, the apparent dichotomy between these two groups of models is perhaps best considered relative, rather than absolute.

Attentional Models

HEILMAN'S ATTENTIONAL–INTENTIONAL MODEL.   The model of unilateral neglect proposed by Heilman and his colleagues (Heilman & Valenstein, 1979; Heilman & Van Den Abell, 1979) is based on four central assumptions. The first of these is that unilateral neglect may arise as a consequence either of a failure to *attend* (i.e., to detect) or a failure to *intend* (i.e., to initiate an appropriate movement) toward stimuli in the contralesional hemispace. The second assumption is that there is a natural seam along which the processing of spatial information is divided between the two cerebral hemispheres (though see Bradshaw, Spataro, Harris, Nettleton, & Bradshaw, 1988). This seam defines the boundary between left and right hemispaces, which may themselves be defined in terms of retinal-, head-, or body-centered coordinates. The third assumption is that the RH has a preeminent role in mediating space-related behavior. In particular, whereas the LH is concerned only with the contralateral hemispace, the RH attends and intends to both hemispaces. The final assumption is that failures of attention and intention in unilateral neglect are attributable to hypoarousal of the damaged hemisphere.

Although these assumptions continue to form the basis of Heilman's model of unilateral neglect, more recent formulations (e.g., Heilman et al., 1993) have been embellished to account for new and apparently incompatible data. For the purposes of this chapter, however, we will restrict our discussion to the four key assumptions made in the original model. Heilman and his colleagues have also endeavored to explain unilateral neglect on the basis of a detailed analysis of neurophysiological data obtained from studies of nonhuman animals. Such considerations are beyond the scope of the present chapter, in which we are concerned primarily (though not exclusively) with data obtained from human neuropsychological studies; thorough overviews of the relevant animal literature have appeared elsewhere (Heilman et al., 1993; Heilman, Valenstein, & Watson, 1994).

Heilman and Valenstein (1979) attributed the manifestations of unilateral neglect to an impairment they called "hemispatial hypokinesia," a defect in attentional orienting and preparation for action in and toward the hemispace contralateral to the lesioned hemisphere. Although several models of unilateral neglect share the assumption that the disorder reflects a deficit in attending to contralesionally located stimuli, the notion that it may, at least in some instances, involve a failure to *respond* to such stimuli in the context of normal *detection* is unique. Watson, Miller, and Heilman (1978) were the first to provide empirical evidence for such a possibility. They trained monkeys to use their left hand to respond to tactile stimuli on the right leg, and their right hand to respond to tactile stimuli on their left leg (i.e., a crossed response task). Subsequently, unilateral neglect was induced by a circumscribed frontal arcuate lesion. The monkeys responded normally using their ipsilesional hand to stimuli delivered on the contralesional leg. However, they often failed to respond using their contralesional hand when stimulated on the ipsilesional side, and occasionally responded (incorrectly) with their ipsilesional hand. Thus, although the animals had demonstrated their capacity to detect tactile stimuli on their neglected side, they were nevertheless unable to initiate a response with the contralesional hand. Similar observations of so-called nonsensory unilateral neglect have been made on monkeys with parietotemporal lesions (Valenstein, Van Den Abell, Watson, & Heilman, 1982).

The findings in monkeys parallel those obtained from human studies of *motor neglect,* in which patients fail spontaneously to use their contralesional limbs, despite demonstrations of adequate power on specific testing (Laplane & Degos, 1983; Ogden, 1988). As already mentioned, motor extinction (Valenstein & Heilman, 1981) denotes a failure to move the contralesional limb when requested to move both limbs simultaneously. In addition, some left unilateral neglect patients are impaired in making eye movements toward the contralesional hemispace (Butter, Rapcsak, Watson, & Heilman, 1988), and others may experience difficulties with *ipsilesional* arm movements directed toward the contralesional hemispace (Bisiach, Geminiani, Berti, & Rusconi, 1990; Heilman, Bowers, Coslett, Whelan, & Watson, 1985; Mattingley, Bradshaw, & Phillips, 1992; Mattingley, Phillips, & Bradshaw, 1994).

There is also ample evidence to suggest that spatial attention is indeed calibrated in hemispatial coordinates in both normal healthy individuals (see Bradshaw et al., 1987), and in patients with unilateral hemispheric damage. Heilman and Valenstein (1979) found that RH-damaged patients erred significantly further to the right when asked to bisect horizontal lines in left hemispace compared with when the same stimuli were presented opposite the body midline. Moreover, there was a trend for patients to err further to the right with midline presentations compared with when lines were located in right hemispace.

Reaction time (RT) studies have shown that patients with left unilateral neglect are slower to respond to visual stimuli on the left than the right of the trunk midline (e.g., Karnath et al., 1991). Similarly, RTs to vibrotactile stimuli are slower in such patients when the stimulated and responding hand is located on the left of the body midline than when it is located on the right (Pierson-Savage et al., 1988).

Although the processes assumed by Heilman to be responsible for mediating attention and intention are subserved by both the LH and RH, the latter hemisphere is said to play a predominant role. More specifically, it is proposed that each hemisphere directs attention and prepares responses in and toward the contralateral hemispace, but that the RH also orients and intends toward the ipsilateral hemispace. It therefore follows that after LH damage, the intact RH should be able to direct attention and intention in and toward either hemispace. However, after RH damage, because the intact LH is only able to attend and intend toward right hemispace, the patient will exhibit left unilateral neglect. The inherent asymmetry in the attentional capacities of the two hemispheres, as conceptualized in this model, is particularly attractive because it accounts for experimental observations on the greater frequency and severity of unilateral neglect after RH compared with LH damage.

One final underlying assumption of the model proposed by Heilman is that unilateral neglect reflects physiological hypoarousal of the damaged hemisphere. We have already discussed the results of neuroimaging studies showing widespread areas of reduced functional activity in the ipsilesional hemisphere of patients with unilateral neglect (e.g., Bogousslavsky et al., 1988; Fiorelli, Blin, Bakchine, Laplane, & Baron, 1991; Perani et al., 1987; Perani et al., 1993). Similarly, it has been found that temporary anesthesia of the RH produces unilateral neglect; Spiers et al. (1990) observed significant deficits in attentional scanning following suppression of RH activity by injection of sodium amobarbital (the WADA test). In this context, it is noteworthy that stimuli presented to the normal RH may be responded to faster, and the RH may exhibit higher levels of arousal or vigilance than the LH. We shall return to consider the possibility that the RH does indeed play a preferential role in mediating arousal.

In terms of anatomy, Heilman et al. (1993) distinguished between two neural circuits underlying space-related behavior, one responsible for mediating spatial attention and the other responsible for preparing the organism for movement. The mesencephalic reticular formation, located in the brain stem, is assumed to provide tonic arousal for those higher level structures incorporated in the two circuits. *Spatial attention* is subserved by the thalamus, through which visual, auditory, and somatosensory information is transmitted to the primary sensory cortical regions and onward via association cortices to specific areas of the parietal (e.g., inferior parietal lobule), frontal (e.g., prefrontal), and limbic (e.g., cingulate) cortices. The *intentional circuit* comprises two pathways; in one, information is transmitted from the supplementary motor area (SMA) through the putamen and globus pallidus (GP) to the ventral thalamus, then back to the SMA (see Chapters 10–12). In the other, transmission is from the prefrontal cortex via the caudate nucleus and GP to medial thalamic nuclei, and back to the prefrontal cortex (see Chapter 10). Interactions between these intentional pathways and the circuit subserving spatial attention are likely to occur in the frontal, cingulate, and parietal cortices. Both the attentional and intentional circuits are also assumed to involve extensive reciprocal connections, thereby permitting certain structures to modulate the output from others.

Perhaps the most critical assumption made in the Heilman model is that space-related behavior is subserved by interconnected networks of brain structures, and that deficits of spatial attention and intention are associated with damage to different parts of these neural circuits. This assumption is certainly consistent with the anatomical data reviewed earlier, which indicate that discrete lesions of a number of cortical and subcortical structures may produce unilateral neglect. The network notion has also been shown to have substantial heuristic power in explaining the neural and cognitive mechanisms underlying many aspects of human behavior (e.g., Damasio, 1990; Goldman-Rakic, 1988b; Rumelhart & McClelland, 1986; and see also Chapters 1, 2, and 8).

MESULAM'S NEURAL NETWORK MODEL.   Mesulam (1981, 1985, 1990) elaborated on the above concept of a complex, interconnected neural network subserving spatial cognition. According to his model, the apparently heterogeneous manifestations of unilateral neglect are best understood in terms of the specific locus of damage. The neural substrate involved in space-related behavior is again conceived of as a network of anatomically distinct but functionally interacting regions. The particular deficits exhibited by individual patients are broadly categorized as reflecting primarily a disruption of perceptual, exploratory-motor, or motivational processes, depending on the locus of cerebral damage. Anatomically, the network now involves three major cortical sites. These are the posterior parietal lobe, the dorsolateral premotor and prefrontal cortex, and the cingulate gyrus. Subcortical structures, including the mesencephalic reticular formation, superior colliculus, striatum, and pulvinar nucleus of the thalamus, are connected with most of these cortical regions and are probably involved in maintaining adequate levels of activation or arousal. It has also been suggested that the superior colliculus and pulvinar are responsible, respectively, for *moving* focal attention from one location to another, and for *engaging* focal attention upon stimuli in particular spatial locations (Posner & Petersen, 1990).

Each cortical site in Mesulam's network is assumed to mediate a different aspect of space-related behavior. Thus, the parietal cortex controls the *perceptual* components of directed attention by providing an internal map or template of space. Evidence from electrophysiological studies suggests that 44% of cells in the lateral intraparietal area of alert monkeys have visual receptive fields that shift *prior* to saccadic eye movements (Duhamel, Colby, & Goldberg, 1992).

Similar techniques in monkeys have revealed the existence of cells in the anterior bank of the parieto-occipital sulcus that respond to visual stimulation in the same spatial location, regardless of eye position (Galletti, Battaglini, & Fattori, 1993). Such cells encode space in craniotopic coordinates, allowing reconstruction of a map of visual space that is independent of eye movements. Cells in the parietal region are also likely to be involved in integrating auditory, somesthetic, and proprioceptive information. Stein (1991) has suggested that the most important characteristics of posterior parietal cells are that they receive sensory, motor, and motivational inputs, and their activity is enhanced when a stimulus is either selectively attended or is the target for movement.

Data from human studies confirm the critical role of the parietal lobe in controlling spatial attention. A visual precue in a particular spatial location will enhance the rate of detection of a target subsequently presented in the same location, even when subjects shift their attention *covertly* (i.e., without head or eye movements, Posner, 1980). Patients with either left or right parietal damage exhibit the beneficial effects of such "valid" precues, even when they occur in the contralesional visual field. However, if such patients are given an "invalid" precue on the ipsilesional side, they are severely impaired in responding to target stimuli on the contralesional side (Posner, Walker, Friedrich, & Rafal, 1984, 1987). In contrast, this severe impairment is not evident when the invalid precue is presented on the contralesional side, followed by a target on the ipsilesional side. Thus, patients with parietal damage have been considered to show an impairment in *disengaging* attention from an ipsilesional location in preparation for a shift toward a stimulus on the (unexpected) contralesional side (Posner et al., 1984, 1987; Posner, Petersen, Fox, & Raichle, 1988; see also Posner & Petersen, 1990). Moreover, the magnitude of this impairment has been found to be correlated with the severity of unilateral neglect as measured by clinical tests (Morrow & Ratcliff, 1988).

The second component of Mesulam's neural network, involving the premotor and prefrontal cortex, is assumed to control the *exploratory-motor* elements of directed attention. We have already reviewed data from both animal and human studies showing that damage to frontal regions may result in unilateral neglect. The question of whether patients with frontal lesions show predominantly exploratory-motor as opposed to perceptual unilateral neglect is still somewhat controversial, though several reports have provided evidence in support of such an anatomoclinical distinction.

In a single-case study of a woman who suffered two focal RH strokes, the first involving the frontal lobe and the second involving the parietal lobe, the pattern of lateralized deficits conformed closely to those predicted by Mesulam's model (Daffner, Ahern, Weintraub, & Mesulam, 1990). Thus, the (initial) frontal lesion (involving the middle frontal gyrus and frontal eye fields) resulted in a severe left-sided impairment on a letter-cancellation task, and prolonged detection times while searching blindfolded with the ipsilesional hand for a target located in left hemispace. However, the patient performed normally on purely "perceptu-al" tests, such as detecting single left-sided visual, tactile, and auditory stimuli. Thus, it was argued, after her frontal lesion the patient exhibited an impairment in exploratory-motor but not perceptual tasks. In contrast, after her second stroke, which involved the inferior parietal lobule, she now showed contralesion-al extinction of visual and auditory stimuli on bilateral simultaneous stimulation, in addition to the reemergence of her previously recovered left-sided deficit on the letter-cancellation task. Thus, she now exhibited unilateral neglect in both the perceptual and motor domains.

Similar results were obtained by Liu, Bolton, Price, and Weintraub (1992), who examined two RH-damaged patients. The first had a parietal lesion and showed contralesional extinction in the visual, tactile, and auditory modalities, in addition to left-sided omissions in cancellation and reading tasks, both of which indicated left unilateral neglect of moderate severity. However, the same patient was unimpaired on an exploratory task performed while blindfolded. The second patient had a frontal lesion and performed normally on tests of extinc-tion, but showed severe left-sided deficits on tasks of cancellation, reading, and exploration.

Bottini, Sterzi, and Vallar (1992) also examined two patients, one with dam-age to the frontal cortex and BG, the other with damage restricted to the tem-poroparietal region. Among other tasks, the patients were given two cancellation tests in which they were required either to cross out targets manually, or to name targets pointed to by the examiner. The patient with frontal–subcortical damage made many left-sided omissions in the former condition, but performed normal-ly in the latter, suggesting that she was able to perceive and identify targets in left hemispace, but could not initiate or execute the appropriate manual actions to cancel them. This suggestion was further supported by her normal perfor-mance on another purely perceptual test involving detection of lateralized anom-alies in illusory figures. In contrast, the patient with temporoparietal damage showed a severe left-sided perceptual deficit on the visual illusion test, and a comparable degree of impairment on both versions of the cancellation task.

In drawing a distinction between the manifestations of unilateral neglect ex-hibited by patients with lesions of frontal and parietal regions, it is well to heed Mesulam's (1990) cautionary note:

> Such dichotomies between action and perception become blurred at this [high]
> level of the nervous system. Complex perception depends on the ability of sense
> organs to act as tentacles or feelers for exploring the external world and continu-
> ously updating an internal perceptual schema. . . . From a behavioral point of
> view, a sensory representation is necessary for the accurate guidance of ex-
> ploratory movements just as exploratory movements are necessary for realigning
> sensory receptors and updating perceptual representations. (p. 600)

As we shall see, the complex interactions between perception and action form the basis of another model of spatial attention and unilateral neglect to be considered shortly; for the moment, we shall simply suggest that perception is something that is not easily divorced from implicit action.

The final cortical region in Mesulam's network, the cingulate gyrus, subserves the *motivational* components of directed attention. Clearly, not all stimuli in our environment are equally worthy of focal attention. There must therefore be some mechanism for specifying the behavioral relevance of individual stimuli, such that processing of selected items can be enhanced while selection of irrelevant items can be inhibited or suppressed (Posner & Dehaene, 1994). There is evidence from anatomical studies of the monkey brain that distinct neuronal networks connect the frontal eye fields and posterior parietal regions with the cingulate gyrus (Morecraft, Geula, & Mesulam, 1993). Single unit recordings in animals have also shown that the motivational relevance of stimuli modulates the activity of parietal and prefrontal structures, which are themselves responsible for directing spatial attention (Bushnell, Goldberg, & Robinson, 1981; Goldberg & Bushnell, 1981).

There is also evidence from PET activation studies in normal healthy humans that the anterior cingulate region is involved in the process of target detection and selection for action (Posner et al., 1988). However, the evidence does not indicate whether this region also provides in our species motivational relevance to attended stimuli. Although there have been no large group studies of unilateral neglect in humans with discrete limbic lesions, several individual cases have been reported following mesial frontal lesions involving the cingulate gyrus and SMA (Damasio, Damasio, & Chui, 1980; Heilman & Valenstein, 1972). There is also anecdotal evidence in support of a motivational component to directed attention; Mesulam (1985) described a patient whose performance on a cancellation task improved when he was offered a monetary reward. However, because the patient had an extensive RH lesion, it is impossible to ascertain whether the deficit was attributable specifically to limbic pathology.

Clearly, any comprehensive model of unilateral neglect must acknowledge the heterogeneity of its manifestations. The model proposed by Mesulam is attractive because it emphasizes, perhaps more clearly than that of Heilman, the apparent association between specific behavioral impairments and the locus of neural damage. In terms of such associations, Mesulam's formulation is but one of many possible alternatives (see, for example, Cappa & Vallar, 1992; Posner & Petersen, 1990; Rizzolatti & Berti, 1993; Rizzolatti & Camarda, 1987; Vallar, 1993). However, one possible inadequacy of the models proposed by both Mesulam and Heilman is that they assume a relatively strict hemisphere–hemispace mapping. In patients with RH damage, unilateral neglect should therefore be restricted to the contralesional (left) hemisphere, with processing in the ipsilesional (right) hemispace remaining intact.

However, both clinical observations and empirical data have shown that unilateral neglect may involve the whole of extracorporeal space, rather than just

the contralesional hemispace. Thus, patients are impaired in detecting and responding to stimuli that occupy a *relative* contralesional position in space, regardless of the absolute position of the stimulus array with respect to some centrally defined midline. As an alternative, therefore, it has been suggested that
unilateral neglect is best conceptualized as a pathological imbalance between
laterally opposed attentional vectors, creating an attentional gradient that is
manifest across the *whole* of extracorporeal space, based on a spatial scale that
changes as a function of specific task demands.

KINSBOURNE'S "VECTORIAL" MODEL.   Rather than assuming a strict mapping between hemisphere and hemispace, Kinsbourne (1987, 1993) has emphasized the
directional nature of space-related behavior. According to his model, each hemisphere is responsible for contraversively directing attention in the horizontal or
azimuthal plane. In healthy right-handers (at least), the rightward (LH) vector
prevails over the leftward (RH) vector. This LH preeminence is in a sense opposite to that attributed to the RH in the models proposed by Heilman and
Mesulam, and marks another important point of departure from them. Although
the anatomical and physiological mechanisms underlying Kinsbourne's model
are yet to be specified in detail, they nevertheless generate easily testable predic-
tions with respect to spatial performance in both normal and brain-injured subjects. In the latter case, a lesion of the RH should release the opposing LH from
inhibition, thereby inducing a pathological rightward-orienting bias, and left unilateral neglect. In contrast, although a lesion of the LH will release the opposing
RH from inhibition, because the attentional vector controlled by the latter is
only relatively weak, it is unlikely to produce a severe right unilateral neglect.
This is consistent with data reviewed earlier on the higher incidence and severity
of unilateral neglect following RH as opposed to LH lesions.

The second critical element of Kinsbourne's vectorial model is that the
strength of the attentional vectors controlled by either hemisphere can be modulated by functional activation of that hemisphere. Thus, tasks that selectively engage LH (e.g., language) functions tend to induce a rightward attentional bias,
whereas tasks that selectively activate RH (e.g., face processing) functions may
induce a leftward bias. Support for this prediction has been obtained from clinical studies. For example, Heilman and Watson (1978) presented RH-damaged
patients with two different cancellation tasks. In one, patients had to detect line
segments with a specific orientation from within an array of differently oriented
distracters. In the second task, patients had to detect target words from an array
of different distracter words. Consistent with Kinsbourne's model, which pre-
dicts a reduction in the severity of left unilateral neglect with nonverbal stimuli
(activating the RH), patients made significantly fewer omissions in the line-
cancellation task compared with the word-cancellation task. However, at least
one subsequent investigation failed to find any difference between the performances of such patients on verbal and nonverbal cancellation tasks (Caplan,

1985). It is likely that factors other than those assumed to be operating also play a role in determining performances on such tasks, for example, the arrangement of stimuli in a spatially ordered versus random pattern (Weintraub & Mesulam, 1988), the inherent complexity of stimulus material (Leicester, Sidman, Stoddard, & Mohr, 1969), and the relative salience of ipsilesionally located stimuli (Robertson & North, 1993).

Another important prediction arising from Kinsbourne's model is that patients with unilateral neglect should not only be impaired in detecting and orienting toward stimuli occupying a relatively *contralesional* position within an array, but they should also exhibit a bias for attending to *ipsilesionally* located stimuli. This prediction highlights a distinguishing characteristic of Kinsbourne's model, namely, that attentional processes are assumed to be disrupted in *both* directions in the horizontal plane, either as a *contralesional impairment* or as an *ipsilesional bias* (Kinsbourne, 1993). In contrast, the models proposed by Heilman and Mesulam assume that although attention in and toward the contralesional hemispace is impaired, attention in and toward ipsilesional hemispace remains normal.

Visual RT studies have provided compelling evidence in support of the notion of an ipsilesional attentional bias in patients with RH damage. De Renzi, Gentilini, Faglioni, and Barbieri (1989) used a computer display to present upper- and lower-case letters in four horizontally adjacent positions, each of which was located progressively further to the right of a central fixation point (i.e., entirely within the right hemifield). Subjects were required to locate and make a dichotomous decision on a prespecified target from within this array. It was found that RTs of patients with left unilateral neglect (determined by clinical screening tests) were fastest to targets presented at the rightmost extremity of the array, and became progressively slower for targets presented closer to the central fixation point. In contrast, the RTs of normal controls and RH-damaged patients *without* unilateral neglect remained constant across target positions. The authors concluded that the ipsilesional orienting bias observed in patients with unilateral neglect was uncontrollable and compulsory, and arose as a consequence of an imbalance between contralaterally opposed attentional vectors controlled by the two hemispheres.

In a similar study by Làdavas, Petronio, and Umiltà (1990), stimuli were presented in one of two horizontally adjacent locations within the right visual field. RTs were obtained from two groups of RH-damaged patients, those with and those without left unilateral neglect. The former group was faster to respond to stimuli in the right than the left relative position, whereas the latter group showed the opposite pattern (i.e., faster responses to targets on the left). Moreover, between-group comparisons revealed that patients with unilateral neglect were slower than those without the disorder to respond to left-sided stimuli, but *faster* to respond to right-sided stimuli. A signal-detection analysis showed that the different patterns of RTs for the two groups resulted from differences in sensitivity, or processing efficiency, rather than from differences in response bias.

Even more compelling evidence in favor of an attentional gradient that operates independently of egocentric coordinates was obtained in a study by Làdavas (1987), who examined the RTs of RH-damaged patients in a task that dissociated egocentric and gravitational or environmental (i.e., those relating to physical space) coordinates. Patients' RTs were faster to (environmentally determined) right-sided compared with left-sided stimuli, regardless of whether they were sitting upright and facing the display, or reclining at an angle of 90° on their left or right sides. Similar findings were obtained by Calvanio, Petrone, and Levine (1987) and by Farah, Brunn, Wong, Wallace, and Carpenter (1990).

The ipsilesional attentional bias shown by RH-damaged patients may persist even under conditions in which patients are allowed unrestricted viewing of a stimulus array. Some early studies (Campbell & Oxbury, 1976; Colombo et al., 1976; Costa, Vaughn, Howitz, & Ritter, 1969) documented the tendency among RH-damaged patients to adopt a position preference for stimuli on the ipsilesional side when choosing stimuli from an array of alternatives (e.g., in Raven's matrices). Similarly, Gainotti, D'Erme, and Bartolomeo (1991) found that RH-damaged patients tended first to identify objects appearing on the right (ipsilesional) side in the Overlapping Figures Test. This tendency apparently occurred in patients without unilateral neglect, in addition to those with moderate and severe symptoms; the extent of the lateral bias in reporting objects increased as a function of severity of unilateral neglect.

More recently, it has been shown that left unilateral neglect patients exhibit an ipsilesional bias when viewing composite and chimeric face stimuli (Mattingley, Bradshaw, Phillips, & Bradshaw, 1993). The strength of this ipsilesional bias remained virtually unchanged on follow-up testing one year postinjury, despite the fact that clinical evidence of left unilateral neglect was now absent in most patients (Mattingley, Bradshaw, Bradshaw, & Nettleton, 1994a). It has also been found that the ipsilesional attentional bias changes according to the type of stimulus material presented. For instance, RH-damaged patients exhibit a stronger and more consistent bias when viewing horizontally oriented, shaded rectangles than when viewing chimeric faces (Mattingley, Bradshaw, Nettleton, & Bradshaw, 1994). Thus, although the ipsilesional bias in such individuals is chronic and pervasive, its magnitude and consistency are perhaps best conceptualized as being variable and task-specific.

Studies of visual search performance for discrete targets in patients with left unilateral neglect have shown that focal attention may be captured by distracters occupying ipsilesional positions in the visual field (Eglin, Robertson, & Knight, 1989). Removing ipsilesional targets, or rendering them invisible by blindfolding the patient, reduces the magnetic attraction they exert upon focal attention. Thus, Mark, Kooistra, and Heilman (1988) found a reduction in the number of left-sided omissions made by patients with left unilateral neglect on a line-cancellation test, under conditions in which patients erased targets, compared with a standard condition in which targets were marked with a pencil. Similarly, Làdavas, Umiltà, Ziani, Brogi, & Minarini (1993) asked patients to point to or remove colored tokens distributed over a sheet of paper, with the aid of vision and while blindfolded. Under visual guidance, patients with left unilateral neglect detected more targets on the contralesional side when removing tokens than when pointing to them. These patients also detected more contralesionally located targets when blindfolded than in the presence of vision, once again illustrating the detrimental effects of ipsilesional stimuli. Finally, Hjaltason and Tegnér (1992) found that the magnitude of rightward errors on horizontal line bisection in left unilateral neglect was reduced (but not eliminated) by asking patients to bisect illuminated lines in darkness.

In summary, the model of unilateral neglect proposed by Kinsbourne accounts for the apparent anisotropy of attention that exists across the horizontal dimension of environmental space in affected individuals. In contrast, the models of Heilman and Mesulam assume that patients with unilateral neglect have a deficit in directing attention in and toward the contralesional hemispace, and that "this attentional deficit *induces* an ipsilesional attention bias" (Heilman et al., 1993, p. 313, italics added). Thus, Kinsbourne's vector model assumes that an ipsilesional attentional bias is the *cause* of the manifestations of unilateral neglect, whereas Heilman has suggested that such a bias occurs as a *consequence* of a contralesional attentional deficit. As with most explanations of unilateral neglect, Kinsbourne's model is not without its shortcomings. For instance, the neuroanatomical basis for the "opponent processors" (Kinsbourne, 1993, p. 64) assumed to control the attentional vectors remains largely unspecified. In this context, it is noteworthy that in the PET study on visual attention in normals described earlier, Corbetta et al. (1993) found that the *direction* of attention shifts made by normal subjects within either hemifield was relatively unimportant in determining the patterns of regional activation.

Representational Models

Despite the apparent persuasiveness of certain attentional models of unilateral neglect, there are also compelling empirically and theoretically based arguments to suggest that the disorder may stem from a failure of neural mechanisms responsible for constructing representations of the spatial environment. Although early theorists hinted at the possibility of a representational deficit underlying unilateral neglect (Zingerle, 1913; cited in Bisiach & Berti, 1987), interest in such an explanation was renewed following the now classic study of Bisiach and Luzzatti (1978). Two patients with left unilateral neglect were asked to describe a scene with which they were familiar (the Piazza del Duomo in Milan; see Figure 3) by *imagining* the scene from two perspectives. When imagining themselves looking across the piazza at the cathedral they reported land-

Figure 3   Facing the cathedral in the Piazza del Duomo in Milan (above), and looking across the square from the steps of the cathedral's main door (below). (From Bisiach & Berti, 1989).

marks that occur on the right in the real scene, but largely omitted landmarks on the left. The same selective reporting was evident when the patients were asked to imagine the piazza from the opposite perspective, looking *from* the cathedral. Thus, the patients reported previously neglected landmarks that now occurred on the right of the imagined scene, but omitted those on the left.

Since the study of Bisiach and Luzzatti (1978), numerous reports have confirmed that patients with left unilateral neglect on standard clinical tests may also be impaired in describing from imagination left-sided details of home plans (Meador, Loring, Bowers, & Heilman, 1987), suburban streets (Meador et al., 1987), towns (Halligan & Marshall, 1988; Rode & Perenin, 1994), and states (Barbut & Gazzaniga, 1987). Such patients may also be unable to compare the angles between hands of an analog clock set at two different times, when the angles occur on the left half-face. In contrast, their performance when comparing imagined angles on the right half-face is unimpaired (Grossi, Angelini, Pecchinenda, & Pizzamiglio, 1993; Grossi, Modaferri, Pelosi, & Trojano, 1989).

BISIACH'S REPRESENTATIONAL MODEL.   On the basis of his original study, Bisiach proposed a *representational* model of spatial cognition, according to which space (both real and imagined) is topographically structured across the two cerebral hemispheres. In patients with unilateral neglect, damage to one hemisphere is assumed to destroy an analog representation of the contralesional side of space. Such a model has considerable theoretical appeal. The fact that researchers frequently refer to spatial attention being "allocated" to particular locations or objects in space disregards a critical intermediate step in the perceptual process, namely, that attention is not allocated to objects or locations *themselves*, but rather to *representations* of objects and locations as they are made available via elementary perceptual processes (Farah, Wallace, & Vecera, 1993). This simple observation is perhaps the cornerstone of representational models of unilateral neglect.

Tests of anorthoscopic visual perception in patients with left unilateral neglect have also revealed representational deficits. For example, Bisiach, Luzzatti, and Perani (1979) showed patients pairs of cloudlike stimuli, with shapes that differed at their left or right extremity. Under anorthoscopic conditions, stimuli must be mentally reconstructed into a coherent whole from a series of smaller "snapshots" (Rock, 1981). With normal viewing conditions, patients could make accurate same–different judgments on vertically aligned pairs of stimuli when the critical anomaly appeared on the right, but not when it appeared on the left. More importantly, when the same stimuli were then moved leftward or rightward behind a mask with a narrow vertical viewing slit, so that they could never be viewed as a single simultaneous whole, left unilateral neglect patients continued to make errors with stimulus pairs in which left-sided details differed, but performed adequately when stimuli differed on their right extremities. Similar find-

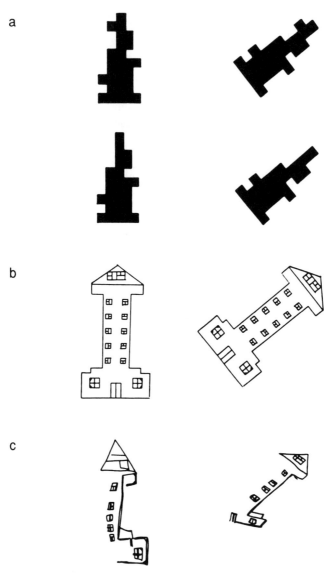

Figure 4   (a) DISPLAY USED TO TEST FOR OBJECT-CENTERED UNILATERAL NEGLECT. THE PAIR OF STIMULI ON THE LEFT HAVE THEIR PRINCIPAL AXES ALIGNED SUCH THAT THE CRITICAL DIFFERENCE BETWEEN THEM FALLS ON THE LEFT OF THE MIDSAGITTAL PLANE. THE PAIR OF STIMULI ON THE RIGHT ARE ROTATED 45° CLOCKWISE, SO THAT THE CRITICAL DIFFERENCE ON THE LEFT SIDE OF THE OBJECT NOW FALLS ON THE RIGHT OF THE PATIENT'S MIDSAGITTAL PLANE. (FROM DRIVER & HALLIGAN, 1991.) (b) EXAMINER'S MODELS, AND (c) EXAMPLES OF COPYING PERFORMANCES FROM A RIGHT-HEMISPHERE-DAMAGED PATIENT WITH LEFT OBJECT-CENTERED UNILATERAL NEGLECT. (FROM HALLIGAN & MARSHALL, 1994.)

ings were reported by Ogden (1985b) in both RH-damaged and LH-damaged patients.

Caramazza and Hillis (1990) reported a woman with left parietal damage and right unilateral neglect dyslexia. She made reading and spelling errors on the *ends* of words, regardless of their orientation on the page (i.e., when printed horizontally, vertically, or even *backwards*). Thus, it was argued, her unilateral neglect occurred at the level of a normalized or canonical word representation, which is itself independent of the spatial orientation of the word input. A similar mechanism may operate for visual objects other than words. For example, Driver and Halligan (1991) asked a patient with left unilateral neglect to compare pairs of vertically elongated nonsense shapes whose contours differed either on the left or right of their principal (i.e., vertical) axes. The patient could detect differences on the right side of the shapes' principal axes, but failed to detect differences on the left. These left-sided omissions were still evident when the shapes were rotated 45° clockwise, so that now part of the left side of the *object* was located on the right of an imaginary vertical axis on the page (see Figure 4). Thus, the patient appeared to exhibit unilateral neglect of an *object-centered* representation that was independent of the spatial orientation of the stimulus on the page. Manifestations of object-centered unilateral neglect may also be apparent in patients' copying and spontaneous drawing (Figure 4).

The data discussed so far suggest that the neural disorder underlying unilateral neglect may indeed reflect a lateralized breakdown of spatial representation. However, it might be postulated that an attentional explanation could suffice, with the addendum that attention may be mobilized on representations derived from both real and *imagined* spatial arrays. It is noteworthy in this context that the mechanisms underlying visual inspection are thought to be similar in perception and imagery (Kosslyn, Holtzman, Farah, & Gazzaniga, 1985), though a recent case study of an RH-damaged patient revealed intact processing of *external* spatial arrays, despite left unilateral neglect of visual *images* (Guariglia, Padovani, Pantano, & Pizzamiglio, 1993). This latter result suggests that different mechanisms underlie visuospatial perception and visual imagery.

Nevertheless, the main proponents of the representational account have criticized attempts to render it in terms of an underlying attentional deficit (see Bisiach & Berti, 1987). Indeed, Bisiach (1993) suggested that if anything, attention should be viewed as being superordinate to representation: "attention is a concept rooted in, and abstracted from, the intrinsic dynamics of representation and not a process acting upon it" (p. 456). Parenthetically, Bisiach and Berti (1987) also maintained that unilateral neglect is only a subcomponent of a broader syndrome involving misrepresentation of personal and extrapersonal space.

In light of these considerations, it is likely that any comprehensive model will need to explain how representational and attentional deficits interact to form the heterogeneous manifestations of unilateral neglect. It has already been hinted that the two kinds of model should perhaps not be considered mutually exclusive. Instead, the explanatory value of each might apply to a different level or domain of cognitive processes (see Bisiach, 1993, and Brewer, 1992, for more detailed arguments of this kind). In the visual modality, one useful heuristic might be to contrast patients' performances on so-called preattentive and attentive tasks (Treisman & Gelade, 1980). Models embracing this dichotomy typically assume that the former operate in parallel and without capacity limitations, whereas the latter involve serial operations and are in some way limited in capacity (Theeuwes, 1993). Thus, elementary shape, orientation, color, and so on, are initially registered in separate representations. Such preattentive processes have the function of parsing the visual field on the basis of primitive features, establishing "regions of interest" for subsequent focal attentive processing. Thus, a lateralized impairment of early visual representations might result in a failure to summon focal attention to objects located contralesionally. In any case, if we believe (see above) that perception involves implicit action, clearly the latter in turn demands the mediation of some form of directed attention. Indeed, aspects of this approach are developed further in the next section.

RIZZOLATTI'S PREMOTOR MODEL.   The account outlined above is only one of several potential ways in which attentional and representational explanations might be combined to account for the manifestations of unilateral neglect. Another possibility is that premotor aspects of response elaboration (i.e., those processes involved in the programming of goal-directed movements) may interact with processes involved in directing attention in space. As we shall see later, recent experimental observations have emphasized the importance of interactions between spatial attention and action. Indeed, an organism's attentional capacity is likely to be determined by demands for the coherent control of action (Allport, 1989; Neumann, 1990). Successful goal-directed action requires specification of parameters derived from a representation of movement space (de Graaf, Sittig, & Denier van der Gon, 1991). The premotor theory of unilateral neglect proposed by Rizzolatti and his colleagues (Rizzolatti & Berti, 1990, 1993) posited that spatial attention is subserved by anatomically independent circuits, each of which subserves a different region of space.

In both monkeys and humans, we can distinguish between three separate regions of space, each defined with respect to its proximity to the body; *personal* space is delimited by the body itself, *peripersonal* space is that region that lies within reach of the upper limbs, and *extrapersonal* space is the region extending beyond the reach of the upper limbs. In the monkey, the frontal eye fields and parietal area PF control attention in extrapersonal space, whereas inferior Area 6 and parietal Area PG control attention in personal and peripersonal space (Rizzolatti & Gallese, 1988). The combined activity of neurons in each circuit forms a representation based on information extracted from a specific region of space, and this is used to program goal-directed movements in that region.

Although the premotor theory was developed on the basis of neurophysiological data obtained in animal studies, it is consistent with some recent findings in humans. For example, Halligan and Marshall (1991a) have provided dramatic evidence suggesting a dissociation between mechanisms underlying representation of different spatial regions. Following an extensive RH lesion, their patient erred significantly to the right on a line bisection task performed in peripersonal space, but was relatively accurate when bisecting lines that had been appropriately enlarged and positioned some distance away in extrapersonal space. In contrast, five left unilateral neglect patients tested by Cowey, Small, and Ellis (1994) made significantly larger rightward bisection errors with lines positioned in extrapersonal space than with lines positioned in peripersonal space, again suggesting a dissociation between representations of different space sectors, albeit in the opposite direction to that reported by Halligan and Marshall (1991a). Dissociations between unilateral neglect of personal and peripersonal space also seem to be a relatively consistent finding (Bisiach et al., 1986; Guariglia & Antonucci, 1992). It is noteworthy, however, that other large group studies have failed to find evidence of dissociations between lateralized impairments in peripersonal and extrapersonal space (Le Fever, 1992; Pizzamiglio et al., 1989), perhaps because they required patients to search for several discrete visual targets, or because they did not require exploratory limb movements. In any case, if dissociations between lateralized impairments in peripersonal and extrapersonal space in humans are indeed analogous to those found in nonhuman animals, it is likely that they should arise only from damage restricted to a small number of neural circuits. Because cerebrovascular accidents leading to unilateral neglect in humans are often extensive and rarely conform to neuroanatomical boundaries, the dissociations that have been found may reflect as yet unspecified differences in task demands.

One final point of interest concerns the compelling findings of I. Robertson and North (1992), who examined the potential therapeutic effects in a left unilateral neglect patient of right- and left-hand movements performed in either left or right hemispace. Passive cuing techniques, such as asking the patient to direct attention to his left arm or to read out a changing number displayed on the left side, failed to reduce the number of omissions on a cancellation task. In contrast, asking the patient to move the fingers of his left hand while it was positioned in left hemispace significantly reduced the number of omissions, even when the patient was unable to see these finger movements. Movements by the left hand in right hemispace, and by the right hand in left hemispace, had no such beneficial effect. Thus, therapeutic effects were achieved only in conditions where both the moving effectors and space of operation were contralateral to the side of the lesion. The authors suggested that combined activation of personal and peripersonal spatial representations via movement was necessary for this patient to overcome his left unilateral neglect, a proposal that is consistent with Rizzolatti and Berti's (1990, 1993) premotor theory of unilateral neglect.

Current Issues in Unilateral Neglect

We have so far introduced the major models of unilateral neglect and provided an overview of some of the evidence garnered in support of each. In the
next section, we examine the adequacy of these models within a broader framework, focusing in particular upon issues that are likely to be of central importance for a comprehensive understanding of disorders of space-related behavior.

The Interrelatedness of Spatial Attention and Action

As we have seen, the traditional (and to some extent prevailing) view among
many (though not all) theorists is that unilateral neglect is a disorder involving
the acquisition and processing of perceptual information, arising perhaps from a
failure to represent stimuli in the spatial domain, or to mobilize attentional resources to specific regions of the spatial environment. As a consequence, relatively few studies have been devoted to examining potential anomalies in the motor or response characteristics exhibited by patients with unilateral neglect. This
state of affairs is unfortunate, because careful examination of the spatial and
temporal characteristics of motor performance may provide fresh insights into
the neural mechanisms that may be disrupted in patients with unilateral neglect
(e.g., Goodale, Milner, Jakobsen, & Carey, 1990). Although the extent to which a
given deficit may be attributable to perceptual or motor factors should be considered *relative,* rather than absolute, it is important to realize that intentional responses are rarely initiated in a perceptual vacuum, any more than perception
usually occurs in an entirely response- or intention-free context.

We have already discussed the existence of nonsensory unilateral neglect in
monkeys with frontal arcuate lesions (Watson et al., 1978). In these animals, the
impairment involved a failure to respond to stimuli using the contralesional
limb. In humans, the analogous phenomenon of motor neglect involves a disturbance of spontaneous movement of the contralesional limbs that cannot be attributed to hemiplegia. However, patients may also exhibit impairments of movement when using the ipsilesional limb. Evidence from both animal and human
studies suggests that the *direction* of planned movements in peripersonal space
is an important parameter in motor control (e.g., de Graaf et al., 1991; Georgopoulos, 1990). Moreover, those structures that are thought to be involved in
mediating visuospatial attention, such as the parietal lobe (Posner & Petersen,
1990), are also considered to play a key role in coordinating goal-directed action
(Goodale & Milner, 1992; Mel, 1991; Stein, 1991).

In humans, *temporal* performance measures such as RT have documented
impairments in directional movement control among patients with RH damage.
For example, using their right hands, such patients were found to take significantly longer to initiate leftward compared with rightward movements of a han-

dle along a fixed linear track (Heilman, Bowers, Coslett, Whelan, & Watson, 1985). Mattingley, Bradshaw, and Phillips (1992) used a visually cued, sequential movement task in which RH-damaged patients made movements toward discrete targets with their ipsilesional hands. Whereas patients without clinical evidence of unilateral neglect performed like controls, those with left unilateral neglect were considerably slower to initiate leftward compared with rightward movements. Interestingly, although these movement anomalies were direction-specific, they occurred in both left *and right* hemispaces, demonstrating that the absolute position of the responding limb with respect to the body midline was *not* of critical importance. Thus, the results provide support, in the motor domain, for Kinsbourne's vectorial model, but are inconsistent with models in which analog aspects of hemispace are considered to be significant in mediating spatial cognition. Mattingley et al. (1992) also found that unilateral neglect patients with *frontal* lesions (but not those with restricted posterior damage) were slower to *execute* leftward movements. These findings are consistent with Mesulam's assumption that the motor aspects of space-related behavior are subserved by the frontal cortex and striatum (Mesulam, 1981, 1990). Of course, it must be noted that the finding of an association between anterior damage and impairments in the execution of contralesional limb movements does not necessarily imply that such impairments are themselves only "motor" in nature (i.e., involving a deficit in the programming of goal-directed movements, independent of (earlier) perceptual processing, Bisiach, 1993).

In a recent study of movement kinematics (Mattingley, Phillips, & Bradshaw, 1994), patients with clinically mild unilateral neglect performed a predictable movement task with roughly the same efficiency as healthy controls, whereas their more clinically impaired counterparts were particularly slow to execute leftward movements. These movements were characterized by a reduced peak velocity, impaired force production, and inefficient on-line control. Interestingly, there was also an abnormal emphasis on terminal visual guidance of *rightward* movements made by both clinically severe and mild patients. This latter impairment is again consistent with Kinsbourne's vectorial model, in which a strong ipsilesional bias is assumed to prevail following RH damage.

The use of *spatial* performance measures has been somewhat more prevalent in studies documenting direction-specific anomalies of movement. For example, Meador, Loring, Baron, Rogers, and Kimpel (1988) found that patients with left unilateral neglect exhibited hypometric movements (i.e., movements of insufficient amplitude) toward the contralesional side when asked to reproduce horizontal displacements with their ipsilesional arms. Bisiach et al. (1990) used a modified version of the horizontal line bisection test, in which patients with left unilateral neglect operated a pulley device to indicate the midpoint of line stimuli. There were two different modes of response, one in which the patients held and moved a pointer from either end of the line to the perceived midpoint (congruent condition), and a second in which they pulled the pointer via a loop attached to laterally positioned pulleys, such that the pointer itself moved in a direction opposite to the direction of hand movement (incongruent condition). In the latter condition, it was argued, the directions of attention and action were set in opposition, "decoupled" so to speak, thereby permitting relatively independent measurement of these two components. Several patients responded in a manner that suggested that they had difficulty with contralesionally directed hand movements, such that the magnitude of these movements in the horizontal dimension was hypometric. Interestingly, these patients had lesions involving the frontal lobe, again suggesting that damage to this region may result in lateralized impairments that are most salient in the motor or premotor domain (Mesulam, 1985).

The notion that "input-" and "output"-related processes may be dissociated has also been applied in cancellation tasks. Tegnér and Levander (1991a) used a 90° angle mirror to dissociate direction of visual attention from direction of arm movement in Albert's line-cancellation task. As in the study by Bisiach et al. (1990), there were two conditions, one canonical (standard) and the other noncanonical (mirror-reversed). Some left unilateral neglect patients only canceled targets that appeared on the right, even though in the mirror-reversed conditions this meant directing their hands into left hemispace. In contrast, in the same mirror-reversed condition, a separate subgroup of four such patients failed to direct hand movements into left hemispace, despite the fact that the targets to be canceled appeared on the "good" right side. Again, of these four patients, three had lesions involving frontal cortex, whereas the fourth had subcortical damage.

In a study involving RH-damaged patients *without* clinical evidence of unilateral neglect, it was found that the trajectories of pointing movements made by the ipsilesional hand to targets on both sides of the body midline were systematically distorted toward the ipsilesional (right) side (Goodale et al., 1990). These distortions were apparent when patients pointed directly at a discrete visual target, and when they were required to bisect the gap between two horizontally separated targets, though in the former condition patients were able to correct their terminal errors to within normal limits (see Figure 5). Similarly, a recent case study of a patient with a right lenticulo-capsular lesion, and no evidence of left unilateral neglect on clinical tests, also found that the trajectories of ipsilesional hand movements toward objects in right hemispace were distorted toward the right (Chieffi, Gentilucci, Allport, Sasso, & Rizzolatti, 1993). This bias, which was maximal in the early stage of the movement trajectory, was exacerbated by the presence of a distracter object situated to the right of the target. This latter finding is consistent with the observations made in the kinematic study of Mattingley, Phillips, and Bradshaw (1994), showing that processes involved in target acquisition and movement guidance toward ipsilesionally located targets may be overemphasized in RH-damaged patients.

Finally, patients with left unilateral neglect may exhibit hypometric saccadic eye movements into contralesional hemispace (Girotti, Casazza, Musicco, &

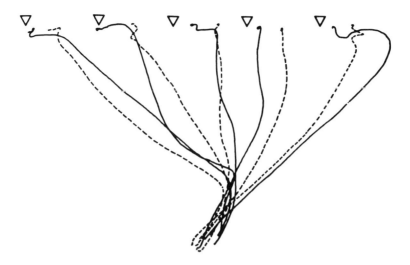

Figure 5 SAMPLE TRAJECTORIES OF REACHES MADE BY A RIGHT-HEMISPHERE-DAMAGED PA-
TIENT USING THE IPSILESIONAL UPPER LIMB. CORRECTIONS TO IPSILESIONALLY SKEWED MOVE-
MENT TRAJECTORIES ARE MORE PRONOUNCED WHEN THE PATIENT POINTS TO DISCRETE VISUAL
TARGETS (SOLID LINES) THAN WHEN HE ATTEMPTS TO BISECT THE GAP BETWEEN TARGETS (BRO-
KEN LINES). (FROM GOODALE, MILNER, JAKOBSON, & CAREY, 1990).

Avanzini, 1983; Heilman, Watson, & Valenstein, 1980). In one study, Butter et
al. (1988) recorded eye movements made toward and away from visual targets in
a patient with a right frontal lesion and left unilateral neglect. Soon after his
stroke, the patient often failed to make leftward eye movements, in response to
either left-sided *or right-sided* (in crossed response conditions) targets. In addition, leftward eye movements were often hypometric and contained many abnormal staircase saccades. It has also been shown that left unilateral neglect patients may restrict exploratory eye movements to the right sides of simple line
drawings, even when there is strong contextual information to suggest the presence of additional information on the left (Karnath, 1994a; see Figure 6).

Hemispheric Asymmetries

The observation that unilateral neglect is more severe and longer lasting after RH damage than after LH damage implies an underlying asymmetry in the
relative contributions of the two hemispheres to space-related behavior. As we
have seen, this asymmetry is incorporated into the models of Heilman and
Mesulam, who assume that the RH controls attention and action in both hemispaces, whereas the LH controls attention and action only in left hemispace.
The PET study on normal healthy subjects by Corbetta et al. (1993) supported

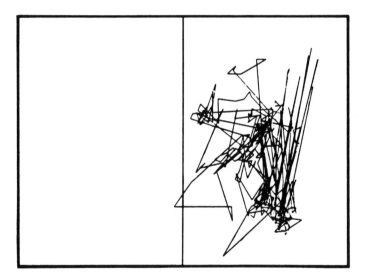

Figure 6   EYE MOVEMENT TRACINGS (BELOW) FROM A PATIENT WITH LEFT UNILATERAL NE-GLECT, DURING VISUAL EXPLORATION OF A LINE DRAWING (ABOVE). (FROM KARNATH, 1994A.)

these models in showing that the right superior parietal cortex is activated by attention shifts in either left or right hemispace, whereas the left superior parietal cortex is only activated by such shifts in the right hemispace. Alternatively, Kinsbourne suggested that the rightward (LH) vector prevails over the leftward (RH) vector, and that selective activation of one hemisphere may enhance the strength of the vector it controls. In this section, we examine evidence for functional asymmetries between the two hemispheres and suggest that they probably contribute in essentially different ways to spatial cognition.

We have thus far considered only the *directional* aspects of spatial attention, which are likely to be concerned with selecting locations and objects in space for further processing. However, it is also important to consider the states of *alertness, arousal,* or *vigilance* in modulating space-related behavior. An adequate state of alertness is likely to facilitate the processes involved in selective attention. Similarly, impaired alertness is likely to reduce the efficiency of selection, and may exacerbate an existing deficit in selective attention. There is substantial evidence to suggest that the RH plays a major role in maintaining general arousal or alertness. In normal healthy subjects, rCBF is particularly high in right frontal areas during a variety of cognitive tasks that demand sustained attention and vigilance. Also in normals, lateralized warning stimuli presented to the RH prior to the appearance of a central target have been found to reduce RTs significantly more than warning stimuli presented to the LH (Heilman & Van Den Abell, 1979). After sectioning of the corpus callosum in humans, vigilance and response speed are superior when stimuli are presented to the isolated RH than when they are presented to the isolated LH (Dimond & Beaumont, 1973; Làdavas, Del Pesce, Mangun, & Gazzaniga, 1994). Patients with RH damage exhibit prolonged simple RTs to unwarned auditory stimuli compared with LH-damaged patients (Howes & Boller, 1975). RH-damaged patients are also more adversely affected by a secondary task than patients with LH damage, and they show greater costs when visual targets are presented without a precue, irrespective of the spatial locations of such stimuli (Posner, Inhoff, Friedrich, & Cohen, 1987).

On the basis of these findings, we can conclude that the cerebral hemispheres differ in their contribution to maintaining a state of alertness, and that the greater severity and persistence of left unilateral neglect may be attributable in part to the specialization of the RH for maintaining a state of alertness (Heilman, Bowers, Valenstein, & Watson, 1987). Data from animal studies have suggested that alertness is maintained by the noradrenergic system. Arising in the locus coeruleus, this system in rats is more disrupted by lesions of the RH than lesions of the LH (see Posner & Petersen, 1990). In monkeys, noradrenergic innervation is strongest in the posterior parietal lobe, pulvinar, and superior colliculus, structures that are thought to form the neural substrate for attentional orienting in space (Posner & Petersen, 1990). Drug-induced blocking of the noradrenergic system in humans using clonidine has been found to decrease the cost of invalid spatial cuing (Clark, Geffen, & Geffen, 1989), a result that may represent a reduction in alertness and an increase in distractibility.

Another functional asymmetry between the cerebral hemispheres concerns their preferred spatial scale of operation. There is abundant evidence to suggest that the RH tends to operate on a global spatial scale, whereas the LH operates on local regions. Early evidence for a distinction between local and global processing came from experiments in which subjects were presented with a large alphanumeric character (e.g., the letter S) composed of smaller characters (e.g., the letter *H*); under these circumstances, normals respond faster to the global configuration (the S) than to the local elements, an effect known as *global precedence* (Navon, 1977). This effect occurs even when subjects are instructed prior to individual blocks of trials to attend only to local or global letters, suggesting that global precedence is an automatic feature of the perceptual system, and is not susceptible to endogenous (i.e., self-generated) control. Evidence of a hemispheric asymmetry for local and global processing comes from studies such as that of Sergent (1982), who in normals found a right visual field (LH) advantage for responding to the local features of Navon-type stimuli, and a left visual field (RH) advantage for responding to their global features.

The asymmetry between deficits in local and global processing of stimuli after unilateral hemispheric damage has long been recognized by clinical neuropsychologists (Lezak, 1983; Milberg, Hebben, & Kaplan, 1986; Walsh, 1985). Patients with RH lesions tend to exhibit global distortion and misplacement of individual elements when asked to copy a visual stimulus such as the Rey Osterreith figure, whereas those with LH lesions reproduce accurately the global configuration but omit many of the details (see Chapter 10, Figure 4). Similarly, RH-damaged patients are likely to break the outer configuration in the Block design subtest of the Wechsler Adult Intelligence Scale, whereas those with LH damage are likely to have difficulty in correctly placing and orienting individual blocks. These clinical observations have been supported by experimental evidence on the performances of patients with unilateral hemispheric lesions (without unilateral neglect) presented with Navon-type figures. Those with RH damage are impaired in responding to global compared with local features relative to controls, whereas those with LH damage are impaired in responding to local compared with global features relative to controls (Robertson & Lamb, 1991).

Several aspects of the performances of patients with unilateral neglect suggest that these individuals may have deficits not only in the directional control of space representation and selective attention, but also in flexibly modulating the spatial scale at which such processes operate. Thus, in addition to showing a strong ipsilesional orienting bias, individuals with RH damage and left unilateral neglect may show specific deficits reflecting an overemphasis on processing local features of stimulus arrays, in the absence of complete global information. Many of the performances of such individuals are in fact consistent with this conjecture. For example, the rightward errors shown by such patients on horizontal line bisection tests may reflect a bias to restrict focal processing to the ipsilesional end of the line (Marshall & Halligan, 1990). Focal cuing to the left end of the line prior to bisection may reduce patients' rightward bisection errors, but often transforms them into substantial *leftward* errors (Mattingley, Pierson, Bradshaw, Phillips, & Bradshaw, 1993; Riddoch & Humphreys, 1983), suggesting a failure to perceive the global extent of the stimulus line.

In contrast, left unilateral neglect patients are relatively accurate in determining the midpoint of figures such as circles and squares (Halligan & Marshall, 1991b; Marshall & Halligan, 1991; Tegnér & Levander, 1991b). Moreover, rightward bisection errors made by left unilateral neglect patients are reduced by positioning a vertical extent that projects above and below the right end of the line, thereby increasing the dimensions of the stimulus as a whole (Halligan & Marshall, 1994b; Mattingley, Bradshaw, & Bradshaw, in press). Another recent investigation has shown that global motion of an unstructured visual background toward the contralesional side may also reduce, and in some cases reverse, rightward errors made by RH-damaged patients on horizontal line bisection (Mattingley, Bradshaw, & Bradshaw, 1994). Thus, despite the apparent failure of global processing in RH-damaged patients, certain stimulus arrays may nevertheless be used to exploit residual or suboptimal global processing.

However, many of the other deficits exhibited by patients with left unilateral neglect imply relatively *intact* global processing. Findings that such patients invariably produce a full circular clock face but fail to position individual numerals on the left (Bisiach et al., 1981; Halligan & Marshall, 1994b; see also Figure 2), and that they may indicate apparently normal perception of the global configuration of a stimulus array by demarcating its outer extremities, but fail subsequently to detect its local left-sided constituents (Halligan & Marshall, 1993b), are consistent with the notion of *preserved* global processing mechanisms. Similarly, the fact that patients often copy only the right sides of individual items in a horizontally distributed multielement array (see Figure 7) suggests normal perceptual parsing of the global scene, but a failure to direct focal attention to the left when reproducing each local object (Halligan & Marshall, 1993b; Marshall & Halligan, 1993). Finally, although patients with left *neglect dyslexia* may omit letters at the beginning of individual words (e.g., COWBOY → boy), they more commonly make errors of substitution (e.g., CABIN → robin) in a manner that preserves word length (Ellis, Flude, & Young, 1987). This again implies relatively intact processing of the global dimensions of such stimuli, which is nevertheless insufficient to guide focal processing of the initial letters of the word.

What are the implications of the local–global processing distinction for the models of unilateral neglect discussed in this chapter? It will be recalled that the models proposed by Heilman and Mesulam assume that the RH is dominant for

 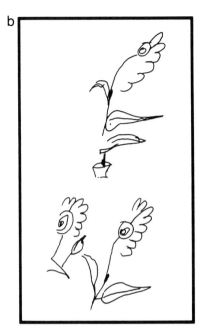

Figure 7   (a) EXAMINER'S MODEL OF TWO FLOWERS DEPICTED EITHER AS PARTS OF A SINGLE OBJECT (ABOVE) OR AS TWO SEPARATE OBJECTS (BELOW). (b) COPIES BY A PATIENT WITH LEFT UNILATERAL NEGLECT. (FROM MARSHALL & HALLIGAN, 1993.)

spatial attention and, in particular, that it subserves this function in both left and right hemispace. It is possible that this bilateral capacity reflects the RH's specialization for processing the global aspects of stimulus arrays, in addition to its likely role in maintaining a state of alertness. On the other hand, the suggestion that the LH has a strong rightward orienting bias coupled with a propensity for processing local features of spatial arrays is consistent with Kinsbourne's vectorial model. Thus, the local–global processing distinction is not necessarily incompatible with the accounts of unilateral neglect reviewed above.

Finally, the local–global distinction also provides a useful clue for explaining the apparent differences in the severity and incidence of left and right unilateral neglect. Because clinical tests such as target cancellation and line bisection require that focal attention be guided by an intact global representation, patients with RH damage are likely to show a more striking pattern of lateralized omissions than patients with LH damage. Patients with right unilateral neglect probably lack the "tightly constricted focus of attention" (Kinsbourne, 1993, p. 79) that is characteristic of left unilateral neglect, and may as a consequence show more subtle and perhaps less debilitating impairments of space-related behavior.

Space- versus Object-Based Attention

Until recently, unilateral neglect has been considered an inherently spatial disorder; indeed, the title of the present chapter reflects this traditional view. However, there is considerable evidence from studies of attention in normals, and more recently from studies of brain-lesioned patients, to suggest the existence of both space- and *object*-based mechanisms in visual attention. Although a thorough discussion of object-based models of attention is beyond the scope of the present chapter, we will consider briefly the relevance of some recent findings of object-based attentional effects in normal and brain-injured subjects. We will argue that some of the manifestations of unilateral neglect are best conceptualized in terms of an impairment in both space- and *object*-based attentional mechanisms.

As we have already outlined, the principal evidence upon which most models of spatial attention is based comes from cuing paradigms of the kind developed by Posner and his colleagues (e.g., Posner, 1980; Posner, Snyder, & Davidson, 1980). It will be recalled that, in such paradigms, a precue facilitates responses to a subsequent target presented in the same location, and inhibits responses to targets presented at locations other than that defined by the precue. It is assumed that the precue acts to orient attentional resources to a specific location, and that the costs and benefits afforded by invalid and valid cuing reflect the specific location of the attentional focus at the moment the target occurs. Thus, spatial attention has been considered to be analogous to a spotlight or zoom lens, which can be moved overtly or covertly through space and focused upon specific regions according to task demands.

We have already considered studies showing that patients with parietal lesions are impaired in disengaging focal attention from ipsilesional locations in preparation for a shift toward more contralesional locations (Posner et al., 1984, 1987). Such a deficit has been cited as one of the major underlying causes of the lateralized component of unilateral neglect (Morrow & Ratcliff, 1988). Another piece of evidence in support of purely space-based models of attention comes from the finding that normal subjects take longer to shift their focal attention over large as opposed to small distances (Egly & Homa, 1991). Similarly, in visual search tasks left unilateral neglect patients are slower to detect contralesional targets when the distracters in a display are distributed over a wide area compared with when the same distracters occupy a smaller area of the display (Eglin, Robertson, Knight, & Brugger, 1994). Thus, the contralesional search deficit exhibited by such patients becomes worse as attention is moved over a progressively larger spatial area.

The dominance of space-based models of attention may have delayed the development of alternative models in which *object properties* play an integral role in attentional selection. Recent evidence, however, has shown that visual attention is indeed modulated according to particular object attributes. In a pioneering study, Duncan (1984) tested normal subjects with briefly presented displays consisting of an outline box, upon which was superimposed a slanted line. The box could be either short or tall, with a gap on the left or right side; the line could be either dashed or dotted, with a positive or negative slope. When asked to make judgments on superimposed box and line stimuli, subjects identified two features belonging to the *same* object (e.g., a short box with a gap on the right) just as readily as they identified a single feature of that object. However, there was a decrement in performance when subjects had to identify two features from *different* objects (e.g., a short box with a dotted line). An explanation of these findings in terms of a space-based model of attention is unlikely because the box and line stimuli occupied overlapping spatial locations. Instead, the results indicate a cost when attending to aspects of separate objects that is not apparent when attending to different aspects of a single object, thereby supporting an object-based explanation.

In another study, Baylis and Driver (1993) presented normal subjects with ambiguous stimuli which, as with the vase-face figure of Rubin, appeared either as a central object against a left- and right-sided flanking background, or as two laterally displaced objects separated by a central background. By manipulating stimulus properties, subjects were induced to attend either to the single central object or to the two flanking objects. Comparing the contours between adjacent elements was easier when they were seen as the outer edges of the single central object than when they were seen as the inner edges of the two flanking objects. Once again, these results suggest a cost in attending simultaneously to two objects. Object-based effects have also been reported in selective attention tasks, in which the effect of competition from irrelevant flanking information is detrimental to identification of a central target when the flanks and target appear to form parts of the same object (e.g., Baylis & Driver, 1992; Driver & Baylis, 1989; Kramer & Jacobson, 1991).

There is now also compelling evidence that object-based attentional mechanisms play a role in some of the manifestations of extinction and unilateral neglect. We have already discussed how perceptual grouping of two laterally adjacent visual stimuli (presented simultaneously) may reduce extinction of the contralesional item, in comparison with conditions in which such grouping is absent (Ward et al., 1994). We have also discussed findings, in patients with left unilateral neglect, of impairments in processing the contralesional sides of discrete visual objects (Driver & Halligan, 1991) and words (Caramazza & Hillis, 1990). Such patients may also be impaired in detecting contours demarcating figure-ground boundaries on the contralesional sides of perceived objects (Driver, Baylis, & Rafal, 1992; Halligan & Marshall, 1994c; Marshall & Halligan, 1994).

In one particularly elegant study, Egly et al. (1994) examined the relative contributions of space- and object-based attention in normal subjects, and in brain-damaged patients without clinical evidence of unilateral neglect, using a modified spatial precuing paradigm. The display consisted of two adjacent rectangular boxes, which appeared simultaneously either to the left and right, or above and below, a central fixation mark. Subjects were precued by a luminance increment to one or other end of a rectangular box; their RTs to detect a target at the cued end were compared with those obtained when the target appeared at the opposite (uncued) end of the same box. Under these conditions, any difference in RT between cued and uncued ends should reflect *space-* but not *object*-based effects, because there was no requirement for subjects to shift their attention between separate boxes. RTs to the uncued end of one box were also compared with those obtained from either end of the box in which no precue appeared. The ends of the uncued box were the same distance from the precue as the uncued end of the cued box. Normal individuals showed an RT cost for shifting attention to uncued locations within the same box, indicating a spatial component of selection, and an additional cost of shifting attention between objects, reflecting an object-based component. Of even greater interest, however, was the finding that although RH-damaged patients showed the familiar contralesional spatial disengage deficit, the cost shown by normals when shifting attention between objects was equivalent for contralesional and ipsilesional targets. In contrast, patients with left-sided lesions showed, in addition to a spatial disengage deficit, an abnormally large cost for between-object shifts to targets in the contralesional field. The finding of normal object-based costs in patients with RH damage is consistent with evidence from studies of patients with left unilateral neglect, indicating intact object segmentation processes (Driver et al., 1992; Driver & Halligan, 1991). This contrasts with the performances of LH-damaged patients, who were impaired in making between-object shifts to contralesional targets.

To summarize, both space- and object-based mechanisms are likely to play a role in attentional selection. However, attentional deficits resulting from unilateral damage probably depend on which cerebral hemisphere is involved. This functional asymmetry, like the arousal and local–global asymmetries suggested above, may underlie the different manifestations of unilateral neglect exhibited after LH versus RH damage. It may also help to explain some of the impairments observed in patients with Balint's syndrome, whose deficits arise from damage to *both* cerebral hemispheres. Whereas the LH lesion in such patients may be responsible for producing simultanagnosia, a characteristically *object-based* impairment, the RH lesion may lead to *space-based* deficits, such as optic ataxia and spatial disorientation (Egly et al., 1994).

Spatial Coordinate Frames

Whether the manifestations of unilateral neglect reflect principally space- or object-based selection deficits, it is important to establish the spatial coordinate frame within which it is manifest. This exercise has in fact proven to be remarkably difficult because, as we have already seen, the actual distribution of impairments shown by individual patients differs according to task demands. Moreover, unilateral neglect may exist simultaneously in several different coordinate frames. Thus, rather than postulating a single coordinate frame centered, for example, on the body's midsagittal axis, it is perhaps more sensible to consider the different coordinate frames that are likely to be necessary for controlling spatial behavior.

One factor that is likely to play a key role in determining the particular coordinate frame in which location and object information is represented is the demand for the coherent control of goal-directed action. If our sense organs always occupied a fixed position relative to each other, and our effectors always operated from the same locations in space, then the transformation of sensory information into a spatial representation for guiding movement would be relatively straightforward. However, such a transformation is rendered much more complex by the fact that both our sense organs and our limbs change their positions with respect to each other, and with respect to the surrounding environment, as we make postural adjustments and locomote to different locations. Thus, the brain must integrate information from many sources simultaneously in order to maintain an up-to-date representation of the spatial environment.

Evidence from physiological studies of monkeys and from artificial neural network modeling suggests that there are several distinct populations of neurons that encode spatial locations in different coordinate frames (Andersen, Snyder, Li, & Stricanne, 1993). In the visual modality, a *head-centered* representation combines information about the position on the retina of a distal stimulus with information about the position of each eye within the orbit. A *body-centered* representation combines retinal, eye, and head position information, the latter being determined in part by afferent information from the neck muscles (Karnath, Christ, & Hartje, 1993). An *environment-centered* representation, which permits coding of information in world-based or *allocentric* coordinates, incorporates, in addition to retinal and eye position information, otolith and vestibular information regarding the position of the head in space. In patients with unilateral neglect, representations in any or all of these coordinate frames may be disrupted.

Physiological studies in monkeys suggest that cells in the posterior parietal cortex encode spatial information in at least two coordinate frames, one head-centered and the other body-centered (Andersen et al., 1993). Thus, there is sound prima facie evidence that damage to this region should impair the processing of spatial information in one or both of these coordinate frames. Indeed, one interpretation of unilateral neglect in humans is that the conversion of perceptual information into a body-centered reference frame is skewed ipsilesionally. Compelling evidence for a distorted representation of egocentric space in unilateral neglect comes from a recent study by Karnath (1994b). In one task, patients with left unilateral neglect were required to reach out to an imaginary point in space directly opposite their body midlines. In a second task, the same patients were required to indicate verbally when a spot of light projected on a wall in front of them was similarly aligned. There was a strong correlation between patients' performances on the two tasks; in both, their midline judgments were displaced approximately 15° toward the ipsilesional side. Left unilateral neglect patients have also been shown to err significantly toward the ipsilesional side in judging the position of stimuli in an auditory lateralization task (Bisiach et al., 1984).

If patients with unilateral neglect do indeed have an ipsilesionally skewed representation of egocentric space, then manipulations intended to bring the coordinate frame back into proper alignment might help to ameliorate any lateralized impairments. Indeed, Karnath and his colleagues observed that prolonged RTs by left unilateral neglect patients to stimuli presented in the contralesional visual field were reduced by turning patients' trunks by 15° toward the contralesional side, so that such stimuli now fell on the ipsilesional side of the body midline (Karnath et al., 1991). Similar beneficial effects were obtained, with head and body aligned, during vibration of the left posterior neck muscles (Karnath et al., 1993). In normals, such stimulation induces a transient ipsilateral shift of the subjective midline, probably as a consequence of proprioceptive input from muscle spindles in the stimulated region that mimic the effects of real trunk turning. In patients, contralesional stimulation may induce a compensatory shift in egocentric coordinates toward the impaired side. Longer lasting, though still transient, benefits are also produced by caloric and optokinetic stimulation (Pizzamiglio, Frasca, Guariglia, Incoccia, & Antonucci, 1990; Rubens, 1985). These maneuvers, which we consider more thoroughly in a later section on recovery and rehabilitation, may also have the effect of inducing a compensatory recalibration of the egocentric coordinate frame (Vallar et al., 1993).

It is likely that the behavioral manifestations of unilateral neglect can indeed be traced back, in the first instance, to a distortion of spatiotopic coordinates of the sort described above. However, this explanation cannot account for instances of object-centered unilateral neglect (e.g., Caramazza & Hillis, 1990; Driver & Halligan, 1991), in which patients may detect all the component objects in a horizontal array, but apparently fail to perceive the contralesional side of each. In this case, the representations of individual objects appear to be disrupted independently of their absolute location in a body-centered coordinate frame. In fact, in vision, the more abstract object-based description are widely considered to be derived from an initial spatiotopic representation (Marr, 1982). Thus, there may in fact be a hierarchy of coordinate frames, each more abstract than its predecessor, from the initial registration of the physical environment to a canonical representation of objects in terms of their principal axes (Behrmann & Tipper, 1994).

If the hierarchical model of spatial coordinate systems described above is correct, it may help to explain some of the heterogeneous manifestations of unilateral neglect. Clinically, a disruption in registering the immediate layout of the spatial environment may result in a patient ignoring objects on the contralesional side of the body. However, when objects within the ipsilesional hemispace are represented at progressively more abstract levels in the hierarchy, the contralesional portions of their representations may now also be disrupted. Such an account helps to explain why unilateral neglect may occur in several coordinate frames simultaneously (Calvanio et al., 1987), and why the coordinate frame(s) in which the disorder is manifest may alter according to task demands; those requiring access to environment-based descriptions, such as pointing at discrete targets in space (Harvey, Milner, & Roberts, 1994), are unlikely to require object-based representations. Conversely, tasks in which veridical object information is critical, such as picking up rods of different diameters (e.g., Chieffi et al., 1993), require intact object representations for successful performance. It may also be relevant that in a recent study (Vallar, Rusconi, & Bisiach, 1994) substantial improvements were observed in cancellation test performances of left unilateral neglect patients following caloric stimulation; in contrast, there was no improvement when the same patients were required to detect left-sided anomalies in pictures of discrete objects. Thus, vestibular stimulation may compensate for a skewed environmental representation (as suggested by Karnath, 1994), but have little or no effect upon object-based representations. Further research is needed to clarify these conjectures.

"Unconscious," Implicit or Tacit Processing

We have so far considered the range of impairments that may comprise the clinical picture of unilateral neglect. We have also seen that contemporary theorists view unilateral neglect as a disorder of high-level cognitive processes involved in mediating space- and object-based perception, attention, and action. It will be recalled that although unilateral neglect patients may exhibit primary sensory impairments such as hemianopia and hemianesthesia, these clearly are not present in all such individuals. The question therefore arises, To what extent is apparently neglected information processed? Obviously, there is some point at which stimuli fail to influence the patient's spontaneous, overt behavior. But as we have seen in other disorders such as agnosia (Chapter 5) and amnesia (Chapter 8), measurements of patients' spontaneous behavior often belie the existence of some degree of partial, covert, or implicit processing. Perhaps not surprisingly, recent investigations have suggested that unilateral neglect is no exception to the frequently encountered manifestations of "unconscious" processing in neurological disorders.

Vallar et al. (1991) obtained electrodermal skin conductance responses from an RH-damaged patient with left unilateral neglect and left hemianesthesia. Although the patient often failed to report somatosensory stimuli delivered to the left index finger, for more than half of the unreported stimuli a *normal* skin conductance response was obtained. Thus, the patient's hemianesthesia could not be attributed entirely to a primary sensory impairment. Rather, the existence of normal autonomic responses in the absence of verbal report suggests a defect in high-level access to intact sensory processes. In another study, Vallar et al. (1991) recorded cortical somatosensory and visual evoked potentials (EP) to contralesional and ipsilesional stimuli in patients with unilateral hemispheric damage. In both LH- and RH-damaged patients, verbal responses and EPs to ipsilesional stimuli were normal. LH-damaged patients with hemianopia and hemianesthesia, but *no* unilateral neglect, failed to report contralesional stimuli; moreover, they showed no discernible EPs to such stimuli. In contrast, RH-damaged patients *with* left unilateral neglect failed to report contralesionally presented somatosensory and visual stimuli, but exhibited *normal* EPs. Once again, these results demonstrate intact early sensory processing of contralesional stimuli in patients with left unilateral neglect, in the context of an inability to report their presence.

The studies by Vallar and colleagues confirm that patients with unilateral neglect may exhibit normal, low-level processing of somatosensory and visual stimuli, although the output of such processing appears not to reach awareness. The findings of other studies have suggested that even more complex information is extracted from apparently neglected stimuli. For example, there is evidence for intact color perception (A. Cohen & Rafal, 1991) and figure-ground segregation (Driver et al., 1992) of apparently neglected stimuli. Both these processes, which occur *preattentively* (i.e., prior to perceptual awareness), are thought to create separate maps specifying regions of interest for guiding later *attentive* ("conscious") processing. Thus, it may be that preattentive processes remain intact in many patients with unilateral neglect, and that although the output from such processes is unable to guide focal attention contralesionally, the extraction of elementary features may nevertheless exert some implicit effect upon patients' behavior.

Perhaps the first systematic study of implicit processing in RH-damaged patients was performed by Volpe, LeDoux, and Gazzaniga (1979). They examined the performances of four patients with right parietal tumors and contralesional extinction on a tachistoscopic task that involved the presentation of pictures or words either individually to one or the other visual field, or simultaneously to both. In the latter instance, stimuli could be either the same (e.g., drawings of an apple in each visual field) or different (e.g., drawings of an apple and a comb). Patients performed with a high degree of accuracy when asked to name stimuli presented individually to either visual field. As expected, on double simultaneous presentations, patients had difficulty naming the left-sided (contralesional) stimulus when the items were different; indeed, two patients failed on all trials to name left-sided stimuli, and insisted that no stimulus had in fact occurred there. However, despite their poor performance in *naming* left-sided stimuli in double simultaneous presentations, the patients performed well above chance when asked to make same–different judgments. Using a similar paradigm, Karnath and

Hartje (1987) obtained comparable results in a patient with left unilateral neglect. In addition, they found that cuing the patient to attend to the contralesional visual field during bilateral presentations failed to improve verbal identification.

The results of these studies have been interpreted by many as indicating the existence of implicit processing of apparently extinguished visual stimuli. More specifically, it has been conjectured that such stimuli may be processed relatively normally, but that the outcome of this processing may not reach consciousness (Volpe et al., 1979). We shall return shortly to consider the issue of *how much* information might be extracted from apparently neglected stimuli. However, it is important at this point to raise a caveat in interpreting the results reported by Volpe et al. (1979) and Karnath and Hartje (1987). Farah, Monheit, and Wallace (1991) demonstrated in an elegant study that the *amount* of visual information required for same–different matching is considerably less than that required for identification, and that a dissociation between performances of patients on these two response measures may therefore not reflect implicit processing of extinguished stimuli. To illustrate this point, Farah et al. (1991) replicated in normals the dissociation found previously in patients, by degrading the visual stimulus presented in one visual field, and recording responses for free identification and same–different matching. They then equated the amount of visual information needed by comparing same–different matching with a forced-choice identification task, and found that the original dissociation disappeared. They also extended their findings to RH-damaged patients with visual extinction. The familiar dissociation of Volpe et al. was obtained, but as with normals, performance on same–different matching and forced-choice identification was virtually identical.

In view of the findings of Farah et al. (1991), is it still possible to maintain that patients extract physical and perhaps even semantic information from apparently extinguished (or neglected) visual stimuli? We can probably answer this question, at least tentatively, in the affirmative. In a widely reported study, Marshall and Halligan (1988) presented a left unilateral neglect patient with a pair of drawings of a house, in one of which red flames were depicted emerging from the left side. These drawings were presented one above the other; the patient was asked to describe what she saw and then to indicate whether the two drawings were the same or different. Her spontaneous descriptions suggested that she was unable to detect any systematic difference between the stimuli, and she consistently indicated that the two stimuli were the same on forced-choice testing. However, when asked in which house she would prefer to live, she consistently selected the house *without* the flames. These results, obtained from stimuli presented centrally in free vision and with unlimited viewing time, do seem to provide strong evidence for tacit processing of neglected information. Parenthetically, it is interesting to note that the patient of Marshall and Halligan (1988) showed a consistent preference for the nonburning house, but was unable to make accurate same–different judgments, whereas the patients reported by Volpe et al. (1979) and Karnath and Hartje (1987) could not identify contralesional stimuli but made *very accurate* same–different judgments.

The results obtained by Marshall and Halligan (1988) have been interpreted by some as suggesting intact processing of semantic information on the neglected side, despite an impairment in the ability to judge physical similarity. It has even been suggested that the readiness with which such patients confabulate when required to select their "preferred" house reflects the extent of "preconscious" processing of unreported information. For example, after indicating his preference for the nonburning house, and despite claiming that the two drawings were the same, a patient tested by Manning and Kartsounis (1993) explained his preference thus:

> There is an extra fireplace, it's the heating, it's warmer and I wouldn't have to buy a wood-burning stove. (p. 212)

When tested the next day, he gave another confabulatory response:

> There are other, extra fireplaces (pointing to the *nonburning house*) and two chimneys. (p. 212)

However, the evidence provided from these "burning house" studies is not necessarily indicative of intact *semantic* processing of neglected information. Bisiach and Rusconi (1990) found that two patients with left unilateral neglect consistently selected the *burning* house as the one in which they would prefer to live, suggesting that their preferences were based on something other than the semantic import of the flames themselves.

More convincing evidence for semantic processing of unreported stimuli was obtained in a study by Berti and Rizzolatti (1992). They examined visual priming of ipsilesionally presented target pictures following brief presentation of contralesional primes. In normals, priming with physically identical or semantically related items facilitates the speed of responses, even when the exposure duration of the prime is too brief to be identified consciously by the subject. Berti and Rizzolatti (1992) asked patients to indicate as quickly as possible whether the target they had seen was an animal or a fruit. When the prime and target were physically identical, or when they were physically different but from the same category, patients' RTs were significantly faster than when the prime and target were from different categories. Moreover, there was no significant difference in RT between identical and categorically related prime–target pairs. Thus, despite denying perception of left visual field primes, these patients were nevertheless able to process stimuli to a categorical level.

Several subsequent studies have confirmed the existence, at least in some patients, of semantic priming effects induced by otherwise neglected stimuli. For example, Làdavas, Paladini, and Cubelli (1993) observed significant effects of semantically related word primes presented in the contralesional visual field upon ipsilesional targets. Apart from denying their presence, the patient could not read left visual field primes when explicitly requested to do so, nor could he decide on their lexical status or categorize them as living or nonliving. The patient also performed at chance level when asked simply to judge whether primes were present or absent. In another study, D'Esposito, McGlinchey-Berroth, Alexander, Verfaellie, & Milberg (1993) observed similar effects with contralesionally presented word and object primes, and centrally presented targets. Interestingly, priming and discrimination differed as a function of lesion site. Patients with right parietal lesions showed normal priming effects with contralesional stimuli, but were impaired in forced-choice discrimination of these items; their ability to discriminate ipsilesional items was normal. In contrast, patients with frontal or subcortical lesions showed either normal discrimination and priming in *both* visual fields, or normal priming and *impaired* discrimination in both fields. It was suggested that patients with posterior lesions cannot direct focal attention contralesionally for explicit stimulus identification, but can extract enough information from primes to activate corresponding semantic representations. Normal priming and discrimination in patients with anterior damage may indicate that their clinical impairments (e.g., lateralized omissions on cancellation tasks) reflect a deficit of intentional or exploratory-motor behavior. Finally, poor discrimination in *both* visual fields may reflect a generalized impairment of arousal, which, as we have already seen, is a common phenomenon in patients with RH damage.

Recovery and Rehabilitation

Although the more florid clinical manifestations of unilateral neglect, such as head and eye deviation toward the ipsilesional side, and failure to orient attention contralesionally even to highly salient stimuli, have been considered in most patients to be transient (Colombo, De Renzi, & Gentilini, 1982; Wade, Wood, & Hewer, 1988), it is unfortunately the case that in a significant proportion of cases, residual impairments of a more subtle, but nevertheless debilitating, nature continue to impinge upon activities of daily living (ADL) (Campbell & Oxbury, 1976; Kinsella & Ford, 1985; Webster, Rapport, Godlewski, & Abadee, 1994). Although the behavioral manifestations of unilateral neglect in the acute stages postinjury are well documented (Bisiach & Vallar, 1988; Mesulam, 1985; Heilman, Watson, & Valenstein, 1993), relatively few studies have focused on the time course and extent of recovery among such patients. Of those studies on recovery that have been conducted, many have involved either qualitative descriptions (e.g., Lawson, 1962) or clinical tests with limited or undocumented sensitivity and reliability (e.g., Campbell & Oxbury, 1976; Denes, Semenza, Stoppa, & Lis, 1982; Hier, Mondlock, & Caplan, 1983). In this section, we consider the physiological and behavioral correlates of recovery in patients with unilateral neglect, and provide an overview of recent rehabilitative interventions.

Anatomical and Physiological Studies

Hier et al. (1983) studied the patterns of recovery from a number of neurological and cognitive impairments in 41 patients with RH damage. They found that recovery from left unilateral neglect (defined in this instance simply as "failure to attend to left-sided visual and auditory stimuli," p. 345) was more rapid among patients with hemorrhagic as opposed to ischemic cerebrovascular events. This may have been due to the smaller volume of hemorrhagic compared with ischemic lesions, or the predilection of the former for subcortical as opposed to cortical regions (Hier et al., 1983). It was also found that patients with relatively small lesions (involving less than 6% of the total RH volume) recovered more rapidly from unilateral neglect (this time as measured by omissions on the Rey Complex Figure Test) compared with those whose lesions involved more than 6% of RH volume. Similar findings were obtained by Levine, Warach, Benowitz, and Calvanio (1986), who reported a positive correlation between lesion volume and persistence of unilateral neglect. Parenthetically, this study also found that the presence of cortical atrophy prior to stroke was associated with more severe unilateral neglect, a finding that may relate to another of the observations made by Hier et al. (1983), namely, a trend toward faster recovery in patients below 60 years of age.

Cappa, Guariglia, Messa, Pizzamiglio, and Zoccolotti (1991) examined 24 patients with left unilateral neglect (as measured by a letter-cancellation test) that had persisted for a variable period (from 2 months to almost 3 years) postinjury. Consistent with the studies of Hier et al. (1983) and Levine et al. (1986), a significant correlation was found between lesion size and severity of unilateral neglect as measured by omissions on the letter-cancellation test. However, the magnitude of this correlation was quite small, suggesting that a substantial proportion of the variability in severity of unilateral neglect could *not* be predicted from lesion size. Moreover, in contrast to the findings of Levine et al. (1986), no relationship was found between the degree of cortical atrophy and severity of unilateral neglect. Although the reasons for this discrepancy are unclear, one possibility is that the different methods employed to measure cortical atrophy (Levine et al. used sulcal width, whereas Cappa et al. used ventricular size) may have yielded different estimates of tissue loss.

Stone and colleagues (Stone et al., 1991; Stone, Patel, Greenwood, & Halligan, 1992) provided evidence for a difference between the rates of recovery from unilateral neglect of LH- and RH-damaged patients. Stone et al. (1991) examined a group of 44 patients with unilateral hemispheric stroke. At 3 days postinjury, approximately half of all LH- and RH-damaged patients exhibited unilateral neglect on standard cancellation tests, a finding that is consistent with previous reports on the relative prevalence of unilateral neglect in the acute stages after LH and RH damage (Ogden, 1985a, 1987). However, three months later, the performances of almost all the LH-damaged patients had returned to normal, whereas those of roughly one-third of RH-damaged patients indicated persistent unilateral neglect.

In a separate study (Stone et al., 1992), 68 patients with unilateral neglect at 3 days postinjury were reexamined at 10 days, 3 weeks, 6 weeks, 3 months, and 6 months. The most marked improvements occurred within 10 days postinjury and reached a plateau within 3 months. Moreover, the extent of recovery was significantly greater in patients with LH damage than in those with RH damage at each retesting session, with the exception of the 6-month follow-up where the difference was not statistically significant. This latter finding was attributable to the fact that although patients with LH damage showed no improvement between 3 and 6 months, patients with RH damage continued to show some degree of improvement over the same period.

Taken together, the results of these studies indicate that recovery from unilateral neglect is more rapid and complete in patients with LH damage. The persistence of unilateral neglect in patients with RH damage has been suggested to underlie the relatively poor outcomes in terms of recovery from hemiplegia and functional abilities, among such individuals several months postinjury (Denes et al., 1982). Nevertheless, the finding that small, incremental improvements may continue several months after injury in patients with RH damage suggests that the discrepancy between LH- and RH-damaged patients may not necessarily persist in the *extended* long-term (i.e., 1 or 2 years postinjury).

As mentioned earlier, although data from structural neuroimaging techniques provide information on the locus and extent of permanent tissue damage, they cannot detect changes in rCBF and metabolism, which are frequent concomitants of primary neural damage. However, the use of SPECT and PET has enabled researchers to obtain estimates of local changes in these physiological indices following stroke, and to examine the relationship between such indices and the severity of unilateral neglect. Thus, Vallar et al. (1988) used SPECT to examine two patients with RH subcortical damage. In the acute phase postinjury, these patients exhibited left unilateral neglect on the same battery of clinical tests as that used by Perani et al. (1987). At this time, rCBF was profoundly reduced, especially in anterior and middle cortical regions. After a mean recovery period of 3.5 months, the same patients now exhibited no signs of unilateral neglect on clinical tests. This change was paralleled by significant improvements in cortical blood flow. On the basis of these findings, Vallar et al. (1988) concluded that the occurrence of cognitive deficits such as unilateral neglect immediately after subcortical stroke is attributable to "functional de-activation" (p. 1275) of remote cortical regions (i.e., diaschisis; see Chapter 1).

Fiorelli et al. (1991) used PET to study cerebral metabolism in six patients with acute unilateral hemispheric damage and predominantly motor neglect.

Patients exhibited widespread hypometabolism of the ipsilesional cortex in addition to a moderate level of metabolic reduction in the thalamus. In one individual who had a small subcortical infarct involving the caudate and lentiform nuclei, motor neglect had resolved but she continued to exhibit unilateral neglect in the visual and auditory modalities in association with significant hypometabolism of the ipsilesional cortex. This latter finding suggests that specific behavioral deficits may have different rates of recovery, and implies that behavioral improvements may not always be accompanied by detectable increments in brain metabolism. One final point of interest to be derived from the study by Fiorelli et al. (1991) is that metabolism in the *undamaged* hemisphere was also found to be lower than that of controls in four patients, two of whom showed statistically significant hypometabolism. If functional hypoarousal of the contralesional as well as the ipsilesional hemisphere can be considered a feature of most patients with unilateral neglect, then recovery may depend upon restitution of function in *both* cerebral hemispheres.

Evidence in support of this suggestion was provided in a recent study by Perani et al. (1993), who used PET to examine two patients with RH damage and left unilateral neglect. One patient showed severe hypometabolism in both RH and LH in the acute phase postinjury. A follow-up examination 8 months later revealed complete recovery from hemianesthesia and hemianopia, and only minor residual unilateral neglect on a cancellation test. A repeat PET study at this time revealed residual hypometabolism in the right fronto-temporo-parietal areas, but a return to virtually normal metabolic values in the undamaged left hemisphere. In the second patient, although PET was not performed acutely, a study at 4 months postinjury revealed severe and widespread metabolic reductions in both the damaged (right) and undamaged (left) hemispheres, which were accompanied by severe and persistent unilateral neglect. These data would appear to support the suggestion that the undamaged hemisphere plays a role in mediating recovery from unilateral neglect, although undamaged regions of the ipsilesional hemisphere must also be involved in the recovery process (Crowne, Yeo, & Russell, 1981; see also Heilman et al., 1993, for further discussion of this point). Indeed, it is possible that the relative contributions of the two hemispheres to recovery may change over time, with contralesional structures playing a predominant role in the acute phase, whereas functions subserved by undamaged regions of the ipsilesional hemisphere may emerge in the later stages postinjury (Perani et al., 1993).

Behavioral Studies

Clearly, the rate of recovery from unilateral neglect must depend upon the instruments used to assess individual patients. Because investigators have used very different tests, estimates of the duration and severity of unilateral neglect postinjury vary widely. For example, Wade et al. (1988) found, in a consecutive series of 62 stroke patients, that 24% of testable individuals showed unilateral neglect on a cancellation task shortly after injury, and that most of these performed normally on follow-up assessment 3 months later. In a more recent study, Marsh and Kersel (1993) found that 22% of unilateral stroke patients continued to show signs of unilateral neglect after 90 days on at least one test from a battery that included cancellation, reading, and line bisection.

Periods of recovery from the behavioral manifestations of unilateral neglect have been reported by numerous authors (e.g., Robertson, 1993), and will not be reviewed in detail here. One important finding to arise from these studies, however, is that although there is some tendency for patients to show milder deficits as a function of increasing time postinjury (e.g., Marsh & Kersel, 1993; Zoccolotti et al., 1989), the correlation between severity of unilateral neglect and time since injury has generally been found to be small (Zarit & Kahn, 1974; Zoccolotti et al., 1989). This latter result is likely to be attributable to the observation that although unilateral neglect resolves gradually in many patients, there is also a substantial proportion of individuals in whom there is little or no improvement, even over the extended long-term.

Relatively few studies have been devoted to obtaining quantitative behavioral measures of spontaneous recovery from unilateral neglect. Campbell and Oxbury (1976) used a battery that included several standardized and quantifiable tests to measure recovery in RH-damaged patients with and without signs of left unilateral neglect in the acute stage postinjury. They found that those individuals with unilateral neglect did not improve significantly at 6-month follow-up on tasks of visual recognition and visuoconstruction. Moreover, these patients continued to show a strong right-sided position preference when selecting visual stimuli from an array of alternatives, despite the fact that most no longer exhibited signs of unilateral neglect on a drawing task. On the basis of these findings, Campbell and Oxbury (1976) suggested that the resolution of unilateral neglect may reflect the development by patients of strategies with which to compensate for a persistent attentional disorder, which may nevertheless reemerge on novel or complicated tasks.

Karnath (1988) examined the performances of three patients with RH damage, two of whom exhibited unilateral neglect on clinical tests, on tasks that involved either same–different matching or naming of geometrical figures presented tachistoscopically to either visual field. For both patients in whom testing was conducted in the acute stage postinjury, naming performance in unilateral right and left visual field presentations did not differ. However, on double-simultaneous presentations, patients were significantly impaired in naming stimuli in the contralesional visual field. This pattern of performance was considered to represent an ipsilesional attentional bias, because patients improved their performances with left visual field stimuli when required covertly to shift attention to this side. Follow-up examinations of two patients 3 and 4 months later revealed the same pattern of impairment in naming contralesional stimuli on double-simultaneous presentations.

In a more recent study, Mattingley, Bradshaw, Bradshaw, and Nettleton (1994a) also found a persistent ipsilesional attentional bias in RH-damaged patients, in this case tested 12 months poststroke. On acute testing, their patients showed left unilateral neglect on clinical tests, in addition to a strong rightward bias in free-viewing of composite and chimeric face stimuli. On follow-up, the same individuals continued to exhibit a rightward attentional bias, though there was no longer any evidence of left unilateral neglect on clinical tests.

Collectively, the results of these studies point to several broad conclusions. First, there is considerable heterogeneity in the performances of unilateral neglect patients during recovery. Second, some patients may continue to exhibit subtle impairments months and even years postinjury; although their performances on standard clinical tests may suggest that they have returned to normal, testing with more sensitive or novel paradigms can reveal persistent deficits. As a corollary, because unilateral neglect is likely to reflect a number of specific impairments, the extent of recovery will be critically dependent upon the task(s) used to assess patients. Finally, the locus of cerebral damage is clearly of critical importance in determining the nature and extent of recovery from these specific impairments.

We have so far concentrated upon the physiological and behavioral correlates of spontaneous recovery from unilateral neglect. In concluding this chapter, we shall consider briefly the efficacy of rehabilitative interventions in patients with unilateral neglect. Although clinicians have long been concerned with issues of rehabilitation, those involved in research have only recently acknowledged that rehabilitation studies may provide fresh theoretical insights into this complex disorder (Robertson, Halligan, & Marshall, 1993). In this section, we will consider separately those interventions that have been shown temporarily to ameliorate unilateral neglect, and those that have been aimed at producing long-lasting therapeutic effects.

Temporary Interventions

Several studies have shown that providing contralesionally positioned spatial cues permits some patients transiently to overcome their attentional bias. In the RT study of Posner et al. (1984), although patients with unilateral parietal damage exhibited *abnormal costs* associated with *invalid* cuing of contralesional targets, they showed *normal benefits* with *valid* cuing of such targets. Thus, patients' responses were normal when they were first cued to attend to a contralesional location.

In the task of horizontal line bisection, the substantial rightward errors associated with left unilateral neglect may be reduced or eliminated by having patients report a cue positioned at the left end of the line prior to placing their transection mark (Nichelli, Rinaldi, & Cubelli, 1989; Riddoch & Humphreys, 1983). Similar effects have been noted when such patients are first required to mark with a pen the left end of the line, even if the act of marking leaves no visible trace (Mattingley, Pierson, Bradshaw, Phillips, & Bradshaw, 1993), and when patients move a cursor from the left as opposed to the right end of lines presented on a computer display (Halligan & Marshall, 1989a,b). Dynamic cues, both lateralized (Butter, Kirsch, & Reeves, 1990) and nonlateralized (Mattingley et al., 1994b) exert an even stronger effect. From a rehabilitative perspective, however, there are two problems with these cuing techniques. First, their effects are transient, probably lasting no more than a few seconds after they are introduced (Halligan, Donegan, & Marshall, 1992). Second, they often induce ipsilesional unilateral neglect, such that patients now appear to be overattracted to the *contralesional* side (e.g., Mattingley et al., 1993; Robertson, 1989). This latter observation is consistent with the suggestion that patients with left unilateral neglect have an abnormally constricted attentional focus, and that discrete spatial cues serve merely to relocate this focus without improving patients' global appreciation of the stimulus array.

Several workers have examined the effects of vestibular stimulation upon unilateral neglect. Irrigation of the contralesional ear with iced water, or the ipsilesional ear with warm water, induces substantial improvements in unilateral neglect after RH damage (Cappa, Sterzi, Vallar, & Bisiach, 1987; Rubens, 1985). Exposure to a rapidly moving background pattern induces optokinetic nystagmus, which may also help to overcome unilateral neglect in RH-damaged patients (Pizzamiglio et al., 1990). Neck muscle stimulation on the contralesional side has also been shown temporarily to reduce the severity of left unilateral neglect (Karnath et al., 1993). As discussed earlier in this chapter, although the precise mechanisms for these effects are unknown, they may modulate the activity of systems responsible for maintaining and updating egocentric coordinate frames (Vallar et al., 1993). Unfortunately, the effects induced by these techniques are again rather limited in duration, typically lasting only a matter of minutes posttreatment. Unilateral neglect patients have also been shown to benefit from manipulations that modulate visual input. One technique is to occlude the ipsilesional eye using an eye patch; the rationale for this maneuver is based on Kinsbourne's assumption that unilateral brain damage releases the intact hemisphere from inhibition, thereby inducing an ipsilesional orienting bias. Because orienting is likely to be controlled to some extent by brain stem structures, most notably the superior colliculus (SC), and because this structure receives most of its input from the contralateral retina, patching the ipsilesional eye should reduce the level of activation of the SC on the undamaged side, thereby reducing its tendency toward pathological ipsilesional orienting. A recent study of the effects of ipsilesional eye patching in patients with left unilateral neglect revealed significant benefits in 11 out of 13 individuals tested (Butter & Kirsch, 1992), though these effects were highly variable and were restricted to the period for which the eye was occluded. In a separate manipulation, Fresnel prisms were placed over patients' spectacles and oriented such that visual information from the contralesional periphery was brought into central vision (Rossi, Kheyfets, & Reding, 1990). Beneficial effects were observed, perhaps as a consequence of a compensatory recalibration of the egocentric coordinate frame. Once again, however, the reported benefits occurred only while patients were actually wearing the prisms, and these did not generalize to ADL.

Finally, there is preliminary evidence that pharmacological interventions directed at specific neurotransmitter systems may modulate the attentional deficits of unilateral neglect. For example, the dopaminergic agonist bromocriptine has been shown temporarily to ameliorate some of the manifestations of unilateral neglect in humans (Fleet, Valenstein, Watson, & Heilman, 1987). Given our growing understanding of the roles played by neurotransmitter systems in modulating attentional functions, it is likely that pharmacological treatments will play a major (and perhaps predominant) role in the management of patients with attentional disorders such as unilateral neglect in the future (Robbins, 1994).

Long-Term Interventions

The study of Lawson (1962) represents one of the first published reports of an attempt to rehabilitate two left unilateral neglect patients. Simple visual scanning and perceptual anchoring techniques (e.g., finding the left edge of a printed page) were used to assist patients with ADLs such as reading. Although patients showed some improvement over the months of treatment, in the absence of suitable control procedures the relative contribution of spontaneous recovery cannot be ascertained. Several studies in the 1970s and 1980s were also aimed at improving visual scanning in patients with unilateral neglect. For example, Weinberg and colleagues (Weinberg et al., 1977, 1979) used a scanning machine, in addition to a number of modified reading, cancellation, and tactile tasks, with groups of left unilateral neglect patients. Patients in the treatment groups improved significantly relative to control patients who received only general occupational therapy, though these improvements tended to be restricted to tasks that were similar to those used in retraining. In a subsequent study employing similar techniques, Webster et al. (1984) observed improvements in wheelchair navigation among three patients with left unilateral neglect. More recently, a large randomized trial of computer-based rehabilitation in patients with unilateral neglect was conducted by Robertson, Gray, Pentland, and Waite (1990). Following intensive therapy using similar techniques to those of Weinberg et al., a blind follow-up examination revealed *no* significant improvement of the treatment group over a group receiving nonspecific computerized tasks.

It is important to make several comments about the results of these early rehabilitation studies. First, the rationale for most of the interventions was symptom-driven, in the sense that treatment strategies were aimed at overcoming the overt manifestations of unilateral neglect, with little or no consideration of the disordered neural mechanisms underlying the symptoms. Second, the therapeutic effects of such treatments were invariably *task-specific*, with minimal generalization to other ADLs. This should not be surprising in view of our first point, namely, that interventions that focus upon ameliorating the clinical *effects* of brain damage are unlikely to change in any significant way their underlying *causes*. Finally, all the early interventions involved many hours of practice. Considering the rather circumscribed (and occasionally nonexistent) benefits, it is probably sensible to conclude that such treatment strategies are unlikely to be an effective use of valuable therapy resources.

The distinguishing feature of recent attempts to rehabilitate patients with unilateral neglect is that they have been theory-driven. For example, Robertson and colleagues (Robertson & North, 1992; Robertson, North, & Geggie, 1992) adopted a series of interventions based on the premotor theory of unilateral neglect proposed by Rizzolatti and colleagues (Rizzolatti & Berti, 1990, 1993). It will be recalled that this theory assumes that the allocation of attention to specific regions of space is determined by neural representations for planning goal-directed actions. Thus, motor activation may be an important factor in helping to mobilize attentional resources contralesionally. In a single-case study of a patient with left unilateral neglect, Robertson and North (1992) found that having the patient wiggle the fingers of his partially hemiparetic left hand significantly improved cancellation test performance relative to a control condition involving left-sided perceptual cuing. Subsequent manipulations revealed that such movements were effective even when the hand was out of sight, as long as it was located in left hemispace. Neither left-hand finger movements in right hemispace, nor *right*-hand movements in left hemispace significantly improved performances. Consistent with the premotor theory, these findings emphasize the important interaction between processes involved in spatial representation and action. In another study involving left unilateral neglect patients, Robertson et al. (1992) found that a 10-hour training program requiring motor activation of the contralesional limb in left hemispace effected lasting improvements on both clinical tests and in ADLs.

In another recent study, Làdavas, Menghini, and Umiltà (1994) examined a group of RH-damaged patients in whom the severity of left unilateral neglect had remained stable for some time prior to treatment. These workers based their intervention on two theoretical principles: (1) that patients with unilateral neglect arising from parietal lesions have an underlying deficit in disengaging attention from an ipsilesional location in preparation for a contralesional shift (Posner et al., 1984, 1987), and (2) that the disengage deficit in such patients is due to a failure of *automatic* attentional orienting, but that voluntary orienting is normal provided the frontal lobes remain intact. Patients received 30 hr of treatment, which consisted either of general therapy, *overt* (i.e., with eye movements) attentional retraining, or *covert* (i.e., without eye movements) attentional retraining. Patients in the latter two groups were trained to shift attention leftward or rightward to peripherally located agents in response to centrally presented valid or invalid precues. Following treatment, patients in *both* experimental groups made significantly fewer left-sided errors on visual cancellation tests, whereas those in the control group showed no such improvement. Because the same degree of improvement was found in both overt and covert orienting groups, this study demonstrated clearly that a deficit in attentional shifting, as opposed to visual scanning per se, is an important factor underlying unilateral neglect. This study also provided compelling evidence in support of our earlier suggestion that treatments aimed at the underlying *causes* of unilateral neglect hold the most promise for lasting therapeutic gains.

Summary and Conclusions

We have seen that unilateral neglect is a heterogeneous disorder affecting perceptual, attentional, and motor processes, with manifestations characterized by their distinctly lateralized quality. Clinically, the disorder manifests as a failure to orient toward or respond to stimuli in the contralesional side of space, coupled with a tendency preferentially to orient toward stimuli in the ipsilesional side of space. There are several other disorders that seem to have at least a superficial relationship to unilateral neglect: *Extinction* involves a failure to respond to one of two horizontally adjacent stimuli delivered simultaneously, despite normal responses to either stimulus presented in isolation. Like unilateral neglect, the phenomenon of extinction is mediated by the relative timing, perceptual attributes, and probability of occurrence of the relevant stimuli. In contrast, *allochiria* involves mislocalization of stimuli from the contralesional to the ipsilesional side, implying relatively preserved perception of contralesional stimuli in the context of an impaired ability to correctly localize such stimuli. *Anosognosia, anosodiaphoria, somatoparaphrenia,* and *misoplegia* all involve misperception or misrepresentation of contralesional body parts. Although dissociations between these impairments are common, at least one conceptual framework considers them as different manifestations of a common underlying disorder (Bisiach & Berti, 1987).

Unilateral neglect typically occurs with pathologies of abrupt onset such as acute hemorrhage or ischemia. It is more frequent, severe, and persistent after RH than LH damage; in the former, the most commonly affected region is the inferior parietal lobe, though the disorder is often observed after lesions of other cortical and subcortical sites. Such observations imply that attentional processes are subserved by a network of brain regions involved in the elaboration and integration of information from each sensory modality; at the highest level, the inferior parietal lobule, prefrontal, and cingulate cortices are likely to play a key role in such operations. Intentional processes are subserved by neural circuits that encompass the SMA, BG, thalamus, and prefrontal cortex. The notion that several interconnected networks control attention and movement in space may help to explain the heterogeneous manifestations of unilateral neglect; thus, depending upon the locus of damage, patients may show predominantly perceptual, exploratory-motor, or even motivational deficits.

Although the disorder has traditionally been considered to reflect an impairment either of representational or of attentional mechanisms, this may be a false dichotomy; both representation and attention are likely to play a role in spatial cognition, and disruption of either may result in the distinct lateralized impairments found in patients with unilateral neglect. To the extent that space is coded in purely egocentric coordinates, unilateral neglect may arise because the contralesional hemispace is deprived of attentional and intentional resources. One reason the disorder is more frequent, severe, and longer lasting after RH than LH damage may be because the RH mediates such processes in both hemispaces, whereas the LH does so only in the contralateral hemispace. However, space is coded not only in body-centered, but also in head-, environment-, and even object-centered coordinates; indeed, unilateral neglect has been observed in all of these coordinate frames.

Perhaps, therefore, the disorder is best conceptualized in terms of an attentional gradient whose spatial distribution changes as a function of specific task demands. According to one conceptualization, the two cerebral hemispheres are mutually inhibitory, each controlling contraversive attentional shifts in the azimuthal plane, with the rightward (LH) vector being stronger than the leftward (RH) vector. RH damage releases the opposing LH from inhibition, thereby inducing a strong rightward orienting bias and left unilateral neglect. Release of the weaker RH from inhibition after LH damage induces only a relatively weak leftward bias, which is unlikely to produce clinically apparent unilateral neglect.

Further asymmetries in hemispheric functioning may help to explain some of the diverse manifestations of unilateral neglect. Arousal and vigilance are mediated predominantly by the RH; thus, patients with left unilateral neglect after RH damage typically exhibit generalized arousal deficits in addition to a distinctly lateralized impairment. Mechanisms underlying the perception of global and local elements in a display are subserved by the RH and LH, respectively. RH damage may therefore induce a breakdown of global processing, and an overemphasis upon local processing of ipsilesional stimuli, whereas LH damage is likely to have the opposite effect. Finally, the two hemispheres may contribute differently to space- and object-based attention; whereas damage to the RH disrupts processes involved in spatial selection, damage to the LH disrupts both space- and object-based mechanisms.

We also noted in this chapter that attentional and motor processes are closely (perhaps inextricably) linked. Indeed, an organism's attentional capacity is not arbitrarily defined; rather, it is determined by demands for the coherent control of action. Another of the models proposed to account for unilateral neglect suggests that brain areas responsible for encoding representations of different regions of space are also concerned with coordinating attention and movement within those regions. Thus, unilateral neglect may dissociate according to which space sectors (personal, peripersonal, or extrapersonal) are affected. Moreover, rehabilitative techniques may use movement as an effective means of ameliorating lateralized attentional deficits. On the other hand, patients with unilateral neglect also exhibit disordered limb and eye movements. These impairments may be restricted to the contralesional hemispace but more often involve contralesional movements directed from any spatial location. When reaching toward targets in front of the body, patients' trajectories are distorted ipsilesionally. In addition, patients may undershoot contralesionally located targets, and they may be slow to initiate and execute contralesionally directed movements.

Despite the wide range of deficits characterizing unilateral neglect, it is also clear that certain attributes of apparently neglected stimuli are nevertheless processed normally. There is evidence in some individuals of intact color perception and figure-ground segmentation of contralesional stimuli. The presentation of word and picture primes in the neglected visual field has also been found to facilitate responses to subsequent target stimuli presented on the ipsilesional side, in a manner that suggests preserved semantic knowledge of otherwise neglected stimuli. The confabulatory responses sometimes observed in such patients may also reflect partial semantic knowledge of unreported contralesional information. The extent to which apparently neglected information is processed differs between patients, probably as a function of such factors as the presence or absence of adequate sensation, and the locus and extent of brain damage.

In the final section of this chapter, we considered the factors underlying recovery from unilateral neglect, and the efficacy of rehabilitation. Although unilateral neglect has traditionally been viewed as a transient disorder it is clear that some patients, particularly those with cortical atrophy and with large RH lesions, may continue to suffer lateralized impairments for months and even years postinjury. Behavioral recovery is paralleled physiologically by restitution of normal levels of cerebral blood flow and metabolism in both the damaged and undamaged hemispheres. However, even in those individuals who do show clinical improvements, more sensitive tests frequently reveal residual impairments.

Temporary amelioration of unilateral neglect has been demonstrated by cuing patients to direct focal attention or limb movements toward the contralesional side. Maneuvers affecting the vestibular system, such as caloric stimulation, act transiently to recalibrate the egocentric coordinate system, though such an effect does not overcome object-based attentional deficits. Manipulating visual input by patching the ipsilesional eye, or by attaching Fresnel prisms to patients' eyeglasses, also provides only partial and transient amelioration of symptoms.

Pharmacological agents that target specific neurotransmitter systems may eventually be the treatment of choice in patients with unilateral neglect, as they are in other disorders of attention and movement, such as Alzheimer's disease, Huntington's disease, Parkinson's disease, and Tourette's syndrome (Chapters 9, 12–14). In the meantime, despite the early failures of interventions aimed at inducing permanent improvements in patients with unilateral neglect, recent techniques, involving contralesional limb movements and retraining of voluntary attentional orienting, have been successful in achieving long-term therapeutic gains.

Further Reading

Bisiach, E. (1993). The Twentieth Bartlett Memorial Lecture: Mental representation in unilateral neglect and related disorders. *Quarterly Journal of Experimental Psychology, 46A,* 435–461.

Bisiach, E., & Vallar, G. (1988). Hemineglect in humans. In F. Boller & J. Grafman (Eds.), *Handbook of neuropsychology* (Vol. 1, pp. 195–222). Amsterdam: Elsevier.

Driver, J., & Mattingley, J. B. (in press). Selective attention in humans: Normality and pathology. *Current Opinion in Neurobiology.*

Halligan, P. W., & Marshall, J. C. (1994). Toward a principled explanation of unilateral neglect. *Cognitive Neuropsychology, 11,* 167–206.

Halligan, P. W., & Marshall, J. C. (Eds.). (1994). Spatial neglect: Position papers on theory and practice. *Neuropsychological Rehabilitation, 4,* 97–240.

Heilman, K. M., Watson, R. T., & Valenstein, E. (1993). Neglect and related disorders. In K. M. Heilman and E. Valenstein (Eds.), *Clinical Neuropsychology* (3rd ed.) (pp. 279–336). New York: Oxford University Press.

Jeannerod, M. (Ed.). (1987). *Neurophysiological and Neuropsychological Aspects of Spatial Neglect.* Amsterdam: North-Holland.

Làdavas, E., Menghini, G., & Umiltà, C. (1994). A rehabilitation study of hemispatial neglect. *Cognitive Neuropsychology, 11,* 75–95.

Posner, M. I., & Petersen, S. E. (1990). The attention system of the human brain. *Annual Review of Neuroscience, 13,* 25–42.

Rafal, R. D. (1994). Neglect. *Current Opinion in Neurobiology, 4,* 231–236.

Rizzolatti, G., & Gallese, V. (1988). Mechanisms and theories of spatial neglect. In F. Boller & J. Grafman (Eds.), *Handbook of Neuropsychology* (Vol. 1, pp. 223–246). Amsterdam: Elsevier.

Robertson, I. H., & Marshall, J. C. (Eds.). (1993). *Unilateral Neglect: Clinical and Experimental Studies.* Hove, U.K.: Lawrence Erlbaum Associates.

Chapter 7

The Acallosal Syndrome

Commissurotomy Studies

Until now, we have discussed cognitive or behavioral deficits consequent upon accidental, traumatic, or pathological changes to localized brain regions. In this chapter we shall consider the effects of a unique surgical procedure, commissurotomy, usually undertaken to limit the spread of otherwise intractable epilepsy. The effects captured the imagination of the public after the publication of Sperry's dramatic observations in the 1960s (see e.g., Sperry, 1968) and the award to him of the Nobel Prize for Physiology and Medicine in 1981. Thus it seemed that each hemisphere could function as a largely independent unit, a mind with its own sensations, perceptions, cognitive processes, experiences, and memories. The impact was not restricted only to the brain sciences, but spread to psychiatry, philosophy, education, and popular culture, often with gross oversimplifications and exaggerated claims about how to release the latent powers of an underutilized and underappreciated hemisphere—the right.

There are in fact two important theoretical issues: hemispheric specialization and cerebral duality. With respect to the latter, there has been speculation for one and a half centuries on the possibly duplex nature of our consciousness, though it is only comparatively recently that the issue has become amenable to proper scientific study. Thus Wigan (1844) noted that each cerebrum is a distinct and perfect whole as an organ of thought, and that a separate and distinct process of thinking may be carried out in each hemisphere. Later Fechner argued that commissurotomy (surgical separation of the hemispheres by severing the commissures), were it possible, which he thought not, should lead to the duplication of a person (Zangwill, 1976); by the turn of the century, the skeptical McDougall bargained with the famous neurophysiologist Sherrington to undertake the operation to resolve the issue should he, McDougall, ever become incurably ill. The challenge was never taken up, and the issue is still partly unresolved; indeed, to the casual observer, commissurotomy patients seem surprisingly normal. Another seeming paradox was that while according to classical lesion studies one might have expected the surgically isolated left hemisphere (LH) to be totally normal with respect to language, and the right to be totally word deaf, word blind, and mute, this was not what actually emerged. Of course we now know, even from lesion studies, that the right hemisphere (RH) plays not an insignificant role in language, and that hemispheric asymmetries are quantitative, relative, and a matter of degree, rather than qualitative and absolute. Nevertheless we shall encounter intriguing differences between the findings from lesion studies and those from commissurotomy patients. Of course many language deficits with LH lesions may reflect not so much RH incompetence, but rather the maladaptive attempts by the damaged LH to maintain functional operation; with commissurotomy patients, the RH can be tested in *isolation* from the left (Nebes, 1990a). Thus we can now directly—in theory—test the relative competence of the two (putatively) intact hemispheres, rather than having to infer it from the effects of brain damage. Indeed we can now test *abilities* rather than *disabilities*. In practice, however, the findings are potentially confused by varying locations and degrees of preexisting morbidity, including epilepsy, which might have led to substantially abnormal brain reorganization, by varying ages at which the commissurotomy operation was undertaken, and when testing took place, and by varying amounts of extracallosal damage that can result from a major brain operation. Factors such as these may explain why phenomena, apparent in unilateral-lesion studies, and predictable in split-brain patients, often do not occur. Thus there is little evidence of unilateral neglect (see Chapter 6) in such patients, with apparently full awareness of both sides of space (Plourde & Sperry, 1982), and either hemisphere able to direct attention contraversively, even though the RH may be the more activated (Làdavas, Del Pesce, Mangun, & Gazzaniga, 1994); however, Heilman, Bowers, and Watson (1984) did find neglect in (visual or tactual) line bisection in a partial commissurotomy patient, with either hand erring towards its own hemispace. Likewise, prosopagnosia (see Chapter 5) seems to be absent in either disconnected hemisphere.

The Forebrain Commissures

There may be as many as 10^9 axons in the corpus callosum and the anterior commissure, the only two direct links between the neocortices of the two hemispheres, and the only pathway through which the higher functions of perception, cognition, learning, and voluntary motor coordination may be unified (Trevarthen, 1990). The corpus callosum on its own is the most prominent commissural structure in the cerebrum, and the largest fiber tract in the brain, with up to 2×10^8 axons interconnecting both homologous and nonhomologous cortical areas (Gordon, 1990). Like the anterior and hippocampal commissures (see

Figure 1), it originates from the midline of the rostral end of the neural tube adjacent to the area of closure of the anterior neuropore (Jeeves, 1990), and may therefore be vulnerable to midline defects during embryogenesis, leading to naturally occurring (congenital) instances of the acallosal syndrome (see below). The anterior commissure largely interconnects temporal cortices, and may, to a greater or less extent, be capable of substituting for the corpus callosum in the event of surgery or congenital absence of the corpus callosum (though see Seymour, Reuter-Lorenz, & Gazzaniga, 1994, who argued that the anterior commissure in humans, unlike the situation in other primates, does *not* transfer visual information). In the event of callosal agenesis, there may even be hypertrophy of the anterior commissure. The midbrain hippocampal commissure plays an uncertain role and cannot be sectioned. Gloor, Solanova, Olivier, and Quesney (1993) noted that although an enormous reduction in the ventral hippocampal commissure has occurred in human phylogeny, leading to its near or total disappearance, its dorsal regions are well developed, crossing the midline under the rostral part of the splenium and the caudal part of the body of the corpus callosum. There is evidence, moreover, from depth electrode electroencephalographic (EEG) recordings of temporal lobe seizures, of in some instances contralateral spread of activity between mesial temporal structures via the dorsal hippocampal commissure. Thus Gloor et al. concluded that, despite claims to

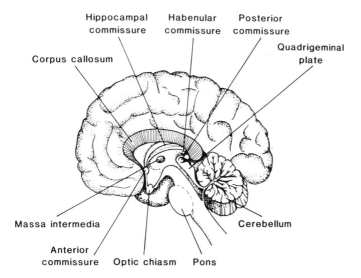

Figure 1 THE MAJOR COMMISSURAL SYSTEMS AND ASSOCIATED STRUCTURES. (FIGURE ADAPTED FROM AN ORIGINAL IN R. W. SPERRY (1964, JANUARY). THE GREAT CEREBRAL COMMISSURE. *SCIENTIFIC AMERICAN, 210*, 42–52. COPYRIGHT © 1964 BY SCIENTIFIC AMERICAN, INC. ALL RIGHTS RESERVED.)

the contrary, we possess a functional dorsal hippocampal commissure that may play a role in memory. Finally, some but not all individuals possess the interthalamic massa intermedia; it may be more common in females.

The corpus callosum, in an anterior-posterior direction, closely matches, topographically, the gross architecture of the cerebral cortex. The anterior third (genu) interconnects frontal structures and plays a largely motor function; the middle third (trunk) interconnects the parietal (somatosensory) and temporal regions, while the posterior third (splenium) interconnects occipital, visual regions (Pandya & Rosene, 1985). There is little or no interconnectivity between primary sensory areas, but homotopic and heterotopic connections are abundant between association cortices, suggesting that the corpus callosum plays a major role in integrating information at various levels of abstraction. There are wide individual variations, together with recurrent suggestions of sex or handedness or both differences in the relative sizes of different regions of the corpus callosum (Cowell, Kertesz, & Deneberg, 1993; Holloway, Anderson, Defendini, & Harper, 1993; Witelson, 1989), possibly accounting for the pervasive claims of sex and handedness differences in cognitive processes (Bradshaw, 1989); indeed similar sex differences are also apparent in the corpora callosa of rats (Bradshaw & Rogers, 1993).

Prior Animal Studies

Bykov (1925, cited in Trevarthen, 1990) showed that transfer of conditioned responses from tactile stimuli, from one flank of a dog to the other, was stopped by hemisphere disconnection. However, it should be noted that the commissures do not always provide a channel of the highest fidelity for interhemispheric transmission of information. Thus interpaw transfer of difficult tactile discriminations in intact cats is not perfect (Berlucchi, 1990), though such interpaw and intermanual transfer of shape and texture discriminations may be completely abolished after commissurotomy in cats and monkeys (Myers & Sperry, 1953; Sperry, 1961). Indeed in young children, whose commissures may be incompletely myelinated, there may also be incomplete intermanual transfer of various tactual discriminations (Burden, Bradshaw, Nettleton, & Wilson, 1985). Interocular transfer of visual learning may similarly be imperfect in normal, laterally eyed animals, where each eye projects to the contralateral hemisphere, and in medially eyed animals where the optic chiasm is sectioned so as to prevent bilateral ocular projection; interocular transfer is then abolished with commissural section in cats and monkeys (see Figure 2), except for simple, gross discriminations (e.g., of brightness and color), which may be mediated subcortically. Similarly, although both forepaws of commissurotomized cats can be directed to targets seen by only one hemisphere, nevertheless, fine coordination of the two hands of split-brain monkeys is typically impaired.

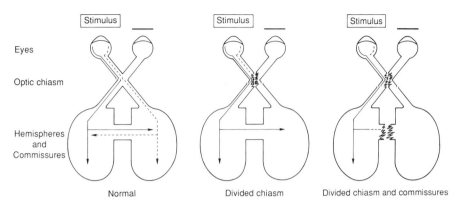

Figure 2 Flow of visual information, with monocular stimulation, in the normal intact animal, in one with a divided chiasm, and in one where both chiasm and commissures are divided. In the first two preparations both hemispheres can receive appropriate information from a single eye, either directly (Normal) or via the corpus callosum (Divided Chiasm). With division of both chiasm and commissures, only the ipsilateral hemisphere can receive input from a single eye. (From Bradshaw, 1989, p. 59.)

The Human Acallosal Syndrome

The earliest systematic studies of the effects of commissural section in humans are those of Akelaitis (1943). Very few behavioral deficits were observed, such that in jocular despair Lashley (1950) declared that the corpus callosum served only to keep the hemispheres from sagging. Testing procedures at the time were far from subtle, but more importantly we now know that, quite apart from the anterior commissure, enough of the splenium may have been spared to permit transfer of some visual information. As more of the splenium is preserved, disconnection symptoms become ever more subtle.

The Caltech series, with the operations performed by Bogen and Vogel (1962) and described by the Nobel laureate Sperry, dates from the 1960s (Gordon, 1990), and is notable for the ingenuity of many of the experimental designs. The classical picture resulted from, typically, complete commissurotomy, including the corpus callosum, anterior, and hippocampal commissures, and demonstrated a dramatic disconnection syndrome. An initial left-side apraxia to verbal commands such as "wiggle your toe" would quickly resolve, due perhaps to increasing ipsilateral motor control, or shifting inhibitory organization. Words or pictures flashed tachistoscopically to the right visual field, which projects to the left cerebral hemisphere, would be reported easily; this was not the case for such material presented to the left visual field (i.e., the RH), for which an expressive aphasia persisted. The left hand could correctly select such objects from within a bag, or even point to the correct name, if short. However, it could not name such objects as it palpated out of view—a left-hand anomia. It could nevertheless retrieve these objects when named or described verbally; the RH might be mute, but it is not word deaf, and can clearly understand simple spoken and even written instructions. One hand could not mimic an (unseen) gesture molded by the examiner on the other; nor indeed can callosally-immature young children (Burden et al., 1985). If the two hands separately felt unseen objects, the patient could not discriminate between them, except perhaps on the basis of any major physical differences (e.g., of weight or hardness, as between a brick and an orange), which might provide sufficient gross information for ipsilateral projection or subcortical mediation. Likewise, two simultaneously flashed pictures, one to each visual field or hemisphere, could not be discriminated, and only the ipsilateral hands could identify them from an array of alternatives by pointing. The right hand was superior for drawing fine, detailed featural information, and the left for the overall gestaltic interrelationships and general outline, in a manner analogous to the left–right dissociations noted (Chapter 10) in constructional apraxia. Similar hand differences appeared with respect to strategies adopted in tactually matching a viewed, unfolded, two-dimensional plan to an unseen, palpated, three-dimensional object; this led to the general conclusion that the RH has the synthetic ability to integrate unrelated parts into a meaningful whole, whereas the LH specializes in the analysis of component features (for review, see e.g., Bradshaw & Nettleton, 1981).

These conclusions were extended with the chimeric technique (Levy & Trevarthen, 1976) where the left half of one object and the right half of another were abutted at their respective midlines, upon which patients fixated when the stimuli were tachistoscopically flashed. They typically reported seeing only a single, complete object, one half of the stimulus being suppressed, with completion across the midline of the other, "seen," object, in a manner similar to perceptual completion in a migrainous scotoma or the blind spot. After viewing such chimeric material, subjects either reported what they had seen, or pointed to the "correct" item among an array of alternatives. Items corresponding to the chimeric half in the left visual field were typically preferred, unless verbal encoding or a spoken response was required (e.g., choose an item whose name rhymes with the target, or name the target). Then the right half was selected. Generally, the right half predominated when phonological aspects were stressed, and the left half when physical similarity was the important criterion. Levy and Trevarthen concluded that each hemisphere has its own cognitive strategy, the RH synthesizing over space, the LH analyzing over time, the RH noting visual similarities, the LH conceptual similarities, the RH perceiving form, the LH detail, the RH lacking a phonological analyzer, the LH a gestalt synthesizer. Parallels with the distinction between apperceptive and associative agnosia are clear; indeed Levy and Trevarthen showed similar dissociations with nonchimeric testing of their commissurotomized patients—the LH matching by function, and the RH by appearance. However, we would caution that hemispheric specialization is relative rather than absolute, a matter of quantitative degree rather than qualitative difference. Indeed interactive processes between hemispheres in the normal undivided brain may further modulate such strategic differences, in some cases reducing them, at other times enhancing them (Bradshaw, 1989; Zaidel, Clarke, & Suyenobu, 1990). Moreover, as we shall shortly see, even in the divided brain, undivided subcortical processes may provide for a surprising degree of interhemispheric cross-talk.

Zaidel's "Z" Lens Technique of Prolonged Visual Lateralization

Zaidel (1990b) in the 1970s attempted to break free from the constraints of brief, eccentric visual presentations. Using a specially designed scleral contact lens system, he devised a method permitting free scanning within a single visual field–hemisphere system. In confirmation of the earlier tachistoscopic studies, he found that the LH was better at the feature-analytic task of figure disembedding, and that the right could read and even write (e.g., by moving cut-out letters) as long as the material was short and syntactically simple. The right, however, was phonologically deficient, with very limited short-term memory and an inability to rehearse. It could nevertheless recognize a large auditory vocabulary, one which in fact was larger than one in the visual modality. It had problems with rhymes and homophones and seemed to access written meaning directly rather than by grapheme–phoneme conversion. It performed best at a semantic-processing level, but its associations were looser, "longer," and more connotative (rather then denotative) in comparison to those of the LH. Its language profile did not correspond to any one stage in first-language developmental acquisition, nor to any specific pattern of LH aphasia. Thus Zaidel noted the abnormality of a pattern of muteness (apart from some clichés, automatisms, and oaths), with substantial auditory comprehension and of a smaller visual than an auditory vocabulary. RH language may approximate closest to (anterior) aphasic alexia, though its muteness matches global aphasia. Its reading resembles that of deep dyslexia (see Chapter 3), as both demonstrate semantic paraphasia, better concrete than abstract word reading, and both lack grapheme-to-phoneme conversion. However, deep dyslexics, though very possibly employing their RH (Coltheart, 1985), are far less impaired than the disconnected RH. Nevertheless, the RH, according to Zaidel, takes turns and responds (nonverbally) fairly appropriately in conversation, though with some deficits at a pragmatic level, and is very passive, rarely generating any spontaneous behavior, linguistic or otherwise.

Often, during interrogation of the RH, the LH would attempt to take over and complete the task in a confabulatory fashion, unless otherwise occupied by a simultaneous concurrent task. Thus, the left hand may commence (correctly) writing PEN (for *pencil*, shown to the left field), and the right hand reaches across and wrongly, though plausibly, completes as PEN . . . NY. Another example of such confabulation, though this time from conventional tachistoscopic presentation by Gazzaniga (1988), involved flashing a picture of a chicken claw to the LH (via the right field), and a snow scene to the RH; the patient had to choose related pictures (from an array of alternatives) one by each hand. The left hand chose a shovel and the right a chicken. The verbal LH, oblivious of the snow scene, when asked why, replied that the chicken claw goes with the chicken, and you need a shovel to clean out the chicken shed. Generally, Zaidel believes that the RH shows characteristically human levels and modes of consciousness, and has the same desires, ambitions, morality, beliefs, and knowledge base (apart from language and presumably computation) as the verbal left, when appropriately interrogated (though cf. Cazzaniga, 1985).

Subcortical Transfer of Information across the Divided Forebrain?

Even in the early, "classical," period of split-brain studies, Trevarthen and Sperry (1973) reported that commissurotomized patients were accurate at judging the relative motion or visual alignment of stimuli simultaneously presented in opposite visual fields. The technique employed prolonged, peripheral viewing with maintained central fixation, and Trevarthen and Sperry appealed to midbrain (collicular) mechanisms sensitive to orientation in space and to peripheral rather than central (foveal) vision. Such attentional and spatial retinotectal pathways, unlike the high-resolution retinogeniculostriate form detectors, were thought to be unaffected by the commissurotomy operation. (Recent studies, e.g., by Mishkin, Ungerleider, & Macko, 1983, have led to revisions in the neuroanatomy.) Indeed, apparent motion between two alternatively flashing dots was still reported by such patients, even when the dots lay in opposite visual fields. Later research confirms such conclusions; Holtzman (1984) reported that patients could direct eye movements to specific target locations in one visual field on the basis of a cue flashed to the other. Sergent (1987) noted that patients could determine the alignment of arrows flashed in opposite visual fields, or whether or not an arrow in one field pointed to a target in the other, or if dots flashed in opposite visual fields summed to an odd or even total, or if a pair of digits in opposite fields summed to a value less or greater than 10. Sergent (1990) also noted that although patients were poor at judging whether or not digits in opposite fields were identical, they could decide which was the smaller and could estimate the relative orientation, location, and size of objects across the midline. Ramachandran, Cronin-Golomb, and Myers (1986) reported that patients could discriminate leftward from rightward apparent motion across the visual fields, and Lambert (1991) found that the semantic category of a word in the left field influenced (i.e., "primed") the processing of words flashed to the right for categorization, an observation to which we shall shortly return.

Response speeds in these tasks were generally too fast for sophisticated or automatic cross-cuing strategies to account for the apparent transfer of information between the divided hemispheres. It had earlier been noted (Gazzaniga & Hillyard, 1971) that patients could name numbers flashed to the left visual field, though the response time increased with the size of the digit flashed, something that did not occur with right-field presentations. It seemed that the subject engaged in silent counting from 0 to 9, until coming to the number that somehow "stuck out." Thus the LH presumably emitted subvocal signals to the RH, which signaled the left to stop and respond on reaching the correct number—another instance of covert processing. Similarly, if a flash could occur in one of four quadrants of a circle, whose center was fixated by the subject, the latter could say where (top or bottom) a left-field flash had occurred by monitoring subsequent eye movements. Again, also as long as the ensemble of possible events was known in advance, a patient could say "sphere," "cube," or "pyramid" in response to such pictures flashed to the left field or RH. The patient apparently did this by looking at a wall clock (if sphere), or a door (if cube), and by responding "pyramid" in default of either of the other two events, a faculty that would be lost in an environment where such cues were absent. Moreover, it has frequently been reported that, during verbal responding, the LH may apparently detect frowns initiated by the RH at an incorrect answer, and utter a correction.

As mentioned above, however, response times may sometimes be too fast to permit explanations in terms of sophisticated cross-cuing strategies. It soon became apparent, moreover, that emotional information could somehow cross from the "isolated" RH to the LH, coloring the latter's speech, even though the LH was unaware of what the RH had experienced (Sperry, Zaidel, & Zaidel, 1979). Generally, contextual, associative, or connotative information, rather than categorical details, seemed to cross (Cronin-Golomb, 1986; Myers & Sperry, 1985). However Sidtis (1985) reported that the LH could correctly "guess," in terms of broad, conceptual categories, information otherwise available only to the right. This was particularly true of partial commissurotomies, the patient seeming almost to play a game of 20 Questions with himself, until the correct answer "popped out." This *conduite d'approche* phenomenon is of course reminiscent of the groping toward the correct meaning that can occur in deep dyslexia (see Chapter 3). Later, evidence appeared of information transfer, in the absence of cross cuing, even in cases of total commissurotomy. Thus surprise stimuli (e.g., a letter during digit naming) could be reported in the left visual field, and colors and even pictures could be matched across visual fields (Cronin-Golomb, 1986; Sergent, 1986). Interhemispheric semantic priming also proved possible in the absence of the forebrain commissures (Sidtis, 1985), with facilitation of a lexical decision to the second word in the right visual field, when the first word, flashed to the opposite field, was of the same semantic class. Similarly, Lambert (1993) found that patients, like normals, were subject to *negative* priming, where responding is slower when a (right-field) attended target is identical to or related to a previously ignored distractor in the opposite (left) visual field. Again, left-field, RH information may somehow influence processing of attended right-field, LH words, and one is reminded of the tacit or implicit knowledge that is apparently preserved in so many clinical cases of alexia, agnosia, or amnesia. Clearly, cross-cuing cannot be invoked under these circumstances. The forebrain commissures may normally be necessary to transfer detailed, complex information at a conscious, explicit level, while subcortical systems may mediate and transmit less detailed, more general, and connotative information at an unconscious, implicit level (and see Sergent, 1986, 1987). The apparent lack of a disconnection syndrome in the coherent, everyday behavior of such patients outside the laboratory in any case contrasts with the traditional (and far more contrived) laboratory demonstrations of mental disunity at the perceptual level. However, careful observation of everyday behavior reveals problems in programming precise, synchronous, or consecutive movements of the hands, a memory deficit with poor interhemispheric cooperation in retrieval and maybe even storage, and an impoverishment of verbal and emotional experience (Zaidel, 1990b). There is also a failure to sustain adequate comprehension of an extended text during reading, conversational deficits, deficits of pragmatics rather than of phonology and syntax, conversational and social inappropriateness, an exaggerated politeness, a tendency to confabulate, and defective appreciation of prosody and metaphor. Such characteristics are suggestive of RH damage or inaccessibility (see Chapter 4), but Zaidel noted that an appreciation of humor, a traditional RH perquisite, is maintained.

At this point we should note Gazzaniga's observation (e.g., Seymour et al., 1994) that although the early classical studies found what they set out to (i.e., hemispheric isolation in perception and information processing), later, largely "West Coast" studies set out with the opposite goal, one of reunification. Seymour et al. (1994) claimed that in the end all such latter studies derive from only two patients, one of whom may not have had complete commissural section, and that commissurotomy patients cannot compare stimuli across the divided hemispheres, except perhaps for very gross visual forms that may be mediated by the superior colliculi, or where there are some residual fibers. They even suggest that *were* there any evidence of an ability, for example, to judge which of two digits in opposite visual fields were the larger, it could be because each hemisphere, independently and without any knowledge of material presented to the other, has a predisposition to respond that is determined by the magnitude of the stimulus presented to it; the hemisphere more disposed to respond then does so. Such an apparent integration then would depend upon hemispheric differences in response readiness, whereby one hemisphere's motor command comes to dominate that of the other, *without any actual transfer of stimulus information.* Seymour et al. (1994) therefore concluded in favor of the traditional view that the corpus callosum is the major route for interhemispheric transfer and integration. We believe that this dispute between the "traditional East Coast" and the "West Coast challenge" will not be readily resolved.

Independent Processors, Wills, and Streams of Consciousness?

The classical disconnection picture is that such patients possess no increase in processing capacity, by virtue of the possession now of two independent processors, except maybe in certain highly artificial circumstances where automatic responding is possible. Thus patients, unlike controls, may show no response-time increment in performing a given task via one visual-field or hemisphere system if at the same time they undertake another task concurrently via the other system; nevertheless, they are much slower overall than controls in performing the single baseline task. This situation of underadditivity is of course curiously reminiscent of what occurs in Parkinson's disease (PD) (Chapter 12). Parkinson's patients, unlike controls, are little slower when performing two tasks concurrently than one alone; they are also little slower in a choice than in a simple reaction time (RT) task. They are indeed already much slower than controls in the baseline conditions of single tasks and simple RT, as if these situations already demand all their processing resources. Perhaps commissurotomy patients, unlike controls, cannot rapidly alternate, "at will," between hemispheres, or, conversely maintain prolonged activation of a single hemisphere. Gazzaniga (1987) however argued for some independent hemispheric processing capacity, on the basis of superiority (over controls) in a memory task requiring visual retention, when the information is distributed between the two hemispheres. Thus with a complex visual array spread across the two fields, controls tend detrimentally to combine it, although patients to their advantage can separately address each half. This is not of course an increase in total capacity, merely an artificial breakdown into two more efficiently manageable subsystems.

Are there two parallel, independent streams of consciousness? If consciousness corresponds to the monitoring of a chosen, selected action sequence performed at a nonautomatic level, given that patients normally can only undertake a single such coherent sequence, any other sequence being fragmentary and automatic, we may wish to conclude in favor of a single consciousness stream (Bradshaw, 1989). We are all of us of course capable of perceptually processing multiple sensory inputs, but again only one such input ever ultimately reaches conscious awareness. Commissurotomy patients may only behave *as if* they have two minds under extremely contrived laboratory conditions, despite the situations (above) where the right hand seeks to complete what the left is writing, and the largely apocryphal stories of one hand trying to interfere with or restrain the other. However, there are occasional reports of such an alien-hand syndrome (Jason & Pajurkova, 1992) after (usually anterior) callosal damage or commissurotomy, though it is not common or evident in most patients, most of the time. Usually under such circumstances the left hand (in dextrals) is involved, with "unwanted" or "unintended" hand and arm movements that nevertheless seem directed to some purpose. The alien hand seems to act autonomously, carrying out complex movements against the subject's verbally reported will, which can interfere with the development of an intended action begun by the other hand. The patient is fully aware of the abnormal, unpleasant, and frightening behavior of the alien hand, but cannot inhibit it; it seems to operate according to its own independent program or will. The "normal" (usually right) hand tries to restrain the alien (usually left) hand. It is worth noting, incidentally, that such callosal damage, together with additional premotor or supplementary motor area (SMA) injury, can lead to synkinesis (Armatas, Summers, & Bradshaw, 1994), whereby the patient, in making a movement with one upper limb, cannot inhibit simultaneous mirror-symmetrical responses in the other. In this context the recent report of Trojano, Crisci, Lanzillo, Elefante, and Caruso (1993) is of some interest. Their patient exhibited alien-hand effects after an ischemic lesion localized to the mesial frontal region and involving the SMA and anterior fibers of the corpus callosum. Compulsive, unwanted groping, grasping, and interference by the left hand, and a feeling of *main étrangère,* gave way to mirror writing of numbers. This evolution, the authors suggest, could reflect initial functional rearrangement of intrahemispheric motor pathways and then of callosal interhemispheric transfer, as mirror movements probably relate to nontransformed motor programs generated in the contralateral hemisphere. They noted that an acute, fleeting, clinical presentation of the alien-hand syndrome may result from (anterior?) callosal lesions on their own, whereas a more chronic and apparent manifestation requires additional frontomesial damage.

According to Puccetti (1985), even normal people must have two ongoing streams of consciousness; thus if split-brain patients have two (which we would anyway dispute), and hemispherectomies (where one hemisphere, or, strictly, one cerebral cortex has been surgically removed) have only one, then in normals there must have been two to start with, the commissurotomy operation merely aiding the demonstration of a preexisting duality. The RH must therefore know the existence of the LH, hear it talking, and try to correct it, even though callosal inhibition may normally conceal, from each hemisphere, introspective access to the conscious contents of the other. Cross-cuing phenomena certainly suggest independence of thought, but that thought need be no more conscious than our own self-corrections in everyday life. In a real sense we are all of us a social mosaic of modular minds, a vector of conflicting drives, feelings, and wills from a multiplicity of processors all operating in parallel, even though unifying attentional mechanisms may normally select a single response. Indeed it has long been recognized that our information-processing attentional bottleneck occurs late at the level of response selection (which we would equate with consciousness), rather than at an early perceptual level. We only become conscious of those action sequences that are finally selected and monitored, requiring executive overview, and cannot be run off automatically and in parallel with others. In any case, we believe that Puccetti's logic itself is faulty; we can tear up a hologram into successively smaller fragments, each of which preserves ever more degraded replicas of the original scene; yet these replicas surely were not present in the original hologram, merely awaiting release by a destructive analog of commissurotomy.

The Congenital Acallosal Syndrome

Occasionally, individuals may be born with partial or complete absence of the corpus callosum. More and more instances of such "experiments of nature" are emerging with the development of new imaging techniques (Jeeves, 1990). Ten percent of such cases are quite asymptomatic, coming to light by chance at necropsy, and others being recognized because of other neurological problems. Indeed any symptoms in life may be more likely to be due to other, associated, central nervous system anomalies, rather than the absence itself of the corpus callosum. Retardation need not occur; indeed the condition is even compatible with a high IQ, though very often there may be a gross discrepancy, in either direction, between verbal and performance IQ. Epilepsy and learning difficulties or behavioral disturbances are common reasons for presenting.

Callosal deficits in such cases vary in degree, depending on the time after conception that the presumed causative insult occurred. The splenium or rostrum are more likely to be absent, and the anterior commissure is usually still present, and is often hypertrophied due to the enclosure of heterotopic callosal fibers. The anterior commissure in fact is important in the extensive functional reorganization (see below) that seems to occur in such cases.

Jeeves (1990) noted that it is of interest to ask the following questions:

1. Can such individuals integrate information presented separately to the two hemispheres?
2. Does learning in one hemisphere transfer to the other?
3. Is the extent and nature of lateralization normal?

The general finding is that congenital acallosals demonstrate few major perceptual changes, though there may be subtle deficits in refined, voluntary bimanual control. This dissociation may reflect the earlier fetal growth of the anterior compared to the posterior callosal axons; consequently, the later developing splenium may remain plastic for a longer period, so allowing for bilateral, cortical reorganization.

At a detailed level of analysis, Jeeves (1990) reported that the intermanual matching of objects, shapes, and letters is intact, if slowed. Unlike commissurotomy patients, congenital acallosals can name such materials palpated by the left hand. Interhemispheric visual transfer time is greatly increased, and values vary with stimulus intensity, a phenomenon not found with surgical patients or controls. Congenital acallosals again are slower in cross-matching color and form, and have problems (which they hotly deny!) in integrating chimeric faces. However, unlike commissurotomy patients, they do not point to the choice alternative corresponding to the left half of the chimera, and name the item corresponding to the right half. They have problems with bimanual coordination, exhibiting a tendency toward synkinesis. They show loss of anticipatory grasp and prehension responses during reaching, due perhaps to loss of inhibition of competing ipsilateral pathways; thus uninhibited, uncrossed output may maintain control of arm movement to the detriment of the crossed output necessary for normal grasping responses by the fingers. Transfer of tactile learning may be impaired, and there may be subtle language or spatial deficits, though these could stem from extracallosal damage. Hemispheric specialization, however, appears grossly normal, with respect to field, ear, and hand asymmetries. Jeeves noted that with respect to possible compensatory mechanisms, bilateral representation of function seems unlikely, anterior commissural and subcortical connections may possibly be standing in for defective callosal pathways, ipsilateral pathways may now be employed, and subtle behavioral strategies such as visual scanning may also be adopted.

What then do the commissures actually *do*? They are clearly not indispensable, but may play the following roles (Jeeves, 1990):

1. Interhemispheric integration, updating, and information transfer
2. Inhibition of ipsilateral pathways to avoid unnecessary competition
3. Inhibition during development so as to facilitate hemispheric specialization
4. Regulation and switching of attention and arousal levels
5. Facilitation of information processing not only at an *interhemispheric* but even at an *intrahemispheric* level (Lassonde, Sauerwein, McCabe, Laurencelle, & Geoffroy, 1988). Indeed, commissurotomy does not just stop seizure activity spreading to the *other* hemisphere, but also reduces it at the *initial* focus.

Case History: Alien-Hand Syndrome

Jason and Pajurkova (1992) provided an interesting illustration of conflict in motor control in a 41-year-old male who had suffered a hemorrhage of an aneurism at the junction of the right pericallosal and callosal marginal arteries, with invasion of blood into the genu and body of the corpus callosum and the inferomedial frontal lobes:

This may be illustrated by the patient's behaviour when he was given a piece of paper and an envelope, with instructions to fold the paper and put it in the envelope. When he was presented with this task 13 days post-onset of lesion of the corpus callosum and inferomedial frontal lobes, the right hand began trying to fold the paper, but the left hand held the paper down and pulled at it. The two hands each pulled at the paper, trying to take it from the other hand. The left hand cooperated in folding the paper one time, but then pulled the paper away from the right hand and crumpled it. When asked at that point what he was doing he replied, "I've got to fold it one more time yet." He ended up by dropping the paper onto the floor with his left hand, 88 sec after the initial command. Greater cooperation between the hands was observed 26 days post-onset, but the two hands did continue to struggle with each other. Fifty-seven days post-onset, he folded the paper twice so that it would fit into the envelope, exhibiting great difficulty and fumbling, but no real conflict. The left hand then pushed the paper away, and the right hand retrieved it. He finally succeeded in completing this task, and was able to seal the envelope with relatively good cooperation of the two hands.

Further examples of conflict between the two hands will be given to illustrate the different ways in which this behaviour could be manifested.

Thirteen days after onset, the patient was holding a small rubber ball in his left hand, and was asked to throw it to the examiner. The right hand reached to take the ball, but the left hand would not let go. A struggle ensued between the two hands for 45 sec. The right hand finally succeeded in wresting the ball from the left, and then he threw the ball to the examiner.

Fifty-seven days post onset, the patient was sitting at a table and was asked to stand up. He started to stand up, pushing up on the chair with the right hand, but was impeded because the left hand continued to hold onto the chair. He stood there like that for a few seconds, before the left hand finally let go, 8 sec after the initial command. Asked to move the chair out of the way, he replied, "Sure," and began to push the chair away with the right hand. At the same time, however, the left hand pulled it back. The left hand succeeded in pulling the chair back and the patient with some hesitation then sat down. He looked upwards in apparent frustration and annoyance. He stood up, but then once more the two hands began pushing the chair in opposite ways. He said, "My right hand wants to go one way, my left hand will want to grab it the other way." He still had not succeeded when the examiner finally intervened, 36 sec after the initial command.

On several occasions the patient was asked to clap his hands. Even in early stages, there were times when he could perform this simple task well. For example, 12 days post-onset he clapped his hands correctly, with both hands moving in concert. At other times, however, he could not do this. Later in the same session, for example, only the right hand made clapping motions. In fact, the right hand moved as if clapping while the left hand was at the patient's face. Demonstration by the examiner did not help.

Fifty-seven days post-onset, the right hand assumed the position before the left. He made a few weak claps, with the left hand grabbing the right on each clap. After demonstration, he began to clap, but using his right hand only for the actions. Asked to clap with both hands, he commented, "the right one's doing all the work." He began to clap with both hands moving. He continued for a few seconds, whereupon the left hand moved to scratch his face while the right hand kept on and made a further clapping motion.

Apart from a slight rustle of clothing, on no occasion could we detect any sound of one hand clapping.

On the same day, while drawing a complex visual figure from memory with his left hand, he commented, "My left hand doesn't go where I want it to. . . . What I wanted was to make a straight line from there to here (pointing with left hand), and it doesn't want to cooperate." The examiner suggested that he try letting it do what it wanted to do. He replied, "Oh, I've tried to do that but it doesn't work. I've tried that before on my own. It just doesn't want to. It does all kinds of goofy things."

Fifty-three days post-onset, at one point during the Picture Completion subtest of the WAIS [Wechsler Adult Intelligence Scale], the patient said that there was nothing missing in the picture and that it looked normal. At the same time, however, his left hand kept tapping at the drawing in the place corresponding to the missing item. The drawing was left in front of him, and only after one minute did he give the correct response verbally.

Further examples were observed during drawing and writing tasks. For instance, the patient was asked to draw a clock with the left hand 159 days postonset. While writing the numbers he exhibited considerable difficulty and made the comment, "I want to draw a straight line. My hands want to make circles." In fact, the number he was making required a straight line. He proceeded to draw a curve with the left hand, and then erased it with the right. Sixty-one days post-onset, the patient was printing with his left hand. When trying to print the word "CAT," he did so with great hesitation, and with tense trembling of the left hand as had been observed during the pointing task. Having completed the "C" and the first vertical stroke of the "A," the left hand moved over as if to print the next letter. At this time the hand muscles tensed and the hand trembled again. The patient said, "Why do you want to run over there for, you fool?" (talking to the left hand). He eventually printed two T's, giving the second T an extra vertical line to make it look like an elongated "Π".

Patient: "It seems late at night if I want to go to close the door, this hand wants to close it (moving right hand) and my left one wants to open it (moving left hand). No matter if it's half way shut or almost shut, I'll try and close it with this one (right), and this one here (left) wants to open it wide open again."
Examiner: "What is that *you* want to do when you're doing that?"
Patient: "Close the door!"
Examiner: "And your right hand will close it . . . "
Patient: "The left one opens it. . . . Even like going to put your pants on: my right pant leg I can put in good; the left one I have to lift my leg to put it—get it started in my pant leg, and after I get it half way in I want to pull it out. I don't know what for. This hand wants to do everything right (shows right hand); my right foots wants to do everything right. My left foot and my left hand want to do the opposite of what my right one does all the time."

"I wanted to go to bed. And I get to the top of the stairs, and the right side of my body would go around the corner and start downstairs. The left side would keep grabbing the door jamb. And there was no way in the world I could let go of that door jamb to start downstairs. It just wouldn't let go. So I'd have to walk all the way to the front entrance, make a big U-turn, come all the way back to the top of the stairs, and do the same thing. I'd done it about 12 times, and I just couldn't go down those stairs. No way in the world."

Examiner: "What happened if you tried to keep on walking in spite of the fact that the left hand was holding on?"

Patient: "It wouldn't work, it makes you stop. You actually have to stop" (Jason & Pajurkova, 1992, pp. 244–253).

Summary and Conclusions

For 150 years there was speculation on the psychological and cognitive effects of surgical separation of the two hemispheres by commissurotomy, but it was not until the 1960s that proper documentation was made. The findings have since been interpreted and oversimplified to fit many popular misconceptions. A recurring problem, moreover, has been the extent of inconsistencies between the traditional clinical and the commissurotomy data. However, although in theory we can now test *abilities* rather than *disabilities,* in practice long-standing preexisting morbidity, such as severe epilepsy, may cloud the issue.

The corpus callosum, which topographically matches the gross architecture of the cerebral cortex, is the most prominent commissural structure, though some of its functions, predominantly visual, can in its absence be taken over by the anterior commissure. The role of the other commissural systems is unclear. Animal studies had shown that commissurotomy could prevent the interhemispheric transmission of complex visual and tactual discriminations, though simpler, gross information may still be transferred presumably via subcortical pathways. The classical picture with human patients is of a dramatic disconnection syndrome, especially in the early postoperative stages. One side of the brain seemed literally not to "know" whatever was presented, visually or tactually, to the other, though often it was necessary to employ subtle and ingenious lateralizing methods, such as the divided visual field technique. Generally, it was found that the LH enjoyed major, if not exclusive, rights to verbal processing, and specialized in the analysis of component features, while the RH generally employed a synthetic strategy in integrating unrelated parts of a configuration into a meaningful whole. With the chimeric testing technique, the right half of a configuration generally predominated when functional or phonological aspects were important, and the left half when physical properties were critical. Parallels with associative and apperceptive agnosia were clear.

Zaidel's scleral contact lens technique permitted longer and more naturalistic interrogation of a single isolated hemisphere, and demonstrated the considerable, though nonstandard, linguistic capacities of the generally mute RH. The latter, moreover, was found to be "fully human," with essentially similar beliefs, knowledge base, and aspirations to its more communicative partner. However, these and other studies more recently have found a curious "spread" of information between the surgically divided hemispheres, presumably via subcortical pathways. Initially ascribed to sophisticated cross-cuing, the phenomena are reminiscent of the covert, or implicit, processing that is becoming apparent in so many other clinical contexts (e.g., acquired dyslexia, prosopagnosia, amnesia, etc.). Generally, what seems to transfer is emotional, contextual, associative, or connotative information, rather than categorical details.

Not surprisingly, the philosophical issues of independent processors, wills, or streams of consciousness in the two cerebral hemispheres have engaged popular interests. The issue is far from resolved, and in certain cases, usually highly artificial, and with certain patients there may be behavior *as if* there is coexisting competition, though not a state of mutual ignorance; nevertheless there is no evidence of an increase in independent processing capacity, and there are many external (and internal) factors that tend toward unification. Conversely, even normal individuals may be subject to multiple, simultaneous, competing inclinations, and in this respect may be no different from the commissurotomy patient. However, the argument, that if commissurotomy (or hemispherectomy) unmasks a preexisting duality, then there must be two individuals to begin with even in every normal person, is seen to be fallacious or at best trivial.

Individuals born without a corpus callosum (congenital acallosals) permit us to study the developmental consequences of such abnormal organization. Many are grossly asymptomatic, and even more subtle techniques than with surgical commissurotomy, may be necessary to demonstrate behavioral anomalies. However, studies of individuals naturally or surgically acallosal reveal that the commissures, although not indispensable, are important in interhemispheric integration, regulation, and inhibition, as well as in the interhemispheric transfer of information. Given the enormous relative size of the corpus callosum, it must play an important role in primate evolution.

Further Reading

Berlucchi, G. (1990). Commissurotomy studies in animals. In F. Boller & J. Grafman (Eds.), *Handbook of neuropsychology* (Vol. 4, pp. 9–47). Amsterdam: Elsevier.

Gordon, H. W. (1990). Neuropsychological sequelae of partial commissurotomy. In F. Boller & J. Grafman (Eds.), *Handbook of neuropsychology* (Vol. 4, pp. 85–97). Amsterdam: Elsevier.

Jeeves, M. A. (1990). Agenesis of the corpus callosum. In F. Boller & J. Grafman (Eds.), *Handbook of neuropsychology* (Vol. 4, pp. 99–114). Amsterdam: Elsevier.

Nebes, R. D. (1990). The commissurotomized brain: Introduction. In F. Boller & J.

Grafman (Eds.), *Handbook of neuropsychology* (Vol. 4, pp. 3–8). Amsterdam: Elsevier.

Trevarthen, C. (1990). Integrative functions of the cerebral commissures. In F. Boller & J. Grafman (Eds.), *Handbook of neuropsychology* (Vol. 4, pp. 49–83). Amsterdam: Elsevier.

Zaidel, E. (1990). Language functions in the two hemispheres following complete cerebral commissurotomy and hemispherectomy. In F. Boller & J. Grafman (Eds.), *Handbook of neuropsychology* (Vol. 4, pp. 115–150). Amsterdam: Elsevier.

Disorders of Memory

The Amnesias

*H*ow is it that, at 80, we may still remember what we did at 8? How are memories *formed*—by local changes at the synapse, or cell nucleus, or both, or more globally in terms of network reorganization? Are these two really alternatives, or merely different explanatory levels in a processing hierarchy? Do we have different mechanisms of memory and forgetting for the ability to ride a bicycle (procedural memory), to remember a new telephone number long enough to place the call (working memory), to spell *ceiling* (semantic memory), to name the capital of Portugal (general knowledge) and to recall what happened at a wedding (autobiographical) (McKenna & Warrington, 1993)? We still do not know the answers to these fundamental questions (Rose, 1992). Indeed, our ignorance similarly extends to the nature and causes of clinical loss of memory in humans.

Parkin and Leng (1993) noted that loss of memory—amnesia—may stem from closed or penetrating head injury, cerebral infection, subarachnoid hemorrhage, hypoglycemia, hypoxia, tumor, metabolic disorders, and other causes. They define the amnesic syndrome as a permanent, stable, global disorder of memory due to organic (rather than psychogenic) brain dysfunction, which occurs in the absence of any other extensive perceptual or cognitive disturbance. (A syndrome is a set of symptoms that persistently co-occur and are therefore probably due to breakdown of a common underlying system.) Although it is not clear that there is a *single* amnesic syndrome—and indeed we shall describe several systems that seem to make their own characteristic contribution to memory—Parkin and Leng claimed that certain commonalities occur in amnesia, constituting what they call the amnesic syndrome:

1. An intact short-term (STM), working, or immediate memory, as measured by the digit span

2. Intact semantic memory and other intellectual functions
3. Severe, permanent anterograde amnesia (i.e., for acquiring new information), both with recall and recognition tests
4. Some degree of retrograde amnesia (i.e., loss or inaccessibility of previously learned material)
5. Intact procedural memory, for learning or retrieving skills
6. Damage to some parts of the mesial temporal lobes (especially hippocampus), and diencephalon (especially dorsomedial thalamic nucleus and mamillary bodies), which are all parts of the limbic system.

In this chapter we shall address these functional and structural aspects of memory, but note that controversy dogs any systematic approach to amnesia, to a greater degree than perhaps in any other neuropsychological domain. Disputes continue at both functional and structural levels. Indeed at neither a macro- nor a micro- (neuronal or synaptic) level do we yet know how memories are established and maintained, though a neuronal mechanism known as long-term potentiation (LTP) is believed to play a crucial role. The effect occurs when repeated transmissions of impulses across the synapse linking two neurons result in a positive feedback effect, making future transmissions easier. For LTP to occur, a signal, the retrograde messenger, must probably be sent back from the postsynaptic neuron to strengthen the connection. Horgan (1994) reviewed evidence that the retrograde messenger may be nitric oxide, a soluble and highly reactive gas that rapidly combines or disappears, and which may therefore also affect *adjacent* neurons and synapses. Indeed, nitric oxide might account for certain other disparate phenomena. As discussed in Chapter 5, neurons gather into columnar structures and layers in certain cortical regions; the width of these columns and layers is approximately that of the effective range of a diffusing nitric oxide signal. Moreover, excitotoxicity involving the neurotransmitter glutamate has been hypothesized to account for the massive neuronal cell death that can occur with oxygen deprivation, and even in Huntington's disease (HD) (Chapter 13). It seems probable that nitric oxide might mediate such cell death, as timely administration of nitric oxide inhibitors is effective in reducing stroke damage.

Twenty-five hundred years ago in Athens, Plato distinguished between failures in

1. *Registration,* which he likened to putting a bird into a cage.
2. *Retention,* where the bird has to be kept there, alive.
3. *Retrieval,* where the bird must be successfully caught again.

These distinctions are still fundamental, and the first two map closely onto the following:

1. *Learning,* the process of acquisition, which may be measured by the rate of recall improvement as a function of the number of successive re-

presentations of the same material, and which may be particularly sensitive to temporal lobe damage.

2. *Memory,* the persistence of learning in a state that permits subsequent retrieval, which may be assessed in terms of how much may be recalled some time after a single presentation, and which may be particularly sensitive to frontal damage.

If amnesia involves a deficit at the level of registration or acquisition, *anterograde* deficits (i.e., for memories of events occurring after the time of injury) should be far more prominent than *retrograde* (i.e., for memories of events that preceded illness or trauma). In other words, memories predating the disorder should be more or less intact, and not so memories that occurred thereafter. Again, temporal lobe amnesics approximate closest to the picture. If, however the amnesic deficit lies at the level of retrieval, registration and retention should be intact. It is difficult, however, to demonstrate intact registration or retention if one cannot retrieve. One possibility is to compare retrieval performance on free recall versus recognition. If both are equally disrupted, encoding or storage problems or both may be more likely than problems with retrieval; retrieval problems might be the more likely diagnosis in the event of failure only to recall, with improved performance in forced-choice recognition tests. However, how do we compare recall versus recognition? We shall later see that covert, implicit memory may be demonstrable in amnesics via forced-choice testing, despite their denial of any overt explicit knowledge of the material. In any case, it may be rash to assume an either–or distinction; both storage and retrieval may well suffer decrement to a greater or less degree in amnesia.

Cognitive psychology recognizes in general three types of event memory (see e.g., Della Sala, & Logie, 1993):

1. *Sensory information storage,* for which as yet we know of no corresponding pathology, and for which only recently has there been any success in anatomical localization (Sams, Hari, Rif, & Knuutila, 1993).
2. *Primary, working, or short-term memory,* the output system
3. *Secondary, associative, or long-term memory,* the system about whose pathology we know most, and upon which we shall largely concentrate.

Before addressing the second and last of the above three systems in detail, we should note that the exact nature and the temporal parameters of all three are still subject to considerable theoretical and empirical dispute. In particular there has not always been agreement between animal and human studies; the latter include both normal subjects and the amnesic. Nor is it always appropriate to group together working and STM, the former term typically applying to human cognition, and the latter often extending to the animal and physiological domains with longer time frames.

Working Memory

Working memory permits the integration of immediate sensory input with stored long-term memories. Because it involves control processes in addition to storage functions, its former label of STM has increasingly given way to working memory, though, as we saw, a distinction is sometimes made between the two. Its operation is essential to a wide range of cognitive tasks (e.g., reading, counting, calculating, and telephoning). It mediates between memory and action, and according to Baddeley (1992) consists of a central controller or executive with a group of slave systems. The central executive is responsible for the allocation and coordination of the brain's attentional resources, and the sequencing of information from the slave systems. Of the latter, the two most important are the following:

1. An inner-speech mechanism, responsible for the silent rehearsal of outline information, (i.e., the articulatory or phonological loop)
2. A visuospatial "scratch pad"

The whole system allows us to develop conscious, introspective thoughts, ideas, and plans for the future. Indeed, the entirely unconscious contents of long-term memory (LTM) are only of use if they can be appropriately assembled in working memory to meet current environmental demands. Along with language, working memory seems uniquely to characterize the human condition. Given the general architecture of the brain as a distributed processing system, we should not seek strict localization of working memory, but according to Wilson, O'Scalaidhe, and Goldman-Rakic (1993) prefrontal structures seem to be closely involved in at least two regions. These regions are in fact in receipt of projections from the two visual systems described earlier by Mishkin, Ungerleider, and Macko (1983). Thus information on shape and color, elaborated in the inferotemporal cortex, projects on to the inferior convexity of the prefrontal cortex, while information on spatial aspects, elaborated in the posterior parietal cortex, projects on to the principal sulcus and the arcuate area of the dorsolateral convexity of the prefrontal regions. Indeed Baddeley's (1992) central executive, coordinating the phonological and visuospatial subsystems, is likely to reside in the prefrontal cortex (Della Sala & Logie, 1993).

At a clinical level, patients with dorsolateral prefrontal damage exhibit deficits in attention, set, and the ability to integrate past events and future requirements on a flexible moment-to-moment basis; according to Funahashi, Chafee, and Goldman-Rakic (1993) the patients' working-memory work space is disrupted. They are deficient in using knowledge for everyday life, despite being relatively intact in performance on conventional intelligence tests and LTM memory tasks. Such patients, and animals, show working-memory deficits when assessed in delayed response tasks, or where familiar material has to be reorganized or combined in novel ways; they behave as if "out of sight, out of mind." Indeed, during delayed response tasks, when the individual must withhold a response during an interval, different single cells in the prefrontal cortex fire on stimulus presentation, during the retention interval, and on responding. Using positron emission tomography (PET) neuroimaging, Raichle (1993) found that in normal subjects holding lists of pseudowords in working memory, prefrontal regions were active, along with the supplementary motor area (SMA); so too were various other regions, depending upon whether the individuals were good (premotor and cingulate—an articulatory strategy?) or bad (occipital and cerebellum—a visual strategy?) at remembering such stimuli. Otherwise, the phonological loop has, not surprisingly, been localized to the inferior parietal cortex on the left, close to the junction with the superior posterior temporal lobe (Della Sala & Logie, 1993); this region, corresponding to the supramarginal gyrus, is of course close to the site commonly identified in conduction aphasia, and its locus was confirmed by Martin, Shelton, and Yafee (1994). These authors in fact identified a second region for *semantic* working memory in posterolateral frontal and adjacent anterior parietal regions. Paulesu, Frith, and Frackowiack (1993) and Jonides et al. (1993) similarly identify the metabolic bases for the phonological buffer store in the left inferior parietal regions, with Broca's area (left inferoposterior frontal) operating during inner speech and the actual *rehearsal* of the contents of the phonological store. There is however much more uncertainty with a number of proposed loci for the visuospatial scratch pad; thus there might be one for the "visual" and another for the "spatial" aspect, the latter perhaps residing in (right) parietal and the former in temporo-occipital regions.

A general conclusion therefore is that working memory depends upon a variety of networked brain areas, depending upon task, stimuli, and strategy, but that the prefrontal regions are involved in an overall executive or supervisory fashion, coordinating and controlling working-memory functions. Working memory is brief, labile, and easily disrupted, unlike LTM, and again unlike the latter it probably depends upon ongoing neural activity rather than structural change. However, contrary to earlier opinion, information possibly may not proceed serially from STM or working memory to LTM. Thus Squire, Knowlton, and Musen (1993) in their review described an acquired amnesic with a severely deficient digit span, but normal verbal long-term memory, as measured by paired-associate learning and the learning of word lists. Although this would suggest that information can bypass STM before entering LTM, or that the inputs are arranged in parallel, it is instead still possible that there is more than one kind of STM or working memory for different types of material (e.g., involving phonological or semantic codes).

Traditionally, STM or working memory for language enables us to maintain an otherwise disconnected sequence of words long enough to extract the syntactic connectivities and associated meaning (see Baddeley, 1986). However, R. Martin (1993) reported a patient who, despite a verbal memory span of only two items, and an inability to learn or maintain verbatim sentences, multisyllabic words, or new phonological forms, nevertheless could achieve normal sentence comprehension. Thus the phonological storage capacity critical to STM may after all play only a limited role in language processing, specifically, perhaps in maintaining and learning phonological forms as when learning a new language. Otherwise, normal language comprehension may instead occur "on-line," with very little reliance on temporary storage, unless one has to backtrack because the initial on-line process has for some reason failed. It would be interesting to see if the patient described above could be taught a new tongue. Indeed R. Martin (1993) mentioned several other patients with phonological storage deficits who still show excellent comprehension even of complex syntax. The "articulatory loop" cannot therefore simply serve to enable us to hold sentence words in memory long enough for us to work out the syntax.

Regions Involved in Long-Term Memory

Leaving aside motor learning and classical conditioning where, in addition to cortical structures, the basal ganglia (BG) and cerebellum may play major roles, there are three main regions involved in the higher cognitive aspects of memory (see Figure 1):

1. The medial temporal lobe, hippocampus, amygdala, and associated limbic regions (the latter constituting a medially located ring of cerebral tissue that surrounds the diencephalon and brain stem), which together make up part of the old circuit of Papez.
2. The thalamus and mamillary bodies—the diencephalon
3. The basal forebrain—the ascending cholinergic pathways originating in the substantia innominata (nucleus basalis of Meynert), which are implicated in Alzheimer's disease (AD).

Frackowiack (1994) noted that metabolically abnormal regions (in terms of neuroimaging) in relatively pure amnesics include the hippocampus, thalamus, cingulate cortex, and basal frontal cortex, all regions involved in a network for explicit LTM. Notably, the hippocampus is active only during LTM formation, not during working memory involvement.

The *anterior medial cortex, hippocampus, and amygdala* constitute a parallel, distributed, associative network that is not itself a memory bank, but that provides a mechanism for storing and consolidating new memories and retrieving old information. It probably receives not so much "encapsulated" memories *per se,* but rather the addresses for accessing all the distributed, multimodal components of memory; hence its important role in associative responding, recognition, and partial information and cuing. The medial temporal cortex maintains the labile component of LTM. The later and more consolidated stage of LTM is

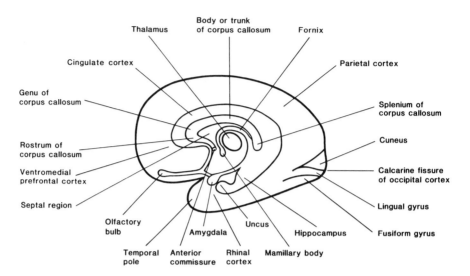

Figure 1   COMPOSITE DIAGRAM OF THE APPROXIMATE RELATIVE POSITIONS OF MAJOR SUB-
CORTICAL STRUCTURES. THE FIGURE DOES NOT CORRESPOND TO ANY ONE SAGITTAL SECTION.

represented elsewhere (e.g., in the anterior temporal cortex). The idea of consolidation is supported by the phenomenon of permanent retrograde amnesia for
events occurring during a varying period of time prior to a concussive blow or
electroconvulsive shock. Some form of structural change, as a result of associating stimuli on the basis of temporal contiguity, is thought to occur at a synaptic
level in the course of consolidation (Miyashita, 1993); this process is further
thought to take place under the influence of afferents from the medial temporal
region, especially the hippocampus, where damage is sufficient to cause major
memory impairment (Zola-Morgan & Squire, 1993). Indeed, as the late James
Olds is reported once to have observed, the hippocampus has become the
Rosetta stone of the brain for neuropsychologists of memory.

The medial temporal lobe also contains the amygdala, a limbic structure
known to be important for other kinds of memory (e.g., conditioned fear, affective memory, and stimulus valence—Zola-Morgan & Squire, 1993), the emotional aspects of a stimulus that are so important in learning. The cingulate regions
that connect to the amygdala and hippocampus are similarly important in emotion, personality, and memory, modulating autonomic responsivity and attention.
Damage here leads to reduction in depression, anxiety, and obsessive–compulsive behaviors (see Chapter 14), though at the cost of slowed thought and action, impaired social judgment, and reduced affect and interests. LeDoux (1994)
noted that the subiculum of the hippocampus communicates with the lateral nu-

cleus of the amygdala, suggesting that in this way contextual information may acquire emotional significance. Indeed the amygdala may play an essential part in modulating the storage and strength of memories.

The hippocampus, though a phylogenetically old part of the brain and technically within the limbic system, is nevertheless in many ways the highest level of association cortex (Wilson & McNaughton, 1993). It plays an essential role in memory encoding, serving as a rapid associative system. It was also thought to be involved in the internal representation of space (O'Keefe & Nadel, 1978); certainly the output of many neocortical processing streams converge in the hippocampus, relating to the spatiotemporal aspects of object and place. However, Zola-Morgan and Squire (1993) noted that spatial memory is merely another, if very clear, example of a broader category of declarative memory (see below), which also includes memory for word lists, faces, odors, and tactual impressions.

The *diencephalic regions* include the mamillary bodies and the anterior and especially the dorsomedial thalamic nuclei, which are closely involved in the Wernicke–Korsakoff syndrome (see below). Damage here is sufficient, as with hippocampal injury, to cause massive amnesia. This may in part stem from disconnection of the temporoparietal stores or addresses of verbal and nonverbal information from supervisory frontal mechanisms controlling attention, planning, and the selection of strategies, goals, and action sequences. It is unclear whether, as in the Wernicke–Karsakoff syndrome, damage to the dorsomedial thalamic nucleus is critical for amnesia to occur, or whether there must also be damage to the mamillary bodies; indeed, it is possible that several regions must be conjointly affected (Zola-Morgan & Squire, 1993). In any case when we discuss Wernicke–Korsakoff amnesia and the effects of diencephalic damage, we shall note that in the former case there is usually also frontal involvement, which of course is absent with purely diencephalic damage.

The *basal forebrain* is the primary source of cholinergic innervation of the cortex. It contains the medial septal nuclei, the diagonal band of Broca that projects to the hippocampus via the fornix, and the nucleus basalis of Meynert (within the substantia innominata), which projects widely to frontal, parietal, and temporal cortices (Zola-Morgan & Squire, 1993). Thus the basal forebrain connects with the medial temporal cortex. Lesions (e.g., from ruptured aneurisms of the anterior communicating artery) disrupt both the other LTM circuits and the orbitofrontal and medial frontal cortices involved with drive, planning, and social judgment. Such lesions are therefore accompanied with persistent memory impairment and personality changes. However, the nucleus basalis of Meynert may be more important for attention than memory per se, and the effects upon memory of basal forebrain damage may be due to disruptions to information processing in the hippocampus and other medial temporal lobe structures. The basal forebrain is a primary initial site of damage in AD (see Chapter 9). Unfortunately, although anticholinergics like scopolamine prevent memory formation, anticholinesterases like physostigmine have proved of limited benefit in ameliorating AD, probably because of the extensive damage elsewhere.

Temporal Lobe Amnesia

The temporal lobes are particularly vulnerable to ischemia, anoxia, metabolic abnormalities, epileptic seizure activity, and viral infection (Parkin & Leng, 1993). Indeed the temporal lobes and associated hippocampus are selectively targeted by the neurotoxin domoic acid, from contaminated mussels. These structures are important in long-term potentiation and consolidation, a process of possibly indefinite duration. Thus they may mediate initial or temporary storage and the transfer of information to other structures for longer term representation. The excitatory neurotransmitter glutamate seems to play a major role in this consolidation process; it is noteworthy that this substance, which is itself neurotoxic in quantity (see Chapter 13), is released on cell death consequent, for example, upon anoxia, though protection against its neurotoxic effects may be afforded by barbiturates or hypothermia.

The temporal-hippocampal axis may be more important for declarative memories (i.e., about facts and events that can be put into propositional form) compared to such procedural memories as skills. The latter can continue to be acquired normally, though the patient has little understanding or knowledge about how, when, or where the learning experience occurred. Anterograde amnesia for such "episodic" (i.e., autobiographical) or "declarative" events is severe, but if by massive repetition or overtraining it proves possible to get memory levels up to normal, then the subsequent rate of forgetting may itself be normal (and see also Bauer, Tobias, & Valenstein, 1993). In any case, *recognition* memory seems much less impaired than *recall*. Semantic memory (e.g., for meaning of words) is relatively intact, especially for *premorbidly* experienced new words, but even postmorbidly some new vocabulary can be acquired (e.g., *psychedelic, jacuzzi, charisma*—Parkin & Leng, 1993). Similarly, implicit memories (see below) are relatively unaffected; the patient continues to show perceptual learning when given progressively less incomplete figures for recognition, and learns how to solve such problems as the Tower of Hanoi and Missionaries and Cannibals. There is continued improvement in mirror reading. Although there is little evidence of frontal deficits, unlike patients with diencephalic damage or the Wernicke–Korsakoff syndrome, there may be more variable levels of retrograde amnesia than with the latter syndrome. Kopelman (1993) speculated that although temporal structures on the left may mediate semantic storage, anterior temporal structures on the right may be more important for autobiographical memories.

Unilateral hippocampal stimulation leads to major temporal lobe activity and hallucinations that are typically verbal if the left side is involved, and non-verbal images with stimulation of the right hemisphere. Similarly, temporal-hippocampal damage on the left results in verbal memory deficits, although damage on the right is associated with problems in recognizing shapes, patterns, and faces, and deficits in spatial memory, maze performance, and ability to recall melodies and tunes. Bilateral damage, however, leads to effects far worse than the algebraic sum of individual, unilateral lesions, indicating that one side can at least partially stand in for the other. Sodium amytal (amobarbital) injection via the internal carotid artery to the side opposite to preexisting unilateral hippocampal damage leads to severe amnesia for material encountered during the last 5 or so minutes that the barbiturate lasts.

The classic case in the study of the lasting consequences of bilateral medial temporal lobectomy is that of H. M. (Scoville & Milner, 1957), who has now been studied for nearly 40 years. His surgically induced lesions in 1953, to alleviate intractable epilepsy, extend up to 10 cm back from the temporal poles and include the uncus, amygdala, peri-amygdalar and peri-rhinal cortices, hippocampus, and parahippocampal gyrus. Bauer et al. (1993) believe that the hippocampus and hippocampal gyrus are probably key elements. H. M. exhibits severe anterograde amnesia (with no improvement), though little retrograde amnesia. He is still highly intelligent, with a good vocabulary and sense of humor. He is dimly aware of important events after the operation, such as his father's death and the current U.S. president, and can draw a plan of his current house and neighborhood. Classical conditioning is intact, as is his STM span and procedural learning, such as mirror drawing, though he denies ever having previously seen the apparatus. Likewise he shows normal improvement in the difficult conceptual puzzle, the Tower of Hanoi, again denying ever having seen it before. Temporal lobe amnesics typically retain some insight into their problem, and may repeatedly describe their state as one of a continual awakening from sleep or a dream, in a present without a past. "Every day is alone, whatever enjoyment I've had, and whatever sorrow I've had" (Milner, Corkin, & Teuber, 1968, p. 217).

Diencephalic Amnesia and the Wernicke–Korsakoff Syndrome

Wernicke in 1881 described an encephalopathy caused by various forms of poisoning, including alcoholism, which involved gait ataxia (imbalance), peripheral neuropathy ("burning" feet), ocular disturbances (ophthalmoplegia), and altered mentation and confusion (Parkin & Leng, 1993). Korsakoff, a few years later in 1889, added severe amnesia to the above syndrome along with confabulatory tendencies in trying to make sense of a now confusing world, euphoria, anosognosia, irritability, and aggression. So was born the Wernicke–Korsakoff syndrome.

The syndrome most commonly stems from chronic alcoholism, though thiamine deficiency is the primary cause. Alcohol interferes with the gastrointestinal transport of thiamine, and chronic liver disease, itself consequent upon alcoholism, diminishes the liver's ability to store thiamine (Parkin & Leng, 1993). Thiamine, or vitamin B1, is important in glucose metabolism and the synthesis of neurotransmitters, especially acetylcholine (which we saw mediates the basal forebrain's contribution to memory) and the inhibitory gamma-aminobutyric acid (GABA). Bilateral diencephalic damage is a typical consequence of thiamine deficiency, especially to the mamillary bodies or the dorsomedial thalamic nucleus or both. It is debated which structure is the most important, and the current consensus seems to be that both may be equally so. However, it now appears that there may also be involvement of the orbitofrontal cortex and the hippocampus. The latter in any case receives input from the mamillary bodies and thalamus. Moreover, thalamic lesions disconnect the frontal systems, which are, as we saw, important for coordinating and imposing cognitive structure upon the posterior cortical systems subserving semantic memory (Bauer et al., 1993).

Diencephalic damage on its own may stem from a primary thalamic hemorrhage, or a nonhemorrhagic infarction. Low-velocity penetrating injuries as from a fencing foil, tumors, and so on, have also been implicated. They all lead to amnesia as a primary symptom, while infarction of the paramedian arterial territory in particular may result in hypersomnolent apathy and amnesia (Parkin & Leng, 1993). Generally, diencephalic damage causes amnesic symptoms essentially similar to those of the Wernicke-Korsakoff syndrome, and the two will be considered together.

A characteristic finding is for amnesia of an arbitrary, autobiographic, or episodic nature, involving how, when, or where something was experienced or learned. This loss is even more pronounced than in temporal-lobe amnesia, and is accompanied by loss of memory for temporal order and sequencing (Parkin & Leng, 1993). Such patients show excessive interference from previously learned material, and are therefore unduly susceptible to proactive interference (PI) (Bauer et al., 1993). Thus they show reduced release from Wickens's type PI. This has been interpreted as an encoding deficit with reduced semantic processing, with processing occurring only at a more superficial level, in the absence of a rich associative network. Interestingly, important or significant information (e.g., the location of alcohol) may be spared amnesic oblivion, perhaps because it becomes integrated within the semantic network. If the information can in fact be encoded, maybe by overlearning, there may, as with temporal lobe amnesia, be a more or less normal rate of forgetting. Again, as with temporal lobe amnesia, recognition is typically better than recall, and retrieval cues may help, suggesting a retrieval deficit in addition to an encoding deficit. Further evidence for the latter comes from the presence of a marked temporal gradient in retrograde amnesia, with oldest material best preserved. This is demonstrable by tests involving "famous faces" and "public events" (see e.g., Kopelman, 1993; Sagar, 1990). Such a gradient, however, is not merely an artifact of subclinical disease and encoding deficits getting progressively worse, as a Korsakoff patient who had previously written his autobiography successfully, later forgot much of the detail therein (Butters & Cermak, 1986). Similarly, nonalcoholic patients with sudden onset of diencephalic damage also show such a gradient, implying a true storage deficit. Thus, older memories may be more practiced, perhaps as a consequence with multiple entries or representations; alternatively they may be more "lawful" or "semantic." Interestingly, recognition testing may not demonstrate a temporal gradient, and according to Kopelman (1993), retrograde amnesia as a whole is more extensive in Korsakoff and AD patients, due perhaps to frontal and anterior temporal damage, whereas basal-forebrain and limbic-diencephalic patients do not show it as much. Other features of the Korsakoff disorder include plausible confabulations ("honest lying") of the sort typically absent in temporal lobe amnesia, and perhaps consequent upon an attempt to make sense of a confusing environment; absence of insight, again unlike temporal lobe patients; meager spontaneous conversation, perhaps in the absence of much material in memory, but normal intelligence, alertness, and procedural memory. Indeed, classical conditioning is preserved, along with skills, maze, and perceptual learning. Thus if gradually more informative fragments of a word or picture are shown (see Figure 2) until recognition is achieved, there are savings on retest after an interval, as with temporal lobe patients. The same applies with use of incomplete words, and there is normal repetition priming even over considerable intervals. In the Wernicke-Korsakoff syndrome there is in fact evidence of frontal pathology as shown by computerized tomography (CT) scans, and behavioral measures such as the Wisconsin Card Sorting Test, which assess attention, set, and executive functions. Frontal pathology is thought to affect access or retrieval more than storage per se, as cues may help recall, and there is little evidence of a temporal gradient in retrograde amnesia; instead, there is a generalized clouding, impoverishment, or disorganization of recall across the life history. Conversely, with temporal pathology, where cues are of little assistance, there may be a true storage loss.

How can we explain diencephalic or Korsakoff amnesia? Is it a problem of too superficial a processing level at *encoding*? Patients are often fooled by irrelevant paraphernalia in face recognition tasks. Perhaps reduced information-processing capacities *lead* to a strategy of having to concentrate only on immediate, superficial attributes or aspect. Is it a problem of *retrieval* because of increased interference and competition? Cues, salient stimuli, unusual experimenter and test locations often help, highlighting the "episodic" aspect of a deficit within a spatiotemporal context. Thus as we shall see, patients can often say if they have seen a picture before, but not where or when (i.e., the episodic aspects). Is it a problem of *storage,* as indicated by the temporal gradient in retrograde amnesia? Or is it simply a problem of *declarative* rather than procedural memory? Although we might have to accept the possibility that all these factors may contribute, with consolidation failure perhaps being less significant than in

Figure 2   AMNESIC PATIENTS MAY SHOW SAVINGS, ON SUBSEQUENT RETESTING, IN RECOGNITION OF FRAGMENTED (a) WORDS OR (b) FIGURES.

temporal amnesia, we should note, in the context of the last possibility, that a taxonomy of memory is yet to be established. Although the patterns of amnesia seem intuitively sensible, we should not perhaps try to force all its manifestations into a single, too strict classification. The aphasias, alexias, and agnosias, as we have seen, are themselves heterogeneous, and so too maybe is amnesia. Indeed the apraxias, aphasias, and agnosias are themselves in a very real sense also amnesias. Language, object recognition, and memory should all perhaps be studied via ecologically more valid or practically more realistic real-life tests, including, in the case of memory, tests of remembering to do future things such as taking medication or keeping appointments. It is far from clear how we can accommodate the current range of taxonomies proposed by experimental studies of memory in normal, healthy individuals (Bauer et al., 1993; Parkin & Leng, 1993; Squire et al., 1993). One approach (see Tulving, 1989) separates procedural memory (phylogenetically ancient, involving skills, implicit knowledge, and knowing "how") from declarative (phylogenetically recent, incorporating semantic and episodic content, explicit knowledge, and knowing "that"). Declarative memory, which may be fast, flexible, and perhaps unreliable, is seen to depend largely upon medial temporal lobe and diencephalic mechanisms, whereas nondeclarative memory, which is slower, less flexible, but maybe more reliable, is far more heterogeneous, involving the neostriatum for skills and the cerebellum for classical conditioning. Within declarative memory, we may further contrast episodic (frontal?) and semantic (temporal?) aspects; the former involve arbitrary

b

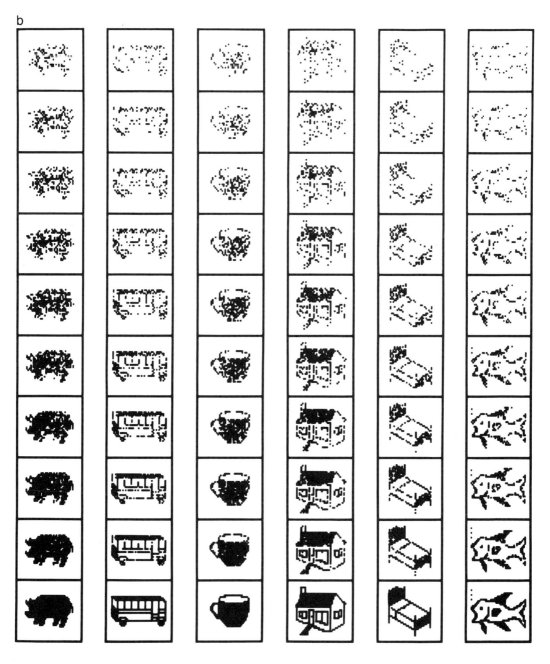

Figure 2 (continued)

autobiographical events with a spatial and temporal context, and are badly affected in amnesia, whereas as we saw, semantic memories, involving general knowledge of lawful world aspects, are possibly less affected by amnesia, and may build upon episodic memories by a process of generalization. If so, given the extreme vulnerability of episodic memory even in the healthy individual, then for semantic memories to be successfully "distilled" from repeated episodic events, something (tacit or implicit memory?) must be successfully preserved despite apparent loss at an explicit level. We shall return shortly to this specific problem. This issue also begs the question as to whether such implicit or tacit memories are merely explicit memories of a subthreshold signal strength, or are different in kind. Nor is it always clear how the conscious–unconscious divide is to be accommodated, especially as much of semantic memory is "cognitively impenetrable" (Fodor, 1983), as with our innate understanding of grammar. This contrasts with the introspective accessibility of much of our autobiographical episodic knowledge. The resolution perhaps is to note that although the episodic–semantic distinction is hotly debated, it is still conceptually and clinically useful, and not to try to force all clinical findings into a preconceived taxonomy.

Preserved Implicit Covert or Incidental Memory

In the 19th century, French neurologist Claparède performed a demonstration that would not pass a modern ethics committee. One week on a ward round, when shaking hands with his patients, he concealed a drawing pin in his hand. Next week the amnesic patient refused, unlike the others, to shake hands, but could not say why. She probably confabulated. More recently, it has been shown that amnesics exhibit preserved repetition priming over quite lengthy intervals in word-completion, perceptual identification, and free-association tasks (Mayes, 1992; Schacter, 1992; see also Schacter, Chiu, & Ochsner, 1993). Such preserved capacities are increasingly being demonstrated by indirect tests, requiring patients to do something that can *indirectly* reflect the occurrence of a prior event (a percept, memory, etc.). Examples include successful perception of fragmentary or briefly presented material after some previous experience, rather than consciously, intentionally, or deliberately reporting on the target material of interest. As Jacoby and Kelley (1992) observed, although the responses in conceptually driven or explicit direct testing are under the control of the subject, with data-driven, implicit, or indirect testing, responses are stimulus-driven and automatic. Performance is influenced by but is not directly dependent on a particular action. Normal healthy individuals show similar "unconscious" memories with masked priming and cryptomnesia (unconscious plagiarism, A. Brown &

Murphy, 1989). The latter, where material is remembered without the conscious experience of previous encounter, is, in an interesting way, the mirror image of confabulation, where the amnesic has an erroneous feeling of remembering. Déjà vu, in temporal lobe epilepsy, similarly dissociates the feeling from the actual mnemonic experience or content. Note that implicit memories may not merely be too weak to enter consciousness, as manipulations affecting them like typeface in priming experiments do not affect explicit memory performance; similarly, manipulations affecting the latter (e.g., increased study time) do not affect implicit memory performance (Bauer et al., 1993).

Moscovitch, Vriezen, and Goshen-Gottstein (1993) categorized implicit memory tests into item-specific and procedural; the former involve for example, a certain face, name, or object, with improved performance on the second or subsequent occasion that it is encountered, compared to the first. The concept of priming is traditionally invoked in this context. With procedural (skill) learning, tasks such as tracking, reading unusual scripts, or solving puzzles have been employed, and the patient shows normal levels of improvement without awareness of having encountered the situation earlier.

Moscovitch et al. (1993) further subdivided item-specific memory tests into perceptual and conceptual. With *perceptual* tasks the patient may have to complete word fragments (e.g., _T_I_G, i.e., STRING) or stems (e.g., S T R_ _ _). Latencies and accuracies improve on repetition, even though explicit recognition levels may remain at chance. Other tasks include identification of degraded line drawings or fragmented pictures, and repetition priming for same–different judgments of faces or objects. Effects may persist for a year or more if solutions are few or unique, so we cannot invoke the temporary activation of preexisting representations. Indeed, the slow rate of decay may be no faster than that shown by normal individuals. The fact that the phenomenon occurs with lexical (word–nonword) decisions, and for reading time of lists of nonwords, indicates that there need not be any preexisting representations of the material, as with real words. Other tasks include joining dots to create patterns. It is noteworthy that although deeper levels of processing may assist performance in explicit memory tasks, the same does not occur with implicit memory tests. Occipitotemporal cortex may be involved with reactivation of perceptual records that however are viewpoint independent. Thus priming effects still occur with changes in orientation of the object or its angle of view, size, or mirror reflection.

With *conceptual* tasks, the target is now supplied in response to a conceptual cue, whereas the specific perceptual information contained within the target is withheld (e.g., "Name a musical instrument containing a reed"). Later, the patient is asked to spell homophonic words like *read* or *reed,* being given only the spoken form, which does not distinguish between them. In this example, a bias toward *reed* would be expected. In another example, subjects would be given highly associated prime-target pairs; later they would be given only the prime

and asked to free associate. Typically, they would respond with the original target. Temporal lobe damage on the left seems to eradicate such conceptual priming (Blaxton, 1992, cited by Moscovitch et al., 1993).

Procedural tests of implicit memory, according to Moscovitch et al. (1993),
measure a general performance change due to a previous experience, which is
not specific to the items then presented, and does not need conscious, explicit
retrieval of the earlier episode. They propose two subdivisions, sensorimotor and
learned rules. *Sensorimotor* tests include such tasks as pursuit rotor and bimanual tracking. Interestingly, patients with HD (see Chapter 13) show little evidence
of implicit memory on such tasks, but show the normal pattern with stem-com-
pletion tasks (above), whereas AD patients (see Chapter 13) show the reverse
pattern (Butters, Heindel, & Salmon, 1990). Thus the BG may be involved in
sensorimotor tasks. Unlike conceptual or perceptual priming, where a single exposure may suffice, many practice trials are usually necessary to demonstrate
implicit memory with such procedural tests.

Where instead *learned rules for ordered sequential operations* are involved,
such as the Tower of Hanoi puzzle, the frontal lobes may be necessary for planning, hypothesis formation, development of strategies, and response monitoring.
The same may apply with maze learning, though good performance may also depend upon an extra, *explicit,* component.

One of the most remarkable instances of implicit or covert memory was recently reported by Vargha-Khadem, Isaacs, and Mishkin (1994), in a case of
childhood global anterograde amnesia consequent upon a pineal tumor. The patient was able to retrieve postmorbid memories via writing, without being able to
provide any oral account whatsoever of the content of his written reports. As the
authors noted, his memory retrieval thus has some of the characteristics of "automatic writing." Such a dissociated form of episodic memory clearly presents a
new challenge to our understanding of how memory is organized at both a functional and a structural level.

Basal Forebrain Amnesia

At present we have comparatively little information on the effects upon
memory of specific damage to the basal forebrain, out of the context of AD (see
Chapter 13), where many other structures are also usually affected. Bauer et al.
(1993) briefly noted the presence of extensive anterograde and variable retrograde amnesia, problems with temporal order, free and wild confabulation, especially with concomitant orbitofrontal damage, a tendency to apathy, again if
frontal structures are also affected, and general similarities to Wernicke-
Korsakoff syndrome. This is not surprising, as we noted the frequent concomitant presence of basal forebrain involvement in the latter syndrome.

An Overview of the Structures and Functions Involved in the Amnesic Syndrome

Clearly memories are not permanently stored in the hippocampus, diencephalon, or basal forebrain. These structures may mediate initial storage, manage consolidation, bind together distributed sites in the cortex perhaps via synchronized firing, and perhaps provide access to information stored elsewhere. The hippocampal system and its limbic (especially amygdalar, see e.g., LeDoux, 1994) interconnections may provide the emotional aspects and the meaningful personal context relative to individual experience (Bauer et al., 1993); hence damage here may permit retention of general (semantic) properties without personal (episodic) context. Hippocampal activation may permit the coordination of all the widely distributed elements involved in an experience. Later, this aspect may drop out so that ultimately we only remember the generalities rather than the personal aspects. Diencephalic damage may prevent effective use of the temporal-hippocampal system; damage to either structure may result in largely similar amnesia as they both participate in a larger functional system, such that their separate contributions cannot be easily discriminated behaviorally (Bauer et al., 1993; Zola-Morgan & Squire, 1993). The particularly severe deficit in Korsakoff patients with respect to memory for episodic aspects and temporal order may reflect additional concomitant frontal damage. In any case, both diencephalic and medial temporal lobe structures project to the ventromedial and dorsolateral frontal cortices, and this linkage plays an important role in guiding behavior both during information encoding and retrieval. Zola-Morgan and Squire (1993) in fact note the following connectivities and projections:

1. The neocortex to parahippocampal and peri-rhinal cortices to entorhinal cortex to dentate gyrus and elsewhere in the hippocampus to subiculum and entorhinal cortex and back to the neocortex

2. Subiculum of hippocampus to fornix to mamillary bodies to mamillothalamic tract to anterior nucleus of thalamus to cingulate cortex

3. Direct projections from hippocampus to anterior nucleus of thalamus

4. Dorsomedial nucleus of thalamus receives input from both the amygdala and peri-rhinal cortex, and from the subiculum of the hippocampus and parts of the peri-rhinal cortex. It projects to the orbitofrontal cortex.

5. The basal forebrain cholinergic system may modulate and nonspecifically activate the hippocampal-neocortical system.

Thus rather than a single structure or even a single system subserving memory, there is a number of different processes involved in memory, and a number of interacting and partially overlapping systems. Damage anywhere in the system as a whole may result in grossly similar behavioral deficits, and it may in fact be beyond the limits of experimental ingenuity to isolate the exact behavioral contribution of each link or node in the interlocking subsystems. Normally, autobiographical memories are likely to be distributed widely in the brain; thus preexisting memories are not likely to be destroyed by injury, merely degraded as to content or made inaccessible, temporarily or permanently. Indeed Conway (1993) described cases where memories, apparently lost for years, suddenly return. Conversely, Kapur, Ellison, Smith, McClellan, and Burrows (1992) reported a closed-head-injury patient with extremely dense retrograde amnesia, such that most of her previous life's events were apparently lost, but with very mild anterograde amnesia, such that new autobiographical information was well acquired. She nevertheless described a feeling of "emotional" recognition when reencountering people from her past. The anterior portions of both temporal lobes were affected, together with the ventral surfaces of both frontal lobes.

Inevitably, most of our detailed information on the amnesic syndrome comes from controlled laboratory studies; however, patients usually present because they or their caregivers are concerned about problems in their daily life. Such problems are less in terms of forgetting *what happened,* and more in terms of forgetting *to do* something in the future (to keep an appointment, etc.). For obvious reasons there has been little systematic study in normal individuals of such *prospective* memory dysfunction, though see Gruneberg, Morris, and Sykes (1988, pp. 348–376). According to Dalla Barba (1993), the far more naturalistic and ecologically valid prospective memory involves forming an intention to do something, and later remembering that there *was* such an intention, the nature of its content(s), *when* to perform it, and that you have *done* so, so that the action is not repeated. While intuitively it would seem that frontal structures and Shallice's (1988) supervisory attentional system are likely to be involved, there has been even less formal study of the pathology of prospective memory in clinical populations. Prospective memory for highly habitual and routine actions may be better preserved than that for one–off actions. It may closely parallel retrospective autobiographical or episodic memory, and both may be similarly affected in classic amnesia and mild dementia. Confabulatory *intentions* may even occur, which the patient may actually attempt to carry out, though with frontal damage the frequent inability to translate intentions into actions may lead to new compounding problems. Generally, however, frontal patients do show problems with prospective memory and intention to action (Dalla Barba, 1993).

Neuroimaging studies have also recently indicated a role for (dorsolateral) prefrontal structures in episodic memory in normal healthy individuals (Shallice et al., 1994). In a verbal task, activity was recorded during *acquisition* on the *left* dorsolateral prefrontal cortex and the retrosplenial area, and during *retrieval* on the *right* dorsolateral prefrontal cortex (DLPFC) and the precuneus bilaterally. The authors review other similar recent findings, suggesting a right hemisphere (RH) prefrontal involvement in episodic retrieval that depends upon memory monitoring and verification (and see also Tulving, Kapur, Craik, Moscovitch, & Houle, 1994, and Tulving, Kapur, Markowitsch, Craik, Habib, & Houle, 1994, for similar findings). Such conclusions, in particular the general idea that the (DLPFC) is important for the executive functions of working memory and the organization of supervisory thought processes, returns us to the theme elaborated at the beginning of this chapter. This region may control hippocampal function, though it is noteworthy that no evidence was obtained by Shallice et al. (1994) for selective activation of mesial brain structures other than the thalamus.

Case History: H. M. and the Lasting Consequences of Bilateral Medial Temporal Lobectomy

H. M. was born in 1926. . . . [He] experienced . . . his first major seizure on his 16th birthday. After high school . . . he was having on the average 10 petit mal seizures per day and 1 major seizure per week. Unsuccessful attempts had been made over a 10-year period to control these seizures with large doses of anticonvulsant medications. . . .

In 1953, when H. M. was 27 years old, Dr. William Beecher Scoville performed bilateral, medial temporal-lobe resection. He approached the brain through two 1.5 inch supraorbital trephine holes. By inserting a flat brain spatula through each hole, he was able to elevate both frontal lobes thereby exposing the tips of the temporal lobes. They in turn were retracted laterally in order to permit access to the medial surfaces, where electrocorticography was carried out in order to assess the activity of the uncus, amygdala, and hippocampus—structures that are often implicated in epilepsy. . . . An incision was then made that bisected the tips of the temporal lobes, and he resected the medial half of the tip of each temporal lobe. Next, Dr. Scoville removed by suction all of the gray and white matter medial to the temporal horns of the lateral ventricles, sparing the temporal neocortex almost entirely. The removal was bilateral, and it is said to have extended 8 cm back from the tips of the temporal lobes. It included the prepyriform gyrus, uncus, amygdala, hippocampus, and parahippocampal gyrus, and must have produced an interruption of some of the white matter leading to and from the temporal lobes. H. M. was awake and talking during the operation.

The operation reduced the frequency of H. M.'s major seizures to the point where now his attacks are infrequent, and he may be free of generalized convulsions for as long as a year. The minor seizures persist . . . [and] do not noticeably disturb him. . . . Since the time of his operation, H. M. has had a profound anterograde amnesia that is especially salient because his overall intelligence and neurological status are relatively well-preserved. H. M.'s global amnesia has put marked limitations on his daily activities and accomplishments. . . .

He [has to be] reminded to shave, brush his teeth, and comb his hair. He spends much of his time doing difficult crossword puzzles and watching television. . . . when he is observed to be acting differently, the nurses question him by running through a list of possible complaints, such as toothache, headache, stomachache, until they hit upon the correct one. He will not spontaneously say, for example, "I have a headache," causing one to wonder whether he knows what is wrong. . . . If he is asked to sit in a particular place, he will do so indefinitely. . . .

He sometimes remembers that his parents are dead, and mentions that he is all alone. . . .

[His] overall IQ ratings have been consistently in the average range, whereas performance IQ ratings fall in the average to superior ranges. . . .

H. M. is able to appreciate puns and linguistic ambiguities, and although he does not usually initiate conversations himself, when someone begins a conversation with him, he talks readily and in general communicates effectively. . . .

He is able to draw an accurate floor plan of the house where he lived during the postoperative years from 1960 to 1974, showing all the rooms in their proper location. He believes that he still lives there, and can recognize the floor plan, drawn by someone else, when it is presented with four foils. In the nursing home where he lives he can find his way from the ground floor to his room, which is one flight up. These latter two instances of preserved spatial capacities occur in environments where H. M. has had thousands of learning trials, however. . . .

He still exhibits a profound anterograde amnesia and does not know where he lives, who cares for him or what he ate at his last meal. His guesses as to the current year may be off by as much as 43 years, and, when he does not stop to calculate it, he estimates his age to be 10 to 26 years less than it is . . . nevertheless he has islands of remembering, such as knowing that an astronaut is someone who travels in outer space, that a public figure named Kennedy was assassinated. . . . He is comfortable dealing with the products of new technology, such as computerized tests and portable radios with headphones. . . . Later experiments [have] provided further evidence of his ability to acquire new motor skills . . . and certain perceptual skills, such as reading briefly-presented words and mirror reading and the cognitive skills required to solve the Tower of Hanoi puzzle (Corkin, 1984, pp. 249–259).

Summary and Conclusions

Amnesia can stem from many causes, but may be defined as a permanent memory disorder consequent upon organic (rather than psychogenic) brain dysfunction, which occurs in the absence of any other extensive perceptual or cognitive disturbance. Although there may be no single amnesic syndrome, nevertheless STM, working, or immediate memory, procedural memory, and semantic memory are typically intact, in the presence of severe, permanent anterograde and some retrograde amnesia. Damage is usually present in the medial temporal lobes (especially hippocampus), or the diencephalon (especially dorsomedial thalamic nucleus and mamillary bodies) or in both. Although amnesia may in principle stem from deficits at the levels of registration, retention, or retrieval, in practice it is not always easy to differentially assess these processes.

Working memory, which is brief, labile, and easily disrupted and mediating between perception, memory, and action, is essential to most aspects of cognitive functioning at a conscious level. Although the brain's architecture as a distributed processing system may dissuade us from seeking strict localization of working memory, nevertheless, prefrontal structures may be closely involved in a central executive, coordinating phonological, and visuospatial subsystems, which are themselves distributed, network fashion, elsewhere in the brain. However, as yet much less is known about the neuropsychology of working or STM than of LTM.

Apart from classical conditioning and motor learning, in the higher cognitive aspects of LTM three main regions are involved—the medial temporal cortex (including hippocampus and amygdala), the diencephalon (including thalamus and mamillary bodies), and the cholinergic basal forebrain (with pathways ascending from the nucleus basalis of Meynert, situated within the substantia innominata). The medial temporal lobe and associated structures, especially the hippocampus, initially maintain the labile component of LTM and provide a mechanism for storing and consolidating new memories and retrieving old information. The system may be particularly important for declarative or autobiographical memories that can be put into propositional form. Acquisition and retrieval of procedural or skill memory, and tacit or implicit learning and problem solving, are more or less unimpaired, whereas recognition memory is far less affected than recall. There is usually some retention of insight.

Damage to the diencephalic regions is sufficient, as with hippocampal injury, to cause massive amnesia; this may be partly due to disconnection of the temporoparietal stores or addresses of verbal and nonverbal information, from the supervisory frontal mechanisms that control attentional and executive functions. The exact role of the mamillary bodies in the associated Wernicke–Korsakoff syndrome, however, is still unclear. This syndrome is primarily due to thiamine deficiency, and although the mamillary bodies and the dorsomedial thalamic nucleus are damaged, there may also be involvement of the orbitofrontal cortex and hippocampus, which receives input from the mamillary bodies and thalamus. The symptoms of diencephalic damage are essentially similar to those of the Wernicke–Korsakoff syndrome; they involve, even more than with temporal lobe amnesia, a severe deficit in the recall of arbitrary, autobiographical, or episodic information, with preservation of tacit, implicit, procedural, and skill learning. In the latter instances different structures are known to be involved. However, plausible confabulation is far more prominent than with temporal lobe amnesia, along with loss of insight. An encoding deficit, additional to a retrieval deficit that can be partly alleviated by cuing, is evident from the marked temporal gradient in retrograde amnesia, as shown by "famous faces" and "public events" tests. Nor is such a gradient merely an artifact of early subclinical disease progress during encoding. Our problem in understanding the disorder relates to the somewhat artificial mode of laboratory testing, compared to the range of very real deficits experienced in everyday life. Thus "future memory," to *do* something, has been little studied. Another problem relates to attempts to reconcile the current range of taxonomies proposed by experimental studies of memory in normal healthy individuals, with the range of deficits encountered clinically.

Damage to the basal forebrain, as in AD, affects the primary source of cholinergic innervation of the cortex, disrupting information processing in the hippocampus and other medial temporal lobe structures. There may be extensive anterograde and variable retrograde amnesia, problems with temporal order, and florid confabulation.

The hippocampus, diencephalon, and basal forebrain may mediate initial storage, manage consolidation, "bind" together distributed sites elsewhere in the cortex, and provide access to information stored elsewhere. However, the various structures are themselves closely interconnected, damage in one part affecting function elsewhere, such that the *general* picture of amnesia does not *greatly* differ as a function of location of damage within the system. Particular deficits, moreover, as a function of localized damage, may reflect the involvement of *other* interconnected subsystems elsewhere (e.g., frontal). There is therefore a number of different processes involved in memory, and a number of interacting and partially overlapping systems; damage anywhere in the total network can cause broadly similar deficits.

Further Reading

Baddeley, A. (1990). *Human memory*. Hillsdale, NJ: Erlbaum.

Bauer, R. M., Tobias, B., & Valenstein, E. (1993). Amnesic disorders. In K. M. Heilman & E. Valenstein (Eds.), *Clinical neuropsychology* (3rd ed.) (pp. 523–602). Oxford: Oxford University Press.

Cohen, N. J., & Eichenbaum, H. (1993). *Memory, amnesia and the hippocampal system*. Cambridge, MA: M.I.T. Press.

Moscovitch, M., Vriezen, E., & Goshen-Gottstein, Y. (1993). Implicit tests of memory in patients with focal lesions or degenerative brain disorders. In F. Boller & J. Grafman (Eds.), *Handbook of neuropsychology* (Vol. 8, pp. 133–174). Amsterdam: Elsevier.

Parkin, A. J., & Leng, R. C. (1993). *Neuropsychology of the amnesic syndrome*. Hove, U.K.: Lawrence Erlbaum.

Petri, H. L., & Mishkin, M. (1994). Behaviorism, cognitivism and the neuropsychology of memory. *American Scientist, 82,* 30–37.

Reber, A. S. (1993). *Implicit learning and tacit knowledge: An essay on the cognitive unconscious*. Oxford: Oxford University Press.

Schacter, D. L., Chiu, C.-Y. P., & Ochsner, K. N. (1983). Implicit memory. A selective review. *Annual Review of Neuroscience, 16,* 159–182.

Smith, M. L. (1989). Memory disorders associated with temporal lobe lesions. In F. Boller & J. Grafman (Eds.), *Handbook of neuropsychology* (Vol. 3, pp. 91–106). Amsterdam: Elsevier.

Squire, L. R., & Butters, N. (Eds.). (1992). *Neuropsychology of memory* (2nd ed.). Hillsdale, NJ: Lawrence Erlbaum.

Squire, L. R., Knowlton, B., & Musen, G. (1993). The structure and organization of memory. *Annual Review of Psychology, 44,* 453–495.

Tulving, F. (1989). Remembering and knowing the past. *American Scientist, 77,* 361–367.

Zola-Morgan, S., & Squire, L. R. (1993). Neuroanatomy of memory. *Annual Review of Neuroscience, 16,* 547–563.

Dementia and Alzheimer's Disease

*L*ewis (1993) noted that anyone acquainted with Alzheimer's disease (AD) will find a familiar ring in the words of Jonathan Swift in *Gulliver's Travels* written well over 250 years ago:

> When they [come] to fourscore years . . . they have no Remembrance of anything but what they learned and observ'd in their Youth and middle Age, and even that is very imperfect. . . . In talking they forget the common Appellation of things, and the names of Persons, even of those who are their nearest Friends and Relations. For the same reason they can never amuse themselves with Reading, because their Memory will not serve to carry them from the beginning of a Sentence to the end . . . neither are they able . . . to hold any Conversation (farther than a few general words). (Lewis, 1993, p. 265–267)

Lewis also noted that Swift himself suffered progressive memory loss in his later years, protracted bouts of walking, a progressive aphasia, and the inability to recognize acquaintances.

AD is perhaps the most common form of dementia. The latter has been called "the disappearing mind," "the 36-hour day," "the funeral that never ends," "emotional widowing," and "the disease of the century." Strictly, dementia is a symptom of many diseases, rather than a disease in its own right. The enormous recent increase in its prevalence stems from the fact that many more people nowadays are surviving to old age, and dementia, largely a disease of aging, is being unmasked by these increased survival rates (Whitehouse, Lerner, & Hedera, 1993). Although loosely definable as an unusual loss of function over and above that associated with "normal" aging, it can only be diagnosed from behavior—when consciousness is unclouded by drowsiness, delirium, or stupor—and not from computed tomography (CT) scan, electroencephalogram (EEG), or positron emission tomography (PET) imaging, even though these procedures may help identify possible causes.

The disorder often has an insidious onset without a clearly defined cause, and no clear threshold can be set on the continuum between normality and dementia; the latter can be said to occur when global cognitive impairment is sufficient to interfere with normal social functioning. Thus dementia may be seen as a loss of function in multiple cognitive abilities in someone previously of normal intellect, and occurring in clear consciousness (Whitehouse et al., 1993).

The criteria, according to the *Diagnostic and Statistical Manual of Mental Disorders* (DSM-III-R) (APA, 1987), include the following:

1. Loss of intellectual ability, with resulting social and occupational handicap
2. Memory impairment at all levels of encoding, storage, and retrieval
3. One or more of impaired thinking and judgment, and aphasia, apraxia, agnosia, constructional difficulties, and personality changes.

However, unlike isolated aphasia, apraxia, or agnosia, dementia usually worsens. Because patients all differ in their premorbid brain architecture, experience, locus of pathology, and their reaction to impairment, commonalities may be hard to find. These may however include impairments in memory, problem solving, and affect, with perhaps changes in attentional processes (e.g., the ability to hold, or to shift, attention, or to attend to more than one event at a time) mediating much of this decline. However, cognitive decline is not necessarily an indicator of dementia, if we are to distinguish the latter from "normal, benign" senile forgetfulness. Nor indeed is cognitive decline or outright dementia even an inevitable consequence of aging, though both are certainly age related, being rare below 65, and rapidly increasing thereafter. As far as dementia is concerned, there may be about 2% annual increase after 75, with around 25% being significantly demented by 80. Early cases may have some, tragic insight into their disorder, with consequent depression, though fortunately anosognosia usually supervenes quite rapidly. Dementia is usually accompanied by a reduced sensitivity to the environment, with a consequent tendency to perform repetitive, stereotyped movements or actions, and a deficit that is greater for controlled (i.e., conscious, deliberate, or planned) processing than for automatic actions or responses. Chiu (1994) asked whether the term *dementia* should perhaps more properly and charitably be replaced by *dysmentia*. Be that as it may, no one behavioral test unambiguously discloses it, though a surprisingly accurate rough-and-ready approach is to ask for the naming of as many animals as possible within 1 min: 9 or less may indicate definite dementia, while 16 or more is an indicator of quite unimpaired functioning (Huppert, 1991). Good health, education, high initial intelligence, and a wide range of interests all seem to help provide protection, though we shall later ask whether it is a matter of "mental jogging"—"if you don't use it you'll lose it"—or instead the protection provided by initial spare processing capacity which is itself associated with life style of the sort described above.

So far we have used the term dementia in an undifferentiated fashion. Elsewhere in this book we shall frequently refer to subcortical dementia, by contrast with the more familiar "cortical" dementia, such as AD, with which this chapter will largely be concerned. Not everyone accepts the concept of a subcortical dementia, or of a distinction between the two manifestations (for reviews, see e.g., Cummings, 1990; Devinsky, 1992; Huber & Shuttleworth, 1990). Indeed, diseases usually extend beyond subcortical structures into the cortex, especially the frontal lobes responsible for a variety of "executive" and "strategic" functions involving attention, the establishment, maintenance, and shifting of set, sequencing of behaviors, and so forth. Moreover, such prototypical "cortical" dementias as AD also involve much subcortical damage. Indeed, cortical and subcortical structures are intimately interconnected, and as we saw in Chapter 2, aphasia can result from subcortical damage. The distinction may still however be anatomically and heuristically useful. Lesions inducing subcortical dementia are generally located in nuclei synthesizing important modulatory or activating neurotransmitters (see below) for example, dopamine (DA), serotonin, noradrenaline (NA), or acetylcholine (ACh), or in structures constituting targets of such systems, or in the tracts transporting such transmitters. Vascular lesions leading to small, deep infarctions (lacunar state), or affecting the basal ganglia (BG), thalamus, or adjacent white matter, may also cause subcortical dementia.

The symptoms of subcortical demential include problems with memory (recent more than remote, both for encoding and retrieval), bradyphrenia (slowed information processing), cognitive deficits (involving the ability to manipulate previously acquired information, calculation, etc.), visuospatial deficits (as assessed by picture arrangement, block design, and object assembly tests), personality and mood changes including depression and apathy, and frontal "executive" deficits as described above, and stemming from damage to striatofrontal circuitry. There may also be slowing in movement initiation and execution, along with adventitious movements (tremor, chorea, and dyskinesia), though myoclonus is more characteristic of cortical dementia.

Causes of Dementia

An enormous range of circumstances are associated with the disorder (Whitehouse et al., 1993). Some major ones include the following:

1. Primary degenerative diseases, for example, AD, Huntington's disease (HD), Parkinson's disease (PD), and so forth
2. Vascular disorders, for example, multi-infarct dementia (MID)
3. Metabolic and endocrine disorders that often are reversible (e.g., disorders of the thyroid, pituitary, calcium metabolism, diabetes, and kidney or liver failure)

4. Infection (e.g., Creutzfeldt-Jakob's disease, kuru, AIDS, meningitis, encephalitis, neurosyphilis, etc.)

5. Toxins such as alcohol, drugs, metal poisoning, insecticides, and iatrogenic disorders (over- or inappropriate medication)

6. Anoxia, cardiac arrest, carbon monoxide poisoning

7. Psychiatric disorders, depression, schizophrenia

8. Nutritional problems and avitaminosis

AD and MID are the most common forms of dementia, and are relatively easy to distinguish from the rest. In MID, prevalence as with most cardio- and cerebrovascular disorders is higher in males. Onset is typically abrupt, being associated with a cerebrovascular accident. Progression is steplike, with episodic exacerbations between plateaus, and initial symptoms tend to be patchy, local, and focal, often with headache and giddiness (see e.g., A. Jones & Richardson, 1990). In these respects MID is somewhat atypical of the other dementias.

Alzheimer's Disease: General Considerations

This disorder, the most common and most devastating form of degenerative dementia, still has no single agreed cause (though cf. recent genetic advances, below), any more than in the mid-19th century when it was variously ascribed to apoplexy, falls on the head, menstrual problems, masturbation, and senility in general. Over half of all cases of dementia may be ascribed to AD, with maybe one sixth to MID, and the rest to the other causes outlined above, many of which are to a greater or less extent treatable. At the moment, AD most certainly is not. Both of the epidemiological indices, prevalence (frequency, in percent, of all current cases for a given place and time), and incidence (frequency of additional new cases per year), show an exponential increase of AD with age, a doubling every 5 years in fact, though fortunately there is no evidence of any change in incidence rates in recent decades. Depending upon criteria, somewhere between 5 and 20% of the population over 65 may exhibit some degree of AD, and perhaps 50% over 85 (see e.g., A. Jones & Richardson, 1990). Unlike MID, it is somewhat more common in females. Just as in PD, the increasing prevalence, with age, of AD seems to reach a plateau around 90 years (T. Wernicke & Reischies, 1994).

AD was first described by Alois Alzheimer in 1906, as a presenile dementia associated with neuritic (senile) plaques and neurofibrillary tangles (see below), which he found at autopsy in the brain of his 56-year-old demented patient. Until recently, AD was indeed considered a *presenile* dementia, though nowadays it is generally recognized that there are no *clinical* differences between presenile (onset before 60) and senile (post-60) forms. Indeed clinical signs of AD may appear only when a threshold number of such plaques and tangles has been exceeded. Thus normal aging may differ from AD only in degree, given that some plaques and tangles are usually present at autopsy, even in the brains of very elderly patients who had shown no clinical signs of dementia. Aphasia, apraxia, and agnosia however do not necessarily occur in normal aging, as they do in AD, and A. Jones and Richardson (1990) argued that AD is a true disease process, or even a collection of different if perhaps related disorders, rather than merely an extension or acceleration of normal aging. Indeed West, Coleman, Flood, and Troncoso (1994) compared the regional pattern of neuronal loss in the hippocampus associated with normal aging and with AD and concluded that the neurodegenerative processes are qualitatively different, with AD a distinct pathological process.

Criteria for Diagnosis and Manifestations of Alzheimer's Disease

There are perhaps three criteria for the diagnosis of AD:

1. Dementia of the "cortical" type (i.e., associated with problems in the areas of memory (especially), language, object recognition, and praxis, together with a deterioration in general intellectual and cognitive function)

2. An insidious onset with uniformly progressive deterioration, even though it may have apparently been initially unmasked after a period of stress or a traumatic event like the death of a spouse or a major relocation

3. Exclusion of all other possible causes, such as stroke, infarct, drugs, and iatrogenic factors, avitaminosis, and so forth. Like the other dementias, it can only be firmly diagnosed after biopsy or more usually autopsy. Until then only a provisional diagnosis is possible, by exclusion of other possible causes, of "probable" AD. Nevertheless, brain imaging (CT, magnetic resonance imaging [MRI], PET) may help. Thus, thin cortical gyri and enlarged sulci and ventricles may be apparent. The width of the medial temporal lobe—a structure essential, as we saw, for memory—is characteristically greatly decreased (A. Jones & Richardson, 1990), and this may in fact almost prove diagnostic. Blood flow and metabolism, likewise, are commonly decreased early in the disorder over the parietotemporal cortex.

Early behavioral manifestations of AD include forgetfulness and impairment of recent episodic memory, but not of old long-term memory (LTM) or of the short-term memory (STM) span. Subsequently, there may be transient confusion, deficits in intellectual and general cognitive functioning, problems with concurrent tasks and the manipulation of data in working memory, and difficulties with judgment and abstract reasoning, such as calculation, the interpretation of proverbs, and the detection of absurdities in pictures. Sensory deficits rarely manifest, and motor disorders only appear late in the progression of the disorder. After a year or two problems will emerge with language, including word-finding difficulties, and with new environments and later even with familiar ones; questions or stories will be repeated; there will be carelessness in dress or hygiene, untidiness, periods of lethargy, indifference, and apathy alternating with restlessness, errors of judgment, rigidity, impatience, argumentativeness, aggression, and disinhibition. There may also be early loss of insight and of interest in current events, inability to grasp new ideas and to discern the significant from the trivial. Although problems with novelty and failure to adapt to changed circumstances are often early alerting factors, with preserved performance during simple, familiar routines, instrumental disorders such as aphasia, apraxia, and agnosia typically come later. Initially, apart from dementia, the neurological examination is grossly normal. Later, there may be mild hyperreflexia and release of primitive reflexes such as the snout and palmomental (Morrison-Bogorad, Rosenberg, Sparkman, Weiner, & White, 1993). The EEG may be normal or diffusely slowed. Language impoverishment develops, with progressive simplification of vocabulary and syntax, and problems with analogy and metaphor. Personality changes often manifest as an extreme caricaturization of premorbid character, beyond the level of eccentricity; thus meticulous individuals may become frankly obsessional. Rarely is there significant change or reversal, and preservation of prior personality may be more associated with vascular dementia (MID). Indeed, some sufferers of AD with gross intellectual deterioration may still perform quite well socially, even though there is major deficit in such "executive" functions as concept formation, problem solving, set, and attention. Later, paranoia, hallucinations, delusions, and sleep disturbances may accompany noradrenergic and serotonergic dysfunction. According to Gilley (1993), psychotic symptoms generally tend to correlate with greater levels of cognitive impairment, faster cognitive decline, earlier onset of the disorder, and motor symptoms. Other behavioral disturbances include depression, apathy, and inappropriate behaviors. Depression can itself cause cognitive impairment and may be a consequence of organic brain syndromes per se. There is debate (Gilley, 1993) about whether depression is associated with rapid cognitive decline in AD or early onset, and even whether it may *precede* clinical manifestations of the disorder as perhaps in PD (see Chapter 12). Apathy in AD is probably associated with damage to prefrontal–BG circuitry (see Chapter 10). Purposeless or inappropriate (socially or sexually) behaviors also probably stem from prefrontal involvement and disinhibition, although extrapyramidal damage may underlie increasing muscle tone and myoclonus and gait disorders. Eventually, the patient becomes bedridden, mute, and unresponsive, with death typically occurring 5–10 years after clinical onset, by pneumonia consequent upon lack of respiratory and pharyngeal control. Fortunately, except perhaps in the early stages of the disorder, or with early-onset familial variants, there is usually lack of awareness (anosognosia) for the condition.

Memory Dysfunction in Alzheimer's Disease

Memory disorders appear among the very first symptoms and are about universal in AD (Nebes, 1989). They occur, of course, to a lesser extent, in normal aging. They are, initially, typically more apparent at the episodic than at the semantic level, for more recent than for remote memory, and for those assessed via explicit rather than by implicit techniques. They involve problems in the spatiotemporal location of autobiographical events, especially for recently acquired information, though eventually only the autobiographical highlights may—temporally—remain (e.g., name, spouse's name, job, etc.). Although initially the immediate STM span is relatively little impaired, recall after a short retention interval is greatly affected, all the more so, in fact, if the material has to be reordered in memory before recall (e.g., reverse-order recall). Functioning in everyday life is impaired by forgetting appointments ("future memory") and the location of objects. There will be numerous attempts to cope, with "aide-memoire" notes littering the house usually to no avail. There will be frequent iterations in conversation, confabulation, false recognition, misidentification, and a general inability to integrate past and present so as to anticipate the future. There will be a severe acquisition deficit, with failure to improve, unlike controls, when given good encoding strategies. However, the subsequent rate of forgetting, if material can be acquired, may then be normal. Patients tend not to cluster or group related items by category when material is presented in random order, unlike controls; nor do they benefit, on recall, if the material *had* been originally presented in clustered or grouped form, rather than scrambled, again unlike controls.

Many of the memory problems suggest a deficit in *controlled* (compared to *automatic*) processing, suggesting a drastic diminution in processing resources. Thus there will be problems in recalling a digit string backwards, despite preservation of *implicit* memory, skills, and language, which require the least controlled and most automatic level of processing. Thus once again we meet the important distinction between, on the one hand, an automatic, unconscious, effortless running-off of overlearnt skills, and on the other hand, the attentional, conscious, effortful, and controlled acquisition of new skills, problem solving, and so on, which all require processing resources and attentional deployment. Processing resources and the deployment of attention of course themselves demand frontal involvement, especially of the cholinergic nucleus basalis of Meynert, a subcortical forebrain structure known to be an early target of AD. Thus registration deficits in AD may reflect problems in supervisory attentional control, via a damaged cholinergic modulatory system, together with poor use of imagery, mnemonics, and encoding organization. Consequently, performance in STM tasks concurrently undertaken along with other secondary tasks is badly decremented in AD. Similarly, recall, a "controlled" nonautomatic operation, is more affected in AD than is recognition, which is more automatic. Likewise, the largely conscious (or potentially so) contents of episodic memory are more affected than the largely unconscious semantic aspects, whereas procedural memory and skilled behavior, which are least conscious and effortful and most automatic, are preserved longest. Semantic information, in so far as it relates to language (see below), may merely become simply less accessible. Thus AD patients show the normal pattern of faster naming of a visually presented final word that completes an auditorily presented sentence, when that word fits the meaning of the sentence. Patients also show normal semantic priming effects, and the normal pattern of faster responses to "typical" stimuli (e.g., "carrot") when deciding if they belong to a target category (e.g., "vegetables") (Nebes, 1990b). Again, the problem with AD may be more one of memory access, especially for controlled rather than automatic information processing, with relatively well-preserved information at a storage level. Indeed, the two fundamental deficits in AD seem to intertwine at various levels:

1. An *attentional* deficit, involving frontal and parietal structures
2. A *memory* deficit, involving temporal, limbic, and forebrain structures.

All of these areas are known to suffer damage in AD, and to be at least partly under cholinergic modulatory control.

Language Deficits in Alzheimer's Disease

We have already addressed some issues relating to language in the context of memory and retrieval. Language problems *per se* manifest early in the course of the disease, though they usually follow the initial signs of memory dysfunction (Nebes, 1989; Zec, 1993). Semantic aspects tend to be lost before the more automatic phonological and syntactic aspects. Indeed, late in the disease there may be preserved (and automatic) correction of another's faulty grammar, without comprehension of what was said. Anomia and word-finding difficulties appear first, followed by a deficit resembling transcortical sensory aphasia (see Chapter 2), speech being fluent and paraphasic, circuitous with little content, with impaired auditory comprehension, and relatively intact repetition. Later this gives way, with increasing impairment of repetition, to a Wernicke-like aphasia that progressively extends to a Broca-like syndrome, with eventual mutism. Semantic errors, on confrontation naming of an object, tend to involve giving the name of the superordinate category (e.g., a picture of a file evokes the response "tool"); there is loss of access to specific details. Indeed, patients may stress irrelevant or inappropriate aspects on confrontation, saying the file is "pretty," or that a table and chair should be grouped together "because both have four legs," rather than because both belong to the category of furniture.

With free associations, normal people typically offer *paradigmatic* associations (e.g., association by meaning as with, boy → girl). Patients with AD however tend to *syntagmatically* associate, associating by syntax (e.g., boy → plays) (Nebes, 1989). With interpretation of proverbs and metaphors, patients, like people with RH damage (see Chapter 4), tend to make *literal* interpretations, losing the connotative aspects (e.g., Make hay while the sun shines → When the farmer is harvesting . . .).

Other Deficits

Visuoperceptual and constructional disabilities manifest in midstage rather than early in the disorder. Face recognition is affected, as are constructional abilities, reflecting parietal involvement. Apart from constructional apraxia (and see Chapter 10), ideomotor and ideational apraxia is apparent, though the patient may be relatively less affected while operating at home with familiar objects in a familiar environment (Zec, 1993).

Neuropathology in Alzheimer's Disease

Later in the disease, there is widespread cortical atrophy and ventricular enlargement, with loss of the larger cortical neurons, especially frontally and temporally, and loss of dendritic branching. With normal aging, possibly by way of compensation, or as a result of storing a lifetime's experiences, there tends to be dendritic growth. There is massive cell loss in the cholinergic nucleus basalis of Meynert and the nucleus of the diagonal band complex, which between them provide modulatory and activating cholinergic input to the hippocampus and much of the cortex. Other similar modulatory neurotransmitter losses also occur (see e.g., Moore, 1990):

1. *Noradrenergic.* The locus coeruleus in the isthmic tegmentum is composed of noradrenergic neurons that are located in the central gray and reticular formation, and, being pigmented, contain neuromelanin. It projects to the cerebellum, septum, amygdala, spinal cord, brain stem sensory nuclei, cortex, and thalamus.

2. *Serotenergic.* The raphe nuclei extend from the midbrain to the caudal medulla and produce serotonin (5-hydroxytryptamine, 5-HT). The largest and most rostral of them, the dorsal raphe nucleus and the central superior nucleus, project predominantly to forebrain structures in a similar though more widespread fashion than the noradrenergic neurons of the locus coeruleus. Passing through the medial forebrain bundle, they project to the thalamus, striatum, brain stem, hypothalamus, cortex, and basal forebrain.

3. *Dopaminergic.* As we shall see in Chapter 12, the substantia nigra

(SN) (pars compacta) has a nigrostriatal projection to the striatum concerned with movement and a mesolimbic projection to the basal forebrain. Another mesolimbic projection arises from the ventral tegmental area, which also has a mesocortical projection to much of the mesial frontal lobe. Innervated areas of the mesolimbic projection include the nucleus accumbens, septal nuclei, the olfactory tubercle, the amygdala, entorhinal, piriform, and cingulate cortices and the hippocampus.

In addition to loss of such modulatory and activating neurotransmitter function in AD, there is also massive cell loss in:

1. The *hippocampus*—memory
2. The *amygdala*—emotion and personality
3. *Olfactory areas*—problems in the sense of smell appear early in AD (and also in PD) and may be partly diagnostic; see A. Jones and Richardson (1990) for review, though Serby, Larson, and Kalkstein (1991) are critical of the hypothesis that infection via the olfactory system could provide a route for the spread into the brain of an infective agent.
4. The *entorhinal limbic system*—which connects to the hippocampus, and so, if damaged, leads to the latter's effective isolation and disconnection (Zola-Morgan & Squire, 1993).

Cell loss also occurs extensively in cortical areas:

1. The *parietal* cortex—leading to spatial problems and anosognosia
2. The *temporal* cortex—leading to agnosia and aphasia, as well as memory deficits
3. The *frontal* cortex—leading to strategic difficulties, executive and social problems.

Two types of lesions are particularly characteristic of AD: neuritic (senile) plaques and neurofibrillary tangles (R. Clark & Goate, 1993; Goedert, 1993). Of course both are only evident at autopsy.

Neuritic plaques occur extracellularly in the vicinity of neurites (axons and dendrites) and are small (15–20-μm diameter) spherical accumulations of cellular debris that consist of β-amyloid protein cores (6–10-nm diameter fibrils) in the extracellular spaces. (This protein also appears within the walls of cerebral blood vessels, and we shall shortly see that recent genetic studies have shown that it plays a fundamental role in the etiology of AD.) The plaques are surrounded by degenerating neural processes, reactive microglia, and astrocytes and occur especially in the terminals of cholinergic neurons of the hippocampus and cortex, and in the amygdala. They occur predominantly in the neuropil layers, regions full of synapses, and eventually vanish, leaving an empty space. The number of plaques tends to parallel the degree of cognitive decline, though some are present in the brains of the "normal" elderly, perhaps indicating a preclinical condition. Aluminum complexes may lie at the center of these plaques, showing up with a silver stain, though it is possible, at least with plaques if not with tangles, that the aluminum may be an artifact of the technique used (Landsberg, McDonald, & Watt, 1992). Indeed the whole role of aluminum in food and water as a risk factor in the etiology of AD is hotly debated (Martyn, 1992), and will be discussed further.

Neurofibrillary tangles are abnormal, paired double-helical filaments (20 nm in total diameter) within the cytoplasm itself of the cell bodies and the apical dendrites, especially of the pyramidal cells. The most common histological feature of AD, they are made up of the microtubule-associated protein tau in an abnormal phosphorylated state. It is possible (R. Clark & Goate, 1993 though see Goedert, 1993) that exposure to fragments of the amyloid precursor protein leads to hyperphosphorylation of tau, which then polymerizes to form stable paired helical filaments. Such tangles are particularly common in the hippocampus, where granular vacuolar degeneration may also be found, consisting of clear, round cytoplasmic zones 4–5 μm across with a centrally stained core. Like plaques, tangles are not unique to AD, occurring also in kuru, a viral disease associated with ritual cannibalism in New Guinea. The level of cognitive decline correlates even higher with tangles than with plaques. Primary projection areas are little affected, with progressively worse involvement out into association cortex, which is of course the most highly evolved in humans. Parietal, temporal, and frontal regions are heavily affected (see Figure 1), together with limbic structures, the raphe nucleus, the SN, and above all the cholinergic basal forebrain. The latter includes the septal nuclei and the nucleus basalis of Meynert (within the substantia innominata). Cholinergic input to the cortex is reduced through loss of the biosynthetic enzyme choline acetyltransferase. However, only limited benefits have been noted from treatment with anticholinesterase inhibitors such as physostigmine, due probably to the concomitant loss also of the other modulatory neurotransmitter systems—NAergic, serotonergic, and DAergic. Indeed atrophy of the cholinergic nucleus basalis of Meynert may not even be the ultimate cause of AD dementia, as severe atrophy of this structure in such spinocerebellar syndromes as olivopontocerebellar atrophy (see Chapter 14) is not associated with corresponding dementia (Kish et al., 1988). We might then have to look instead to local cortical neuronal loss, plaques, and tangles in explaining the dementia of AD, and the reduction in levels of the excitatory neurotransmitter glutamate in the pyramidal cell region.

The Role of β-Amyloid in the Etiology of Alzheimer's Disease

The β-amyloid protein (a unique 39–43 amino-acid peptide deriving from the much larger amyloid precursor protein) occurs in the blood vessels as well as the brain itself (Selkoe, 1991). It accumulates in the elastic lamina of brain

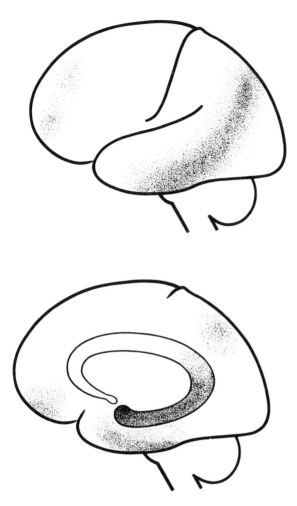

Figure 1 Distribution and severity (lateral, top, and mesial, bottom), of degenerative processes in typical Alzheimer's disease. (Figure redrawn from B. Kolb & I. Q. Whishaw, 1990.) *Fundamentals of human neuropsychology* (3rd ed.) (New York: W. H. Freeman, p. 828).

blood vessels, and may be revealed by congo red or thioflavine staining. It has the property of self-aggregating into 8-nm fibrils. The gene ultimately responsible for its coding is found on chromosome 21; an extra copy of this chromosome is also the cause of Down's syndrome, and individuals with Down's syndrome typically exhibit signs and symptoms of AD by their fourth decade. If β-amyloid protein is injected into the hippocampi of rats, plaques and tangles result similar to those occurring in AD, an effect which is stopped by the injection of the neurotransmitter substance P. The therapeutic possibilities are self-evident, though there are suggestions (Regland & Gottfries, 1992) that β-amyloid may instead have a secondary role in the disease process, even as a protective reactant when brain cells are injured. However, as Morrison-Bogorad et al. (1993) observed, physical trauma and ischemia may result in the deposition of β-amyloid. There are also suggestions (Breitner et al., 1994) that people taking nonsteroidal anti-inflammatory drugs for arthritis may have a lower incidence or slower progression of AD, which could indicate an inflammatory autoimmune response being provoked by the β-amyloid protein (for review, see Schnabel, 1993). Other observations note that aging is associated with increasing vulnerability of mitochondrial DNA, leading to problems with oxygen metabolism, production of excess free radicals, which can lead to amyloid plaque formation, and increased levels of the excitatory and neurotoxic neurotransmitter glutamate (Growdon, 1992). Glutamate of course is involved in the long-term potentiation of memory in the hippocampus (see Chapter 8), and may act as an "excitotoxin" in AD just as in HD (see Chapter 13). Consequently, glutamatergic neurons decrease in number, especially in the hippocampus, where neuritic plaques and neurofibrillary tangles form. Thus a combination of antioxidants, anti-inflammatories, and glutamate antagonists, together with choline agonists, may provide the ultimate therapeutic key. Other additional possibilities are to add the nerve growth factor gene specific for ACh to hobbled retroviruses, which can alter the host cells' DNA, so as to place the gene within the brain cells, or to employ cholinergic neural transplants as in PD, though now to the basal forebrain (Dunnett, 1991). However as Growdon (1992) observed, pharmacologic potentiation of cholinergic neurotransmission has not substantially improved patients' memories. Nerve growth factor is a neuropeptide promoting and regulating the survival of specific neurons during brain development. Normal aging may lead to reduced levels of the substance, and even to AD. The factor induces the rate-limiting enzymes for the synthesis of ACh and NA. For this reason its infusion may slow the development of AD (Kennedy & Whitehouse, 1993), though it may also promote the expression of the amyloid precursor protein.

The Inheritance and Genetics of Alzheimer's Disease

Some cases of AD are sporadic; others are clearly inherited. People with first-degree relatives with AD are thought to have a 50% chance of getting AD by their mid-80s, if they survive that long (Breitner, 1990). This suggests autosomal dominant inheritance with full penetrance by the ninth decade. To date, at least four genetic loci have been identified that seem associated with AD (J. Martin, 1993):

1. Chromosome 21, which is linked to early-onset forms, Down's syndrome, and a mutation in the amyloid precursor protein gene
2. Chromosome 14, which is also linked to early onset
3. Chromosome 19, with a locus close to the gene for apolipoprotein E, which is involved in cholesterol transport into the brain, and may cause later onset AD, both inherited and spontaneous (see below)
4. A fourth locus, not yet identified except by exclusion of the above.

In 1993 a gene was found on chromosome 19, with an allele that leads to increased risk for AD after 65 years of age (Corder et al., 1993). It codes for a substance known as apolipoprotein E4 (apoE4), which ferries cholesterol in the bloodstream. ApoE4 is highly concentrated in the plaques and tangles of AD patients. The gene lies on the region of chromosome 19 that is associated with β-amyloid and AD. People without the gene seem to have around 20% chance of developing AD by 84 years; those with one copy (maybe 25% of the population) have a nearly 50% chance by 76 years, and those inheriting the gene from both parents (between 2 and 3%) have over 90% chance by 68 years. The apoE4 form may bind to β-amyloid and anchor the deposits in the brain and within the neurites themselves, instead of in a more diffuse and apparently benign fashion (Hardy, 1994). If so, an intervention strategy may not be too hard to find. Findings and conclusions such as those just described have now been confirmed and extended by Mayeux et al. (1993). However, as Roses et al. (1994) noted it is not necessary to inherit an apoE4 allele to develop AD, nor, at any given age, does inheritance of one or two such alleles predict the diagnosis of the disorder; it merely provides a biological risk factor that even then may only be quantitatively applied to risk estimates when proper epidemiological data are available. The inheritance of apoE4 alleles affects the rate of the process recognized as AD, and we might all be at risk of it if we lived long enough. ApoE4 testing, should it ever become commonplace, will never be diagnostic or even predictive of when or if an individual will develop AD; its main use will be to subdivide patients by biological risk for studies of the rate of disease progression and the testing and use of possible therapeutic agents.

Strittmatter et al. (1994) suggested that maybe it is not so much the case that apoE4 *causes* late-onset AD, but rather that it is the *absence* of the *other* variants, apoE2 and apoE3, which *normally* protect against AD. Thus apoE3 binds to tau protein, possibly slowing the initial rate of tau phosphorylation and self-assembly into the paired helical filaments. Garruto and Brown (1994) reviewed evidence that aluminum might promote, in vitro, the phosphorylation and aggregation of tau protein, and may also induce the accumulation of amyloid precursor protein in rats' brains. Harrington et al. (1994) reported that renal dialysis patients, whose treatment exposes them to greatly increased levels of aluminum, exhibit AD-like changes in the processing of tau protein, with increased concentration of insoluble hyperphosphorylated tau. As Garruto and

Brown (1994) observed, the intracerebral biochemical similarity described in AD, renal dialysis patients, and patients suffering from amyotrophic lateral sclerosis or parkinsonism dementia of Guam (where neurofibrillary tangles occur with very high aluminum concentrations) invites speculation about a shared pathogenesis. Zinc, too, has been very recently implicated in the formation of β-amyloid plaques in vitro (Bush et al., 1994), and in further cognitive deterioration in AD patients (see e.g., Kaiser, 1994, for review).

Although the etiological relationships between tau, β-amyloid, and aluminum are still unclear, how do tau and β-amyloid relate to the cholinergic deficit found in AD? Wallace et al. (1993) reported that induced lesions of the cholinergic nucleus basalis of Meynert elevate the synthesis of β-amyloid precursor protein in the cortex. Lesions causing reductions in cortical noradrenergic and serotonergic innervation similarly induce the synthesis of β-amyloid precursor protein. Consequently, the association of the amyloid peptide with senile plaques may be caused by loss of subcortical innervation. At a more general level, Roberts et al. (1994) found that head injury can promote overexpression of the β-amyloid precursor protein with subsequent deposition of β-amyloid; they concluded that head injury may be an important etiological factor in AD.

There have been recurring hints that females on estrogen therapy are considerably less likely to develop AD (see e.g., Henderson, Paganini-Hill, Emanuel, Dunn, & Buckwalter, 1994, and Mestel, 1993). Estrogen is known to maintain or promote cell growth in the nucleus basalis of Meynert, and also to promote learning. Men, who have a reduced incidence of AD, go on throughout their lives producing testosterone, some of which metabolizes to estrogen, whereas postmenopausal women have neither.

There have also been recurring hints that smoking somehow protects against AD, just as may also be the case the PD (Baron, 1986). However, as Hardy and Roberts (1993) observed, and see also Riggs (1993), such apparent protective effects from smoking could be due to genetic differences between smokers and nonsmokers, which also influence AD and PD survival. Thus, as we saw, apoE regulates the clearance of plasma lipoproteins and so is also involved in coronary heart disease. Patients with the critical genotype will also be at increased risk for heart disease, and so will be culled from the population of smokers. Consequently *surviving* smokers will tend to have fewer individuals with the critical apoE4 genotype compared to nonsmokers, and thus the risk for apoE4-related AD will be increased in nonsmokers.

Katzman (1993) offered a model of AD that is based on analogy to certain cancers (see Figure 2). At time A, initiating factors (e.g., associated with apoE4) lead to brain changes including β-amyloid deposition. At time B this latent phase accelerates, perhaps in response to certain other factors that convert this latent phase into a malignant cascade of events that include the development of plaques and tangles, loss of synapses, and cell death. The preclinical phase precedes the development of the first cognitive symptoms. Eventually, a threshold of

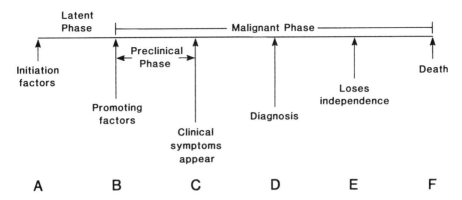

Figure 2 A MODEL OF ALZHEIMER'S DISEASE THAT IS BASED ON ANALOGY TO CERTAIN CAN-CERS, WITH RESPECT TO INITIATING AND PROMOTING FACTORS, AND THE PRE- AND POSTCLINI-CAL PHASES OF THE DISORDER AND ITS PROGRESSION. (FIGURE REDRAWN FROM AN ORIGINAL IN KATZMAN, 1993.)

damage is reached and clinical symptoms manifest at time C, even though the criteria for clinical diagnosis (time D) may not yet have been met. Thereafter there is gradual development of loss of independence (time E), and, eventually, death (time F).

Does "mental jogging" somehow protect us from this melancholy procession of events? It has long been known that a high initial intelligence is an excellent predictor of longevity, and people with a tertiary degree have half the incidence of AD at any given age (Katzman, 1993). However, as the prevalence of the disorder doubles every 5 years, a high initial intelligence and/or sustained mental activity may merely postpone the onset by that amount, perhaps because of the extra neuronal spare capacity in such individuals (see Figure 3). Perhaps the most optimistic alternative explanation is the old finding that an enriched environment is associated with an increase in the number of (juvenile) rats' visual cortical synapses (Turner & Greenough, 1985), coupled with newer studies (Merzenich, 1987) that even in adult primates there can be massive and almost immediate cortical reorganization in response to changed environmental circumstances. We are however still a long way—much further than is the case with PD (see Chapter 12)—from finding a useful therapy for AD. It has long been known that anticholinergics like scopolamine, when given to healthy young volunteers, cause transient AD-like confusion and memory deficits. Unfortunately little therapeutic benefit has been reported from a wealth of studies employing precursors for the synthesis of ACh, inhibitors of acetylcholinesterase such as tacrine, or muscarinic receptor agonists. The best option may be to use a nerve growth factor to prevent degeneration of basal-forebrain cholinergic neurons.

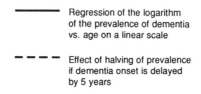

Regression of the logarithm
of the prevalence of dementia
vs. age on a linear scale

Effect of halving of prevalence
if dementia onset is delayed
by 5 years

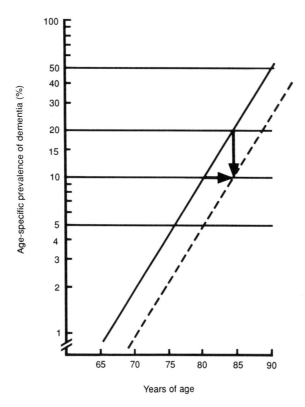

Figure 3   Age-specific prevalence of dementia (expressed logarithmically). (Unbroken line). If its prevalence doubles every 5 years, clearly anything that halves the prevalence will postpone dementia onset by about 5 years (broken line). (Figure redrawn from an original in Katzman, 1993.)

Summary and Conclusions

Dementia, "the disease of the century," is being unmasked by our increasing survival rates, though it is not an inevitable consequence of aging. With an insidious onset and no clear threshold between normality and pathology, dementia involves a loss of function in multiple cognitive abilities in someone of previously normal intellect, and without clouding of consciousness. Impairments in memory, problem solving, and affect occur, due partly perhaps to an underlying and developing attentional disorder. However, no one behavioral test unambiguously discloses it, and it may be associated with a wide range of circumstances (e.g., primary degenerative diseases, vascular, metabolic, and endocrine disorders, infections, toxins, and psychiatric and nutritional problems). Thus, although one-quarter of the population over 80 may be significantly demented, in at least some cases the situation is potentially reversible.

AD, however, the most common cause of disorder, with an exponential increase in age, is not reversible. Debate continues as to whether AD, like PD, is merely accelerated aging, or whether it is a disease process (or even a collection of different, if similar, disorders) in its own right. There is also debate about the position of AD in any distinction, valid or otherwise, between cortical and subcortical dementia. Until autopsy, a diagnosis of AD can only be provisional, by exclusion of other possible causes of dementia. At autopsy, thin cortical gyri and enlarged sulci and ventricles may be apparent, together with considerable reduction in the width of the medial temporal lobe, and an abundance of neuritic plaques and neurofibrillary tangles widely present throughout cortical association areas, parietal, temporal, and frontal, and the hippocampus and amygdala.

Early behavioral manifestations of AD include forgetfulness, transient confusion and disorientation, untidiness, periods of lethargy alternating with restlessness, rigidity, and aggression, along with deficits in executive functions. Language impoverishment develops, though aphasia, apraxia, and agnosia typically appear later. Memory deficits are more pronounced for episodic, recent, and explicit aspects than for semantic, more remote, and implicit components. Disturbed autobiographical memory and confabulation are characteristic. Many of the registration deficits in AD may reflect problems in supervisory attentional control, consequent upon damage to subcortical cholinergic forebrain structures and pathways originating mainly from the nucleus basalis of Meynert (within the substantia innominata), though modulatory neurotransmitter losses also occur in noradrenergic, serotonergic, and DAergic systems. Thus deficits in encoding and access, rather than storage per se, may underlie the memory deficits of AD.

Neuritic plaques occur extracellularly in the vicinity of axons and dendrites, and contain β-amyloid protein cores and, possibly, aluminum complexes. Their abundance parallels cognitive decline. Neurofibrillary tangles are abnormal, paired, helical filaments within the cell bodies themselves and apical dendrites, and they contain the microtubule-associated protein tau in an abnormal phosphorylated state. Their cortical distribution generally matches that of the plaques; the damage they cause to the cholinergic basal forebrain may cause the attentional and executive dysfunction of AD, and their abundance in the hippocampus probably underlies the disorder's memory deficits.

The gene that codes for the β-amyloid protein lies on chromosome 21, an extra copy of which is also the cause of Down's syndrome; individuals with this disorder typically show evidence of AD by their fourth decade. A mutation in the amyloid precursor protein gene (on chromosome 21) has been linked to early onset forms of AD, though abnormalities on other chromosomes, notably 14 and 19, have also been associated with the disorder. Thus a gene has been identified on chromosome 19 with a locus close to the gene for apolipoprotein E, which is involved in cholesterol transport into the brain and may cause later onset AD. The apoE4 form of apolipoprotein E may bind to β-amyloid and anchor deposits in the brain and within the neurites themselves. People with the gene seem to have a considerably increased risk of developing AD. Although the relationships between tau and β-amyloid are still unclear, lesions causing reductions in cortical cholinergic, NAergic, and serotonergic innervation may themselves induce synthesis of the β-amyloid precursor protein. On the other hand, anti-inflammatory drugs may slow an inflammatory autoimmune response provoked by β-amyloid, and estrogen therapy and even smoking have been claimed to have prophylactic properties. Whether "mental jogging" also somehow protects us, on the basis of "if you don't use it, you'll lose it," is still to be determined. However, as the prevalence of AD doubles every 5 years, and individuals with a tertiary degree have half the incidence of AD at any given age, a high initial intelligence, itself associated with sustained mental activity, may merely postpone clinical onset by 5 years because of the extra neuronal spare capacity in such individuals.

Further Reading

Clark, R. F., & Goate, A. M. (1993). Molecular genetics of Alzheimer's disease. *Archives of Neurology, 50,* 1164–1172.

Corder, E. H., Saunders, A. M., Strittmatter, W. J., Schmechel, D. E., Gaskell, P. C., Small, G. W., Roses, A. D., Haines, J. L., & Pericak-Vance, M. A. (1993). Gene dose of apolipoprotein E type 4 allele and the risk of Alzheimer's disease in late onset families. *Science, 261,* 921–923.

Gilley, D. W. (1993). Behavioral and affective disturbances in Alzheimer's disease. In R. W. Parks, R. F. Zec, & R. S. Wilson (Eds.), *Neuropsychology of Alzheimer's disease and other dementias* (pp. 112–137). Oxford: Oxford University Press.

Goedert, M. (1993). Tau protein and the neurofibrillary pathology of Alzheimer's disease. *Trends in the Neurosciences, 16,* 460–465.

Hart, S., & Semple, R. (Eds.). (1990). *Neuropsychology and the dementias.* London: Taylor & Francis.

Jones, A. W. R., & Richardson, J. S. (1990). Alzheimer's disease: Clinical and pathological characteristics. *International Journal of Neuroscience, 50,* 147–168.

Jorm, A. E. (1990). *The epidemiology of Alzheimer's disease and related disorders.* London: Chapman & Hall.

Katzman, R. (1993). Education and the prevalence of dementia and Alzheimer's disease. *Neurology, 43,* 13–20.

Kennedy, J. S., & Whitehouse, P. (1993). Alzheimer's disease. In L. Barclay (Ed.), *Clinical geriatric neurology* (pp. 76–89). Philadelphia: Lea & Febiger.

Morrison-Bogorad, M., Rosenberg, R. N., Sparkman, D. R., Weiner, M. F., & White, C. L. (1993). Alzheimer's disease. In R. N. Rosenberg, S. B. Prusiner, S. DiMauro, R. L. Barchi, & L. M. Kunkel (Eds.), *The molecular and genetic basis of neurological disease* (pp. 767–777). Boston: Butterworth-Heinemann.

Nebes, R. D. (1990). Semantic memory function and dysfunction in Alzheimer's disease. In T. M. Hess (Ed.), *Aging and cognition* (pp. 265–296). New York: Elsevier.

Whitehouse, P. J., Lerner, A., & Hedera, P. (1993). Dementia. In K. M. Heilman & E. Valenstein (Eds.), *Clinical neuropsychology* (3rd ed.) (pp. 603–646). Oxford: Oxford University Press.

Zec, R. F. (1993). Neuropsychological functioning in Alzheimer's disease. In R. W. Parks, R. F. Zec, & R. S. Wilson (Eds.), *Neuropsychology of Alzheimer's disease and other dementias* (pp. 3–80). Oxford: Oxford University Press.

Chapter 10

Movement Control I

Cortical Systems and Apraxia

$\mathcal{T}$he *sensory* systems provide us with an internal perceptual representation of the outside world, both in terms of object recognition, especially via the inferotemporal (IT) cortex, and in terms of object localization, via parietal structures (Mishkin, Ungerleider, & Macko, 1983). This involves the close integration of exteroceptors (especially the eyes) and proprioceptors (stretch, tension, and load receptors in muscles and tendons), which signal object position in coordinates involving retinal location, eye, and head turn. *Motor* systems transform neural signals, derived ultimately from the integration of external (perceptual) and internal (learned and motivational or limbic information), into physical energy, transmitting commands via the cortex, basal ganglia (BG), cerebellum, brain stem, spinal cord, and muscles. We can thereby maintain balance and posture, move head, eyes, and limbs, and communicate via speech and gesture. At a cortical level, these motor representations are largely anterior, while the perceptual processes are posteriorly represented.

To a broad and imprecise level of approximation, we can distinguish between the following classes of movement patterns:

1. Reflex—coughing, the knee-jerk response, withdrawal of a hand from a hot surface, and so on. These responses are rapid, involuntary, and stereotyped, with a graded control proportional to the eliciting stimulus. Many "primitive" (infantile, "subcortical") reflexes (e.g., Babinski, snout and rooting) disappear with growing cortical inhibitory control, only to reappear, "disinhibited," in the event of a stroke.

2. Automatic—walking, speaking, and chewing. These largely rhythmic responses, once initiated, lead to stereotyped, repetitive automatic sequences, and combine both voluntary and reflex acts. Thus often only the initiation and termination of the sequence is voluntary, the rest being performed without conscious effort.

3. Semivoluntary—the need to stretch or yawn and some forms of compulsive touching or tics. Typically they are responses to "irresistible" internal sensations or pressures, as with Tourette's (TS) and "restless legs" syndromes.

4. Voluntary, which may be fully intentional, planned, or self-induced (e.g., to lift an object) or may be responsive to an external signal (e.g., an orienting response, a head or body turn to a call, etc.) Purposeful and goal-directed actions such as manipulation, praxis, and speech involve a high level of learning and improve with practice, though they may subsequently subside into rapid and fluent unconscious automaticity.

Most movements occur at the joints and involve the operation of two opposing sets of muscles capable of contracting—and contraction is the only possible active response of a muscle—one on each side of a joint, which thereby acts as a fulcrum. We see this typical *agonist—antagonist* opposition in the biceps and triceps system of the upper arm, permitting flexion and extension. Usually of course several joints and many muscles are involved (e.g., shoulder, elbow, wrist, and fingers) in reaching and grasping, and debate continues as to whether movement control is *ultimately* coded in terms of patterns of contraction (force and duration) of individual muscle systems, or of the desired trajectory of the limb extremity as it moves through space to a target (Schmidt, 1988). In any case prior postural adjustments must also usually be performed automatically before reaching, grasping, and lifting, to avoid shifts in the center of gravity and loss of balance.

Movement control involves a hierarchical organization, anterior to posterior, of the central nervous system (CNS). The anterior regions are concerned with the more strategic aspects, whereas the more posterior components manage the more detailed execution. At many levels there is a somatotopic (topographic) organization, a one-to-one mapping between adjacent muscle systems and adjacent neural representation. Moreover, each level receives its own sensory input, whereas higher levels control input to and output from lower levels.

Movements will normally be triggered by two events, the recognition of an object (as desirable or threatening) via the IT lobes, and a consequent motivational change (approach or avoidance) registered in the limbic system, especially the cingulate cortex, which is a mesially located infolding of the base of the frontal cortex. Information then passes forward to the prefrontal cortex, which in growth and development most distinguishes us from our hominid ancestors of around 6 million years ago (Bradshaw & Rogers, 1993), to establish intentionality, planning, general aims, and procedural schemata. Information then passes back to premotor regions to establish complex motor synergies, or alternative combinations of muscle systems, to achieve a desired response along one of several alternative routes (e.g., large or small signatures, via movements of forearm and shoulder, or fingers and wrist); at this point appropriate motor programs will be selected for specific environmental demands, with commands passing back to the primary motor cortex for realization. At the same time the cerebellum and BG, the "extrapyramidal" motor system, modulate the above pyramidal activity via the brain stem and thalamus. This provides additional automatic computational assistance for the smooth, well-coordinated, and accurate achievement of complex voluntary praxic behaviors, which finally "exit" via the spinal cord and the corticospinal tract. In the meantime, the parietal cortex maintains the spatial maps wherein action occurs.

Prefrontal Cortex

For "primitive" vertebrates like the lamprey, a system is needed that can maintain orientation against the pull of gravity, which can provide propulsion and that can mediate the simple automatisms needed for catching prey, feeding, and reproduction. For these functions, a brain stem and cord provide a few rhythmic activities that are sustained by oscillatory program generators (Steg & Johnels, 1993). Terrestrial quadrupeds require more highly developed control of posture, locomotion, and skilled paw manipulation. They therefore possess newly acquired orienting, righting, and protective reflexes and learned motor schemata for reaching, catching, and manipulation. Simple programs for the simultaneous control of posture and locomotion continue to be maintained in the brain stem and cord; thus kittens with the cord transected shortly after birth show near-normal hindleg gait patterns when held over a variable-speed treadmill, even changing gait from a trot to a gallop. However, the newly evolved cortex coordinates different motor programs into purposive behavior. Humans, who appeared in the last few million years (Bradshaw & Rogers, 1993) possess even more advanced control of posture, equilibrium, and above all manipulation, with the unique ability to predict and plan, to initiate appropriate and to suppress inappropriate emotional responses. Much of this advanced supervisory control and "executive" functioning resides in our enormously evolved prefrontal cortex, which subserves our unique capacity to think, model, and plan (Benson, 1993b). Thus our prefrontal lobes are larger than those found in any other species, relative to the rest of the brain and when corrected for body weight. Indeed, they constitute about one-third of our neocortex (Kimberg & Farah, 1993), yet their function has remained largely mysterious until very recently. (Other areas of massive cortical evolution in humans include language-related inferior parietal and superior and middle temporal regions.) The prefrontal cortex has no simple sensory or motor function, and is not the seat of intelligence per se; specialist modules for automatic and dedicated processing are generally located posteriorly. However, its major executive and supervisory role *deploys* intelligence *usefully*, mediating set and flexibility via an overall grasp of a current situation (Damasio & Anderson, 1993). If it is not the seat of our intelligence, it is certainly the place where our humanness and personality resides, though its exact functions are poorly understood and not easily subsumed under a single principle. It manages behaviors sequentially in space and time and it organizes goals, intentionality, and anticipatory set, which it maintains or changes as appropriate. It plans, prepares, formulates, and oversees the execution of action sequences; it monitors the strategic aspects of success or failure and the consequences (including social) of actions; it applies both foresight and insight for nonroutine behaviors, and provides a sustained activating and motivating level of drive. By ruminating about past activities and contemplating future ones (Benson, 1993b), it provides a supramodal direction and management of more posterior and automatic brain functions. Consequently, while psychometric intelligence may be well preserved after frontal damage, the patient may no longer behave adaptively. This mismatch between measured and real-life intelligence reflects the fact that the latter requires more basic problem-solving rules and a core system of values based on information that is both inherited (instincts and drives) and acquired (by education and socialization, see Damasio & Anderson, 1993). Much of the input to these prefrontal structures comes from the BG via the thalamus. Indeed inputs to the BG from the sensorimotor association and limbic cortical areas remain anatomically and functionally segregated throughout the BG–thalamocortical circuitry; they influence separate regions of the frontal lobe, where lesions (in either the frontal lobe or the associated BG circuits) are associated with a number of discrete syndromes. Indeed, five distinct BG–thalamocortical circuits have been identified (for review, see Cummings, 1993). All five circuits have a similar basic structure of frontal cortex to striatum (caudate or putamen) to globus pallidus internal (GP_i) segment or substantia nigra pars reticulata (SN_r) or both to various thalamic nuclei, and back (in a closed loop) to the regions of the frontal cortex from which the circuit previously had originated. All remain contiguous but anatomically segregated, and also receive projections from and project to other areas. Within each of these five circuits, there are two pathways, a direct one linking the GP_i/SN_r, and an indirect one via the striatum to the GP external segment (GP_e) to the subthalamic nucleus (STN) and on to the GP_i or SN_r, which will be dealt with in detail when we consider the BG. At this point we might merely note that the STN, which also receives topographically organized cortical inputs from the motor, premotor, and prefrontal regions, regulates the output of the GP_i and SN_r. Lesions in the STN cause involuntary movements (dyskinesias) contralaterally. We might also note at this point that there is important dopaminergic input from the SN pars compacta (SN_c) to the striatum (which is lost in and a major cause of Parkinson's disease [PD]), and from the ventral tegmental area to mesolimbic regions. A ventral limbic circuit, which includes the amygdala, orbitofrontal, and anterior temporal pathways, and which influences the striatum, may modulate responsivity in terms of environmental relevance via fear, anxiety, and negative affect; a dorsal limbic circuit that includes the cingulate cortex may be more important for the more hedonic and positive aspects of stimuli and behavior.

Much of the cognitive, motor, and limbic dysfunction of PD and Huntington's disease (HD) (see Chapters 12 and 13) stems from changes in both the BG and the particular frontal circuits affected. The five circuits are as follows:

1. *Motor circuit*, originating in the supplementary motor area (SMA), premotor, motor, and somatosensory cortices and the superior parietal lobule. This circuit, which unlike the next three, passes through the putamen rather than the caudate in the striatum, and projects via ventrolateral thalamic nuclei to the SMA and premotor cortex, is important in preparatory movement activity and the serial processing of movements initiated in the cortex. Proprioceptive information plays a major role in movement control via a comparator that matches the "expected" consequences of an action previously "fed forward" (as an efference copy) with information ultimately fed back from the *actual* consequences. In this context, the BG are important in scaling the amplitude and velocity of movements.

2. *Oculomotor circuit*, originating in the frontal eye fields, which lie just rostral to the premotor cortex in Brodmann's area 8, as well as the dorsolateral prefrontal (see below) and posterior parietal cortices. This circuit, passing through the caudate and terminating again in the frontal eye fields, is analogous to the above motor circuit. It controls saccade generation at a "cortical" level (cf. similar orienting control of saccades by the superior colliculus), and is part of a larger system responsible for the motor aspects of orienting and exploration with the eyes, head, and probably limbs.

3. *Dorsolateral prefrontal circuit*, originating and terminating in the dorsolateral convexity of the prefrontal cortex. This circuit, passing through the caudate, plays an important role in cognitive and executive functions (see below).

4. *Lateral orbitofrontal circuit*, originating and terminating in inferolateral prefrontal regions. This circuit also passes via the caudate, and like the dorsolateral prefrontal circuit is thought to mediate such behaviors as delayed responding and delayed alternation.

5. *Anterior cingulate circuit*, originating and terminating in that region. This limbic circuit, unlike the others, passes via the nucleus accumbens, and also receives input from the hippocampus, amygdala, entorhinal, and peri-rhinal cortices.

Cummings (1993) reviewed three distinct frontal-lobe syndromes, each corresponding to damage to one of the last three circuits described above.

Dorsolateral Prefrontal Syndrome, with Deficits in Executive Functions

The patient cannot generate hypotheses and flexibly maintain or shift set as required by changing task demands (e.g., with the Wisconsin Card Sorting Test). There is reduced verbal and design fluency or creativity, paucity of ideas, poor organizational strategy for learning tasks, poor construction strategies for copying complex designs, and problems in motor programming with reciprocal motor tasks and sequential motor tests. These difficulties are all reminiscent of subcortical dementia (see below). Thus there are memory problems (for recall more than for recognition), as the region is important for memory search and activation, whereas the temporal cortex may mediate the actual storage processes (see Chapter 8). Indeed, there are extensive interconnections between the dorsolateral prefrontal cortex (DLPFC) and the amygdalar-hippocampal regions, but Funahashi, Chafee, and Goldman-Rakic (1993) see the DLPFC as instantiating a form of working memory that they term *representational* memory. Thus Kimberg and Farah (1993) noted that prefrontal damage causes problems in an apparently disparate group of situations—motor sequencing, Stroop, the Wisconsin Card Sorting Test, and memory for context. Patients may be able to perform all the individual movements needed to prepare a meal, but may get them in the wrong order, or perseverate; they are greatly slowed in naming the color of the ink that constitutes the letters of a color word that is incongruent with the to-be-named dye; when sorting cards according to, for example, color and the experimenter shifts to a new category (e.g., shape), which the subject must determine from the changed nature of the feedback ("right," "wrong"), patients exhibit excessive perseveration and inability to employ feedback; finally, DLPFC patients, although not amnesic per se, display loss of contextual information. They also have difficulty in sourcing a correctly recalled fact or its temporal context. Kimberg and Farah (1993) commented that although there *may* be no common underlying process or deficit, nevertheless there seems to be a strong family resemblance, and they noted that Shallice and Burgess's (1991) central executive or supervisory attentional system fails fully to capture prefrontal deficits in contextual memory. They opt instead for a single underlying impairment in the strength of working memory associations that include representations of goals, environmental stimuli, and stored declarative knowledge. The actual *representations* of the goals are presumably unaffected, because patients may still say what *should* be done, while failing to act accordingly. However, after decades of viewing the (dorsolateral) prefrontal lobes as effectively silent, from the middle of this century it has increasingly come to be generally recognized that they operate as a final stage in information processing, where external sensory input and stored information are integrated, in a supervisory or executive fashion, to determine a particular course of action. They are of course uniquely able to do this being in receipt of posterior sensory and memorial information, and being part of a closed-loop system involving the BG, which may act to sequence or schedule response stages. We therefore see the DLPFC as going beyond a sophisticated system of working memory, and as operating to establish, maintain, and shift attention, where appropriate, generating flexible solutions to novel problems, and as planning and regulating adaptive goal-related activity (and see also Dubois, Verin, Teixeira-Ferreira, Sirigu, & Pillon, 1994, and Taylor

& Saint-Cyr, 1992). As we shall see in Chapter 12, patients with PD may show evidence of frontal or executive deficit; it is unclear whether this is a consequence of loss of appropriate input to the system from a defective BG, or whether this system is itself somehow compromised, due perhaps to loss of dopaminergic activation.

Mesial-Frontal and Anterior-Cingulate Syndrome

This syndrome is marked by profound apathy, lack of spontaneous activity, reduced verbal fluency, emotional indifference, and, in the extreme, akinetic mutism or the apathetico-akinetic-abulic syndrome. Interests and ideas may be intact but are not translated into the least demanding action. Bilateral lesions may be necessary for major permanent deficits. According to some authorities, left-sided lesions may be associated more with depression, and right-sided damage with euphoria (Damasio & Anderson, 1993). Note that the cingulate cortex is a mesocortical paralimbic region that modulates limbic outflow to the prefrontal cortex, and which plays an important role in attention, orienting, affects, and the expression of personality.

Orbitofrontal Syndrome

This syndrome is marked by personality changes, apparent lack of insight, social disinhibition, shallow affect, antisocial and "unacceptable" behaviors, tactlessness, childishness, irritability, elevated and labile mood, fatuous euphoria, automatic initiation of gestures, and compulsive utilization of any objects occurring in the environment (Jason, 1990). Despite such behaviors, if interrogated, the patient may be able to describe accurately what should and should not occur in public—while nevertheless proceeding to perform the latter, though Damasio and Anderson (1993) noted that the orbitofrontal patient is rarely truly vicious or psychopathic. Thus, unlike the prior syndrome where the patient "leaves undone that which ought to be done," now the patient commits sins of commission rather than of omission. Nevertheless, frontal lobotomies (excisions) and prefrontal leucotomies (tract severances) in these regions may help excessively shy, anxious, or "inhibited" individuals. As may be expected, hyperactivity as measured by positron emission tomography (PET) in these orbitofrontal regions may be associated with anxiety and obsessive–compulsive disorder (OCD) (see Chapter 14). In the context of insensitivity to future consequences following prefrontal damage, Bechara, Damasio, Damasio, and Anderson (1994) noted that ventromedial prefrontal injury is associated with severely impaired real-life decision making, with otherwise preserved intellect. Such patients seem unable to learn from their mistakes, while their solution of verbally posed social problems and ethical dilemmas is apparently unimpaired. Bechara et al. developed a card-selection task involving gains and losses, and requiring an ability to develop an estimate of which decks are risky and which are ultimately profitable. They found that such patients seem unresponsive to future consequences, and are more controlled by immediate profit-and-loss prospects. They explain such "myopia" for the future in terms of a failure to act upon otherwise-intact knowledge necessary to conjure up scenarios of future outcomes.

In 1868, Harlow reported a patient, Phineas Gage, who hitherto had been an efficient and trustworthy railroad foreman, "a shrewd and smart businessman, very energetic and persistent in executing all his plans of operation," (p. 327–347). A blasting accident with a tamping iron, a bar 1 m long, 4 cm wide, and weighing 6 kg, drove it in below the orbit of the left eye and out through the top of his head (see Figure 1). He "gave a few convulsive motions of his extremities, but spoke in a few minutes, (p. 327). He was able to walk and eventually recovered, but was

> no longer the same Gage, . . . [becoming] fitful, irreverent, indulging at times in the grossest profanity, which was not previously his custom, manifesting but little deference for his fellows, impatient of restraint or advice when it conflicts with his desires, at times pertinaciously obstinate, yet capricious and vacillating, devising many plans of future operation, which are no sooner arranged than they are abandoned in turn for others appearing more feasible (Harlow, 1868, p. 327–347).

Damage in HD and PD affects the corresponding striatal components of these BG–prefrontal circuits, though the consequences are essentially similar. With HD, caudate damage (initially medial, later lateral) leads to mood and personality disorders, irritability, and "explosive disorder," antisocial behaviors, apathy, and cognitive dysfunction. The latter manifests as set-related problems with the Wisconsin Card Sorting Test, deficits in verbal fluency, and poor recall as with the dorsolateral prefrontal syndrome. Early symptoms of personality change correspond to orbitofrontal and anterior cingulate involvement. Dorsal caudate damage results in executive dysfunctions, ventral caudate damage with disinhibition, and destruction in the nucleus accumbens leads to apathy and loss of initiative. With PD, there may only be a motor disorder as the putamen becomes increasingly inoperative, but if the caudate also is involved, cognitive deficits and dementia may manifest. With both PD and HD, depression may occur with dorsolateral or orbitofrontal hypoactivity or both, and OCDs may manifest with HD and TS with orbitofrontal or anterior cingulate hypermetabolism or both.

Premotor Cortex and Supplementary Motor Area

The premotor cortex (premotor area, PMA) lies just posterior to the prefrontal cortex and anterior to the primary motor cortex. The SMA is a continuation, on the anterior mesial surface of the hemispheres, of the PMA, which oc-

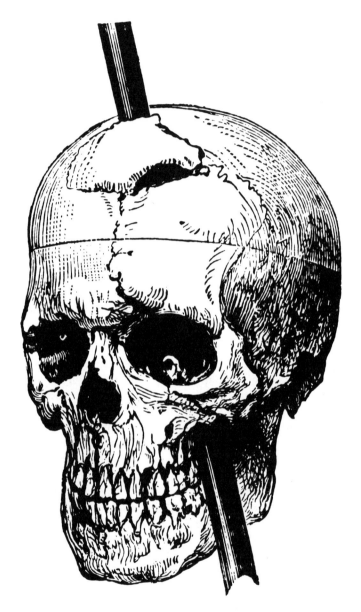

FIGURE 1   THE SKULL, AND PART OF THE TAMPING IRON, BELONGING TO PHINEAS GAGE, WHO SURVIVED THE ACCIDENT. (FROM AN ORIGINAL ILLUSTRATION OF 1868; COURTESY OF MALCOM MACMILLAN.)

cupies the lateral convexity. The two regions lie in Brodmann's area 6 and possess extensive interconnections. Although the SMA receives its main subcortical input from thalamic relay nuclei and the internal segment of the GP, it also receives input from cerebellar nuclei (the contralateral dentate, especially), though the PMA receives most of the latter; the SMA also receives extensive cortical input from the primary and secondary somatosensory areas, the primary motor cortex, and the posterior parietal association cortex, whereas the PMA receives input from the secondary somatosensory and posterior parietal cortex (for review, see e.g., Cunnington, Bradshaw, & Iansek, in press). The SMA outputs to the primary motor cortex (which gets its major ipsilateral input from the SMA, and lesser inputs from the PMA), and also via the corpus callosum to the contralateral motor cortex, PMA and SMA; thus the SMA projects bilaterally to the motor cortex and therefore is intimately involved in bimanual integration and probably (via the corpus callosum) in the inhibition of symmetrical upper-limb movements (synkinesis) and certain "alien-hand" behaviors (Jason & Pajurkova, 1992, and see Chapter 7). The SMA also outputs subcortically back to the striatum of the BG (thereby, as we saw earlier in this chapter, effectively closing the motor circuit) and to the cerebellum.

There are two components to the movement-related potentials that precede voluntary actions. The late component involves a rapidly increasing negative potential beginning about 500 ms before movement onset; it is maximal over the contralateral primary motor cortex. An earlier "readiness" (*Bereitschaftspotential*) potential involves slowly increasing negative activity commencing 1 to 2 s prior to movement onset; it is bilaterally symmetrical and maximal over the vertex and the two SMAs (Mushiake, Inase, & Tanji, 1991). Greater SMA activity precedes noncued or voluntary movements compared to externally cued movements (Shimizu & Okiyama, 1993), and for *internally* cued movements, although activity is greater over the PMA in response to *externally* cued movements (Mushiake et al., 1991). This situation is of course reflected in the nature of the inputs of the two systems—largely internal, via the BG, for the SMA, and external and cerebellar for the PMA. Between them, therefore, the SMA and the BG are involved more in movement sequences, especially if qualitatively *heterogeneous*, synergistic, or multijoint, than in single movements or in the repetition of *homogeneous* movement sequences (Niemann, Winker, Gerling, Landwehrmeyer, & Jung, 1991), as shown by the effects of SMA lesions or of PD that affect the BG. The BG may therefore, along with the SMA, schedule movements, allowing their automatic and coordinated execution (Brotchie, Iansek, & Horne, 1991a,b; Phillips, Bradshaw, Iansek, & Chiu, 1993). Tonic activity (the *Bereitschaftspotential*) builds up over the SMA as the subject prepares to initiate a predictable well-learned response (Ikeda, Lüders, Burgess, & Shibasaki, 1992), and is terminated by a cue in the form of phasic activity from the BG, which performs a synchronizing function and permits release of the upcoming response by the primary motor cortex, and then allows preparatory tonic buildup for the next response element (see Figure 2). PD patients, with disturbed BG output to the SMA, rely heavily on external cues and benefit by their provision (Georgiou et al., 1993), probably via the premotor cortex. Indeed SMA activity is most pronounced for movements fitting into a precued, internally organized timing sequence (Lang, Obrig, Lindinger, Cheyne, & Deeke, 1990), forming a queue of well-learnt and temporally ordered motor commands (Halsband, Ito, Tanji, & Freund, 1993), where the aspect of temporal organization may be more important than motor programming per se (Kornhuber, Deeke, Lang, Lang, & Kornhuber, 1989). Motor programming, *sensu stricto*, may be more the province of the premotor cortex, where lesions may affect multijoint activities of the whole limb.

Activity in the SMA, and mesial frontal motor regions generally, may occur even with *imagined* movements (Roland, Larsen, Lassen, & Skinhøj, 1980). Tyszka, Grafton, Chew, Woods, and Colletti (1994) identified several sites of mesial frontal activity during finger movements; during imagination of the same movements, rostral areas were more active than caudal areas. Stimulation of the

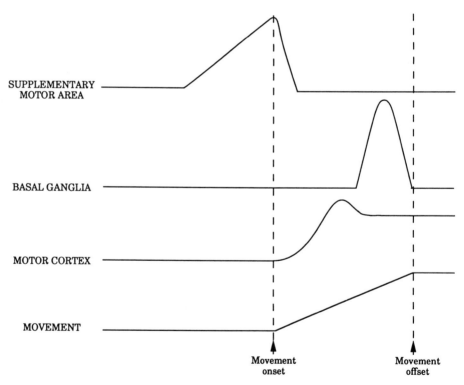

FIGURE 2   SCHEMATIC REPRESENTATION OF PROPOSED NEURONAL ACTIVITY BETWEEN CORTICAL REGIONS AND THE BASAL GANGLIA FOR ONE SUBMOVEMENT IN A SEQUENCE.

SMA produces an urge to perform movements (which is absent with primary motor cortex stimulation, where only the response occurs), an anticipatory feel of an imminent movement, compulsive responding, and response facilitation. Damage to the SMA may also lead to awkward hand postures, clumsy reaching, problems in maintaining appropriate arm muscle contractions with varied loads, and problems with bimanual coordination. Indeed the SMA is thought to play a major role in coordinating the activity of the two hands (Jason, 1990). Both structures, however, the SMA and the PMA, are particularly important in the temporal organization, timing, initiation, and execution of voluntary movements, and their integration into sequences or synergies, and much of the parkinsonian deficit is a consequence of altered BG output to these two regions. Thus increased tonic inhibition of thalamocortical neurons by increased BG output in PD may render the precentral motor areas less responsive to other inputs normally involved in initiating movement, or may interfere with set-related activity (Wichmann & DeLong, 1993).

Primary Motor Cortex (Motor-Sensory Cortex)

The primary motor cortex in the precentral gyrus is responsible for simple contractions of single muscles, but not for higher level planning or synergies. It is one of the last brain regions to be active just before movement onset. Topographically (somatotopically) organized, laterally to medially, it controls the body rostrally to caudally (see e.g., Ghez, 1991a). This representation was predicted by Hughlings Jackson over 100 years ago on the basis of his observation of the orderly sequential spread of epileptiform responses across the body. Like a traveler's "mud map," it is not to scale, with large areas responsible for fine movements (e.g., of fingers and mouth), whereas smaller areas code for coarser arm or leg movements. Columnar organization permits, via lateral inhibition, sharper processing, whereas with laminar organization deeper regions may mediate more specific movements. Its cells code at least two features (parameters) of an upcoming movement, force, and direction. Thus the direction in which a limb is commanded to move represents a weighted sum of all the directions (vectors) signaled by a population of its cells (Georgopoulos, Schwartz, & Kettner, 1986, and see Figure 3). For a given direction, more cells fire that correspond to that direction than fire corresponding to any other direction. Consequently, according to the population coding hypothesis, the commanded direction is the largest weighted sum. Such a system, like a hologram, is of course very tolerant of local injury, as long as the different cells, representing different directions, are distributed randomly (as they are) across the cortex. Reciprocal connections with the parietal sensorimotor cortex provide, via feed forward (reafference), for anticipated changes of sensory input consequent upon an intended movement. Because of the strong sensory as well as motor representation in this region, it is

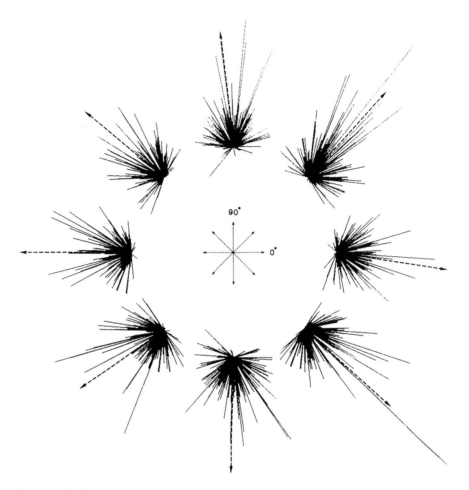

FIGURE 3   THE DIRECTION IN WHICH A LIMB IS COMMANDED TO MOVE REPRESENTS A
WEIGHTED SUM OF ALL THE DIRECTIONS (VECTORS) SIGNALED BY A POPULATION OF MOTOR
CORTICAL CELLS; FOR A GIVEN DIRECTION MORE CELLS FIRE WHICH CORRESPOND TO THAT DI-
RECTION THAN FIRE CORRESPONDING TO ANY OTHER DIRECTION. (FROM A. P. GEORGOPOULOS,
R. CAMINITI, J. F. KALASKA, & J. T. MASSEY, 1983. SPATIAL CODING OF MOVEMENT: A HY-
POTHESIS CONCERNING THE CODING OF MOVEMENT DIRECTION BY MOTOR CORTICAL CELLS.
EXPERIMENTAL BRAIN RESEARCH SUPPLEMENT, 7, 327–336.)

often referred to as the motor-sensory cortex. Damage to the system may lead to
weakness, paresis, an initial flaccid paralysis or hemiplegia, like a limp rag doll,
later giving way to hypertonia, hyperreflexia, and brisk reflexes and spasticity.
Brown (1994) provided a detailed review of the latter, an exaggeration of stretch
reflexes manifesting as hypertonia, with a decrease in their thresholds and an in-

crease in their gain. The result is a velocity-dependent increase in resistance of passively stretched muscles. This resistance may suddenly melt away—the "clasp-knife" phenomenon. The velocity-dependence of the rigidity, moreover, contrasts with the rigidity of PD (Chapter 12). Although spasticity is traditionally ascribed to damage to the pyramidal tract, Brown observed that other tracts are also clearly involved. Damage to descending pathways in the internal capsule may produce more striking spasticity than cortical damage, as such lesions interrupt fibers originating in other cortical areas (and see also Rothwell, 1994). Thus additional premotor and SMA damage may add extensor spasticity to the paralysis typically seen after focal damage to the primary motor cortex. Brown suggested that loss of cortical drive to bulbar inhibitory regions may be the cause of spastic hypertonus after such damage to the frontal cortex or internal capsule. Normal tone is therefore consequent upon a balance between the inhibitory effects on stretch reflexes mediated by the dorsal reticulospinal tract, and facilitatory effects on extensor tone mediated by the medial reticulospinal (and, to a lesser extent, the vestibulospinal) tract. With complete cord transection and loss of all supraspinal influences, hypertonicity may be less marked, as the descending excitatory systems no longer act unopposed; there may also be delayed reorganization, over a year, within the isolated cord.

At this point it may be of historical interest to note that predominantly contralateral motor control was recognized as early as the fifth century B.C. by the Hippocratic writers:

> And, for the most part, convulsions seize the other side of the body; for, if the wound be situated on the left side, the convulsions will seize the right side of the body; or if the wound be on the right side of the head, the convulsion attacks the left side of the body. (Hippocrates, *On Injuries of the Head*, p. 19)

Parietal Cortex

Two major regions are relevant to the control of movement. The *anterior* parietal cortex somatotopically (topographically) codes sensory input. It used to be thought that this representation is fixed and permanent, at least in adult primates; however even here plastic reorganization has recently been found (Pons et al., 1991). In humans, Yang et al. (1994) studied two adult amputees. In one, 8 years after traumatic avulsion of the left brachial plexus, significant intrusions were found of facial localizations into hand territory on the affected hemisphere. In the other patient, studied 11 years after accidental amputation of the right forearm, the affected hemisphere exhibited intrusion of facial and upper arm localizations into the hand region. In both cases displacements exceeded 30 mm. The hand area is of course flanked by that responsible for the face and arm. The authors noted that such findings prove that massive reorganization of sensory maps is possible in the adult human brain, and that new patterns of precisely organized and functionally effective connections can emerge.

Proprioceptive information from muscle stretch receptors (spindles) passes via the spinal cord and medulla (where it decussates) through the lateral lemniscus to the ventral posterior nuclei of the thalamus and on up to the somatosensory parietal cortex. Such proprioceptive information is vital for fine limb control and targeting, and the maintenance of long sequences of simple acts without feedback (Jason, 1990). With damage to this system, old ballistic motor programs may be retained, at least for a period, but new ones requiring fine motor control cannot be acquired. Thus a patient may be able to continue driving his or her old car, but not be able to handle a new one.

The *posterior* parietal lobe massively interconnects with structures within the premotor cortex. These regions integrate information on voluntary head and eye turns (the midbrain superior colliculus performs the same function for more automatic orienting responses), thereby recoding retinotopic information into spatiotopic coordinates. Thus a map is established of the location of events with respect to the body, independent of current head or eye orientation. The somatotopic map of the anterior parietal cortex is thereby integrated with the real world. Damage to the posterior parietal lobe, particularly of the right hemisphere (RH), leads, as we saw (Chapter 6) to the patient ignoring ("neglecting") events occurring in the side of space contralateral to the lesion. It can also lead to a type of constructional apraxia (see below), due in part to disturbances in the ability properly to achieve appropriate spatial integration of composite elements.

Apraxia: A Disorder of Voluntary Gesture and Action

Movement disorders associated with damage to circuits within the BG are characterized by disturbances to the more automatic and even involuntary aspects of movement. Whether they involve such positive symptoms as tremor or rigidity in PD, or chorea in HD, or negative symptoms like akinesia or bradykinesia, or general problems in fluent action sequencing or movement scaling, the patient generally knows what *should* be done. Indeed, if two tasks are simultaneously performed, one overlearned and automatic, and the other deliberate and requiring care, it is the former that typically drops out in PD, while the patient attempts to concentrate upon the harder, less familiar task.

By contrast, apraxia is a disorder of learned or skilled movement, typically associated with left-hemisphere (LH) damage, and which is not attributed to weakness, akinesia, abnormal tone or posture, movement disorders, sensory loss, poor comprehension, inattention, general intellectual deterioration, or uncooperativeness (Harrington & Haaland, 1992; Heilman & Rothi, 1993; Watson, Rothi, & Heilman, 1992). Apraxics, often anosognosic for the condition, rarely present with the disorder as the chief complaint; usually it is noticed later in the context of aphasia, often Broca's or conduction. Indeed a majority of aphasics are also apraxic, though not all apraxics are also aphasic. Moreover, although congenitally deaf users of sign language, such as American Sign Language for the Deaf (ASL), use the same brain areas for ASL as we do for speech (Bradshaw, in press; see also Soderfeldt, Rönneberg, & Risberg, 1994), the regions handling nonlinguistic praxis and language are closely adjacent, and may have recently coevolved in hominid evolution (Bradshaw & Rogers, 1993). Indeed Kimura (1987) suggested that females tend to become both apraxic and aphasic after anterior LH damage, whereas males become so after lesions to either anterior or (especially) posterior regions on that side. This claim is disputed by De Renzi (1989). In any case, it must be emphasized that apraxia and aphasia do often dissociate, and that there are more cases of crossed aphasia than of crossed apraxia (Watson et al., 1992) (i.e., there are more dextral aphasics than dextral apraxics after RH lesions). Thus the RH is more likely to be dominant for language than for praxis, and handedness is more associated with praxis than with language. However, as aphasia is more common than apraxia in dextrals after left-sided lesions, praxis may also therefore be more bilaterally represented than language. Finally, apraxia, as a disorder of skilled movement, is more than just a symbolic deficit, as even nonsymbolic movements, like unfamiliar and intrinsically meaningless sequences, are typically performed poorly by apraxics.

Apraxia involves an acquired deficit in performing complex, purposive, deliberate, skilled sequences, on request and out of context, often with a preserved ability to perform individual components, especially in a naturalistic context (Duffy & Duffy, 1990; Roy & Square-Storer, 1990), though maybe in the wrong sequence. Thus the sign of the cross may be performed automatically on entering a church, but not on command. Apraxia is therefore tested clinically by examination of gesture performance to command or on imitation (e.g., brushing teeth, saluting, or lighting a cigarette). Indeed, there may be an excessive clinical reliance on such familiar movements. There are perhaps six kinds of tests (Watson et al., 1992):

1. Pantomime—"Show me how you would salute"
2. Imitation of pantomime—"This is the way you salute; now *you* do it."
3. Use of object—"Here's a hammer; show me how to use it"
4. Imitation of the examiner using the object (e.g., hammer)
5. Discrimination between object uses that are correctly or incorrectly pantomimed by the examiner
6. Comprehension of gesture—"What am I doing now?"

Pantomime performance on command is typically the hardest, spontaneous evocation under naturalistic conditions is the easiest, with imitation (of an object's use, or of the examiner pretending to use it) somewhere between. Thus deficits in intransitive movements (e.g., making signs) are more characteristic of apraxia than are deficits in transitive movements (i.e., using objects or tools), as objects or tools may provide the naturalistic context wherein to evoke the response automatically. Indeed Heilman and Rothi (1993) distinguished between the following gestures or acts that may be required of a patient in testing for apraxia:

1. Intransitive limb gestures (e.g., waving good-bye, saluting, beckoning, signaling someone to stop, making the sign of the cross)
2. Intransitive buccofacial gestures (e.g., blowing a kiss, sticking out the tongue)
3. Transitive limb gestures (e.g., opening the door with a key, flipping a coin, opening a screw-topped bottle, using a tool)
4. Transitive buccofacial gestures (e.g., blowing out a match, sucking on a straw)
5. Serial acts (e.g., cleaning a pipe, inserting the tobacco, lighting it; folding a letter, putting it in an envelope, sealing and stamping it)

Traditionally, apraxic errors involve the following:

1. Errors of content—substituting one pantomime for another (e.g., hammering instead of using a screwdriver)
2. Errors of spatial production—using a body part as an object (e.g., using the forefinger as if it were the screwdriver, instead of holding the latter and gripping and twisting it in the normal fashion)
3. Errors of temporal sequence—striking a match (or pretending to) and putting it (instead of the cigarette) in the mouth; such switching of elements is particularly apparent with difficult, nonnaturalistic settings.
4. Problems with the rate and speed of movement, such that both single movements and sequences may be prolonged.
5. Problems in initiating movements—a struggle or delay before starting, false starts, and repetitions.
6. Difficulty in temporally coordinating simultaneous movements

Liepmann at the turn of the century distinguished between ideomotor and ideational apraxia (see e.g., Grüsser & Landis, 1991), a distinction that has been resurrected in various guises, and which is still subject to much dispute as to the reality of the distinction, its nature, and its likely localization. According to one approach, ideomotor apraxia involves an inability to perform single movements to command (perhaps with preserved automaticity), whereas ideational apraxia involves deficits in correctly performing action sequences (Rothwell, 1994). Alternatively, at the risk of excessive generalization, we might describe *ideomotor* apraxia as an impairment in knowing *how*, rather than *what*, to do. It is the implementation rather than the concept that then seems to be at fault. Performance is typically worst with transitive limb gestures on command (i.e., with pretend responses), better with imitation of the examiner, and best of all with actual object use. Errors of perseveration, timing, and sequencing are common, together with spatial errors involving hand posture, orientation, direction, and movement of the actual joints. The hand may be used as if it were the tool itself, rather than a guidance system. The patient may however be able to confirm whether or not a model is correctly performing an action. With *ideational apraxia*, a term where Heilman and Rothi (1993) noted there is much confusion, there is perhaps a more fundamental loss of the concept of an object's use, or loss of a whole schema or action plan, so that the patient does not even know what to do, or whether or not a model is correctly performing an action. Although there is therefore typically impaired performance of complex movement sequences, individual elements may be managed appropriately. Heilman and Rothi (1993) emphasized problems in correct action sequencing in ideational apraxia (e.g., putting tobacco in a pipe before rather than after cleaning it) and suggest the term might be reserved for when patients cannot even use the actual object, or when they make major content errors in inappropriate use of an object, or entirely inappropriately pantomime its use (e.g., a toothbrush perhaps being used as a fork, etc.). The distinction between ideomotor and ideational apraxia remains appealing and even heuristically useful, yet it may not be really valid; patients rarely fall neatly into one or the other category, and it is not clear how lesions might localize. In any case, just as with the aphasias, we should probably invoke a network of distributed mechanisms rather than a series of discrete, if interconnected centers, as originally conceived by Liepmann one hundred years ago. Thus McDonald, Tate, and Rigby (1994) asked whether even ideomotor apraxia, on its own, encompasses a heterogeneous group of disorders reflecting damage to different underlying mechanisms. Different performance profiles may reflect deficits at such different processing stages as motor program representation (largely ideational), motor program retrieval (ideational and ideomotor), and motor program implementation (largely ideomotor). In this way we can accommodate such profiles as, for example, an inability to *pantomime* a movement, with preservation of *imitation*, and so on.

According to Liepmann, in dextrals the LH contains the "space–time" engrams or "movement formulae" for controlling purposeful, skilled movements (praxis) on both sides of the body (Rapcsak, Ochipa, Beeson, & Rubens, 1993). He and others based this conclusion upon such observations as the following:

1. Right hemiplegics with LH lesions often demonstrate "sympathetic dyspraxia" of the left hand, whereas left hemiplegics with RH lesions are not similarly apraxic with the right hand.

2. A severe left-hand apraxia often occurs to verbal commands, in the context of imitation and even actual object use, with destruction of (the anterior part of) the corpus callosum. (Such effects tend to be more common or stronger in the early acute phase of disconnection, and where the damage occurred naturally rather than from surgery to relieve long-standing epilepsy, when in fact brain reorganization may already have taken place.)

Thus space–time engrams in the LH were seen to control the right hand directly and the left hand via the commissures. It is now recognized that there are extensive reciprocal connections between parietal and frontal lobes, constituting a distributed network for mediating such complex skilled behaviors as speech and praxis. Anterior premotor damage near Broca's area may affect largely the *execution* side in apraxia (an aspect of ideomotor apraxia), perhaps with preserved *recognition* of an exemplar of correct behavior, whereas posterior parietal damage may destroy the latter capacity itself (an aspect of ideational apraxia). Thus, the supramarginal gyrus on the left may contain the actual engrams or spatiotemporal motor programs, whereas corresponding structures on the right may mediate the more spatial, extrapersonal, or attentional aspects. As we saw, the premotor cortex is involved when skilled movements are made under sensory (usually visual) closed-loop guidance; it is not active when routine tasks are undertaken automatically, without the need for external guidance. Then, however, the closely adjacent SMA assumes an important role in open-loop control, which relies upon internal cueing in scheduling movement sequences. Indeed, the SMA receives input from the GP, the output circuit of the BG, and may thus lie at the interface between, on the one hand, that system's mediation of automatic internally cued response scheduling and control of force production, and, on the other hand, conscious, closed-loop control by the premotor and parietal cortex. The latter two systems in turn receive automatic, subcortical, extrapyramidal backup from the cerebellum for closed-loop visual control. Thus the relationship of the cerebellum to the parietal–premotor system for automatic closed-loop control is similar to that of the BG with respect to the SMA for conscious, deliberate, open-loop sequencing. The superior parietal lobule in turn perhaps acts as a comparator for matching the feed-forward–efference copy, generated by the SMA, to the subsequently obtained proprioceptive feedback, and so ensures that skilled praxic sequences are correctly executed. The inferior parietal lobule, on the other hand, may contain the spatial and temporal memories for extrapersonal, object-related behaviors (Watson et al., 1992). The LH temporal cortex along with the cerebellum (Keele & Ivry, 1990) and the BG are all involved in timing functions.

There is a long tradition of a functional separation between two visual systems. Schneider (1969, and see also Trevarthen, 1970) proposed a cortical–subcortical distinction:

1. The *retinogeniculostriate* pathways subserved object identification and focal vision.
2. The *retinotectal* pathways subserved object localization and ambient vision.

Because this formulation, appropriate maybe for the rat, seemed incomplete at the primate level, Mishkin et al. (1983) proposed an alternative cortical–cortical distinction:

1. A *ventral* occipitotemporal pathway for object discrimination (striate → prestriate → inferotemporal cortex)
2. A *dorsal* pathway for spatial orientation and spatial interrelationships (striate → prestriate → posterior parietal cortex)

Jeannerod and Rossetti (1993) proposed two corresponding cortical visuomotor systems for representing the goal of movement:

1. A pathway involving the inferior and posterior parietal lobule, mediating an object's *extrinsic* properties like its spatial location with respect to the observer's body, its velocity and direction of movement, and so forth. This pathway computes an appropriate trajectory for the hand in extrapersonal space to a given target location, in terms of distance and direction. Thus reaching involves control of the proximal joints of an arm operating within a body-centered coordinate system.

2. A pathway involving the occipitotemporal cortex, and coding an object's *intrinsic* properties, such as shape, size, and orientation, so that it can be "acquired" for manipulation, identification, or transformation. Thus grasping is a highly evolved behavior that relates not to a body-centered coordinate system, but to the object and hand, almost irrespective of hand position in extrapersonal space. Object recognition thereby occurs up to a level sufficient, if not for explicit identification, then certainly for purposive "sizing" or forming of the fingers or grasp appropriate to the size, shape, and consistency of the observed object. As we saw in Chapter 5, such pragmatic or implicit "recognition" can dissociate from explicit object identification, categorization, or memorization. Thus Goodale, Milner, Jakobson, and Carey (1991) reported agnosics who could nevertheless correctly perform the shape of grasp before contacting an unrecognized object.

As Jeannerod and Rossetti (1993) observed, both systems are activated when we reach for an object. From visual signals we prepare two appropriate kinetic codes, one for reaching and one for grasping, and normally the two systems seamlessly merge, with preparations for grasping occurring "in flight" during the hand's reaching trajectory toward the target object. Breakdown of these two semi-independent, automatic control systems, together with the (dys)coordination of concomitant eye and hand tracking movements, is an aspect of apraxia that is now receiving increased attention.

LH damage has long been associated with bilateral deficits in a wide range of motor tasks (e.g., the production of single movements, timing and sequencing responses, memory for sequences, and the perception and production of rapid sequential movement) (Haaland & Harrington, 1994). However, it is not clear what cognitive deficits are general to LH damage, and what are specific to apraxia. Nor is it clear whether the LH is specialized for the following:

1. The control of complex, purposeful movements per se
2. The control of ballistic aspects that are *open*-loop, preprogrammed, and minimally dependent upon sensory feedback
3. The control of the *closed*-loop, sensory-dependent aspects
4. The perception of fine temporal (or even spatial) discriminations, and the production of rapid sequences requiring timing and the coordination of movements, both speech and nonspeech.

Haaland and Harrington (1994), finding that LH damage affects the acquisition of *larger*, rather than *smaller*, targets, emphasized the ballistic, open-loop aspects, and in any case we would suggest an RH involvement in the closed-loop, sensory-dependent aspects. Indeed, Elliott et al. (1993), in their study of hand target aiming, found that a normal right-hand advantage gave way to a left-hand superiority under conditions of spatial uncertainty. They concluded that the RH is important in preparing the spatial aspects of an aiming movement, whereas the left normally mediates movement execution. The ultimate asymmetry (hemispheric, and hand) in a given task is in a real sense the resultant vector of these two component processes (and see also Morgan et al., 1994).

Harrington and Haaland (1992) noted that LH lesions can also produce deficits with single, isolated movements, and so ask whether the sequencing deficits of apraxia are merely the cumulative effects of deficits in programming single movements. They concluded that impaired sequencing occurs over and above separate problems with single movements. They also noted that both clinically apraxic and nonapraxic LH patients have similar problems in scheduling or timing sequential motor programs, but only apraxics have difficulty in reprogramming *heterogeneous*, as compared to merely *repetitious*, hand sequences. They concluded that although LH damage *per se* may lead to deficits in encoding and generating single movements, and in scheduling or timing a simple series of homogeneous actions, only apraxics have difficulty in programming long heterogeneous sequences prior to and during movement. They see apraxia as a breakdown in the parsing of actions into natural rhythmic or spatial groupings, so that the organizing principles for grouping responses are disrupted. The analogy with the parsing, grammar, and syntax of language is obvious, and we remind the reader of our earlier observations concerning the coevolution of language and praxis.

Other forms of apraxia include the following:

1. *Limb-kinetic apraxia*, where pyramidal and corticospinal damage results in loss of fine precise movements in the limb contralateral to the lesion, especially with the distal musculature (fingers). Thus dexterity itself is now affected, whereas in ideomotor apraxia it may be near normal, especially in the ipsilesional arm.

2. *Buccofacial (oral) apraxia*, where there is difficulty in performing skilled movements with face, lips, tongue, cheeks, larynx, or pharynx on command. The

patient cannot pretend to blow out a match, suck on a straw, blow a kiss, etc. Errors of content, sequencing, or muscle systems may occur and may lessen if a real object can be seen or used appropriately.

3. *Truncal apraxia*, where the patient may be asked to bow or to assume a boxer's posture (Heilman & Rothi, 1993).

4. *Constructional apraxia*, which is tested by getting the patient to copy or draw figures, or to arrange blocks in a pattern. LH lesions typically lead to over-simplifications, loss of detail, or poor discrimination of elements and features, whereas RH damage results in loss of outline or gestalt configuration, a haphazard or inappropriate combination of correctly discriminated features, faulty proportions, and spatial interrelationships, and, often (left-sided) neglect (see Figure 4). This classic double dissociation occurs typically with parieto-occipital lesions.

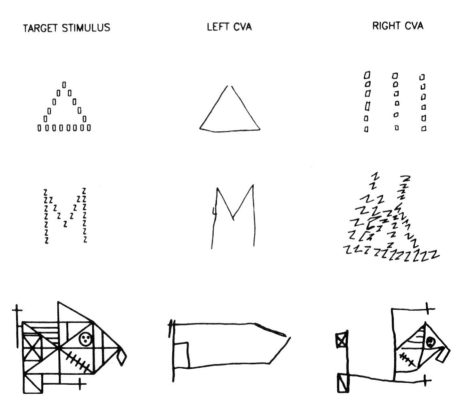

FIGURE 4 DRAWINGS OF TARGET FIGURES COMPOSED OF SUBFIGURES, AND THE REY-OSTERREITH FIGURE, FROM MEMORY BY A RIGHT- AND LEFT-BRAIN-DAMAGED PATIENT (CVA, CEREBRO-VASUCLAR ACCIDENT). (FROM ROBERTSON & LAMB, 1991.)

What role therefore may the RH play in manual praxis, apart from spatial, extrapersonal, or attentional aspects? Normally, the LH may transcallosally inhibit the right, so that the left controls both hands. Release of RH processing after LH damage may be a function of the size of any left-sided lesion. The information stored on the right may be less accessible and require stronger multimodal stimulation than is the case with the LH, according to Rapcsak et al. (1993). Thus the patient, relying upon the RH, may perform well if handling an actual concrete object, which thereby provides tactuo-kinesthetic feedback for the automatic execution of well-established action routines, and far less well if undertaking intransitive gestures by pantomime entirely from memory. Rapcsak et al. (1993) in fact described praxis in a strong dextral after a massive LH stroke destroyed most of that hemisphere. There was severe impairment in pantomime of transitive gestures with the left hand and in reproducing novel, nonsymbolic hand and arm movement sequences, but with relative sparing of overlearnt, habitual actions like actual object use, and intransitive gestures (e.g., sign of the cross). Gesture recognition and discrimination were also preserved. The above authors suggested that the praxis system of the RH is strongly biased toward concrete, context-dependent execution of familiar, well-established action routines; the RH, they say, is critically dependent upon a transcallosal contribution, from the LH, for left-hand control in abstract or context-independent performance of transitive movements, and in learning novel movement sequences. They concluded that in at least some individuals the RH can recognize and discriminate gestures, just as it can understand but not produce speech.

Implicit in our consideration of skilled motor performance, apraxia, and the roles of cortical and subcortical structures in movement control is the concept of the motor program. Originally conceived as a set of muscle commands that are structured in advance, before commencing a movement sequence, and that allow it to be carried out without peripheral feedback, the concept emphasizes open- rather than closed-loop control (Schmidt, 1988). Morris, Summers, Matyas, and Iansek (1994) reviewed evidence for such open-loop control. Thus sufferers from tabes dorsalis lose peripheral sensory feedback but can still perform routine actions such as walking; piano playing and speech may involve sequences that are emitted far too fast for on-line feedback control; anticipatory postural adjustments routinely and automatically occur prior to reaching and lifting; movements may be unexpectedly perturbed by externally imposed loads, but nevertheless can continue for a few hundred milliseconds uncorrected; reaction (i.e., preparation) time is typically (though within limits) a function of the duration of the upcoming movement (execution) time, as if reflecting the time needed to "unpack" the (longer or shorter) motor program. All such demonstrations are at least compatible with the concept of a motor program. However, it is far from clear where such an entity might be located. As we have seen, prefrontal regions seem to specify goal and strategic aspects, whereas the premotor cortex may formulate plans in response to environmental cues—*assembling*, as it were, the motor programs wherever formulated. The SMA seems to link together and sequence commands, as if providing a buffer store for motor programs before their final execution. The BG seem to automatically initiate sequencing, in conjunction with the SMA. The posterior parietal cortex may provide the spatial context within which action can take place, whereas the spinal cord can generate, endogenously, repetitive, rhythmic, serial activity as in walking, stepping, chewing, and maybe even breathing. Does the motor program reside somewhere else, or is it an emergent property of such a functional network? Does the network operate serially, as we may seem to have suggested so far, through the constraints of description, or is it essentially a far more powerful parallel interactive system, each part simultaneously "updating" the rest? We have already suggested just such a system for speech (Chapter 2), and it is likely that similar architectural considerations apply here also, to the motor program. Increasingly, the brain is seen as a system of massively parallel, distributed networks, whose components all interact for a given function, and which may themselves, in different circumstances and contexts, participate in more than one function.

If we do conceive of the motor program as perhaps a higher order construct, consequent upon the functional, as well as the structural, architecture of the motor system as described in this chapter, we must also allow some role for closed-loop feedback contributions, and we must address the problem of how a potentially vast number of individual motor programs, to deal with every different slightly varying environmental contingency, may be stored. We must also accommodate the concept of motor equivalence, the fact that the same end target can be achieved via alternative systems of joints; thus we can produce an idiosyncratic and identifiable signature written large via the forearm on a blackboard, small via the fingers on a check, or even by mouth or the foot. One approach (Morris et al., 1994; Schmidt, 1988) is to invoke the concept of an abstract schema, independent of an actual or particular effector, and independent of a given force, amplitude, speed, and so forth, which may be added in as specific parameters on a particular occasion to suit the circumstances of the movement. The motor schema is therefore an abstract *class* of a domain of actions, which is stored in only the most general form. The motor program is still conceptually and heuristically useful as a concept, but may perhaps more properly be conceived of as a realization of a motor schema that has been "fitted" with the particular parameters to suit the movement. Neither concept, however, should be sought in any one brain region, any more than we should perhaps seek individual phonemes in a register in Broca's area, or morphemes in a lexicon in Wernicke's area. Both regions, together with inferior parietal structures, the SMA, thalamus, BG, and so forth, all contribute, in different ways, to the overall realization of phonemes and morphemes that in any case are themselves meaningful only within a mutual (and in fact much wider) context of speech and language as a whole.

Summary and Conclusions

Motor systems transform neural signals into physical energy, transmitting commands via the (largely anterior) cortex, BG, cerebellum, brain stem, spinal cord, and muscles. Movement patterns range from reflex to automatic, semivoluntary and voluntary, though the distinctions are not always easy to make, and practice can convert a conscious decision to respond in a certain way to unconscious automaticity. Although movements ultimately involve patterns of agonist and antagonist muscle contractions around one or more joint(s), it is unclear whether they are *coded* in those terms, or in terms of the requisite trajectory to be followed by a limb extremity through space. Nevertheless, movement control is hierarchically organized in an anterior to posterior fashion. Anterior regions are responsible for the more strategic aspects, whereas the more posterior components of the motor system manage the details of execution. Thus, intentionality, planning, and procedural schemata are the province of the prefrontal cortex, whereas premotor regions establish complex motor synergies and motor programs, which achieve realization at the level of individual muscles in the primary motor cortex. The extrapyramidal motor system (cerebellum and BG) modulate the above pyramidal activity, while the parietal cortex maintains the spatial maps wherein action occurs.

Our enormously evolved prefrontal cortex exercises a major executive or supervisory role over the more automatic and dedicated processing systems situated more posteriorly. It organizes goals, plans, and action sequences, and monitors the strategic aspects of success or failure, with major input from the BG via the thalamus. Five distinct BG-thalamocortical circuits have been identified, all with a basically similar closed-loop configuration to and from different regions of the frontal cortex. Much of the cognitive, motor, and limbic dysfunction of subcortical motor disorders such as PD and HD stems from changes in the BG and the particular frontal circuits affected—the motor, oculomotor, dorsolateral prefrontal, lateral orbitofrontal, and anterior cingulate circuits. Three distinct frontal-lobe syndromes have also been described, each corresponding to damage to one of the last three circuits above. The *dorsolateral prefrontal* syndrome is associated with deficits in executive functions involving goal-oriented behavior, flexibility, set, fluency, and creativity. With the *mesial-frontal* (anterior-cingulate) syndrome, the patient exhibits profound apathy, lack of spontaneous activity, and emotional indifference. With the *orbitofrontal syndrome*, there are personality changes, apparent lack of insight, social disinhibition, and otherwise unacceptable behavior.

The premotor cortex may be involved in the motor aspects of spatial orientation, preparatory movements and postural adjustments, and choice and operation of an action plan, especially with respect to *externally* generated environmental signals. The SMA, a continuation of the premotor cortex on the anterior mesial surface of the hemispheres, is more concerned with running movement sequences that are *internally* generated or controlled, selecting each next component of a sequence from internal cues. Slow, negative-going potentials (*Bereitschaftspotentials*) build up over the SMA when the next response is readied. Damage to the premotor cortex results in the disssolution of complex behaviors into discrete, unitary elements, while its stimulation may release complex, patterned synergies and gross stereotypies; damage to the SMA may lead to similar effects to those found with the premotor cortex, together with awkward, clumsy postures and problems with bimanual coordination, whereas its stimulation may produce an actual *urge* to respond.

The primary motor cortex (or motor-sensory cortex) is responsible for the responses of single muscles, and is somatotopically organized on a fine scale, with larger areas responsible for finer movements (e.g., of the fingers and mouth). Reciprocal connections with the parietal sensorimotor cortex provide, via reafference, for anticipated changes of sensory input consequent upon an intended movement. Damage leads to weakness, paresis, or hemiplegia, later giving way to hypertonia, hyperreflexia, and spasticity.

In the parietal cortex, *anterior* regions somatotopically code sensory input, including proprioceptive information that passes from the muscle spindles via the lateral lemniscus. Such proprioceptive information is essential for fine limb control and targeting. *Posterior* regions massively interconnect with structures within the premotor cortex, and integrate information on voluntary head and eye turns, recoding retinotopic information into spatiotopic coordinates. Thus a map is established of the location of events with respect to the body itself.

Apraxia is an acquired disorder of learned or skilled movement sequences, of the limbs or the articulators, and is typically found with LH damage. It is frequently, but not necessarily, associated with aphasia, and is most evident when a complex sequence is requested out of context; performance of individual components may be preserved, and the whole action may still be performed automatically or spontaneously, in a naturalistic context. The distinction between ideomotor and ideational apraxia is still often made, the former reflecting an impairment in the implementation rather than in the concept, whereas with ideational apraxia even the concept is said to be lost. However, in practice the distinction may not easily be made. Nevertheless anterior premotor damage near Broca's area may largely affect the *execution* side of apraxia, whereas posterior parietal damage may destroy the actual *capacity* itself. The supramarginal gyrus on the left may contain the actual engrams or spatiotemporal motor programs, whereas corresponding structures on the right may mediate the more spatial or attentional aspects. However, details of the exact interplay between the two hemispheres with respect to praxis in general are yet to be established; one possibility is that LH damage affects the more ballistic, open-loop, or abstract aspects of behavior, whereas the RH may be more involved in closed-loop, sensory-dependent, or concrete aspects, where perhaps context also is important. It is also noteworthy that, according to one study, although both clinically apraxic and nonapraxic LH patients may have similar problems in scheduling or timing sequential motor programs, only apraxics seem to have difficulty in reprogramming *heterogeneous*, as compared to merely *repetitious*, hand sequences.

Further Reading

Benson, D. F. (1993). Prefrontal abilities. *Behavioural Neurology, 6,* 75–81.

Brown, P. (1994). Pathophysiology of spasticity. *Journal of Neurology, Neurosurgery and Psychiatry, 57,* 773–777.

Cunnington, R., Bradshaw, J. L., & Iansek, R. (in press). The role of the supplementary motor area in the control of voluntary movement. *Human Movement Science.*

Damasio, A. R., & Anderson, S. W. (1993). The frontal lobes. In K. M. Heilman & E. Valenstein (Eds.), *Clinical neuropsychology* (3rd ed.) (pp. 409–460). Oxford: Oxford University Press.

Dubois, B., Verin, M., Teixeira-Ferreira, C., Sirigu, A., & Pillon, B. (1994). How to study frontal lobe functions in humans. In A.-M. Thierry, J. Glowinski, P. S. Goldman-Rakic, & Y. Christen (Eds.), *Motor and cognitive functions of the prefrontal cortex* (pp. 1–16). New York: Springer.

Duffy, J. R., & Duffy, R. J. (1990). The assessment of limb apraxia: The limb apraxia test. In G. R. Hammond (Ed.), *Cerebral control of speech and limb movements* (pp. 503–534). Amsterdam: North Holland.

Haaland, K. Y., & Harrington, D. L. (1994). Limb-sequencing deficits after left but not right hemisphere damage. *Brain and Cognition, 24,* 104–122.

Harrington, D. L., & Haaland, K. Y. (1992). Motor sequencing with left hemisphere damage. *Brain, 115,* 857–874.

Heilman, K. M., & Rothi, L. J. G. (1993). Apraxia. In K. M. Heilman & E. Valenstein (Eds.), *Clinical neuropsychology* (3rd ed.) (pp. 141–164). Oxford: Oxford University Press.

Jason, G. W. (1990). Disorders of motor function following cortical lesions: Review and theoretical considerations. In G. E. Hammond (Ed.), *Cerebral control of speech and limb movements* (pp. 141–168). Amsterdam: North Holland.

Jeannerod, M., & Rossetti, Y. (1993). Visuomotor coordination as a dissociable visual function: Experimental and clinical evidence. *Baillière's Clinical Neurology, 2,* 439–460.

Levin, H. S., Eisenberg, H. M., & Benton, A. L. (Eds.) (1991). *Frontal lobe function and dysfunction.* Oxford: Oxford University Press.

Morris, M. E., Summers, J. J., Matyas, T., & Iansek, R. (1994). Current status of the motor program. *Physical Therapy, 74,* 738–752.

Passingham, R. E. (1993). *The frontal lobes and voluntary action.* Oxford: Oxford University Press.

Rapcsak, S. Z., Ochipa, C., Beeson, P. M., & Rubens, A. B. (1993). Praxis and the right hemisphere. *Brain and Cognition, 23,* 181–202.

Roy, E. A., & Square-Storer, P. A. (1990). Evidence for common expressions of apraxia. In G. E. Hammond (Ed.), *Cerebral control of speech and limb movements* (pp. 478–502). Amsterdam: North Holland.

Sungaila, P., & Crockett, D. J. (1993). Dementia and the frontal lobes. In R. W. Parks, R.

F. Zec, & R. S. Wilson (Eds.), *Neuropsychology of alzheimer's disease and other dementias* (pp. 235–264). Oxford: Oxford University Press.

Watson, R. T., Rothi, L. J. G., & Heilman, K. M. (1992). Apraxia: A disorder of motor programming. In A. B. Joseph & R. R. Young (Eds.), *Movement disorders in neurology and neuropsychiatry* (pp. 681–690). Oxford: Blackwell Scientific.

Chapter 11

Movement Control II

Cerebellum and Basal Ganglia

*I*n the previous chapter we discussed the role of the cortex (prefrontal, premotor, motor-sensory, and sensorimotor cortices) in the initiation, monitoring, sequencing, and realization of voluntary movements. However, in addition to deliberate, conscious movement control by the pyramidal system, the extrapyramidal motor system, in particular, the cerebellum and basal ganglia (BG), play a major modulatory and supportive role. Reentrant feedback loops involving these structures are important in the smooth, automatic management of these voluntarily initiated movement sequences. As we shall see, characteristic deficits result from damage to these regions, damage which in the case of the cerebellum tends to be structural, and in the case of the BG, neurochemical.

Cerebellum

This "little brain" (the literal translation from the Latin) forms about 10% of the brain by volume, and contains nearly half of the total number of its neurons, with an information-processing capacity nearly equal to that of the cerebral cortex. Like the BG, it provides a major subcortical reentrant loop in motor pathways, receiving information from the cortex and (in the case of the cerebellum) outputting it to the brain stem. It serves as an integration and predictive control system for a vast array of sensory, motor, and limbic systems, improving the performance of all parts of the brain to which it is reciprocally connected, by feedback and feed-forward computations that assist in regulating behavior. However, compared to the BG, cerebellar input is more exclusively sensorimotor. The structure is important in controlling, and, especially, terminating aimed (particularly visual) movements, whereas the BG, with their wider cortical input, initiate and mediate a wider range of movements. The cerebellum indirectly regulates movement and posture, by adjusting motor output, regulating tone, coordination, timing, and movement synergies, via proprioceptive input from the muscle spindles, Golgi tendon organs, and the vestibular (balance) mechanisms, as well as from the exteroceptors, especially vision. It acts as a comparator, reducing discrepancies between intention and performance, via feed forward (reafference) and feedback, like the parietal cortex but now at a more unconscious level. Damage can lead to the need to think out each next movement deliberately. The cerebellum may have first evolved as a sensory system to visually track objects, with movement control later taken on via its involvement in the sensory consequences of movement (Paulin, 1993), which then led to its (perceptual and productive) role in timing. Rothwell (1994) sees the cerebellum as fulfilling two primary functional roles, timing and learning new movements, with a third role in humans that combines aspects of the first two. This third aspect involves movement coordination by transforming the idea of where to move into the appropriate and necessary changes in joint angle. Thus the cerebellum mediates the coordination of multiple systems of joints, the attainment of target positions of, for example, the hand via alternative routes or trajectories, amplitude scaling to suit external demands, compensation for changing loads, the coordination of reflexes, and the fulfillment of a comparator function in matching feed-forward and feedback information. Clearly we still have some way to go in fully understanding the functional roles of the cerebellum.

Stemming from his observations of the consequences of cerebellar tumors and gunshot wounds to the cerebellum during World War I, Gordon Holmes (1922) described the resultant disorders of movement. He concluded that the fundamental disturbances are hypotonia (diminished resistance to passive limb manipulation), tremor, asthenia (diminished strength), fatigability, and astasia (jerky incoordination) (see e.g., Gilman, 1994). He subdivided astasia into dysmetria (where range of movement is disturbed), directional error, and disturbances in rate of movement. He also described dysdiadochokinesia (see below) together with inappropriate associated movements, saccadic abnormalities, and problems in balance, locomotion, and speech. In advance of his time he used the term *decomposition of movement* to describe the degradation of smoothly performed complex movements into irregular, jerky components (Gilman, 1994). Defective motor coordination is still seen as a principle feature of cerebellar pathology, manifesting as lack of movement smoothness, with multiple peaks in the velocity profiles and oscillation during target acquisition. We shall shortly review evidence that structures within the cerebellum control the timing and sequencing of agonist and antagonist contractions by enhancing phasic activity in the motor cortex. Cerebellar systems may also process kinesthetic feedback, modulate long-latency reflexes, compensate for inherent mechanical instability, control movements requiring multiple joints, and calculate transformations between internal and external geometrical plans for predictive coordination of complex movement (Gilman, 1994).

Although the basic circuitry is essentially similar throughout, three major functional regions, successively evolved, have been identified in the cerebellum (Adams & Salam-Adams, 1992; Ghez, 1991b);

1. The *vestibulocerebellum* (flocculomodular lobe) is concerned with balance and equilibrium, coordination of movements of the head, neck, trunk, and limbs, and with maintaining the body in the three Cartesian planes of space. It is the central integrator of information from the extraocular muscles, labyrinth, visual centers in the pons, midbrain, and higher regions, and receptors in muscles of the neck and trunk. Unilateral lesions result in eye deviations, fixational problems, dysmetria, fragmentation of slow pursuit movements, and jerky microsaccades.

2. The *spinocerebellum* (medial cerebellum or central body) partially overlaps in function with the vestibulocerebellum. It receives, via the gamma system of muscle efferents, major somatosensory inputs, and controls the medial and lateral components of the descending motor system. Although it therefore modulates the ongoing execution of limb movements, damage (chronic or via acute alcoholic intoxication) also leads to gait ataxia. This manifests as a wide-based Donald Duck stance, the trunk slightly pitched forward and arms held stiffly away from the sides, with problems in postural adjustments before reaching out.

3. The *cerebrocerebellar* structure or lateral body gets sensory, motor, and premotor cortical input, and (like the BG) outputs via the (ventral) thalamus to the motor and premotor cortex, including Broca's area. Important in learned and skilled movements of the extremities and the cranial musculature (ocular, lingual, and laryngeal), it plays an important role in planning, preparing, initiating, and timing movement, becoming active 40 ms before the motor cortex. It thus helps achieve precise control of rapid limb movements, fine dexterity, coordination, and sequencing. Damage, according to Adams and Salam-Adams (1992), leads to a delay in motor cortex activation, a reaction time increase, and prolongation of the initial agonist burst (with consequent hypermetria), such that both the preparation and execution phases of a movement are affected. According to Leiner, Leiner, and Dow (1993a,b), this structure, which they referred to as the lateral cerebellum or neodentate, is enormously evolved in humans (being in fact central to hominid evolution with its prefrontal outflow), and provides a general computational network the importance of which transcends control of fine motor sequencing. They argued that it mediates control of syntactic, symbolic, and ideational sequences intrinsic to language and thought. Counting, timing, sequencing, predicting, anticipatory planning, error-detecting and correcting, attention-shifting, adaptation, and learning all come under its purview. Kim, Ugurbil, and Strick (1994) reported that this region was extremely active during attempts to solve a pegboard puzzle, much more so in fact than during simple

movements of the pegs alone. Ivry and Diener (1991) further emphasized the role of velocity and timing mechanisms where discrete adaptive responses must be precisely controlled, whether at a conscious or automatic level. In a very recent study, Manto, Godaux, and Jacquy (1994) added extra weights to the moving limbs of cerebellar patients and found that this increased overshoot, while in normal controls overshoot was reduced. They concluded that the lateral cerebellum modulates not only the timing of agonist and antagonist activity, but also the intensity of contraction, computing the parameters for the braking function of the antagonist. This includes the onset time and the amplitude, based on initial limb position, target location, and the inertia to be overcome. Consequently, cerebellar patients display lower peak accelerations and higher peak decelerations than controls, taking longer to reach peak velocity than to decelerate back to zero (Rothwell, 1994). Thus, as Gilman (1994) noted, the lateral cerebellum is responsible for more than simple timing even in limb control.

Damage to the last two of the above three regions clearly will be particularly detrimental to such rapid skilled movements as speaking, typing, and instrument playing. Errors may take two forms:

1. Errors in direction, intensity, velocity, aim, timing, and smoothness, with premature and delayed stopping (under-and-over-shoot, i.e., dysmetria), and breakdown of the normal triphasic cycle of agonist–antagonist–agonist activation, with the need for further corrections and oscillations. This leads to "action tremor," especially at the *end* of a movement, unlike a parkinsonian resting tremor, due to dysmetria with the consequent need for visually guided corrections. A separate, regular, rhythmic cerebellar tremor may stem from other mechanisms, according to Adams and Salam-Adams, 1992. Rubral tremor, a severe postural cerebellar tremor of between 2.5 Hz and 4 Hz, is proximal rather than distal, waxing and waning and tending to progressively increase in amplitude with prolonged posture. The electromyograph (EMG) shows bursts of 125–250 msc, with alternation of antagonists (see e.g., Hallett et al., 1994).

2. Problems in sequencing, timing, and temporal coordination of movements that involve multiple joints (e.g., the spatiotemporal coordination of hand and finger joints), leading to asynergia. As a consequence there may be a "decomposition" of complex movements into their component elements, with general ataxia (lack of coordination), dysdiadochokinesia (deficits in the rate and regularity of the timing of rapidly alternating movements, e.g., repetitive sequences of back of hand to thigh, fist to chest, and palm to forehead), and speech dysarthria. The patient must now consciously plan each next movement.

Cerebellar neurons are unique in their highly regular repetitive arrangement of a common basic circuit or module. Although all regions perform similar functions, they do so on different sets of inputs, via different connections, controlling balance, limb and eye movements, and muscle tone. The circuits are modifiable by experience and are important in motor learning, adaptation, anticipation, and prediction. Curiously, congenital absence of the cerebellum may produce very few later deficits, with early plastic takeover by the cerebral cortex, as shown by the appearance of deficits with subsequent additional cortical damage.

The Basal Ganglia

The BG are a group of gray-matter structures deep beneath the cortex, and surrounding the thalamus and hypothalamus. In fact, although the caudate, putamen, and globus pallidus (GP) (see below) are diencephalic, the closely related structures—the substantia nigra (SN) and subthalamic nucleus (STN)—strictly belong to the midbrain (Rothwell, 1994). The BG are of very complex anatomy and physiology, and their damage can result in major impairment of speed, sequencing, and fluidity of movement (A. B. Young & Penney, 1993). In many respects the BG parallel the cerebellum in function, although they receive a much wider cortical input, whereas the cerebellum gets input mostly originating from the sensory and motor cortices. Although the cerebellum outputs mainly to the premotor and motor cortex and the spinal cord, the BG output via the thalamus to the prefrontal, premotor and motor cortex, and the supplementary motor area (SMA). Thus while the cerebellum regulates movement execution, especially visually based targeting, the BG regulate higher order cognitive planning and also nonmotor cognitive behaviors, and the planning and execution of more complex motor strategies. The BG may therefore select and prepare movement initiation by disinhibitory release of motor programs. They act like an autopilot (Steg & Johnels, 1993), controlling at a subconscious level the coordination of different motor programs, on behalf of the conscious, volitional, decision-taking cortex. The system operates more in the context of well-learned, semiautomatic behaviors, rather than where a novel, unfamiliar, or difficult task requires conscious and controlled monitoring; where the system is dysfunctional, usually a consequence of neurotransmitter breakdown as in Parkinson's disease (PD), unlike the cerebellum where structural damage is more likely to occur, performance reverts to a deliberate, slow, nonfluent behavior. However, where cerebellar damage affects the execution of each response, BG dysfunction is reflected more in dysfluencies between successive responses. Thus as we shall shortly consider in more detail, it is likely that phasic activity generated by the BG for predictable, well-learned, and internally generated movements provides an internal cue to signal the switch between components in a sequence. This is possibly achieved by terminating sustained activity in the SMA for the impending movement, and turning on the preparatory phase of sustained activity for the next movement (Brotchie, Iansek, & Horne, 1991a,b). Thus in monkeys SMA activity terminates sharply at or before movement onset, whereas with animals rendered parkinsonian by BG damage from MPTP (1-methyl-4-phenyl-1,2,3,6-

tetrahydropyridine) (see Chapter 12), SMA neurons show no abrupt termination of activity, but continue firing. Abnormalities in the termination of movement-related SMA activity are therefore thought to be associated with disruption of BG cuing mechanisms by PD in humans. Normally, BG cues, generated at movement end point, are likely to be distinct for well-learned, predictable movements that require few attentional resources, but to become indistinct for unpredictable new movements that require a considerable amount of attentional resources.

The BG are components of a family of largely segregated, parallel cortico-subcortical pathways. As we saw in Chapter 10, each of the five circuits engages separate portions of the BG and thalamus, and the output of each circuit is directed toward functionally different areas of the frontal lobes (DeLong, 1993). The BG therefore presumably perform similar operations on input from each circuit, the function of the circuit being specified by its inputs and outputs. In the specific context of motor deficit, the latter can emerge from dysfunction anywhere along the BG motor circuit, though its character may vary as a function of the exact site and nature of damage.

The BG motor circuit provides a reentrant pathway from the motor-sensory and somatosensory cortices, the SMA, and premotor cortex, returning inputs to the same regions after intermediate processing via the thalamus (see Figures 1 and 2). The somatotopically organized excitatory cortical input to the striatum (putamen from motor and sensory cortices, caudate from association cortex) is glutamatergic and excitatory. Conversely, output from the BG via the internal (or medial) segment of the GP (GP_i) and the SN pars reticulata (SN_r) to the ventrolateral nucleus of the thalamus (VL) is inhibitory, via the neurotransmitter GABA (gamma-aminobutyric acid); the VL in turn has an excitatory projection back to the cortex (see e.g., Hallett, 1993; Wichmann & DeLong, 1993). In fact all the intrinsic connections within the BG are inhibitory (GABA), except for the pathway from the STN to the GP_i and SN_r, which is excitatory (glutamate). Note that although the caudate and putamen together constitute the striatum (strictly, the neostriatum), the caudate, putamen, and GP constitute the corpus striatum, whereas the putamen and GP together are sometimes referred to as the lentiform nucleus. The nucleus accumbens occupies the ventral anterior striatum adjacent to the putamen, and has recently been shown (Harris & Aston-Jones, 1994; see also Schulteis & Koob, 1994) via its dopamine (D2) receptor activity to be important in the rewarding effects of abused drugs and in opiate withdrawal. Within the striatum, the caudate nucleus and the putamen are separated by the internal capsule. Bands or bridges of gray matter perforating the capsule cause the striated appearance of the striatum, hence its name. The putamen has the same cytological features as the caudate, even though it is physically continuous with the GP (Domesick, 1990). Anatomically and behaviorally, the striatum (together with the GP and thalamus) form an integrated circuit with the frontal lobe. However, it is unclear whether lesions anywhere within this cir-

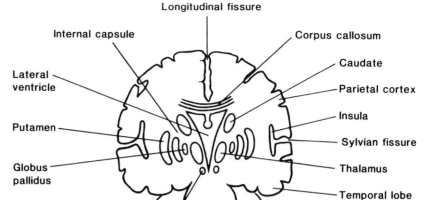

a

Longitudinal fissure

Internal capsule

Corpus callosum

Caudate

Lateral ventricle

Parietal cortex

Insula

Putamen

Sylvian fissure

Globus pallidus

Thalamus

Temporal lobe

Pons

Cerebellum

Subthalamic nucleus

Medulla

Coronal Section

b

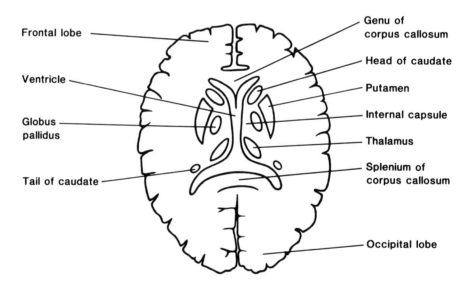

Frontal lobe

Genu of corpus callosum

Head of caudate

Ventricle

Putamen

Globus pallidus

Internal capsule

Thalamus

Tail of caudate

Splenium of corpus callosum

Occipital lobe

Horizontal Section Through The Striatum

Figure 1   STRUCTURES IN THE BASAL GANGLIA: CORONAL (a) AND HORIZONTAL (b) SECTIONS.

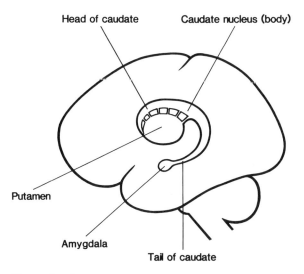

Figure 2   Structures in the basal ganglia: sagittal section.

cuit produce exactly the same behavioral deficits, or whether, as is perhaps more probable, there are subtle differences reflecting local information processing. The prominence of motor abnormalities after lesions of the putamen nevertheless indicates its largely motor role, whereas behavioral disturbances such as abulia (apathy with loss of initiative and spontaneity) after caudate damage indicate a predominantly cognitive function (Bhatia & Marsden, 1994).

Two pathways, direct and indirect (see Hallett, 1993; Wichmann & DeLong, 1993; A. B. Young & Penney, 1993, for reviews) have been identified in the motor circuit between the putamen and the GP_i and SN_r (see Figure 3). The *direct* pathway projects monosynaptically on to the motor portions of the GP_i and SN_r. The net result is that the latter structures can release the VL thalamus from tonic inhibition, so that it can in turn excite the cortex. The direct pathway is therefore a positive feedback circuit, for sustaining or facilitating an ongoing action. Damage to this direct pathway in PD, or late Huntington's disease (HD), results in akinesia or inability to sustain effort. The *indirect* pathway connects in an inhibitory fashion between the putamen and the external (or lateral) segment of the GP (GP_e) which in turn inhibits (via GABA) the STN, which then projects via an excitatory (glutamatergic) pathway to the GP_i and SN_r. The latter two structures are thereby activated, and able to inhibit (via GABA) VL thalamic neurons that otherwise would facilitate unwanted motor programs. There are two additional reciprocal inhibitory interconnections, between GP_e and GP_i and SN_r, and between the latter structures and back to GP_e. In effect, the indirect

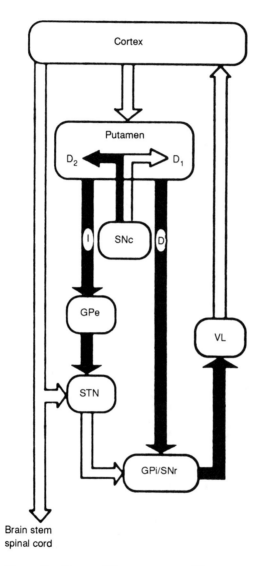

Figure 3   Direct (D) and indirect (I) pathways in the motor circuit of the normal basal ganglia. D_2, D_1, dopamine receptors; SN_C, substantia nigra zona compacta; GP_E, globus pallidus external; VL, ventrolateral nucleus; STN, subthalamic nucleus; SN_R, substantia nigra pars reticulata; GP_I, globus pallidus internal. (From DeLong, 1993.)

pathway suppresses unwanted movements. In PD it is hyperactive, and the patient cannot switch to new programs. In early HD it is dysfunctional, causing chorea and unwanted movements. In the striatum, at synapses located at the base of the dendritic spines of the medium spiny neurons, DAergic neurons originating in the SN zona compacta (SN_c) play a modulatory, gating role—a mechanism that we shall see is affected in PD. This nigrostriatal DAergic modulation differentially affects the direct and indirect pathways, facilitating transmission via the direct and decreasing activity in the indirect pathway, respectively, through D1 and D2 receptors (DeLong, 1993). Normally, activity in the direct pathway increases BG output on to thalamocortical neurons, which in turn modulate the activity of the precentral motor areas, notably the SMA and premotor cortex.

The BG therefore can select (via the direct pathway) and inhibit (via the indirect) specific motor synergies, respectively, via phasic reduction of BG output and via its tonic inhibitory influence on the activity of thalamocortical cells. Moreover, when voluntary movement is facilitated, spinal reflexes will be inhibited. An overactive indirect or underactive direct pathway, as in PD (see Chapter 12), will increase pallidal activity, suppress thalamic and cortical activity, and make movement initiation difficult, while at the same time increasing the responsiveness of spinal reflexes, leading as we shall see to parkinsonian akinesia and rigidity. Conversely, an underactive indirect and overactive direct pathway will lead, via reduced pallidal activity and thalamic release, to involuntary excess voluntary movements, and to chorea, as in HD. Nevertheless, as we shall see in Chapter 13, parkinsonian akinesia and bradykinesia also coexist in that disorder.

As mentioned above and discussed in Chapter 10, the BG influence movement via the SMA and premotor cortex. The motor component of the BG collaborates with the SMA in running learned movement sequences, the involvement of the BG only becoming apparent after the sequence has been well learned and relegated to motor memory. The SMA generates *tonic* premovement neuronal activity (*Bereitschaftspotential*), which ceases immediately when movement commences; the BG generate *phasic* activity during and toward the end of such movements only, however, when movements and sequences are predictable, require few attentional resources, and are easy to perform. Such phasic activity of the BG is used as an internal motor cue to terminate the tonic premovement activity of the SMA, thereby enabling the submovement sequence to be executed. This execution then causes another phasic cue to be generated by the BG, which in turn can extinguish tonic premovement activity for the *next* submovement, enabling it to execute, and so on, until the movement sequence is run off to completion. Thus the phasic cue generated by the BG serves to *synchronize* activity. As we shall see (Chapter 12) such synchronization is not achieved in the event of a defective BG cue, as occurs in PD. For detailed consideration of the issues reviewed above see Brotchie et al. (1991a,b), Georgiou et al. (1993), and Cunnington, Iansek, Bradshaw, and Phillips (in press).

Summary and Conclusions

The cerebellum, like the BG, serves as a major subcortical reentrant loop for the motor system, receiving information from the cortex and outputting it to the brain stem. It acts as an integrative and predictive control system, and as a comparator mediating feedback and feed-forward computations. It is particularly important in visual aiming behaviors as well as adjusting posture and motor output, and controlling timing and movement synergies.

Three major functional regions have been identified in the cerebellum. The *vestibulocerebellum* is concerned mainly with balance, equilibrium, and control of the extraocular muscles. The *spinocerebellum* modulates the ongoing execution of limb movements, and is concerned with gait and postural adjustments. The *cerebrocerebellar structure* (or lateral body) helps achieve precise control of rapid limb movements, dexterity, coordination, and sequencing. Damage, especially to the last two regions, may result in errors in direction, intensity, velocity or timing, and the need for further corrective action. There may be a general "decomposition" of complex movements into their component elements.

The functions of the BG partly parallel those of the cerebellum, though the former structure receives input from and outputs to a greater cortical area. While the cerebellum regulates movement execution, the BG regulate higher order cognitive planning, selecting and preparing movement initiation by disinhibitory release of motor programs. An internally generated cue signals the switch between successive phases. The system operates more in the context of well-learned, semiautomatic behaviors, rather than where conscious monitoring is required. Where cerebellar damage may affect the execution of each individual response, BG dysfunction may result in dysfluencies between successive responses. Thus in PD the BG may not be able to regulate the SMA by appropriately turning off and on sustained activity for each successive phase of a motor program.

The excitatory cortical input to the striatum, the input circuit of the BG, is glutamatergic; output from the BG via the internal segment of the GP and the SN_r, to the VL nucleus is inhibitory, via GABA. The VL nucleus has an excitatory projection back to the cortex. Nigrostriatal DAergic input to pathways within the BG modulates their activity, and, in turn, the activity of the SMA and premotor cortex. With PD, when DAergic activity is reduced, pallidal activity is increased, such that thalamic and cortical activity is suppressed, making movement initiation and execution difficult.

Further Reading

Adams, R. D., & Salam-Adams, M. (1992). Disorders of movement associated with diseases of the cerebellum, particularly the degenerations. In A. B. Joseph & R. R.

Young (Eds.), *Movement disorders in neurology and neuropsychiatry* (pp. 310–318). Oxford: Blackwell Scientific.

Cummings, J. L. (1993). Frontal-subcortical circuits and human behavior. *Archives of Neurology, 50,* 873–879.

Ghez, C. (1991). The cerebellum. In E. R. Kandel, J. H. Schwartz, & T. M. Jessell (Eds.), *Principles of neural science* (3rd ed.) (pp. 626–646). New York: Elsevier.

Hallett, M. (1993). Physiology of basal ganglia disorders: An overview. *The Canadian Journal of Neurological Sciences, 20,* 177–183.

Rothwell, J. (1994). *Control of human voluntary movement* (2nd ed.). London: Chapman & Hall.

Wichmann, T., & DeLong, M. R. (1993). Pathophysiology of Parkinsonian motor abnormalities. In H. Narabayashi, T. Nagatsu, N. Yanagisawa, & Y. Mizuno (Eds.), *Advances in neurology* (Vol. 60, pp. 53–61). New York: Raven Press.

Young, A. B., & Penney, J. B. (1993). Biochemical and functional organization of the basal ganglia. In J. Jankovic & E. Tolosa (Eds.), *Parkinson's disease and movement disorders* (2nd ed.). Baltimore: Williams & Wilkins.

Subcortical Movement Disorders I

Parkinson's Disease

*I*n 1817, James Parkinson described the occurrence in paralysis agitans of tremor, alterations in gait, posture, and speech, and depression, with, he believed, relative unimpairment of the intellect and senses:

> Involuntary tremulous motion, with lessened muscular power, in parts not in action . . . with a propensity to bend the trunk forward and to pass from a walking to a running pace, the senses and intellects being uninjured. . . . The first symptoms perceived are a slight sense of weakness, with a proneness to trembling . . . most commonly in one of the hands and arms. These symptoms gradually increase in the part first affected, and at an uncertain period, but seldom less than 12 months or more, the morbid influence is felt in some other part . . . or limb. (Parkinson, 1817, pp. 1–3)

However, as Golbe and Langston (1993) noted, nearly 2,000 years ago Galen described elements of the "shaking palsy," and similar mention is made in the Ayurvedic medical text of India of 1000 B.C. (Manyam, 1990). Of course as the disorder is largely one of aging, neither authority would probably have encountered in those days many sufficiently elderly individuals at risk.

Parkinson's Disease, Aging, and the Dopaminergic System

Parkinson's disease (PD) is a disorder primarily of late middle life, whose major manifestation is a progressive disturbance in motor function produced by dopaminergic (DAergic) cell loss in the pigmented substantia nigra zona compacta (SN_c), which projects to the striatum, with formation of Lewy bodies. Lewy bodies are the most characteristic pathological marker of idiopathic PD, and are laminated, eosinophilic, cytoplasmic inclusions, 7 to 20 nm in diameter, in neurons still remaining. They possess a dense core of granular material, with the main body composed of ringlike filaments intermingled with degenerating organelles (Gibb, 1993). Their presence also, however, in Alzheimer's disease (AD) and amyotrophic lateral sclerosis indicates that they may be a nonspecific morphological indicator of cell pathology in various neurodegenerative diseases (Stacy & Jankovic, 1992). Pathology may also be found in the ventral tegmentum, the locus coeruleus (another pigmented structure and source of ascending noradrenergic activation), the serotonergic raphe nucleus, and the cholinergic nucleus basalis of Meynert. This may lead to widespread loss of biogenic amine activity and cognitive, affective, and even autonomic disorders, due to damage to sympathetic ganglia. Nonmelanized (unpigmented) cells may be more resistant to damage, though Gibb (1993) noted that the nucleus basalis of Meynert (see Chapter 9) lacks neuromelanin, but is affected by Lewy body disease. The lateral (caudal) portions of the SN_c, which project preferentially to the putamen, are affected more than the medial (rostral) portions, which project to the caudate. The putamen, as we saw, is part of the motor loop, whereas the caudate is involved in more cognitive activities. DAergic neurons in the ventral tegmental area are also affected, with deficits in the medial olfactory area, lateral hypothalamus, amygdala, and entorhinal, cingulate, hippocampal, and frontal cortices (Gupta, 1993). DA is synthesized from the dietary amino acid tyrosine, which is hydroxylated to levodopa (L-dopa) in a reaction catalyzed by tyrosine hydroxylase (Hauser & Olanow, 1993). L-dopa is then decarboxylated to DA. The presence of tyrosine hydroxylase serves as a histochemical marker. After release, DA can be catabolized to homovanillic acid (HVA) by monoamine oxidase (MAO), or can undergo reuptake. There are decreased levels of HVA in the putamen and SN in PD.

Nigrostriatal degeneration in PD must progress to the stage of 80% cell loss before clinical symptoms appear (Snow & Calne, 1992, though Stacy & Jankovic, 1992, opt for a lower figure of 50–60% loss). Until then, there may be maintenance of normal extracellular concentrations of DA by increased activity and DA output by surviving neurons, and by reduction of the reuptake sites responsible for DA removal from the synaptic cleft. Yet later, there may be increase in the number of postsynaptic receptors, before there is failure of further compensation and the disorder manifests clinically. Indeed it is possible that a self-perpetuating vicious circle may be established from the compensatory increase in DAergic activity, if, as is sometimes suspected, the metabolism of DA is itself neurotoxic. Because damage may extend from the nigrostriatal system to many other regions, with considerable individual differences in severity, extent, and progression, it may be asked whether PD is a single disorder or a syndrome with multiple causes. Thus we may distinguish between early and late onset, and between PD presenting primarily with tremor or rigidity and akinesia, and with or without concomitant cognitive dysfunction, dementia, and depression. Tremorous PD generally seems to be the more benign, but Yahr (1993) believes that most differences are quantitative rather than qualitative, and indicative of a single disease process. Nevertheless we must distinguish between "idiopathic" PD, and "PD-plus" syndromes such as progressive supranuclear palsy, Shy-Drager syndrome, and olivopontocerebellar degeneration (see Chapter 14). Indeed, the initial clinical diagnosis of PD is frequently inaccurate, the standard pathological diagnosis at autopsy being loss of SN cells and abundant Lewy bodies in remaining cells. Until then, as we shall see, PD is diagnosable from tremor, akinesia, impaired balance, muscle rigidity, and a positive response to L-dopa. Striatonigral degeneration (see Chapter 14) shares many of these features, but is characterized by an impaired range of eye movements and little response to L-dopa.

The prevalence of PD ranges between 50 per 100,000 in those less than 50 years, 1000 per 100,000 in those 60 or more years old, and 2500 per 100,000 in those 85 years or older (Golbe & Langston, 1993). The mean age of onset is around 60 (Cummings, 1992), and males are twice as commonly affected as females (Shoulson & Kurlan, 1993). Unlike the situation with Huntington's disease (HD) (see Chapter 13), earlier onset seems to be associated with a better prognosis. At the "normal" rate of cell loss in the nigrostriatal system, we might expect everyone to manifest PD by 100 years if they lived that long, but something in PD causes an up to five-times increase in cell loss. Perhaps 1% of the population over 60 has PD, and 2% over 80. Like aging itself, it is progressive, remorseless, and without remissions or exacerbations, though progress can in some cases be very slow. Indeed PD is seen by many as merely accelerated aging; old people certainly show many of the features of PD—stooped posture, shuffling step, poor arm swing, and abnormal long loop reflexes with increased reaction time (RT). However, the incidence of PD seems to decrease after 75, unless this is an artifact of underreporting of illness by the very old. Moreover, disparity for PD in monozygotic (MZ) twins in the very old, differences in the distribution of nigral cell loss between PD patients and the normal elderly, and the fact that L-dopa may not help alleviate the age-related motor dysfunctions of the elderly, argues against aging as being the sole cause of PD (see e.g., Golbe & Langston, 1993; Snow & Calne, 1992, though according to Ollat, 1992, DA insufficiency of the three ascending systems is a major aspect of cerebral aging, explaining many of the associated motor, emotional, affective, and cognitive disorders, especially those analogous to frontal symptomatology; she concluded that there can be considerable beneficial effects from the DAergic agonist piribedil). Untreated, PD progresses to death typically in around 10 years, though current therapy extends life expectancy for at least 14 or so years after the initial symptoms, and indeed longevity may not be substantially reduced. There are no specific radiological, biochemical, or electroencephalographic (EEG) diagnostic tests, diagnosis being purely clinical in terms of the cardinal symptoms (tremor, rigidity, bradykinesia); however, an electromyogram (EMG) may demonstrate tremor before it is clinically evident, and a neuroimaging positron emission tomographic (PET) scan, involving fluorodopa uptake in the putamen, may demonstrate DA deficiency in the nigrostriatal system in idiopathic PD, and in presymptomatic individuals exposed to the neurotoxin 1-methyl-4-phenyl-1,2, 3,6-tetrahydropyridine (MPTP), which specifically targets the nigrostriatal pathway.

Loss of nigrostriatal DA in idiopathic PD leads to overinhibition of the globus pallidus external (GP$_e$) in the *indirect* pathway (see Chapter 11) of the basal ganglia (BG), with consequent disinhibition of the subthalamic nucleus (STN) and thereby increase in GP$_i$/SN$_r$ activity. Loss of nigrostriatal DA similarly leads to reduced activity along the *direct* inhibitory projection from the striatum to the GP$_i$ and SN$_r$ ($_i$, internal, $_r$, reticulata), thereby disinhibiting the output of the BG (Wichmann & DeLong, 1993). The positive feedback characteristics of the internal reciprocal connections between the GP$_e$ and GP$_i$ may further enhance GP$_i$ activity (see Figure 1). Ultimately, according to Wichmann and DeLong, all these changes lead to increased tonic BG output, with consequent overinhibition of parts of the motor thalamus (VL), and therefore reduced activi-

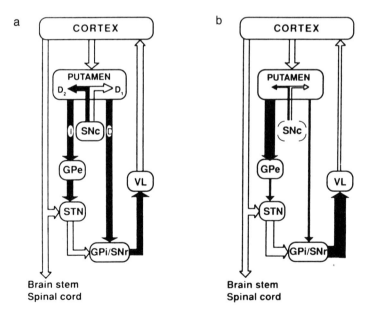

Figure 1 DIRECT AND INDIRECT PATHWAYS IN THE MOTOR CIRCUIT OF THE PARKINSONIAN BASAL GANGLIA (a) COMPARED WITH THE NORMAL BRAIN (b). D$_2$, D$_1$, DOPAMINE RECEPTORS; SN$_C$, SUBSTANTIA NIGRA ZONA COMPACTA; GP$_E$, GLOBUS PALLIDUS EXTERNAL; VL, VENTROLATERAL NUCLEUS; STN, SUBTHALAMIC NUCLEUS; GP$_I$, GLOBUS PALLIDUS INTERNAL; SN$_R$, SUBSTANTIA NIGRA RETICULATA. (FROM DELONG, 1993.)

ty along the thalamocortical neurons. Consequently, phasic responses to passive manipulation are enhanced overall, and are somatotopically less specific than normal, with increased gain of the motor circuit for somatosensory responses.

Positive and Negative Symptoms in Parkinson's Disease

We can divide the cardinal symptoms of PD into positive and negative. Negative symptoms, indicative of loss or impairment of some motor functions, include loss or reduction of spontaneous movement (hypokinesia), a slowness in *initiating* movement (i.e., an RT increase, or akinesia), and a slowness in *executing* movement (i.e., a movement time increase, or bradykinesia). Bradykinesia may be a consequence of disturbance of the normal triphasic pattern of initial agonist burst followed by antagonist and final agonist activity. Patients may lose the normal amplitude-to-velocity relationship, so that large amplitude movements are performed at abnormally low velocities, due to failure to generate an adequate initial agonist burst. Thus PD patients perform large amplitude movements at low velocities, segmented into multiple small-amplitude movements. As we saw, according to Wichmann and DeLong (1993), excessive *tonic* output from the GP_i may lead to increased tonic inhibition of the thalamocortical neurons, with reduced responsiveness of the cortical mechanisms, and therefore insufficient phasic activation of the alpha motoneurons during attempted movements. Superimposed *phasic* reductions in activity during movement execution may not be faithfully transmitted to the cortex, leading to a reduced range and scaling of movement. Increased gain in the motor circuit may lead to near-maximal phasic responses, even with movements of small amplitude, and, ultimately, a reduced range of available velocities. Akinesia may additionally stem from a disturbance of programming or motor planning, in terms of, as we saw of the assembly, sequencing and release of the individual motor programs involved in more complex movements. The subsequent loss of automaticity, and the "frontal" problem in forming and maintaining sets, results in the need to operate at a more conscious, deliberate level.

Other negative symptoms may include micrographia (a progressive decrease in size and speed of writing over a couple of lines), slowed gait, often with freezing or with very small steps (*marche à petit pas*) though a rapid festinating gait may also occur, poor arm swing, postural instability, movements en bloc rather than sequential, a dull, weak, uninflected, and hypophonic voice with slowed speech (again, paradoxical tachyphemia may sometimes occur), a masklike, unemotional expression, reduced blink rate, and slowed, hypometric, or absent saccades. Indeed, changes in fixational eye movements (increased latency, reduced velocity, multiple stepping, hypometria, and fewer spontaneous saccades) closely parallel the motor deficits; they are mediated via damage to the oculomotor circuit between the BG and frontal eye fields, as described in Chapters 10 and 11.

There may occasionally be increases in speed (paradoxical kinesia or kinesia paradoxa) manifesting in, for example, tachyphemia or a festinating gait. A variety of visual cues may be effective in evoking this response in receptive individuals, and often, in the context of gait, they may simulate objects that the patient perceives as having to be stepped over (e.g., a line of cards spaced at intervals equalling or slightly exceeding the walking stride length) (T. J. Reiss, personal communication, 1994). However, a general motor instability may underlie the phenomenon of kinesia paradoxa. Thus as we saw in Chapters 10 and 11, after a movement sequence is commenced with the establishment of submovement execution, the execution of one submovement leads to the generation of a phasic cue by the BG. Such a cue would normally terminate tonic set-related activity for the next submovement in the supplementary motor area (SMA), serving to synchronize activity. However, a defective cue due to PD would not achieve such synchronization; the execution of the next submovement would be ill prepared, with an end point differing from what was initially planned. This would further impact upon an already-disordered cue generated by the BG, which would lead to further errors in preparation for the next submovement, and so on. Festination or hastening would occur in the context of an imposed motor set for fast movement, especially perhaps with simple, repetitive, homogeneous movements like stepping or tapping. Progressive slowing or freezing might occur when initially operating at a normal preferred speed, or with complex or heterogeneous movements such as handwriting (micrographia) or speech production, where recomputation of a program is required. Note that freezing should be distinguished from an akinetic inability to commence, for example, ambulation, and from voluntary cessation of the activity; it involves cessation contrary to the patient's will.

The major positive symptoms in PD (i.e., abnormally excessive responsivity) are rigidity and tremor. Rigidity manifests as resistance to passive movement in both extensors and especially flexors, leading to postural problems; flexion of the knees, hips, elbows, and neck leads to the typical simian (apelike) stooped posture. Other associated postural problems include loss of righting reflexes and stepping responses when pushed, and loss of anticipatory postural adjustments before lifting, and so forth. Rigidity typically starts axially, and spreads distally, to the limbs. It has a smooth, plastic, viscous "lead pipe" quality when the examiner passively bends the patient's limb, though it may interact with tremor to produce a "cogwheel" or "ratchet" quality. It is not the same as spasticity, which occurs with corticospinal tract lesions and is characterized by a clasp-knife type of resistance that is velocity dependent. Subtle rigidity in a limb may also be brought out by voluntary movements of the contralateral limb. Rigidity may be due to loss of nigrostriatal inhibition of descending motor pathways and enhanced reactivity to passive stretch, via heightened responses of the tonic stretch reflex, through facilitation of the alpha and gamma systems (Delwaide & Gonce, 1993). Alternatively, abnormal long-latency reflexes may account for abnormal alpha-motoneuron excitability in PD (Wichmann & DeLong, 1993, and see also Cummings, 1992). Thus in PD the long-latency reflex, which enables us appropriately and automatically to adjust our grip on a slipping wine glass and is amenable to some voluntary control, is less subject to such subjective control and may sit at near maximal gain. In any case, stiffness and rigidity is reinforced by contralateral movement, stress, and anxiety, whereas its reflexive component is evidenced by its disappearance with dorsal root section (Delwaide & Gonce, 1993).

Tremor, which occurs in a majority but by no means all cases of PD (Marsden, 1994), manifests at rest, disappearing when the limb is used, or during sleep. This "resting" tremor, by contrast with cerebellar "intention" tremor that is maximal during movement and targeting, has a typical frequency around 5 Hz; later, it may also manifest as action tremor during voluntary movement, when its frequency typically increases to around 7 Hz. Initially asymmetrical, it starts in an upper limb as a "pill-rolling" movement; it may then progress to the ipsilateral leg and thence to contralateral limbs; lips, tongue, and jaw may also be involved. It may increase in one arm if the opposite arm is used ("overflow"), or under stress (e.g., when performing mental arithmetic). It may be suppressed by an act of will, sleep, or complete relaxation. It may be a consequence of phasic oscillatory activity in the BG themselves, or of increased tonic BG output promoting oscillatory activity in the thalamus, with consequent rhythmic activity in thalamocortical and corticospinal projection neurons (Wichmann & DeLong, 1993). It may be influenced by peripheral kinesthetic information starting in the muscle spindles, and passing up to the cortex and back to the muscle via the thalamus and the long-loop reflex pathways (Delwaide & Gonce, 1993). Tremor is the symptom least responsive to antiparkinsonian medication, and unlike bradykinesia its severity correlates little with cognitive performance.

As Stacy and Jankovic (1992) noted, the traditional division of parkinsonian symptoms into negative and positive, reflecting respectively direct dysfunction of the BG and release phenomena, is now being questioned. Thus Bergman, Wichmann, and DeLong (1990) reported improvement of bradykinesia after STN lesions, suggesting that this traditionally negative symptom should be reassessed.

Deficits in the Internal Generation of Programmed Activity

Patients with PD may also have problems in using internalized, learned information (Georgiou et al., 1994), as might be expected from the review (see Chapters 10 and 11) of the effects of damage to or faulty output to the SMA. Thus patients show reduced *Bereitschaftspotential* (readiness potential) before making regular, predictable, internally governed responses (Dick et al., 1989). They also have difficulty in performing rapid sequential movements according to an internalized timetable, in the absence of external cues (Georgiou et al., 1993). They may not show the usual increase in response time with dual compared to single tasks (though see e.g., Malapani, Pillon, Dubois, & Agid, 1994), choice compared to simple RT, or more difficult or uncued compared to easier or cued tasks (Jones et al., 1994). Such underadditivity suggests that they have to control *all* movements on-line, in a closed-loop mode depending on visual feedback, and with full attentional monitoring (Saint-Cyr & Taylor, 1993) rather than in advance according to internal cues. Indeed patients are unusually susceptible to external visual cues, freezing when trying to walk through a doorway, but walking well along stripes laid out at an appropriate gait frequency or up and down stairs. Patients have difficulty with complex behaviors involving simultaneous or smooth and rapidly sequential movements at several joints, and bimanual alternation (Jones, Phillips, Bradshaw, Iansek, & Bradshaw, 1992). When a patient *does* try to perform two actions simultaneously, frequently one fails—usually the more automatic, subconscious one, whereas the nonautomatic, consciously attended response succeeds.

The frontal and prefrontal cortex, BG, and thalamus are involved (see Chapter 11) in the operation of motor plans or schemata—the selection of appropriate motor programs from memory, and their assembly into an adequate sequence. Each motor program is a set of motor commands needed to execute a simple act. It is stored in a generalized or abstract form, and before a particular operation specific parameters (force, duration, direction, specific muscle groups, etc.) are thought (Schmidt, 1988) to be fed in to it; after execution it receives proprioceptive (from muscles and joints) and exteroceptive (from vision and hearing) information, and information on the operation's success or failure. It is thereby continuously updated during use, and provides each of us with a unique idiosyncratic *modus operandi* (e.g., our particular characteristic writing style, whether by hand, foot, or mouth). As Soliveri, Brown, Jahanshahi, and Marsden (1992) observed, we rarely think about the act of walking, because we can adapt existing motor programs speedily and flexibly to cope with concrete, carpets, or steps of various lengths, and only need consciously to reset the specific parameters when suddenly encountering ice or deep mud. According to some theorists, the PD patient's problem is in rapidly changing specific parameters, and certainly manifests in assembling, switching, and terminating such programs, in performing simultaneous movements, and in the automatic execution of such learned motor plans (Delwaide & Gonce, 1993). As we saw in Chapters 10 and 11, in the normal brain the BG probably provide a cue to synchronize premotor activity, and crisply release each next stage in a learned motor sequence (Cunnington, Iansek, Bradshaw, & Phillips, in press); such a cue is defective or missing in PD, tonic activity in the SMA is no longer sharply terminated by phasic BG output, and movement initiation is impaired in the absence of an additional external cue bypassing the SMA. Cunnington et al. (in press) additionally showed that, in normal healthy individuals, unpredictable cuing such that move-

ments can no longer be sequenced or determined according to an internally generated schedule results in reduced activity in the SMA. Indeed, such activity is
only apparent in PD patients in the absence of external cues, when movements
must be internally determined. In the absence of BG input, as in PD, the SMA
may only be involved in movements that *must* be internally determined; hence
the heavy reliance of parkinsonian patients upon external cues.

Cognitive Impairments

PD is a heterogeneous condition, both clinically and neuropathologically.
Consequently, cognitive deficits vary widely between patients, and even within
the same patient during different times of the day (Dubois, Boller, Pillon, &
Agid, 1991), as a function of medication status, nutritional level, and so on.
While the disorder was originally thought to be largely or entirely motor, with little or no cognitive impairment (see e.g., Parkinson, 1817), greater clinical awareness and the longer survival of patients have recently revealed the extent of cognitive changes during disease progression. Cognitive decline however progresses
slowly, and, as in HD, tends to parallel the motor changes—especially akinesia,
less so rigidity, and perhaps not at all with respect to tremor (Dubois & Pillon,
1992).

Many of the problems associated with speech do indeed reflect difficulties
in respiratory control, again as in HD, and the increasing stiffness of facial, oral,
buccal, and pharyngeal muscles. As previously adumbrated, pitch tends to increase and loudness to decrease (hypophonia), and both parameters become
more variable. There are problems with stress and emphasis, with imprecise articulation, dysprosody, hypokinetic dysarthria, and reduced phrase length, sometimes with tachyphemia and compulsive repetition or palilalia. However, a true
disorder of planning seems to underlie these difficulties (Dubois et al., 1991).
Anomia, word-finding difficulties, and "tip-of-the-tongue" effects are common,
due perhaps to the simultaneous frontal involvement, together with impaired
comprehension of complex grammar, though it should be noted that although up
to one-third of PD patients may have a language disorder, many of these may simultaneously suffer from AD (see below). Brentari and Poizner (1994) described
a congenitally deaf user of sign with PD, who exhibited reductions in speed and
amount of signing, disruptions of timing, distalization of the movement, and prolonged hand-shape changes. They noted however that once again problems lie
more in the domain of execution than in the grammar itself.

Visuospatial difficulties also occur (Levin, 1990) involving reasoning, constructional praxis, and attentional shifting, though maybe somewhat less commonly (Dubois et al., 1991), and perhaps only under conditions of significant
processing demand, or when supervisory attentional systems in the prefrontal
cortex are affected (see below). Moreover, motor disability may bias the evalua-

tion of performance. However, the caudate is known from animal studies to be important for spatial operations, so it is not perhaps surprising that many patients have problems in appreciating the relative positions of objects in space, in route finding, in copying complex figures, and in processing complex gestures (Cummings & Huber, 1992).

Memory disturbance is of course a hallmark of dementia, an issue to which we shall shortly turn. Deficits in visuospatial memory are said to be particularly prominent, though as Dubois et al. (1991) noted it may be simply that such tasks are intrinsically harder than those involving language. Recognition memory for most kinds of material, verbal or visuospatial, is better preserved than recall, and problems become particularly prominent when material must be reorganized in memory. The fact that such deficits are present even in early-onset PD indicates that they are not merely an artifact of aging. Recall even of remote memory, personal or of public events, is affected, and may parallel the development of dementia. Registration, storage, and consolidation may occur relatively normally, problems arising in accessing the information. Frontal or strategic aspects are especially affected, as with delayed responding and delayed alternation, recency discrimination, and temporal ordering. Indeed, memory for temporal order is more affected in PD than in AD, whereas long-term storage and remote memory are more affected in AD than in PD. Similarly, implicit memory, procedural memory for skills, and performance in puzzles such as the Tower of Hanoi, again where striatal involvement is known, are decremented more in PD than in AD; declarative memory however (see Chapter 8) is more affected in AD than in PD. That said, according to Appollonio et al. (1994) parkinsonian memory deficits still primarily affect conscious, effortful strategic aspects of searching long-term memory.

The concept of parkinsonian bradyphrenia or slowness of thought or information processing is controversial (Pirozzolo, Swihart, Rey, Mahurin, & Jankovic, 1993). Clinically, it is said to manifest as apathy, intellectual inertia, and slowness of information processing (Dubois et al., 1991). Cognitively, the last ("psychic akinesia") is clearly the most important aspect, and may be testable with a constant motor but variable cognitive element, as with mental rotation tasks. However, many studies have failed to find it when movement problems are controlled for (Duncombe, Bradshaw, Iansek, & Phillips, 1994). It may perhaps only manifest under high information-processing loads, partly perhaps as a result of defective strategic approaches, again indicating prefrontal involvement. Otherwise it may be a consequence of frequently occurring (in PD) depression or dementia, if they are not controlled experimentally or medically. Treatment itself, of course, whether with anticholinergics, L-dopa, or DA agonists, may lead to abnormal mentation, either affecting cognitive performance detrimentally, as found by Brown et al. (1993) and Owen et al. (1992), or beneficially (Lange et al., 1992). The absence of significant correlations between speed of information processing and bradykinesia in early-onset patients suggests that a DAergic deficit in PD may not determine bradyphrenia (Dubois & Pillon, 1992; Pirozzolo et al., 1993). Indeed, there is now evidence that cognitive dysfunction correlates best with those motor symptoms that respond little or not at all to DAergic medication (e.g., disorders of gait and dysarthria), suggesting that cognitive disorders may instead relate to non-DAergic systems, as indeed seems to be indicated at autopsy (Dubois & Pillon, 1992). Thus PD is also associated with reduced cholinergic, serotonergic, and noradrenergic activity, with damage in the nucleus basalis of Meynert, the raphe nuclei, and the locus coeruleus, especially in the presence of dementia. Other neurotransmitters are also known to operate within the BG and to be affected in PD (Gupta, 1993). These include the enkephalins, opiates, substance P, somatostatin, cholecystokinin, neurotensin, glutamate, aspartate, and gamma-aminobutyric acid (GABA). Thus PD is a multitransmitter deficiency disorder, with motor, mood, and cognitive deficits.

We can attempt to study drug effects directly via "drug holidays," but effects may be contaminated by permanent or long-term drug-induced changes. In more chronic patients whose response to medication may fluctuate, we can compare cognitive performance between "on" (near normal) and "off" motor states, but again findings are contradictory (Levin, Tomer, & Rey, 1992). Although debate continues as to whether asymmetries in side of onset or severity of motor symptoms correlate with cognitive (verbal or visuospatial) function, the best evidence for the influence of DAergic transmission on cognition, perhaps via the mesolimbic or mesocortical "frontal and executive" functions, comes from studies (see below) of MPTP intoxication in young, previously healthy individuals. Here, only the DAergic system is damaged, unlike the other neuromodulatory transmitter systems; memory and attention deficits *are* found, indicating a DAergic role in cognition. In any case, cognitive programs intrinsic to the cortex probably are not so much damaged as deactivated. However, cognitive deficits are far from universal in PD; they are usually moderate rather than severe, and are more evident in older patients and under conditions of cognitive load.

As we have by now repeatedly indicated, cognitive deficits in PD are essentially frontal or "executive" (e.g., problems in goal or strategic aspects and in maintaining and shifting attention) (Bradshaw et al., 1993), problems with manipulating (initiating, maintaining, or switching) mental set as with the Wisconsin Card Sorting Test and Stroop Color and Word Test, problems with fluency and with delayed recognition and responding, with automatic calling-up of responses (Saint-Cyr & Taylor, 1993), with generating solutions to novel problems without the help of external guidance, and with judgments of temporal ordering and relative recency (Pirozzolo et al., 1993). Stam et al. (1993) asked whether those "executive," set-related, and attentional mechanisms that are deficient in PD reside in the frontal cortex itself, which has its own DAergic system originating in the ventral tegmentum, or in the "complex loop" (see Chapter 10) running between the caudate, thalamus, and frontal cortex. Owen et al. (1993) in fact reported subtle differences in the manifestation of set-related deficits between PD and purely frontal syndromes. Thus although loss of DAergic influence upon the dorsolateral prefrontal cortex may play a major role in the cognitive deficits of PD, the lack of disinhibition indicates the relative noninvolvement of the orbitofrontal regions.

Additionally, PD patients usually fail to exhibit loss of concern and insight, and of major tendencies to perseverate. Nevertheless, it must be concluded that problems in attention, in puzzles such as the Tower of Hanoi, in the Wisconsin Card Sorting Test, and in the Stroop Color and Word Test (where the reader must inhibit the strong natural tendency to read the color word, rather than name the dye in which it is written), all suggest the dysexecutive syndrome and defective prefrontal functioning. We see the same disability in parkinsonian problems in predicting target movements in predictable or redundant tracking tasks, where there may instead be excessive dependence upon the environment in the absence of adequate self-generated programming (Dubois et al., 1991). Likewise, we see parkinsonian problems in alternating between different motor programs, any one of which on its own could be satisfactorily performed. Of course, the frontal lobes receive much input from sources other than the BG, so parkinsonian damage to the BG need not result in exactly the same deficits as occur with purely frontal injury, the issue of interest being to determine those specific executive routines that stem from BG disruption and PD. A second issue, already alluded to in Chapter 10, is the extent to which we can still characterize the problem as "executive," rather than perhaps in a narrower sense one of a partial failure of working memory (Baddeley, 1992). In some respects the distinction is perhaps merely semantic, if we view the frontal lobes as mediating a working memory (see Chapter 8) which has limited processing resources, which in turn become even more limited in PD.

Retinal and Visual Changes in Parkinson's Disease

There are now numerous reports of changes in the visual system in PD (for review see e.g., Ikeda, Head, & Ellis, 1994), suggestive of DA deficiency in the retina. DAergic amacrine and interplexiform cells receive input from cone-driven bipolar cells, and release DA to horizontal cells and photoreceptors. Ikeda et al. noted that DA receptors are widely distributed throughout the retina, although degeneration of DAergic cells in the retina in PD is more severe in the fovea than in the periphery. They reported that in PD the visually evoked potential may be delayed, along with changes in the flash- and pattern-evoked electroretinogram.

Dementia and Depression in Parkinson's Disease

The possible involvement of a range of neuromodulatory deficits is compatible with the frequent development of dementia in PD. Clinical observations that dementia may be higher in PD patients manifesting hypokinesia and rigidity rather than tremor suggest that this "subcortical" dementia (manifesting as slowing, forgetfulness, alterations in mood and personality, and inefficiency rather than aphasia, agnosia, apraxia, or amnesia) is not merely a consequence of aging; on the other hand, the greater decrement to performance IQ than to verbal IQ resembles an exaggeration of normal aging. PD patients of course are typically elderly, and may therefore also suffer concomitant AD (see Chapter 9) with its *cortical* dementia. (In Chapter 9 we noted that the distinction between cortical and subcortical dementia is blurred and often disputed.) Indeed, subclinical AD may even be unmasked by PD, and many PD patients are known to exhibit some cortical pathology. The frequency of dementia in PD may possibly reach 20% (Hammond-Tooke & Pollock, 1992). It is far milder and less progressive than the dementia of AD, and in fact is not easily distinguishable from bradyphrenia. The latter term however should be reserved for slowed thinking with unimpaired intelligence, with dementia here referring to reduced levels of voluntary attention, spontaneous interest, initiative, and effort, with increased fatiguability and a slight diminution in memory. As noted by Dubois et al. (1991), the condition is more one of a cognitive disorder of subcortical origin, rather than a true dementia. Clearly the frontal "executive" functions discussed above can come under this category. According to Hammond-Tooke and Pollock (1992, and see also Saint-Cyr & Taylor, 1993) dementia correlates more with severity of motor deficits and age of onset than with disease duration, and also correlates with depression, both perhaps being a consequence of concomitant damage to cholinergic, noradrenergic, and serotonergic systems in addition to those involving DA. Generally, the incidence of dementia in PD is higher in older, and older-onset patients, the response to drug treatment is poorer in demented patients, and the progress of PD is more rapid in such cases.

The frequency of depression again is uncertain, perhaps reaching 40%, but not correlating with age or gender. (It is much lower in AD, the classical dementing disorder.) Debate continues as to whether it is reactive to the disease or endogenous and integral to the disorder, though the latter possibility seems the more likely, given the changes to the biogenic amine systems and the suggestions of a long, prodromal parkinsonian personality of depression, introversion, pessimism, melancholia, rigidity, diligence, stoicism, industriousness, and reduced affect (Stacy & Jankovic, 1992). Interestingly, tricyclic antidepressants and electroconvulsive therapy may help both depression and PD. In PD, depression may be of mild to moderate intensity, and characterized by dysphoria, pessimism, and somatic symptoms (Hammond-Tooke & Pollock, 1992). It correlates poorly with the motor severity of PD, and its relationship to DAergic status is uncertain. The anterior cingulate circuit connecting the BG and frontal structures, as discussed earlier, and otherwise known to display hypometabolism during endogenous depression (Sano & Mayeux, 1992), is likely to be involved, together with concomitant disturbances of the biogenic amine systems. These factors, together with autonomic involvement, account for the other common PD symptoms of seborrhea, sialorrhea, impotence, constipation, and postural hypotension. Although depression tends not to be related to age of onset, duration, or degree of the disorder, there is some evidence (Santamaria & Tolosa, 1992) that patients where depression preceded motor symptoms tend to be younger, less impaired, and to have a family history of PD. However, depressed patients may also show a faster progression of the disorder, despite the fact that otherwise it is *older* patients who may show the faster disease progress.

Disease Progression

We can identify the following stages of PD (Hoehn & Yahr, 1967). [Note all these figures are essentially pre-L-dopa therapy.]:

Stage 1: Unilateral, a single extremity tremor and mild rigidity; mild inconvenience, lasting about 3 years.

Stage 2: Asymmetrical bilateral tremor and rigidity, a mildly flexed posture with limb adduction, facial masking, monotonous hypophonic speech, mild gait disturbance and slowness, decreased spontaneous movement, mild fatiguability. About 6 years total.

Stage 3: Disequilibrium, impaired balance, posture, and righting reflexes especially on turning or if pulled backwards; hesitation, freezing, festination. About 7 years total.

Stage 4: Help may now be needed for walking, dressing, feeding, or bathing. Around 9 years total.

Stage 5: Severe disability requiring use of wheelchair or bed; complete invalidism. Death due to the complications of inactivity after a total of about 14 years.

Therapy with L-dopa modifies the rapidity with which these stages are reached, and there may now in fact be no excess mortality over normals, even though the underlying disease progresses at the same rate.

Management of the Disorder

Narabayashi (1993) reviewed the use of stereotaxic surgery, which started in the early 1950s and before the advent of drug treatment, for the alleviation of PD symptoms. He noted that stereotaxic surgery is still important and can achieve almost total control of rigidity and tremor on the side of the body contralateral to surgery, when medical treatment otherwise fails. Initially the GP$_i$ was targeted, and was found to be effective against muscle rigidity but not tremor. Later, attention turned to the thalamus for the control of tremor. The ventral intermediate nucleus, adjacent to tactile sensory (proprioceptive) neurons, produces rhythmic bursts of unitary activity synchronous with tremor of certain parts of the contralateral side of the body. Lesions here were found to suppress tremor. In the ventrolateral (VL) nucleus of the thalamus, adjacent to the ventral intermediate nucleus, lesioning was found to abolish rigidity, just as with lesioning of the GP_i. The VL nucleus of course receives the main outflow of the BG. Akinesia secondary to rigidity could also be abolished, but in most patients akinesia seems to be primary or independent of rigidity, rather than secondary to it, and to require L-dopa medication. Iacono and Lonser (1994) reviewed posteroventral pallidotomy for the successful control of akinesia, dystonia, tremor, and rigidity, which they say is very similar to thalamotomy. Wichmann and DeLong (1993) similarly reviewed the effects of surgery on the STN and GP_i, noting that inactivation of the STN reduces GP_i output toward more normal levels, leading to reduction of PD motor signs. For a general review of lesion treatment of PD, see Guridi, Luquin, Herrero, and Obeso (1993), and Marsden and Obeso (1994). The latter observed that motor circuits within the BG are thought to facilitate desired and inhibit unwanted movement through their influence on the thalamus, and that stereotaxic lesions in the motor thalamus or GP should therefore impair voluntary movement. They noted that although such lesions may improve rigidity and tremor, they nevertheless seem not to worsen bradykinesia, and they discussed possible reasons for such an apparent discrepancy. They concluded that the motor circuits of the BG participate in a distributed motor system that can operate, albeit imperfectly, in the absence of much of the former's contribution. They suggested that significant external or internally derived events may capture the attention of the nonmotor striatum (i.e., the caudate), which then can interrupt the routine operation of the motor circuit to permit new cortically initiated activity.

Before the L-dopa era of medication, excess mortality from PD was around three times that of the general population. Nowadays there is an almost normal life expectancy, with very effective early symptom control, though the underlying loss of nigrostriatal cells continues inexorably. Anticholinergics have long been used to control tremor, but they are not efficacious against bradykinesia, rigidity, or postural instability, and have unpleasant side effects. In the late 1950s DA therapy was initiated. As DA does not cross the blood–brain barrier, primary treatment is usually oral with L-dopa, a precursor of DA, together with a peripheral decarboxylase inhibitor, for example, carbidopa (Sinemet) or benserazide (Madopar, Prolopa), to prevent peripheral conversion of L-dopa to DA, with unpleasant side effects (Cummings, 1992; Yahr, 1993); carbidopa does not cross the blood–brain barrier. Adding the selective MAO-B inhibitor selegiline (L-Deprenyl) may increase the amount of DA available to receptors, as well as having protective antioxidant effects, slowing disease progression. DA agonists stimulate the postsynaptic DA receptor (D2 especially) directly, rather than facilitating production of DA itself; consequently, their metabolism does not lead to toxic byproducts. They include bromocriptine (Parlodel) and pergolide (Permax). Anticholinergic agents (e.g., benztropine) may also be given to reduce excess cholinergic activity. Side effects of L-dopa therapy can include vivid dreams and nightmares, hallucinations, delusions, confusion, nausea, and orthostatic hypotension. Later, there may be decreased symptom control of PD. This includes involuntary movements and drug-induced dyskinesias, alterations in mentation, and diurnal "on–off" fluctuations or sometimes extremely rapid oscillations between severe parkinsonian immobility ("off") and a mobile dyskinetic state ("on"). There may also be episodes of akinetic freezing, fatigue, and neuroasthenia. These distressing changes, after initial excellent control, occur after about 4 or 5 years of L-dopa treatment in almost half PD patients, and begin with an end-dose loss of effect, progressing to non-dose-related random on–off phenomena (see e.g., Guridi et al., 1993). It is not known whether these phenomena are due to progression of the disease itself, or to drug effects, though the former is suspected. In fact, Utti et al. (1993) reported that L-dopa treatment was not associated with any increase in mortality, and reject any substantial drug toxicity. Moreover, Bravi et al. (1994) concluded that the wearing-off phenomenon that complicates L-dopa therapy in advanced PD may be attributed both to a reduction in striatal DA storage, consequent upon the progressive degeneration of *presynaptic* DAergic terminals, and also to *postsynaptic* changes. Dyskinesias, furthermore, may stem from supersensitive postsynaptic DAergic receptors. In any case, DAergic precursor therapy depends on the integrity of at least some nigrostriatal neurons to metabolize L-dopa to DA and transport it. When most of these cells are lost, DA receptor agonists like bromocriptine may be added.

New forms of treatment being developed are all directed toward restoring striatal DA activity. One option (Hauser & Olanow, 1993) is the neural transplantation of tissue capable of synthesizing and releasing DA into the striatum to compensate for endogenous deficiency of DA. Experimental parkinsonism in rats and monkeys was successfully reversed by transplant of fetal SN cells, though benefits were inconsistent in advanced human patients. Because of ethical, immunological, and infectious considerations, autoimplants (autologous transplants) have been used of cells from the patients' own adrenal medulla; such cells normally produce epinephrine and norepinephrine, though they also produce DA when isolated from the adrenal cortex and the effects of cortisone. Again the procedures have been more successful in experimental animals than in human patients, though Diamond, Markham, Rand, Becker, & Treckiokas (1994) reported that in four PD patients who had undergone adrenal medulla-to-caudate transplants 4 years previously, the disease course became more benign and more slowly progressive. However, they concluded that improvement was insufficient to warrant routine use of the procedure. Other experimental techniques involve introduction of skin cells genetically modified to manufacture DA, cloned brain-derived nerve growth factor specific for DAergic activity, and antioxidants (see below) to counter pathogenic oxidative deamination of MOAs (Yahr, 1993).

Causes of Parkinson's Disease

We still do not know the ultimate causes of PD, except when it is due simply to vascular problems. Such cases are distinguishable from idiopathic PD by the absence of tremor, the presence of brisk muscle stretch reflexes, extensive plantar responses, stepwise progression, and a history of hypertension and stroke (Cummings, 1992). Simple Mendelian inheritance seems not to be a common cause, apart from some comparatively rare familial cases. Thus MZ-twin concordance rates do not support heredity as an important component, though it is possible that PD incidence in cotwins has been underestimated, and insufficient time may have elapsed for PD to develop in the cotwin. Although the fact that MZ and dizygotic (DZ) twin concordance rates are similar may seem to suggest low heritability, it must be remembered that if the mitochondrial genome is involved, inheritance would be exclusively maternal and concordance rates would not differ. Indeed, the possibility of an inherited mitochondrial defect is being discussed, apart from the fact that the mitochondria are very susceptible to mutagenesis and certain toxins, as is the possibility of an inherited predisposition to PD, which will only emerge in the presence of an environmental factor. Thus the fact that PD may develop almost simultaneously across generations in a family argues for an additional environmental contribution. Environmental factors such as industrial waste, toxins in well water, pesticides, and so on, have all been suggested from epidemiological studies. Indeed Hall (1841, cited in Snow & Calne, 1992) noted the similarities between the effects of mercury poisoning and "paralysis agitans"—the first suggestion of an environmental toxin. The best available model for a neurotoxin comes from the effect of MPTP, a congener of illegal narcotics. This substance, which is skin and lung absorbent, a likely route for an environmental toxin, is a lipophilic protoxin that readily crosses the blood–brain barrier; it is rapidly converted by MAO-B to 1-methyl-4-phenylpyridium (MPP^+), its active metabolite, which is concentrated in the mitochondrial matrix, and leads to depletion of adenosine triphosphate (ATP) and neuronal destruction (Stacy & Jankovic, 1992). This sequence of events results in a syndrome that is clinically indistinguishable from PD. It exclusively damages DAergic cells of the SN. Indeed, people exposed some years ago to MPTP and then without symptoms are now in many cases beginning to show parkinsonian symptoms. Some pesticides have a chemical structure similar to the precursor to MPTP, whereas others are known to target the mitochondria (Golbe & Langston, 1993).

Quite apart from MPTP, the Chamorro Indians of the Pacific Island of Guam, who ate flour from the seed of the cycad *Cycas circinalis*, developed a disorder combining PD, dementia, and amyotrophic lateral sclerosis (ALS, or motor neuron disease). A single exposure is said (Snow & Calne, 1992) to have been sufficient to produce the disorder up to 35 years later, supporting the idea that idiopathic PD may be due to an environmental toxin superimposed upon progressive neuronal loss with age. It is also noteworthy that the excitatory neurotransmitter glutamate, thought to possibly act as a neurotoxin, is involved in the Guam disorder, "idiopathic" ALS, and HD (see Chapter 13). PD could therefore be due to an accumulation of damage from an exogenous toxin that is insufficiently metabolized by defective enzymatic degradative pathways. Alternatively, even a *brief* exposure to a nigral toxin might be sufficient, with the resultant compensatory increase in DA turnover in the remaining nigral neurons gradually producing toxic by-products, the resultant nigral damage only accelerating this vicious circle (Golbe & Langston, 1993).

Viruses have also been implicated in PD, and prions, another infective agent, in a number of other neurodenerative disorders. Thus bovine spongiform encephalopathy (BSE) or "mad cow" disease (in cows, and now in some other species where it is spread by eating infected nervous tissue) exactly mimics human kuru in the Fore region of Papua New Guinea. There it was spread by ritual cannibalism (Carr, 1993). Both together with nonfamilial Creutzfeldt-Jakob disease (which is transmitted to people after treatment with contaminated products from human cadavers) and scrapie (a sheep disease of the central nervous system) result in a staggering gait, and are also caused by prions (Prusiner, 1993). Prions in fact cause a variety of degenerative neurological disorders that can be infectious, inherited, or sporadic in origin (Cohen et al., 1994); they differ from all other infectious pathogens in that they are devoid of nucleic acid, and their only known constituent is a protein that is an altered form of a normal one encoded by a chromosomal gene. True viruses, however, are implicated in encephalitis lethargia (EL, or von Economo's encephalitis), a form of postencephalitic PD with other extrapyramidal symptoms (Cummings, 1992) that developed often years after acute infection between 1919 and 1926. Akinesia and rigidity were the most common symptoms, often accompanied by catatonia, tremor, chorea, myoclonus, tics, oculogyric crises, forced gaze deviations, stereotyped thinking, hallucinations, obsessions and compulsions, mood disorders, and personality changes. Although, fortunately, the disease seems now for some reason to have vanished, lesions were found in the SN and widespread throughout the BG, the thalamus, and hypothalamus. The disorder, with its bizarre and sudden leaps from one state to the other, has of course been popularized in the film *Awakenings*, which in fact often underplays the nature of the disorder and the (unfortunately short-lived) relief gained by L-dopa therapy. (The influenza pandemic and that of EL overlapped in time, but were probably unrelated.)

An olfactory dysfunction has often been noted in both PD and AD (see Chapter 9), suggesting that in both neurodegenerative disorders the olfactory system may be vulnerable to environmental agents, toxins, or even viruses (R. L.

Doty, Stern, Pfeiffer, Gollomp, & Hurtig, 1992). From the nasal cavity, and via the nasal epithelium, it is possible that such agents might gain access to the central nervous system; alternatively, olfactory dysfunction may merely reflect parkinsonian degeneration of central processes. Stern et al. (1994) noted that the degree of olfactory dysfunction differs as a function of symptomatology (gait disorder vs. tremor) and age of onset.

There has been much recent evidence of neurotoxicity from oxidative free radicals and oxidative stress in PD brains. Indeed oxidative stress may be a major factor in the neuropathology of adult neurodegenerative disorders (PD and HD, see Chapter 13) along with seizures, stroke, and trauma (Coyle & Puttfarcken, 1993). Moreover, DA itself may be neurotoxic, as free radicals are produced when MAO oxidizes DA. Oxidative stress and free radicals have also been implicated in ALS, and the role of MAO inhibitors like selegiline is to decrease the generation of free radicals by reducing the oxidative metabolism of DA. A maternally inherited defect in the mitochondrial genome may lead to inadequate detoxification of environmental toxins, with consequent nigral damage (Di Monte, 1991). Alternatively, a defect in the regulation of oxidative metabolism of DA may result in the excess generation or defective removal of free radicals (Olanow, 1990, and see also Schapira, 1994) generated by the oxidative metabolism of DA. Total iron content in the brain may increase in PD, with decreased ferritin iron binding in the BG (see e.g., Shoulson & Kurlan, 1993), leading to the formation of highly reactive hydroxyl radicals. Indeed, the possible role of iron in oxidation reactions and free radical formation in PD is discussed in detail by Fahn and Cohen (1992) and by Jenner (1994), who extended the implications to other degenerative diseases of the BG, and even to other neurodegenerative diseases in general, such as AD (see Chapter 9). In this general connection it is noteworthy that a possible protective role for smoking in reduction in the incidence of PD repeatedly reemerges (Baron, 1986), though Riggs (1993) noted that such effects could still stem from a differential survival bias, and Mayeux, Tang, Marder, Côté, & Stern (1994) find that PD *reduces* smoking, rather than smoking protecting against PD. Although the carbon monoxide in inhaled cigarette smoke may possibly help remove oxidative free radicals, nicotine binds to certain cholinergic receptors (anticholinergics can help control PD symptoms); in any case, the prevalence of both smoking and PD may be due to linked individual differences in personality. Thus PD patients are thought to have long had quiet, uncreative rule-abiding personalities of a depressive predisposition and a lack of affect—maybe the earliest preclinical feature with a very long latency. This feature may also predispose them not to smoke.

Summary and Conclusions

PD has been described for nearly 200 years. It is a progressive and predominantly motor disorder of middle life involving DAergic cell loss in the nigrostriatal system, though other DAergic regions and neurotransmitters are also affected. While clearly related to the aging process, more may perhaps be involved than simply accelerated aging. The disorder, at a motor level, results in increased tonic output of the BG, with consequent overinhibition of parts of the motor thalamus and therefore reduced thalamocortical activity. Resultant negative symptoms include slowness in initiating (akinesia) and executing (bradykinesia) movements, and changes in gait, speech, and writing. Rigidity and tremor are the major positive symptoms. However, a deficit in the internal generation of programmed activity, with a consequent overreliance upon external cuing, may underlie many of the parkinsonian difficulties during complex motor and cognitive activities, such as the smooth initiation of each next stage in a sequence. When a patient tries to perform two actions simultaneously, moreover, the better learned, more automatic or subconscious one may fail; indeed it may often be as if all tasks, even the simplest, require full attentional monitoring, and cannot be prepared in advance or run off automatically.

At a cognitive level, deficits vary widely between patients, and even within the same patient at different times of the day, though the overall steady decline, which rarely results in incapacity, tends to parallel the progressive motor decrement. Planning difficulties again seem to underlie many of the difficulties with language, and a mild anomia is common. Visuospatial difficulties may only appear with the imposition of significant processing demands, though difficulties have been reported in appreciating the relative position of objects in space, in route finding, and in copying complex configurations. Deficits in visuospatial memory may occur, though generally recognition is preserved better than recall. Indeed the problem may be more one of access than of registration, storage, or consolidation. Procedural memory is more affected than declarative memory, the reverse of the situation typically obtaining with AD. There is really very little evidence for significant cognitive slowing (bradyphrenia) analogous to the characteristic motor slowing of PD, in the absence of significant dementia or depression.

Many of the cognitive deficits in PD are suggestive of the "dysexecutive" syndrome associated with prefrontal damage (e.g., defective goal or strategic aspects, problems in maintaining and shifting attention or mental sets, problems with delayed responding and with the automatic calling-up of responses). However, unlike true frontal patients, patients with PD lack disinhibition or loss of concern or insight, and do not usually perseverate. It is unclear whether their problems reflect DAergic malfunction in the frontal cortex itself, or in the "complex loop" running between the caudate, thalamus, and frontal cortex, or even concomitant damage to other neuromodulatory systems (e.g., the noradrenergic, cholinergic, and serotonergic). Indeed, damage to all four neuromodulatory systems may underlie the frequent development of dementia and depression in PD. Dementia however is rarely severe, and accompanies hypokinesia and rigidity rather than tremor in the manifestation of positive symptoms. It is often described as subcortical, through the cortical–subcortical distinction is disputed, and the condition should perhaps be described as a cognitive disorder rather than true dementia. Where the latter does indeed manifest in PD, it may be due to concomitant AD. Depression also frequently accompanies PD, and may even exist as a long, prodomal personality trait in individuals later to develop the disorder.

Stereotaxic surgery preceded L-dopa medication in the management of PD, and is still employed in the control of rigidity and tremor. The target of surgery is the interruption of pathways so as to reduce the tonic output of the BG. However, medication with L-dopa, a precursor of DA, together with the administration of various other pharmacological agents, is now generally the treatment of choice, with the aim of restoring nigrostriatal DA levels to something approaching normality. However, side effects are common, including dyskinesias and alterations in mentation, as eventually are "on–off" fluctuations and problems in control. It is not always clear to what extent such problems stem from medication or are intrinsic to the disease process itself. Cell grafting may be the way forward to the future.

Simple Mendelian inheritance does not normally seem to operate in the etiology of PD, as evidenced by discordant MZ twin studies, though an inherited predisposition, possibly via a mitochondrial defect, cannot be excluded. Environmental agents almost certainly play a role, though they too have proved elusive. However, the selective neurotoxin MPTP is currently providing us with an ideal experimental model, and other naturally occurring agents are known to cause more or less similar motor disorders, probably via the excitatory neurotransmitter glutamate. Viruses have also been implicated in EL, which exhibits a number of extrapyramidal symptoms resembling PD. Currently, a promising line of research addresses the issue of neurotoxicity from oxidative free radicals and oxidative stress in such neurodegenerative disorders as PD and HD, whereas DA itself may be neurotoxic. Again, a (maternally inherited) defect in the mitochondrial genome may lead to inadequate detoxification of environmental toxins, with consequent nigral damage. A possible protective role for smoking remains controversial, just as it does for AD.

Further Reading

Dubois, B., Boller, F., Pillon, B., & Agid, Y. (1991). Cognitive deficits in Parkinson's disease. In F. Boller & J. Grafman (Eds.), *Handbook of neuropsychology* (Vol. 5, pp. 195–240). Amsterdam: Elsevier.

Hauser, R. A., & Olanow, C. W. (1993). Parkinson's disease. In L. Barclay (Ed.), *Clinical geriatric neurology* (pp. 155–169). Philadelphia: Lea & Febiger.

Huber, S. J., & Cummings, J. L. (1992). *Parkinson's disease: Neurobehavioral aspects.* Oxford: Oxford University Press.

Jankovic, J., & Tolosa, E. (Eds.), *Parkinson's disease and movement disorders* (2nd ed.) (Chapters 5, 6, 26). Baltimore: Williams & Wilkins.

Joseph, A. B., & Young, R. R. (Eds.). (1992). *Movement disorders in neurology and neuropsychiatry.* Oxford: Blackwell Scientific.

Mahurin, R. K., Feher, E. P., Nance, M. L., Levy, J. K., & Pirozzolo, F. J. (1993). Cognition in Parkinson's disease and related disorders. In R. W. Parks, R. F. Zec, & R. S. Wilson (Eds.), *Neuropsychology of Alzheimer's disease and other dementias* (pp. 308–349). Oxford: Oxford University Press.

Narabayashi, H., Nagatsu, T., Yanagisawa, N., & Mizuno, Y. (Eds.). (1993). *Parkinson's disease: From basic research to treatment: Advances in neurology, Vol. 60.* New York: Raven Press.

Nutt, J. G., Hammerstad, J. P., & Gancher, S. T. (1992). *Parkinson's disease: 100 maxims.* London: Edward Arnold.

Pirozzolo, F. J., Swihart, A. A., Rey, G. J., Mahurin, R., & Jankovic, J. (1993). Cognitive impairments associated with Parkinson's disease and other movement disorders. In J. Jankovic & E. Tolosa (Eds.), *Parkinson's disease and movement disorders* (2nd ed.) (pp. 493–510). Baltimore: Williams & Wilkins.

Schneider, J. S., & Gupta, M. (Eds.). (1993). *Current concepts in Parkinson's disease research.* Seattle: Hogrefe & Huber.

Stern, G. (Ed.) (1990). *Parkinson's disease.* London: Chapman & Hall.

Wichmann, T., & DeLong, M. R. (1993). Pathophysiology of Parkinsonian motor abnormalities. In H. Narabayashi, T. Nagatsu, N. Yanagisawa, & Y. Mizuno (Eds.), *Advances in neurology,* (Vol. 60, pp. 53–61). New York: Raven Press.

Chapter 13

Subcortical Movement Disorders II

Huntington's Disease

George Huntington in 1872 first described

the hereditary chorea, as I shall call it, [which] is confined to certain and fortunately few families, and has been transmitted to them, an heirloom from generations away back in the dim past. It is spoken of by those in whose veins the seeds
of the disease are known to exist, with a kind of horror, and not at all alluded to
except through dire necessity when it is mentioned as 'that disorder.' (Huntington, 1872, p. 317)

Huntington went on to describe the three main peculiarities of the disease: its
hereditary nature, the fact that it manifests usually in adult life, and a tendency
in its sufferers to insanity and suicide. The first cases are believed to have occurred in early 17th-century Europe (Koroshetz, Myers, & Martin, 1992), and
debate continues as to whether all present-day cases of this autosomal-dominant
inherited disorder are direct descendants, or whether any new mutations can or
have occurred. We shall see that very recent findings on the genetic nature of
the disorder have thrown a new light on this particular issue.

Prevalence, Age of Onset, and Inheritance of Huntington's Disease

Estimates of the prevalence of Huntington's disease (HD) vary, ranging from
around 5 or 10 per 100,000 in Northern Europe and the U.S.A. to 17 per
100,000 in Tasmania and 700 per 100,000 in Lake Maracaibo, Venezuela. As
the average age of clinical onset (around 45 years, but with wide variations) is
postreproductive, its prevalence (overall percentage in a population currently
with the disorder) and its incidence (the number of new cases annually) might

be expected to rise, except for the countereffects of declining family sizes. A progressive neurodegenerative disorder, it is characterized by the insidious onset of minor motor incoordination followed by uncontrollable choreiform movements, cognitive deterioration, and affective and psychiatric symptoms with personality change. Memory and speech deficits appear, along with abnormal saccadic control. There is no cure, and treatment is at best palliative, with symptoms progressing relentlessly to death (typically from aspiration pneumonia or cardiorespiratory failure) about 16 years after onset of clinical symptoms. Survival times may however vary considerably, and diagnosis is typically made several years after disease onset. As it is an autosomal dominant trait with complete lifetime penetrance, a child of an affected parent has a 50% chance of developing the disorder, irrespective of sex. Moreover, rare HD homozygotes, both of whose parents have the disorder, do not differ clinically from heterozygotes (Gusella, MacDonald, Ambrose, & Duyao, 1993). Often, of course, homozygotes for a condition are more seriously affected. The HD gene may therefore operate via control processes, rather than in a structural protein or enzyme (Penney & Young, 1993). Homozygous HD cases tend to have an age of onset corresponding to the allele with the larger number of repeats (Gusella et al., 1993, see below). Late onset is typically more benign than early, which is associated with a rigid, akinetic PD-like form with tremor and dystonia, and known as the Westphal variant. The latter, constituting about 10% of HD cases, is associated with severe intellectual impairment, short survival, and paternal inheritance. (Maternally inherited genes are more methylated and are therefore less likely to be expressed, but see below for recent findings concerning sex differences in the number of trinucleotide repeats in the HD gene.)

The gene for HD was first localized to the end of the short arm of chromosome 4 in 1983, but it took 10 years to isolate the gene and identify the disease-associated mutation. This has now been identified as an expanded and unstable trinucleotide (CAG) repeat sequence in a gene (IT15) on chromosome 4p16.3 (The Huntington's Disease Collaborative Research Group, 1993): the nucleotide sequence CAG probably codes for the amino acid glutamine, though how the expanded stretch of CAG repeats produces its calamitous effects is still unknown (Gusella & McDonald, 1994). Such a repeat, like a kind of genetic stutter, is now also known to occur in fragile X syndrome, myotonic dystrophy, spinocerebellar ataxia type 1, and spinobulbar muscular atrophy (Kennedy's disease) (see e.g., Gusella et al., 1993, and J. Martin, 1993). Unaffected individuals have between 11 and 34 copies of the triplet, and people with HD between 37 (or possibly 42) and 86. (Recent reports are emerging which blur, close, or even overlap the "border" region in the vicinity of 35 repeats, see e.g., Kremer et al., 1994.) The number of repeats does not perfectly predict the age of onset, especially with paternal inheritance; indeed, a father's sperm may vary in the number of repeats, whereas the number tends to be constant in the mother's ova (Wexler, 1993). Generally such observations account for the fact that differences in age of onset tend to be fairly stable within families, and for the apparent age-related penetrance of the disorder. Paternal transmissions are associated with more repeats and earlier onset. Occasionally, the number of repeats may apparently decrease, so that in theory HD might miss a generation; more commonly the reverse occurs, even within the upper normal range of repeats, so "new mutations" may after all appear. Effects such as these account for the general phenomenon of anticipation, whereby age of onset tends to be reduced and severity of symptoms increased in successive generations in families susceptible to the disease. However, some other mechanism must also be operating, as otherwise all normal families would be expected eventually to evince HD. The effect of the repeats may be one of excess protein production and/or expression of the gene at the wrong time. It may not be coincidental that HD, myotonic dystrophy, spinocerebellar ataxia type 1, and spinobulbar muscular atrophy all involve approximately the same number of repeats of CAG (though on different chromosomes), the synthesis of glutamine, damage to (different) areas of the motor system, and have an association of paternal transmission and juvenile onset (Wexler, 1993). Findings such as these will now allow direct genetic diagnosis (Kremer et al., 1994), without, as before, the need to use linkage markers and DNA donation from other family members. They also open up the possibility (see below) of early therapeutic intervention, while simultaneously necessitating a revision of the traditional or classical concept of the gene (Sutherland & Richards, 1994). Thus although the latter is still seen as the basic unit of inheritance, and to be subject to mutations that result in changes in the behavior of an organism, mutations are no longer seen as stable. This instability results from changes in the number of times a block of three subunits (nucleotides) is represented (repeated) in the normal DNA sequence. These repeats seem to occur in regions performing a regulatory function, and changes in the number of such repeats may be influenced by the sex of the parent, and may in turn influence the severity of disease symptoms and age of onset. In all the disorders, including HD, where this phenomenon occurs, there seems to be an overlap, in terms of number of repeats, between the upper limit of normality and the lower limit where morbidity might occur (Sutherland & Richards, 1994).

Symptomatology of Huntington's Disease

The first clinical signs of the disorder are typically affective, involving depression, anxiety, irritability, emotional lability, impulsivity, and aggression (Pirozzolo, Swihart, Rey, Mahunin, & Jankovic, 1993). These are soon followed by restlessness, clumsiness, incoordination, forgetfulness, personality changes, altered speech and writing, and saccadic disturbances, leading on to problems with fine motor dexterity, steadiness, and speed. Athetosis and chorea follow, although later still involuntary movements decrease, with the appearance of upper limb flexion and lower limb extension. The disorder shares with PD common negative symptoms of bradykinesia and akinesia, especially in the end stages, though these symptoms probably underlie the superimposed hyperkinetic chorea even in the early stages. Indeed HD and PD differ in their positive symptoms, chorea in HD replacing the tremor of PD. In fact, akinesia and bradykinesia are better indicators of functional ability and better preclinical predictors than chorea, which while often little apparent in the earliest stages, may come to dominate the condition, before generally dying away in the later stages of HD, leaving hypokinetic rigidity. Chorea may initially present as minor tics and twitches, slower than myoclonus, developing into abrupt jerks that occur randomly in place and time, interspersed with periods of quiescence. Distal involvement may initially predominate, with, later, more frequent and pervasive, and higher amplitude movements spreading to the face and eventually to the whole body. They present as a constant, unpredictable flow fitting across the body of what seem to be complex fragments of normal behaviors, like the flicking of cigarette ash from between the fingers, the shrug of a shoulder or raising of an eyebrow—premotor synergies, perhaps. Their involvement of multiple joints resembles voluntary action; indeed, like tics (Chapter 14) they may be briefly suppressible. Their quasi-undulating manner is idiosyncratic to each individual. Slower, choreoathetotic movements are more writhing and flowing, resembling the operation of a string puppet (Koroshetz, Myers, & Martin, 1992). Slowly evolving over time, such hyperkinesias give purposeful acts an exaggerated baroque quality, as if the patient (as indeed may be the case) is trying to smuggle in and conceal choreic activity amidst normal voluntary movements. At their peak in some patients a million or more such movements may occur per day, leading to exhaustion. When slowing down in the disorder's later stages, they become more athetotic, reflecting perhaps the change of one dystonic posture to another. As with the tics of Tourette's syndrome (TS) (see Chapter 14), they may be temporarily suspended with a great effort of will. They usually decrease during sleep and increase with stress and with voluntary movements like walking. Analogous problems occur with eye movements (Koroshetz et al., 1992); saccades are slow, and smooth pursuit movements are often broken by fixations, followed by corrective saccades, although there may be involuntary eye movements to peripheral stimuli. Speech becomes dysarthric, with disturbed rate and rhythm leading to unintelligibility; bursts of words are emitted with pauses in midsentence, often with parkinsonian slowing down and loss of volume toward the end of a phrase. Some of the speech problems stem from incoordinated control of respiratory, tongue, facial, and pharyngeal musculature. Gait becomes "cerebellar," wide-based, and ataxic, resembling that due to alcoholic intoxication, with short, slow steps, irregular rhythm, problems in turning, impaired balance, and sudden falls. The head may be slightly flexed forward or to one side, with extension of the metacarpophalangeal joints and dorsiflexion of the wrist. There may be problems in walking tandem or standing on one foot; misstepping may occur, and loss of the walking rhythm (Koroshetz, Myers, & Martin, 1993). There may be hypotonia and hyperreflexia.

As with PD, HD patients exhibit slower movement execution times (MTs) than controls when performing repetitive tapping tasks. Not only are they slower with simple wrist flexions, but they are especially slow when combining two movements (e.g., hand squeezing and elbow flexing), in a sequential movement task, irrespective of neuroleptic medication (Thompson et al., 1988). Agostino, Berardelli, Formica, Accornero, and Manfredi (1992) found that HD patients were particularly slow at switching from one movement to the next. We (Bradshaw, et al., 1992) have reported, in a sequential button-pressing procedure, that like PD patients, HD patients have particular difficulty in executing movements at very low or very high levels of advance information, and in initiating movements at low levels of advance information. Problems seem to lie in using advance information and in generating appropriate motor programs, especially when sequential movements must be conducted on the basis of internal cues (i.e., planned programmed responses) to guide movement, rather than via external cuing. Indeed, patients with basal-ganglia (BG) disorders, PD, or HD have a general deficit in sequencing complex motor programs, especially when successful performance depends upon some degree of automaticity, and in the absence of external visual cues. HD patients may also have difficulty in controlling the specific parameters of movement, like direction, force, extent, and duration (Phillips, Bradshaw, Chiu, & Bradshaw, 1994). Reduction in processing capacity may mean that HD patients, like PD patients, behave as if all operations also involve a second, simultaneous, concurrent task.

Cognitive Deficits in Huntington's Disease

A cognitive decline parallels—precedes, even—motor decline (Brandt, 1991), correlating closely with the extent of caudate damage that in HD, unlike PD, precedes and is greater than the damage inflicted on the putamen. A "subcortical" dementia develops, resembling that of PD, Wilson's disease (WD), and supranuclear palsy, rather than the "cortical" dementia of Alzheimer's disease (AD); thus there is no real aphasia, agnosia, or apraxia, and there is good orientation as to place and time, with preserved recognition of friends and family, and preserved insight, sense of humor, and social intelligence (Woodcock, 1992). Rather, there are learning and retrieval inefficiencies, an impaired ability to operate upon, organize, or sequentially rearrange acquired information, a mild to moderate intellectual impairment with slowed information processing, reduced alertness, and a reduced cognitive capacity for performing concurrent tasks. As noted elsewhere in this book, the cortical–subcortical dementia distinction has been queried on neuroanatomical, conceptual, and empirical grounds, but survives as an operationally useful concept.

The cognitive deficits involve the construction and execution of mental plans, rather than the loss of specific skills; concentration and attention both decline, simple tasks are performed slowly and with abnormal effort, though the fact that vocabulary and syntax may be so little affected leads to an underestimation of the degree of cognitive impairment (Koroshetz et al., 1992). Difficulties with insight and judgment in problem solving and the manipulation of knowledge indicate a frontal lobe deficit. Indeed, the head of the caudate nucleus receives major innervation from the dorsolateral and orbitofrontal cortices. Thus there are problems with planning and set, with perseveration and distractibility, loss of flexibility and fluency, difficulties with sequencing, organization and scheduling, together with both apathy (a failure spontaneously to initiate activities) and disinhibition (Folstein, Brandt, & Folstein, 1990). The frontal involvement is therefore more severe than in PD. Memory deficits appear early; they are visual more than verbal, and involve short- rather than long-term memory, episodic rather than semantic contents, and recall rather than encoding or recognition. Thus there seem to be problems in efficient mental search and retrieval, and in the conscious, deliberate controlled aspects of memory, rather than the automatic. Although the rate of forgetting may be normal, in the short-term at least, an extensive impairment of remote memory without a temporal gradient (see Chapter 8) again suggests a retrieval problem. A retrieval deficit may therefore reflect strategic problems of inappropriate ways of searching (Pirozzolo et al., 1993). However, unlike AD and Korsakoff's amnesia, but as often may be the case with striatal damage, the acquisition of implicit memories and especially skills may be badly affected, and again unlike Korsakoff's amnesia, remote autobiographical memories may be as badly disturbed as the more recent (Folstein et al., 1990). Cognitive deficits extend to problems with calculation, verbal fluency, and some anomia, though speech comprehension is well retained, even after total loss of intelligible production. Syntax, repetition, content, and word usage are well preserved, despite problems in control of syllable and pause duration and timing. Indeed TV watching behavior may well continue even to the late stages, with normal levels of disappointment if expected visitors fail to appear. Visuospatial difficulties make an early appearance, are generally more severe than those involving language, and include problems with egocentric localization, directional sense, and orientation—all deficits that have been reported in rats with caudate lesions. For review, see Brandt (1991).

Often before motor signs clinically manifest, relatives, friends, and colleagues may observe increased levels of irritability, anger and aggression, frustration, explosive outbursts, anxiety, apathy, and withdrawal (Koroshetz et al., 1992). Depression, suicidal tendencies, and panic all occur, and seem not to be merely reactive to HD diagnosis; they also respond well to medication. Later inappropriate sexual, eating, alcohol, and drug-related behaviors reflect frontostriatal dysfunctions, while paranoia and schizophreniclike psychosis, delusions, hallucinations, and confabulation reflect disturbances in DAergic activity; they are therefore amenable to neuroleptic treatment.

A subset of individuals genetically at risk of developing HD may all show some abnormalities in cognitive and motor tests involving spatial performance, direction, memory, and the utilization of advance information (Bradshaw et al., 1992). They may also have problems in performing rapid tapping movements of the tongue and rapid sequential finger movements, and may exhibit dysdiado-chokinesia, or problems in performing sequences of rapidly alternating manual responses. Spouses may report that newly diagnosed patients were restless while sleeping for several years before clinical diagnosis. PET imaging may also reveal abnormally reduced metabolism some years before clinical diagnosis (Penney & Young, 1993). The disease process seems therefore already to have started. While gene-probe testing is now available and can predict HD with nearly 100% accuracy, individuals at risk are understandably often reluctant to submit to the possible confirmation of their worst fears. Provocation with DA agonists is another promising but little-tried procedure, which in view of the probable outcome, however, may more likely be replaced by gene-probe testing.

The Neuropathology of Huntington's Disease

With HD, progressive atrophy of the striatum (caudate before putamen, with the nucleus accumbens least affected) is the hallmark. The medial caudate is the first to show signs of damage, the small spiny neurons being most affected, followed by the dorsal putamen and then the tail of the caudate (Vonsattel, 1992). As we saw in Chapter 11, the role of the striatum seems to be to modify cortical activity and patterns of movement by timing, ordering, and sequencing via a number of parallel loops—motor, oculomotor, prefrontal, orbitofrontal, and cingulate—all of which become more or less faulty in HD, and which between them explain much of its symptomatology. The putamen, contrary to the situation in PD, is less affected. Defective striatal and frontal metabolism, as indexed by PET and regional blood flow, precedes loss of brain tissue (Hasselbalch, et al., 1992), and see Figure 1 for an example of a computerized tomography (CT) scan of the brain in HD. Indeed, according to Aylward et al. (1994) magnetic resonance imaging (MRI)-assessed volumes of not only the caudate but of all BG structures (i.e., including also the putamen and globus pallidus (GP)) are significantly reduced even in gene-marker-positive at-risk individuals without clinical evidence as yet of the disorder. Thus BG volume is reduced before individuals become symptomatic, while subsequently volume reduction, imaged metabolism, neuronal loss, and astrogliosis all correlate with clinical severity. As we saw earlier, the caudate receives extensive nonmotor (i.e., limbic and prefrontal) inputs, and it is dysfunction of this particular BG-frontal circuit that accounts for the characteristic cognitive symptoms of HD. However, the cortex is also in-

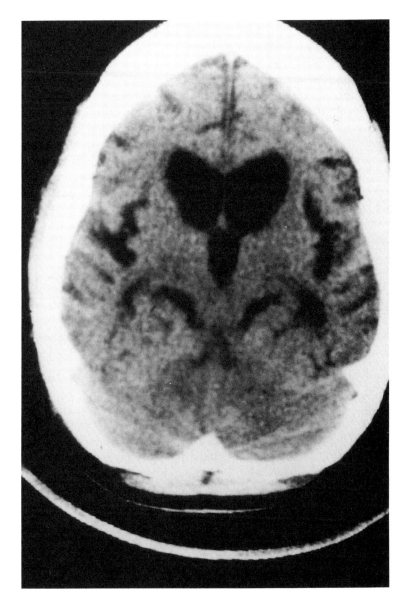

Figure 1 CT SCAN OF BRAIN IN HUNTINGTON'S DISEASE; NOTE THE ENLARGED VENTRICLES CONSEQUENT UPON SEVERE CAUDATE ATROPHY. (FROM MATERIAL KINDLY SUPPLIED BY E. CHIU.)

volved, especially that of the frontal lobe, with 20 to 30% loss by volume, widened sulci and shrunken gyri, correlating with the degree of caudate damage, and resulting in an end state reduction in brain weight of 200 g to 300 g, or 20 to 25% (Vonsattel, 1992). Resultant widening of the lateral ventricles is clearly visible on CT scans, as is the striatal hypometabolism on PET scans, sometimes even before clinical symptoms manifest. White matter, thalamus, and hippocampus are also affected, making HD a kind of genetically programmed multiple system atrophy, though the cerebellum, brain stem, and spinal cord seem to be more or less unaffected. The caudate damage of course accounts for the cognitive and emotional deficits, prefrontal involvement accounts for the disinhibition, apathy, set, and strategy-related problems, whereas hippocampal dysfunction explains the other cognitive and memory deficits. Loss in the striatum of gamma-aminobutyric acid (GABA) neurons projecting to the GP leads to its disinhibition; it then more actively inhibits the subthalamic nucleus (STN), resulting in abnormal movements (Young & Penney, 1993) that reflect the consequent imbalance of DAergic and GABAergic neurons; hence the beneficial symptomatic treatment with DA antagonists. This treatment leads to the remaining putamen neurons of the *indirect* BG pathway increasing activity (Hallet, 1993). Striatal degeneration leads to underactivity of the *direct* pathway, as we saw in PD, and resultant akinesia and rigidity. Indeed Hallet (1993) summarized the effects of various kinds of pathology upon the direct and indirect pathway in the BG as follows:

1. Underactivity of the direct pathway, resulting in akinesia and rigidity, causes these symptoms in both PD and HD.
2. Underactivity of the indirect pathway causes the chorea and unwanted movements of HD and L-dopa-induced dyskinesia.
3. Overactivity of the direct pathway, with excessive and overflow movement, similarly is associated with L-dopa-induced dyskinesia and dystonia.
4. Overactivity of the indirect pathway causes the bradykinesia found in dystonia and PD.

In animal models of HD, overstimulation (with kainic or quinolinic acids) of striatal glutamate receptors can produce HD-like effects. The striatum of course receives excitatory glutamatergic input from the cortex, while according to Penney and Young (1993) damage to the N-methyl-D-aspartate (NMDA) subtype of glutamate receptor in the striatum results in extracellular calcium flooding in. This has led to the neurotoxic or excitotoxic theory. Coyle and Puttfarcken (1993) noted that activation of glutamate-gated cation channels may also be important sources of oxidative stress, and these two semi-independent but mutually reinforcing processes may result in metabolic stresses and neural degeneration not only in HD but also in PD, amyotrophic lateral sclerosis (motor neuron disease) epilepsy, stroke, and hypoglycemia. In any case excessive excitatory glutamatergic input to the medium-sized spiny neurons of the caudate results in loss of inhibitory GABAergic circuits (Wexler, 1993). Storey and Beal (1993) measured concentrations of GABA, glutamate, and other neurotransmitters in the BG of HD patients exhibiting either chorea or bradykinesia and rigidity within 6 months of death. Their findings confirmed that HD chorea is due to loss of striatal GABAergic inhibitory projections to the globus pallidus pars externa (GP_e), with resulting increased activation of its inhibitory neurons projecting to the subthalamic nucleus; later in the disease loss of GABAergic input to the globus pallidus pars interna (GP_i) leads to rigidity and bradykinesia. As we saw above, in the normal brain the balance between inhibition via the direct pathway and excitation via the indirect pathway projecting to the GP_i controls the level of thalamocortical activation.

The "genetic stutter" (see above) now known to be associated with the HD gene may, moreover, lead to excess glutamatergic activation of the caudate, with consequent excitotoxicity. Clearly, future *prevention* of HD may involve pharmacological reduction of such activity in genetically at-risk individuals, perhaps by antagonists to neurotoxins, neurotoxic metabolites or NMDA, whereas future *treatment* in such individuals might involve neural transplant. The Italian ALS Study Group (1993) reported trials of branched-chain amino acids that activate glutamate dehydrogenase, an enzyme involved in the metabolism of glutamate, while the $GABA_B$ receptor agonist baclofen may possibly help by inhibiting the release of glutamate; so far, little success has been reported from clinical trials, though other compounds with more specific effects upon the glutamate system, currently under investigation, may prove more profitable.

Summary and Conclusions

Huntington's disease (HD) has been described for over 100 years, though the disorder is thought to have been present in 17th-century Europe. As the average age of clinical onset is postreproductive, its prevalence could possibly rise, were all other factors to remain constant. A less-common, early-onset form with paternal inheritance is more severe and more rapidly progressive. A progressive neurodegenerative disorder like PD, it is characterized by the insidious onset of minor motor incoordination followed by uncontrollable choreiform movements, cognitive deterioration, and affective and psychiatric disorder. Treatment is at best palliative, and is far less successful than currently is the case with PD. As it is an autosomal dominant trait with complete lifetime penetrance, a child of an affected parent has a 50% chance of developing the disorder. The gene lies near the end of the short arm of chromosome 4, the mutation involving an expansion and instability of the CAG trinucleotide repeat sequence. The cut-off between normality and HD pathology lies somewhere between 34 and 42 repeats. Other motor disorders also seem to involve CAG repeats, though on different chromosomes, and may also involve disorders of glutamate metabolism.

The first clinical signs include restlessness, clumsiness, incoordination, forgetfulness, personality and mood changes, with the subsequent development of problems in fine motor dexterity, steadiness, and speeds. The negative symptoms of bradykinesia and akinesia, particularly in the end stages, are shared with PD, whereas the major positive symptom, corresponding to parkinsonian tremor, is hyperkinetic chorea. The latter commences distally as minor tics and twitches, and spreads to the rest of the body in the form of apparently complex fragments of normal behaviors. They can be temporarily suppressed, or smuggled into or concealed within normal voluntary movements. Later in disease progress they become athetotic and dystonic. As with PD, patients have problems in preparing movements in advance and in controlling movement parameters.

A cognitive decline parallels, or may even precede, the motor decline, and correlates with the extent of caudate damage which in HD, unlike PD, precedes and is greater than that inflicted on the putamen. As with PD, a "subcortical" dementia develops, again with evidence of frontal involvement. There are problems with planning and set, with perseveration and distractibility, loss of flexibility and fluency, difficulties with sequencing, organization, and scheduling, together with both apathy and disinhibition. Thus the frontal involvement is more severe than in PD. There are early memory deficits, visual more than verbal, involving short- rather than long-term memory, episodic rather than semantic aspects, and recall rather than encoding or recognition. However, the acquisition of implicit memories and skills may be badly affected. Visuospatial deficits appear earlier and are more severe than those involving language. A subset of individuals at risk of developing HD may show abnormalities in cognitive and motor tests prior to clinical diagnosis.

Defective striatal metabolism, especially of the caudate, precedes loss of brain tissue. Subsequently there is loss of cortical tissue, especially of the frontal lobe, though white matter, thalamus, and hippocampus are also affected. Loss in the striatum of GABA neurons projecting to the GP leads to its disinhibition; resultant abnormal movements reflect the consequent imbalance of DAergic and GABAergic neurons; hence the beneficial symptomatic treatment with DA antagonists. As in PD, a neurotoxic or excitotoxic theory has been proposed, involving the excitatory neurotransmitter glutamate and oxidative stress. The "genetic stutter," now known to be associated with the HD gene, may lead to excess glutamatergic activation of the caudate.

Further Reading

Brandt, J. A. (1991). Cognitive impairments in Huntington's disease: Insights into the neuropsychology of the striatum. In F. Boller & J. Grafman (Eds.), *Handbook of neuropsychology* (Vol. 5, pp. 241–261). Amsterdam: Elsevier.

Brandt, J. A., & Bylsma, F. W. (1993). The dementia of Huntington's disease. In R. W.

Parks, R. F. Zec, & R. S. Wilson (Eds.), *Neuropsychology of Alzheimer's disease and other dementias* (pp. 265–281). Oxford: Oxford University Press.

Folstein, S. E. (1989). *Huntington's disease: A disorder of families.* Baltimore: Johns Hopkins University Press.

Folstein, S. E., Brandt, J., & Folstein, M. F. (1990). Huntington's disease. In J. L. Cummings (Ed.), *Subcortical dementia* (pp. 87–107). Oxford: Oxford University Press.

Gusella, J. F., MacDonald, M. E., Ambrose, C. M., & Duyao, M. P. (1993). Molecular genetics of Huntington's disease. *Archives of Neurology, 50,* 1157–1163.

Harper, P. S. (1991). *Huntington's disease.* London: W. B. Saunders.

Joseph, A. B., & Young, R. R. (Eds.) (1992). *Movement disorders in neurology and neuropsychiatry* (Chapters 22–24). Oxford: Blackwell Scientific.

Koroshetz, W. J., Myers, R. H., & Martin, J. B. (1993). Huntington's disease. In R. N. Rosenberg, S. B. Prusiner, S. DiMauro, R. L. Barchi, & L. M. Kunkel (Eds.), *The molecular and genetic bases of neurological disease* (pp. 737–752). Boston: Butterworth-Heinemann.

Sutherland, G. R., & Richards, R. I. (1994). Dynamic mutations. *American Scientist,* March–April, 157–163.

The Huntington's Disease Collaborative Research Group (1993). A novel gene containing a trinucleotide repeat that is expanded and unstable on HD chromosomes. *Cell, 72,* 971–983.

White, R. F., Vasterling, J. J., Koroshetz, W., & Myers, R. (1992). Neuropsychology of Huntington's disease. In R. F. White (Ed.), *Clinical syndromes in adult neuropsychology: The practitioner's handbook* (pp. 212–251). Amsterdam: Elsevier.

Chapter 14

Other Movement and Neuropsychiatric Disorders

Tics, Tourette's Syndrome, Obsessive–Compulsive Disorder, Autism, Affective Disorders, and Other Such Disabilities

*U*p to this point we have discussed a continuum of movement disorders, starting with the apraxias, where there is a deficit in the voluntary initiation and control of intentional, cortically mediated movement, and progressing to such essentially subcortical deficits as Parkinson's disease (PD) and Huntington's disease (HD). These latter diseases, primarily of the basal ganglia (BG), involve problems that can be loosely characterized as reflecting deficits of automaticity and response modulation. A further range of disorders, also involving aspects of the striatum, include the tic-related syndromes, obsessive–compulsive disorder (OCD), and autism. We end this chapter with a brief account of the affective disorders.

We can classify movements along a continuum, as previously described, ranging from the automatic to the involuntary:

1. Automatic—learned, and performed without conscious effort or a readiness potential (*Bereitschaftspotential*)
2. Voluntary—intentional, prepared and planned, and exhibiting a *Bereitschaftspotential*
3. Semivoluntary—in response to an "irresistible urge" (e.g., yawning, compulsive touching, the "restless legs" syndrome [Tarsy, 1992], and tics)
4. Involuntary—tremor, chorea, and the abnormal postures of dystonia (see e.g., Jankovic, 1992).

Tics and tic disorders, with which this chapter is partly concerned, lie uneasily in the borderland between semivoluntary and involuntary movements. Although tics cannot be indefinitely inhibited, they nevertheless can be subjected to a considerable range of voluntary modulation. Nor is it always clear where, or if, we should place the boundary between tic disorders generally, and the particular syndrome known as Tourette's syndrome (TS).

Extrapyramidal Movement Disorders

As we have seen, extrapyramidal movement disorders, which to a greater or lesser extent involve the BG, break down into two types (Hallett, 1993):

1. Akinetic, where as with PD, there is a difficulty in generating voluntary movement
2. Hyperkinetic, where involuntary undesired movements occur, as with tics, dystonias, and dyskinesias.

It might be useful at this point to review some of the symptoms (and see Calne, 1992; Hallett, 1993). These include the following, apart from akinesia, rigidity, and tremor, with which we have already dealt:

1. *Dyskinesia*—strictly, any kind of involuntary movement, including tremor, chorea, ballismus, tic, dystonia, or myoclonus, but a term usually employed for complex, choreatic, and dystonic movements occurring after prolonged neuroleptic treatment for schizophrenia (tardive dyskinesia) or L-dopa treatment for idiopathic PD.

2. *Tremor*—repetitive rhythmic movements consistent in time and place, which may be designated as, for example, resting, action, or postural, and which result from alternate or synchronous action of groups of muscles and their antagonists (Fahn, Tolosa, & Marin, 1993).

3. *Myoclonus*—brief (less than 100-ms duration) shocklike muscle jerks of central nervous system (CNS) origin, often recurring repetitively in the same muscles, but (unlike dystonia) without corresponding limb movement. Depending upon frequency, they may merge with tremor.

4. *Tic*—repetitive, stereotyped jerks of limited distribution that can often be voluntarily suppressed at cost of increased inner tension. If multiple and simultaneous, they merge into multifocal myoclonus or chorea or both.

5. *Ballismus*—gross, unilateral, irregular choreic movements, usually of the upper limb, often a consequence of lesions to the contralateral subthalamic nucleus (STN), and ameliorated by dopamine (DA) antagonists.

6. *Chorea*—irregular, generalized movements, apparently purposive but like tics largely involuntary, which flit and flow.

7. *Athetosis*—peripheral dystonic movements, usually iatrogenic.

8. *Dystonia*—sustained contractions, often involving twisting and repetitive movements or abnormal postures. Cocontractions can cause pain.

Strictly, dystonia refers to sustained postural aspects, whereas with dyskinesia it is the movement that is important. However, dystonia may be sustained (athetoid) or shocklike (myoclonic), fixed (tonic), or kinetic (clonic), of "unknown" origin (idiopathic or primary), often with a strong hereditary component, or due to metabolic or toxic disorders, injury, or stroke (secondary), or as an irreversible, iatrogenic side effect of medication (tardive) (see e.g., Chan, Brin, & Fahn, 1991, and Rothwell & Obeso, 1987). It may occur as "overflow," when another body part is voluntarily moved, and may be exacerbated with stress. Tone (resistance to passive stretch) typically is increased, due maybe to cocontraction resulting from lack of reciprocal inhibition. It may be focal, multifocal, segmental, or generalized. Primary dystonia commencing before 20 years of age may involve first the legs, then the trunk, neck, and upper extremities, later becoming generalized. When it commences in adulthood it tends to be segmental, involving the neck (spasmodic torticollis) or one or both upper limbs, or the head (eyes and mouth), after damage to the putamen or globus pallidus (GP) of the BG.

The frequently occurring abnormal lack of reciprocal inhibition between flexors and extensors and excess cocontraction of agonists and antagonists during voluntary movement suggest spinal cord involvement in dystonia, though overflow to other muscles, as noted above, indicates higher level (BG) dysfunction (Ghez, Gordon, & Hening, 1988; Obeso & Giménez-Roldám, 1988); so too does the fact that dystonic difficulties may disappear if the nature of the response is changed, even though similar muscles continue to be used. Thus a patient may not be able to walk forward, but can do so backwards, or up stairs, or hop or run. Dystonia may therefore stem from erroneous selection and modulation of motor programs, perhaps in the supplementary motor area (SMA), due to faulty BG output, and reduced tonic inhibition to the rest of the cortex. Exaggerated tonic reflexes consequently occur, and abnormal flexor postures. Indeed Rondot (1991) proposed that the frequent dystonic attitude of extended elbow, flexed wrist, and extended fingers imitates the posture of locomotion in lower primates. Dystonia is therefore seen as a disturbance of the postural adjustments normally preceding voluntary movements and necessary to maintain balance. In effect, postural mechanisms become dominant and take over from the kinetic components of voluntary movement. Dystonia may be reduced by such sensory tricks as the examiner forcing round even further the patient's dystonically deviated head, leading to a sudden release via action perhaps of the muscle and tendon stretch and load receptors (Christensen, 1991).

Writer's cramp (graphospasm) may be examined as a model example of dystonia, though similar disorders may extend to typing, music, painting, and sport generally (Sheehy, Rothwell, & Marsden, 1988). It may commence only when writing, leading to dystonic postures of the hand (fingers flexed or extended) that impede writing. Later, these postures may extend to other manual activities, and overflow also to other muscle systems. Patients may change hands to compensate, but the problem may then slowly extend to the other hand, or the original hand may show dystonic postures when the unaffected hand writes (synkinesis). Writer's cramp may initially be specific to a particular object (e.g., a pen), and not occur to another (e.g., chalk on a board), before extending also to that. We may ask why focal manual dystonia typically strikes first at an individual's most skilled motor accomplishment, and why, often, there is a history of recent local trauma to the affected part. Fletcher, Harding, and Marsden (1991) in this context review evidence that local peripheral injury can affect BG chemistry, and suggest that this mechanism might interact with the effects of a single autosomal dominant gene.

Jankovic, Leder, Warner, and Schwartz (1991) studied 300 patients with cervical dystonia, comparing their findings with those of an additional series of 266 from another laboratory. They noted that cervical dystonia is a focal dystonia characterized by sustained, involuntary contractions of the neck muscles that result in abnormal movements and head postures. Spasmodic torticollis is commonly used to describe this form of dystonia, but as they observed it is not always spasmodic and does not always consist of torticollis. It is however often associated with other movement disorders, particularly tremor. They found that age of onset was typically in the fifth decade, with a female preponderance, a frequent family history of tremor or other movement disorders, and, quite often, prior (i.e., within about 1 year of onset of clinical symptoms) history of local trauma. Of course a majority of whiplash victims do *not* go on to develop cervical dystonia, but Jankovic et al. (1991) speculated that it might occur in genetically or otherwise predisposed individuals. A functional, genetically determined abnormality of the BG seems to be implicated. In two-thirds of the patients, stress appeared to exacerbate symptoms, a phenomenon that we note occurs commonly with other movement-related disorders of the BG (e.g., parkinsonian tremor, the tics of TS, and the chorea of HD).

Restless legs syndrome (RLS) is a form of dyskinesia that is familial, idiopathic, and present in up to 5% of the population (Tarsy, 1992). Originally described by Thomas Willis in 1685 (Tarsy, 1992) it has since come to be known also as Ekbom's syndrome. It is a symptom complex characterized by subjective lower limb discomfort relieved by movement. It occurs mostly late in the day or at night, after prolonged sitting or lying. The abnormal sensations or paresthesias are deep-seated and usually localized to the lower legs. They are variously described as crawling, creeping, pulling, itching, tingling, cramping, aching, drawing, or stretching sensations within the muscles, and are momentarily alleviated by movement, or massage, before the urge returns. The parallel with tics and chorea is clear. Sudden jerking movements of the legs during sleep (nocturnal myoclonus) are often reported by the patient's spouse. Age of onset is typically in early adult life, with an equal gender distribution and a strong family history (Walters, Hening, & Chokroverty, 1991). Its cause is unknown, the condition is lifelong, mild or severe, static or progressive, remitting or unremitting, and a variety of drug treatments may help, notably DA agonists. In some respects it resembles drug-induced (tardive) akathisia, where there are similar feelings of unease in the legs, a compulsion to move, fidgety restlessness, and squirming repetitive limb movements, which like tics can be briefly controlled by an effort of will. Present in most patients on neuroleptic antipsychotics (DA antagonists), the condition (tardive akathisia) may be acute or chronic, and may be irreversible.

Before proceeding to discuss in some detail tic disorders, TS, OCD, and autism, as further manifestations of BG dysfunction, we shall briefly mention some other extrapyramidal movement disorders. However, as they are either comparatively rare, or do not throw much light upon neuropsychological processes, we shall not consider Wilson's disease (WD), the Shy-Drager syndrome, or progressive supranuclear palsy in any detail. In any case, idiopathic PD constitutes around 75% of the extrapyramidal parkinsonian syndromes (Gibb, 1993).

Wilson's disease (WD) is a systematic disorder of copper metabolism caused by an autorecessive gene on chromosome 13 (Patten, 1993). Biliary excretion of copper is impaired, and the consequent buildup of metal in the brain causes involuntary movements, mental changes, and loss of coordination that superficially resemble HD. It can be controlled by chelating agents. The BG are particularly affected, and WD usually presents between the ages of 5 and 50, with a peak between the ages of 8 and 16. Dysarthria and loss of coordination of voluntary movements are most commonly observed, together with involuntary movements, dystonic postures, and tremor. Intellectual deterioration and behavioral disturbances also occur, and accompanying psychiatric symptoms may be classified into affective, behavioral, personality, schizophreniform, and cognitive. The prevalence of WD may be around 1 in 50,000, and siblings have a 25% risk of developing the disorder.

Multiple system atrophy (MSA) is a group of neurodegenerative movement disorders involving the cerebellum, brain stem, and BG (Pirozzolo, Swihart, Rey, Mahurin, & Jankovic, 1993), and which usually begin after age 50 years, affecting men more commonly than women (Polinsky, 1993). There are three main subgroups:

1. The spinocerebellar degenerative disorders, including *olivopontocerebellar atrophy* (OPCA). There is progressive cerebellar ataxia, dysarthria, parkinsonian rigidity and akinesia, tremor, involuntary movements, hyperreflexia (i.e., hyperactive muscle stretch reflexes), nystagmus, and dysphasia. Positron emission tomography (PET) imaging reveals cerebellar and brain stem hypometabolism. Despite anxiety, depression, and emotional reactivity, intellectual functions are little impaired. According to Berciano (1993), OPCA is a group of heterogeneous syndromes whose only common factor is neuronal loss in the ventral portion of the pons, inferior olives, and cerebellar cortex, with cerebellar ataxia usually the initial clinical symptom, and progressive cerebellar disturbance the disorder's

most outstanding feature. Parkinsonian rigidity and akinesia occurs in about half of the cases.

2. *Shy-Drager syndrome*, a disorder of progressive autonomic failure with parkinsonian motor features of rigidity, akinesia, tremor, ataxia, dysarthria, emotional lability, and cognitive impairment that probably resembles that of PD (Pirrozolo et al., 1993). No definite cause has been identified.

3. *Striatonigral degeneration* (SND), a neurodegenerative disorder most commonly mistaken for PD. Cell loss and gliosis occur in the putamen, caudate, GP, substantia nigra (SN), and subthalamic nucleus (STN) (Gibb, 1993), and there is parkinsonian bradykinesia, rigidity, and tremor, though autonomic functions are unimpaired.

Thus MSA can present as a parkinsonian or cerebellar disorder, or one with autonomic failure.

Progressive supranuclear palsy (PSP, or Steele-Richardson-Olszewski syndrome, see e.g., Lees, 1990), despite its rarity (a prevalence of about one hundredth of that of PD), is becoming a routine diagnostic consideration when PD patients are examined (Golbe & Davis, 1993). Dysarthria is an early symptom, although its most characteristic feature, downgaze paresis (ophthalmoplegia), appears relatively late. Other notable indications are an extremely reduced blink rate (even more than with PD), and nuchal dystonia with hyperextension of the neck. Mean age at symptom onset is around 60, and PSP patients, like patients with PD, are somewhat more likely to come from rural areas or to have been exposed to environmental toxins; there is no direct evidence of a genetic involvement. As might be expected in a disorder with symptoms (e.g., rigidity and bradykinesia) in common with PD (see e.g., Gibb, 1993; Pirozzolo et al., 1993), there is pathology of the SN, GP, and STN as well as the pontine tegmentum and regions participating in ocular motor control, including the superior colliculus. DAergic and cholinergic systems are affected, with resultant subcortical dementia and frontal lobe dysfunction. Thus there are manifestations of forgetfulness, bradyphrenia, perseveration, apathy and depression, irritability, inappropriate mood, an impaired ability to manipulate acquired knowledge, forced grasping, and utilization behavior (Lees, 1990). This contrasts with MSA and the olivopontocerebellar disorders where frontal symptoms are typically absent.

Tic Disorders: General Features

According to Lees and Tolosa (1993), the word *ticq* appears in 17th-century French, meaning "an unsightly muscular caprice," and may be connected with the idea of goats. Indeed the Old German word *ziki* itself means *goat*, while the Italian *ticchio* is related to *capriccio*, itself a form of *capra* or *goat*. Maybe bewitchment by the goat god Pan is the common thread.

Tics are distinguished by three features (Jankovic, 1987): the occurrence of an unusual sensation together with an irresistible urge to move prior to the tic, the ability voluntarily to suppress the tic for variable periods, and finally their occurrence during all stages of sleep. According to Devinsky and Geller (1992) motor tics are involuntary, rapid, nonrhythmic movements. They are, moreover, sudden, repetitive, unpredictable, stereotyped, and jerky, involving individual muscles or groups and apparently without purpose. They last around 100 ms. They involve gestures, expressions, sounds, and even stereotyped phrases. They are commonest in the head, face, and neck, though they fluctuate in time, region, and form, like the choreiform movements of HD (above). Although they are suppressible at will, much more so than other abnormal movements like chorea and dystonia, this results in an increase in tension and may then result in an explosive outburst with transient relief, before once again the urge is experienced (Leckman et al., 1992). Simple motor tics include eye blinking, forced staring, facial grimaces, jerks of the head, neck, or arm, shoulder shrugging, and contraction of the abdominal muscles. Simple vocal tics may include vocal and other movements of the larynx, tongue, lips, and so on, resulting in throat clearing, coughing, spitting, sniffing, gurgling, whistling, sucking, grunting, snorting, clicking, shrieking, and a variety of animal sounds, such as barking, hooting, hissing, growling, and quacking (Devinsky & Geller, 1992). Complex motor tics may run into compulsive behaviors and may include stereotyped actions such as facial expressions, grooming, touching, smelling an object or body part, and aggressive actions against others, as well as hopping, clapping, throwing, gyrating, head banging, pelvic thrusting, pinching, rolling the eyes, and so on. Copropraxia (obscene gestures or actions) and echopraxia (imitating others' behaviors) may also occur compulsively, together with apparently dangerous or potentially self-injurious actions (e.g., while driving, climbing, or flying, though sufferers may still be successful as mechanics, surgeons, or pilots). Complex vocal tics involve words and phrases that may be emitted with an unusual loudness or pitch (e.g., "oh boy," "you know," "shut up," "all right," "what's that?"). As automatisms, they are slipped in at syntactically (thought not semantically) appropriate breaks. One's own (palilalia) or another's (echolalia) words or phrases may be repeated, or obscene, aggressive, or socially inappropriate words used (coprolalia). Unusual rhythms, tones, accents, volume, or speed of speech may be adopted (speech atypicalities), or a phrase may be repeated a set number of times until it feels "just right" (vocal rituals) (see e.g., Leckman et al., 1992). Coprolalia and copropraxia, unlike ordinary swearing and obscene angry gestures, are compulsive, repetitive, and ritualistic and devoid of emotive content. They may be deliberately left incomplete ("fu . . . h") or camouflaged or substituted. Their content (racial, religious, or political) may be partly determined socially, and coprolalia may occur in sign with congenitally deaf users of sign.

Tics generally are exacerbated both by boredom and by stress, with their strength and frequency reduced at intermediate levels of arousal, and with distraction and sleep (Jankovic, 1992), though they can still occur during all stages and phases of sleep. They are commoner in males, and usually start in childhood, around 7 years. Indeed transient tic disorder may occur in up to 15% of children and adolescents and is characterized by simple motor or vocal tics lasting from 2 weeks to 1 year (Devinsky & Geller, 1992). Such tics usually disappear in adolescence, but may briefly recur in adulthood with anxiety, anger, fatigue, or stress, or in old age when former tics may reemerge or new ones appear. In chronic tic disorder, the tics last longer than a year, and the condition merges with that of TS.

Tics, though classifiable perhaps as semivoluntary and occupying perhaps the same hierarchical level as subtle gestures like a frown or scowl (Lees & Tolosa, 1993), unlike true voluntary actions are not preceded by the readiness potential (*Bereitschaftspotential*) (see e.g., Obeso, Rothwell, & Marsden, 1981). They often appear as caricatures of normal behaviors, and the sufferer may try to conceal them by incorporating them into such normal activities, or alternatively may employ a variety of sensory and motor maneuvers to reduce them (*gestes antagonistes*). These include deliberate induction of pain, for example, via pressure or doing something to occupy the affected body part. It is possible, moreover, that in susceptible individuals local irritation (e.g., sore eyes, allergies, etc.) can "sow the seeds" for the functionless perpetuation of initially adaptive responses such as blinking or sniffing. Thus tics may start as voluntary movements, later becoming stereotyped and automatic. Indeed latent tic activity may be reactivated or triggered by environmental stimuli (e.g., hearing a cough or stifling a yawn). The "infectiousness" of yawning in some ways is a near-universal tic, suggesting that tics are semivoluntary stereotypies with a low release threshold involving DAergic (and probably serotonergic, see below) midbrain systems. In fact, manipulation of such mechanisms in animals can lead to responses such as grooming, shaking, licking, rocking, rubbing, and vocalizing (Handley & Dursun, 1992). As we have seen, there are extensive interconnections between the striatal and limbic systems, and electrostimulation of the cingulate in humans can lead to complex, coordinated, stereotyped movements, accompanied by an urge to move. Similarly, von Economo's encephalitis lethargica, as we saw, may affect the striatum and anterior cingulate, with resultant tic behavior, and cingulotomy can help alleviate tic behavior and compulsions. Tics may also be secondary to medications (e.g., amphetamine, cocaine, L-dopa, phenytoin, phenobarbital, and antipsychotics) (tardive tics). Other secondary causes of tics include strokes, head trauma, carbon monoxide poisoning, and degenerative disorders.

Lees and Tolosa (1993) compared tics with mannerisms, stereotypies, and rituals:

1. *Mannerism*—a bizarre way of performing a purposeful act, often involving incorporation of a stereotyped movement (e.g., of the limbs) into goal-directed behavior.

2. *Stereotypy*—a purposeless, voluntary movement of whole body areas performed in a uniformly repetitive fashion, often for long periods, at the expense of other activities. Stereotypies include head nodding and banging, body rocking, and the endless pacing of caged mammals.

3. *Ritual*—a seemingly meaningless repetitive action, of a complex nature, due to an irresistible irrational urge that interferes with normal living but nevertheless relieves inner anxiety. Pirouetting and bowing before opening a door would be an example. The relationship with superstitious and even religious behavior is clear.

Tourette's Syndrome: General Symptomatology

Tourette's syndrome (TS) is an extreme form of tic disorder. Itard (1825, cited in Singer & Walkup, 1991) reported on his patient, a French noblewoman, who developed persistent body tics, barking sounds, and uncontrollable utterances of obscenities. In 1885, George Gilles de la Tourette described patients with multiple tics, involuntary movements, coprolalia, and echolalia, and Charcot rewarded his contribution by bestowing the eponym "Gilles de la Tourette syndrome" on the disease *"maladie de tic"* (Singer & Walkup, 1991). Charcot similarly identified in the condition involuntary, impulsive ideation, repetition of an action a given number of times, and compulsive checking—*"folie du doute"* (Cohen, Bruun, & Leckman, 1988). However, descriptions of "bewitched and possessed" individuals go back centuries. Thus Sprenger (1489, see e.g., Lohr & Wisniewski, 1987, p. 191) described:

> A sober priest without any eccentricity . . . no sign of madness or any immoderate action [who was said to be possessed of the devil]. . . . When he passed any church, and genuflected in honour of the Glorious Virgin, the devil made him thrust his tongue far out of his mouth; and when he was asked whether he could not restrain himself from doing this, he answered: 'I cannot help myself at all, for so he used all my limbs and organs, my neck, my tongue and my lungs, whenever he pleases, causing me to speak or to cry out; and I hear the words, as if they were spoken by myself but I am altogether unable to restrain them.

According to contemporary accounts, the 18th-century English lexicographer Samuel Johnson suffered from what we would now call akathisia, unusual complex compulsions, and various vocal and motor tics. He was said to be "awkward" at 7 years, and later exhibited facial grimacing, finger twirling, head tilting, shoulder shrugging, gesticulations, loud vocalizations, and bizarre rituals on going through doorways. He even commented upon the compulsive life of a lexicographer in his own dictionary. According to Parry-Jones (1992), he was also legendary in his lifetime for his uncouth table manners and gustatory perspiration, his life-long greediness, and uncontrolled eating even in the last year of his life. Such behavior, moreover, was punctuated by intermittent abstention, which, together with his regular utilization of purges, is clearly indicative of bulimia. Bulimia and self-mutilation behavior of a minor sort in fact often coexist with the OCD manifestation of TS; Samuel Johnson indeed is recorded as continually picking at his fingers, cutting at the knuckles, biting the nails to the quick, and repeatedly hitting and rubbing his legs. Chapman (1953, cited by Parry-Jones, 1992) records in modern English Boswell's account of Johnson's gustatory infelicities:

> When at table, he was totally absorbed in the business of the moment: his looks seemed rivetted to his plate: nor would he, unless when in very high company, say one word, or even pay the least attention to what was said by others, till he had satisfied his appetite, which was so fierce, and indulged with such intenseness, that while in the act of eating, the veins of his forehead swelled, and generally a strong perspiration was visible. To those whose sensations were delicate, this could not but be disgusting: and it was doubtless not very suitable to the character of a philosopher, who should be distinguished by self-command . . . and not only was he remarkable for the extraordinary quantity which he eat, but he was, or affected to be, a man of very nice discernment in the science of cookery. (p. 851)

Sacks (1992) suggested that Mozart too might have suffered from TS; he was hyperactive, given to tics, sudden impulses, odd motor behaviors, echolalia and palilalia, with a love of nonsense words and a head full of melodies—like Johnson, with his head full of words. We can ask, with Sacks, whether Mozart and Johnson were creators and innovators in their fields *despite* TS or *because* of it. TS may therefore be a continuum with two ends: one (stereotypic) extreme involves simple motor tics and vocalizations, iterations, and perseverations, which are largely a nuisance and an irrelevance; the other extreme perhaps involves elaborations, playful mimicry, extravagant impudent inventiveness, and audacious dramatizations, surreal associations, uninhibited inventiveness, incontinent reactivity, stimulus hunger, imagery, and exuberant art. Sacks saw this latter extreme to be a source of inspiration of "ticcy witticisms and witty ticcisisms" in language, music, art, athletics, and games, a usefully harnessed disorder. Indeed jazz and "tic dancing" are extremely popular at Tourette Syndrome Association socials, and Sacks reminded us of his patient who went off his medication because it destroyed his creativity. A recent illustrative case history is provided by Singer and Walkup (1991):

> An 11-year-old child of an upper middle class family was referred for evaluation of spitting, cursing, and copropraxia. This patient had the onset of ocular tics at age 7, subsequently replaced after several months by a transient horizontal head tic and sniffing. At age 9, he began to hit himself, smell his hands, kiss objects, and kneel. After washing his mouth with soap failed to deter spitting sounds, he was referred to a child psychologist who interpreted his symptoms as an indica-

tion of severe psychopathology. Symptoms led to punishment by parents, discipline by teachers, and ostracism by peers. Intensive psychotherapy failed to diminish his movements and vocalizations. Finally at the age of 11, concurrent with the onset of vocalizations including socially unacceptable words and common four-letter obscenities, he was diagnosed as having Tourette's syndrome. Initiation of tic-suppressing therapy with haloperidol diminished tics to a socially acceptable level, but produced unacceptable lethargy, dysphoria, and excessive weight gain. Substitution of pimozide for haloperidol eliminated the side effects without negatively affecting the control of his tics.

During the first half of this century, TS was primarily studied by psychoanalysts, in the light of its profanity, uncontrolled sexual impulses, and aggression. It was not, however, until the successful use of the DA antagonist haloperidol by Seignot (1961) that modern pharmacological and neurochemical studies began (Devinsky & Geller, 1992). TS represents a spectrum of familial, involuntary motor and vocal tics together with neuropsychiatric manifestations that make it relevant for neurologic, psychiatric, and pediatric disciplines (Singer & Walkup, 1991).

The first symptoms of TS involve an impulsive hyperactivity that only later, with the onset of tics, become distinguished from non-TS behavioral disorders (Devor, 1990). The first tics typically involve the face, especially the eyes, then progress rostrally to caudally, maybe never getting to the feet but often stopping at the upper torso, legs, and arms. Vocal tics are rare as initial symptoms; latest of all to appear, and then only in a minority of cases, is the socially disabling coprolalia, along with echolalia, palilalia, and echopraxia (Bruun & Budman, 1992). Simple tics appear first and complex tics like jumping, hopping, and touching appear later; thus the spatiotemporal evolution of tics seems to mimic the topographic representation and maturation of the CNS. The syndrome starts between 5 and 8 years of age for around two-thirds of cases, with one sixth developing symptoms before and one sixth after then, up to the age of 15 when very few new cases appear. Indeed, its symptoms may remit after adolescence (Park, Como, Cui, & Kurlan, 1993). Prevalence estimates depend upon criteria, and whether clinic or community-based assessment is made (Fallon & Schwab-Stone, 1992). The lifetime risk may however approximate 40 per 100,000 individuals (Singer & Walkup, 1991). It is perhaps three times more common in boys than in girls, though the closely related OCD is more common in girls, suggesting modulation by sex hormones (Leckman et al., 1992). Indeed, adrenal androgens first appear around 7, when tics commonly manifest, and while androgen antagonists help OCD, androgenic steroids as used by athletes can increase tic frequency in TS. The disorder appears in all socioeconomic classes and racial groups. Its etiology is unknown (though see below), and diagnosis can only be made on the basis of signs, symptoms, and clinical course. Indeed, correct diagnoses are often long delayed, and may in fact initially be made by parents, friends, or relatives, or based on information in the lay press, TV, or educational media.

According to *Diagnostic and Statistical Manual of Mental Disorders (DSM-III-R)*, both multiple motor and one or more vocal tics must have been present at some time during the illness, though not necessarily concurrently; the tics must occur many times a day, usually in bouts, nearly every day or intermittently throughout a period of more than 1 year; the anatomical location, number, frequency, complexity, and severity of the tics must change over time; the onset must occur before age 21, and occurrence must not be due exclusively to psychoactive substance intoxication or known CNS disease (e.g., HD or postviral encephalitis). Typically, the patient will be able for a while to suppress them, and they may also be associated with obsessive thoughts and images or compulsive behaviors and urges, and preceded by premonitory feelings, which are relieved by performance of the act. The disorder is often diagnosed by an alert worker in another specialty, (e.g., an ophthalmologist observing excessive blinking in a patient, an otolaryngologist noting excessive throat clearing, or an allergist observing abnormal sniffing, or a psychologist detecting compulsions). However patients often also present with attention deficit hyperactivity disorder (ADHD) (see below). Their IQ distribution parallels that of the general population, and thought disorder rarely occurs. Their emotional responsivity and their experience and perception of affect appears entirely normal. Thus their reduced inhibition of aggressive and sexual impulses appears as a form of emotional incontinence. However, there may be evidence of impaired visuoperceptual performance, visuomotor skills, reading, and mathematical skills, and discrepencies between verbal and performance IQ. Singer and Walkup (1991) also believe that there may be deficits in "executive functions" (see Chapter 10), in terms of capacities to plan and sequence complex behaviors and to organize and sustain goal-directed activities. Although the neurological examination usually shows no focal or lateralizing signs, "soft" signs may include abnormalities of coordination and of fine motor performance, synkinesis, motor restlessness, and sleep disturbances; electroencephalograph (EEG) abnormalities sometimes are reported but are certainly far from universal (Singer & Walkup, 1991).

Hereditary and Environmental Factors in Tourette's Syndrome

Gilles de la Tourette originally suggested that hereditary factors were important, and he has been proven right. Monozygotic (MZ) twins show a very high (at least 50%) degree of concordance, while in dizygotic (DZ) twins and first-degree family members the incidence of tics varies from 15 to 85%, and of TS from 4 to 8% (Devinsky & Geller, 1992, and see also Leckman & Peterson, 1993). A single autosomal dominant gene with variable penetrance seems responsible (Devor, 1990), which may be expressed as either TS or OCD, or possibly as ADHD. There is a very high incidence of TS among certain inbred communities like the Amish. Increased blink rate has been reported in otherwise unaffected relatives, and it may be that in certain family members transient childhood tic and sleep disorder may be *formes frustes* of TS.

Clearly, epigenetic (environmental) factors are also operating prenatally, and influencing intrauterine growth. They may mediate vulnerability to toxins, prenatal stress, severe nausea and/or vomiting during the first trimester of pregnancy, chronic psychosocial stress and exposure to androgens, cocaine, stimulants, and so on. Thus there is not 100% concordance with MZ twins, and in discordant MZ twins the worse affected usually is of lower birth weight. The role of sex hormones is evident from the increased vulnerability of males, despite the (presumed) autosomal dominant inheritance. Androgenic steroids early in fetal life may influence neurogenosis, neuronal migration, neurite extension, synaptic connectivity, periods of programmed cell death, axonal elimination, and synaptic pruning (Leckman & Peterson, 1993). Such effects normally may result in sexually dimorphic brain structures and asymmetries (Bradshaw, 1989). It is therefore interesting to note two very recent studies (Peterson et al., 1993, and Singer et al., 1993) showing anomalies in morphological brain asymmetries in TS, which Witelson (1993) said indicated that the lenticular region (putamen and GP) of the BG is reduced on the left in TS compared to controls, with attenuation of the normal predominance of the left hemisphere (LH) contralaterally. It is unlikely however that the TS gene results in any gross brain dysmorphology, as in HD, though Leckman and Peterson (1993, and see also Stoetter et al., 1992) noted in TS a reversal of the normal relative patterns of cortical and subcortical activity; thus TS patients are said to display reduced metabolic rates in the striatum and ventral cortex, and increased metabolic rates in the somatosensory cortex and SMA, which is the reverse of the normal pattern (and see below). Other evidence of the effects of sex hormones comes from increased incidence of tics with anabolic steroids and reduction with androgen receptor blockers. It is noteworthy that DAergic areas of the SN pars compacta (SN_c) contain androgen receptors. Maternal stress during pregnancy is associated with increased tic behavior, and rats born of mothers stressed during pregnancy show altered amphetamine-induced turning behavior and altered asymmetries in striatal DAergic structures. The viral disease encephalitis lethargica, moreover, was associated with vocal and motor tics, OCD, and damage to DAergic structures like the ventral and midbrain tegmentum.

Anatomical and Neurochemical Aspects of Tourette's Syndrome and Obsessive–Compulsive Disorder

Electrostimulation of the thalamus can, as we saw, induce throat clearing, phonation, exclamations, repetitions, pallilalia, and head and arm swings, whereas thalamic lesions and isolation of certain prefrontal areas may ameliorate TS (L. Cohen et al., 1988; Leckman et al., 1992). PET imaging indicates decreased regional metabolic activity in insular, prefrontal, and cingulate cortices in TS, which is compatible with some degree of behavioral disinhibition, whereas individuals with OCD may have *increased* metabolic activity in caudate, orbitofrontal, and cingulate regions (Rauch et al., 1994), which is again compatible with their behavior patterns. Thus the two closely related disorders, TS and OCD, may involve abnormally correlated activity between frontal and striatal regions, involving the motor, cognitive, and limbic circuits of the BG described above. Not surprisingly, then, DA antagonists help ameliorate TS, whereas DAergic substances like amphetamine exacerbate it, and lead to complex stereotypies in animals like chewing, licking, sniffing, and head-searching behaviors, and learning and attentional deficits (Devor, 1990; Handley & Dursun, 1992). Paradoxically, DA turnover may be decreased in TS, perhaps due to a down-regulatory, compensatory effect consequent upon an initial hypersensitivity (Devinsky & Geller, 1992). The DA hypothesis in TS (and cf. the analogous DA hypothesis in schizophrenia, see Chapter 15) suggests either increased density of DAergic innervation, or excessive levels of DA, or increased receptor sensitivity to it. The latter is suggested by reduced baseline and turnover levels of homovanillic acid (HVA), a major metabolite of DA, though Singer and Walkup (1991) found an increased number of DA terminals originating from either the nigrostriatal or ventral tegmental area. However, Turjanski et al. (1994) found that striatal metabolism of exogenous L-dopa and the density of striatal D2 receptors were normal in TS patients; they concluded that the syndrome may not arise from a primary dysfunction of DAergic terminals. Moreover, serotonin seems also to be implicated. This neurotransmitter, as we have seen, itself modulates the activity of DA. Its role in this respect, and with respect to movement generally, is reviewed by Jacobs and Fornal (1993) and Spoont (1992). Serotonin reuptake inhibitors given to animals can produce "wet-dog shake" behavior and ticlike head movements, especially if the skin is stimulated, while local anesthesia may reduce the behavior pattern (Handley & Dursun, 1992). Clomipramine, an antidepressant serotonin reuptake inhibitor, may exacerbate tics while ameliorating OCD (see below), although DA antagonists may have exactly the opposite effects (and see Montgomery, 1992). Because some serotonin neurons in the midbrain seem to be active in grooming, the idea has arisen of tics as a form of grooming gone wrong. Consequently, Kurlan (1992) viewed the defective genes of TS, via inappropriate DAergic and serotonergic modulation, as leading to the defective operation or development of the BG, frontal, and limbic systems. These systems subserve primitive motor, vocal, and emotional processes involved in grooming and reproductive behaviors, which are further modulated by gonadal steroidal hormones. In this way we can accommodate sex differences and a tendency for improvement after puberty. Copropraxia and coprolalia will then be the inappropriate expression of normally inhibited thoughts and emotions, together with fragments of primitive motor and vocal programs involved in reproductive activity.

In addition to abnormal DAergic and serotonergic dysfunction, there are continuing suggestions (for reviews, see e.g., Suchowersky, 1994) that the endogenous opioid system may also be involved in TS. Thus opiates can help alleviate tics, and withdrawal of opioid therapy can precipitate or exacerbate them. There is also evidence of abnormal dynorphin staining in the GP of TS subjects at autopsy. Finally, there may even be noradrenergic abnormalities (Shoulson & Kurlan, 1993), as the α-adrenoreceptor stimulant cloridine can ameliorate tics.

Obsessive–Compulsive Disorder: General Symptomatology

Two conditions, OCD and ADHD, may persist after the tics of TS have abated in adulthood. Around 50% of children with TS may show ADHD (D. Cohen et al., 1988). Problems occur with respect to attention span, concentration, and impulsivity; there is motoric hyperactivity and problems in maintaining and changing set, and with social disinhibition—all, as we have seen, clearly frontal deficits. The disorder is again common in boys, and is in fact perhaps the commonest childhood disorder. ADHD symptoms usually precede TS symptoms by 2 to 3 years, and may be more common in individuals with more severe tics (Singer & Walkup, 1991). The closest cognitive analog of TS, on the other hand, is perhaps OCD, again a familial disorder with around 2% of the population exhibiting some form of it. Again the BG, DA, and serotonin are involved; it co-occurs with TS around 40% of the time, and may be an alternative expression of the same genetic vulnerability. While tics wax and wane, OCD symptoms tend to be more constant.

The *obsessive* component of OCD relates to the intrusion of senseless, unwanted, unacceptable, and repetitive thoughts, which are perceived by the thinker as repugnant, blasphemous, and obscene, and as invasive of consciousness (and see Cath et al., 1992). Like tics, tremor in PD, and chorea in HD, they increase with stress; they may resemble echophenomena in healthy individuals—the tunes that sometimes one cannot get out of his or her head. In compensation, the sufferer may develop an excessive concern over, for example, dirt or religion, or disgust over sex and excretion, leading to repetitious washing or religious rituals, or proselytizing in connection with religious or moral beliefs. The obsessions of the "idiot savant," the arithmomania, calendrical calculations, and preoccupations with timetables, winners of horse races, and lucky numbers all come into this category. Indeed, we may cease wondering at the perceived island of unusual ability, such as calendrical calculation, in a sea of otherwise intellectual mediocrity in savants if we remember the number of hours per day, seven days a week, spent by such individuals on these kinds of activities. Such obsessions with arithmetic and numbers of course merge into compulsions, and we shall shortly consider them further in the context of autism.

The *compulsive* component of OCD relates to the actual behaviors performed in response to obsessions, or to overt irrational obsessive fears. Indeed, there may well be a continuum between complex tics and compulsions. Such behaviors include repeated checking of locks, vehicle brakes, cooking or ironing appliances, repeated washing, arranging or touching objects, never walking on cracks, entering or leaving a doorway or a passage with a set number of steps so that perhaps the left foot always leads, the need to "even up" both sides of the body symmetrically, to touch posts a certain number of times, or to repeat something until it feels "just right." As children, and possibly at that age universally on the verge of being "at risk" for OCD, most of us remember walking rituals involving flagstones or turning, or compulsive counting. It is possible that pathological fears—phobias—of snakes, spiders, or heights come into the same domain. It is also possible that the OCD symptoms in TS are really *impulsions*, in some way rewarding, while with true, independent OCD the behaviors are more in the way of *compulsions*, performed for anxiety reduction or tension release, to avoid something "bad" happening. Generally, patients "know that it is crazy," but cannot help it, and may even withdraw from society to indulge it. The antidepressant clomipramine, a serotonin reuptake inhibitor, proves highly efficacious against OCD, and for related disorders like nail biting or hair pulling (trichotillomania), even in dogs given to excessive flank lickings; again a "grooming disorder" seems implicated (Kurlan, 1992). The perceived preponderance of kleptomania or compulsive shoplifting in postmenopausal females again may indicate the hormonal involvement in these limbic-frontal-striatal disorders, as well as highlighting the overlap, in this context, between neurology and neuropsychiatry. Generally, OCD may be seen as inadequate suppression of primitive and "instinctual" behavioral routines related to grooming, hygiene, and perhaps diet and territoriality that are hardwired into the BG, and that are normally inhibited by the serotonergic system. The emergence of such apparently psychiatric behavioral problems probably indicates malfunctioning, structurally, neurochemically, or both, of this system; hence they may be treated more or less successfully by drugs that block the reuptake of serotonin (e.g., fluoxetine [Prozac]).

Rapoport (1990) reviewed findings that report high concentrations of serotonin receptors, and of serotonin itself, in the ventral striatum and nucleus accumbens. (Harris & Aston-Jones, 1994, and see also Schulteis & Koob, 1994, reported that D2 receptor activity is important in the rewarding effects of abused drugs and addiction, itself clearly a form of compulsive behavior.) She also reviewed evidence from brain imaging of reductions in the volume of the caudate nuclei in OCD, of higher metabolic rates in the caudate and orbitofrontal cortices of patients with the disorder, and of the benefits of psychosurgery. Here, interruption of the connections between the frontal lobe and the striatum, by cingulotomy or anterior capsulotomy, or stereotaxic lesions within the mediodorsal and anterior nucleus of the thalamus, may help alleviate OCD symptoms. The general picture is of damage, in OCD, to the orbitofrontal loop, which incorporates the orbitofrontal cortex, caudate nucleus, GP, substantia nigra pars reticulata (SN$_r$), and the ventroanterior and mediodorsal thalamic nuclei, according to Rapoport (1990). Indeed two other disorders, postencephalitic parkinsonism and Sydenham's chorea, probably also involve these pathways. Postencephalitic parkinsonism was described by Constantin von Economo in 1917, though its English translation was not available until later (von Economo, 1931); an infectious, neurotoxic disease of the BG and other anterior structures, which came and vanished mysteriously early this century as a pandemic, it is associated with compulsive actions and thoughts, in addition to motor changes. Sydenham's chorea, a disorder of children and adolescents after rheumatic heart disease, is associated with neural degeneration of the striatum, and sudden, aimless irregular movements of the extremities, together with OCD.

Autism

Kanner (1943) reported children who were unable to carry out normal conversations, who were out of contact with other people, confused personal pronouns (you, I), exhibited an obsessive desire for sameness, and had exhibited solitary tendencies from their earliest days, and yet who could handle objects skillfully and often had abnormally good memory. Initially he blamed faulty parenting (perfectionism, with a cold, obsessive, humorless approach), though later he came to acknowledge the condition's innateness. (It is of course possible, if the condition has a heritable component with variable penetrance, see below, that some parents *would* exhibit such autisticlike behavior *themselves*.) Debate has in fact, as in so many other similar domains (e.g., schizophrenia) (see Chapter 15), continued inconclusively on the nature–nurture controversy, though as Gillberg and Coleman (1992) argued, a major organic component seems undeniable. The term *autism* ("aloneness") was coined by Bleuler (1911) to designate a category of thought disorder in various schizophreniform conditions characterized by a love of isolation (αὐτίτης—in isolation). Although autism (or Kanner syndrome) currently is not seen as an aspect of schizophrenia, as we shall see it possesses many parallels to that disorder. Indeed, like schizophrenia, with its complex symptomatology, it is probably not a single disease entity, but represents the final expression of various etiological factors, one of which may be a disorder of the DAergic mesolimbic system (Gillberg & Coleman, 1992). Its criteria currently include early onset (before 3 to 5 years of age, though later onset *has* been reported), with severe abnormalities of reciprocal social relatedness and of verbal and nonverbal communication, restricted, repetitive, and stereotyped patterns of behavior, interests, activities, and imagination. Gillberg and Coleman speculated whether the notorious "feral" children of antiquity (Romulus and Remus, Victor the child boy of Aveyron around 1800, and Kamala and Amala in India in the 1930s) could themselves "merely" have been autistic.

The incidence of autism is about 5 per 10,000 with a high number of first- or only-born (it is possible that parents, after an autistic child, opted against further children), or fourth or later born, both of which suggest that perinatal birth difficulties may play a role. Three to four times as many boys are affected as girls, though the rate is even higher in some studies, and in the closely related Asperger syndrome. The overrepresentation of boys with autism is less evident in the severely mentally retarded, suggesting, according to Gillberg and Coleman (1992), that although boys are perhaps more genetically or otherwise prone to developing the disorder, worse brain damage is necessary before it manifests in girls. Most cases do exhibit some mental retardation, though as we shall shortly discuss "islets" of special ability, often involving rote memory, music, art, visuospatial, or repetitive activity, often are evident. The fact that around one-third develop epilepsy again is indicative of a brain abnormality, and one which is affected by sex hormones, as in TS, as epilepsy commonly manifests at puberty. (Indeed there are continuing suggestions that, as in TS, girls may have a different autistic phenotype than boys; moreover, the condition may worsen at puberty, with increasing destructiveness, aggressiveness, restlessness, and hyperactivity.) However, the abnormality may well have been present from before birth, symptoms only emerging when changing environmental demands, with development, exceed current capacities to cope. Hence, autism, like schizophrenia, even if congenital, will seem to manifest after infancy. Again, as with schizophrenia, there has been debate as to whether to *exclude*, as autistics, individuals with clear evidence of brain dysfunction! In fact, although autism should be distinguished from childhood schizophrenia, TS, OCD, and perhaps Asperger syndrome, it cannot be denied that there are many commonalities. Indeed Asperger syndrome is probably a variant of autism; it is more common and less severe than autism, with possibly greater evidence of heritability and higher intelligence and superficially excellent language skills, but the same deficits of empathy. The male–female ratio is even higher than in autism, and individuals, especially as adults, are described as eccentric, odd, or original, though they have a tendency toward OCD and dangerous fascinations (e.g., involving fire or explosives) and criminality.

Repetitions in speech (echolalia) and imitations of others' actions (echopraxia) are common in autism. Memory for conversations is often extraordinarily good with respect to form (i.e., the actual words used), though with little comprehension of content. Restricted, repetitive, obsessive, stereotyped behavioral activities are the norm, with a fascination for a limited range of objects or activities, an insistence on "sameness," emotional outbursts if any object or routine is changed, however slightly, preoccupations with timetables, weather reports, numbers, and so on, and a generally ritualistic behavior very like OCD, suggestive of DAergic or serotonergic dysfunction. Behavioral stereotypies such as hand flapping or tiptoeing are common, along with a hypersensitivity to mild noise and a reduced sensitivity to pain. Also suggestive of serotonergic dysfunction are sleep disturbances and abnormal circadian rhythms, and anorexia nervosa and bulimia that have been linked both to autism and, as we saw in the case of Samuel Johnson, to TS. Indeed there is a high incidence of autism in hyperserotoninemia (which is familial); disorders of serotonin metabolism again are linked to eating disorders and TS. Of course, as Gillberg and Coleman (1991) observed, a high *blood* level does not necessarily mean a high *brain* level.

As with schizophrenia, there are repeated reports of abnormal lateralization in autism in terms of handedness or visual field or ear asymmetries; a "failure of lateralization" is often suggested, though it is hard to see how this would in itself necessarily cause the disorder. The EEG is often diffusely abnormal, though there may be localized or focal abnormalities, perhaps associated with epileptiform tendencies. Brain imaging has suggested brains that are more symmetrical than normal, or even with *reversed* asymmetry, though this has been disputed, as also have other claims such as abnormal ventricular dilation, and developmental failure (rather than atrophy) of the cerebellum. We have already noted the involvement of abnormal striatal functioning in TS and OCD; Horowitz, Rumsey, Grady, and Rapoport (1988) similarly found, from PET imaging, that autistics show a functional impairment in interactions between the striatum and the frontal cortex. Bauman (1992) provided the most detailed review so far of brain abnormalities. These include reduced positive PET correlation patterns in frontal and parietal regions, together with the thalamus, caudate, and lenticular nuclei and insula, which Bauman suggested indicate impaired functional interaction between these regions. Increased cell packing densities and reduced neuronal size also have been reported in the hippocampus, subiculum, entorhinal cortex, amygdala, mamillary bodies, medial septal nucleus, and anterior cingulate. Such abnormalities may disrupt circuitry and function of the hippocampus and amygdala, and the limbic and sensory association neocortex, affecting motivation, emotion, memory, and learning, as is seen in fact in autism; indeed, such lesion studies in monkeys, as Bauman observed, indicate strikingly autistic behavior, such as purposeless hyperactivity, impaired social interaction, learning deficits, compulsive exploration, withdrawal from socially rewarding situations, loss of fear of normally aversive stimuli, and disturbed cognitive learning with intact (striatal? see Chapter 8) habit learning. Cerebellar abnormalities by CT and MRI have also been reported, along with findings of changes at the neuronal level, though their role in the context of autism is not clear.

However, it again must be emphasized that autism is probably not a single, homogeneous disorder with clearly defined localization of dysfunction. As Gillberg and Coleman (1992) observed, there may be various subtypes, with nonspecific ventricular atrophy, cerebellar atrophy (see e.g., Courchesne, Townsend, & Saitoh, 1994), or left-temporal damage. Thus tuberous sclerosis, neurofibromatosis, fragile X syndrome, Rett syndrome, Williams syndrome, Jubert syndrome (with cerebellar dysfunction), and Moebius syndrome are all associated with autistic symptoms, suggesting that several different brain areas, perhaps forming a functional unit, can be separately targeted.

At a cognitive level, Frith (1991) suggested that autism depends upon lack of early development of a theory of mind; patients cannot intuit what others think and feel, or even that they *do* think and feel, rather than being mere robots. This leads to extreme loss of empathy while a mechanical understanding of the natural (or artificial) world is well preserved. Thus autism is not exclusively a cognitive ("cortical") or a social ("limbic") disorder, but an interaction between them, arising from some other primary dysfunction. As Frith (1991) observed, mental states (knowing, believing, etc.) are not directly observable but have to be inferred. Autistic children cannot do this, or develop a theory of mind like normal 4-year-old children. A typical experimental demonstration involves two dolls; in Sally's basket there is a marble, but nothing in Anne's. Sally leaves the room (and her basket) and so cannot see Anne move the marble to her own basket. When Sally returns, the children are asked where Sally will look for the marble. Normal and Down's syndrome children claim that Sally will look for it where it was before she left the room, while autistic children say that she will look for it where it currently is.

A familial pattern is often apparent in autism, though the pattern may differ depending upon the underlying etiology (e.g., fragile X syndrome, tuberous sclerosis, etc.). Other family members also often have affective or learning disorders or anorexia nervosa (Gillberg & Coleman, 1992). There is high concordance for autism between MZ (and much lower between DZ) twins. It is interesting to note, as we saw in Chapter 13, that the gene for the fragile X syndrome includes the DNA base pair sequence CGA, with number of repeats correlating with the severity of clinical presentation, in a somewhat similar fashion to the situation with HD. However, not all sufferers from the fragile X syndrome have autism (though most do have neuropsychological abnormalities), and most autistics do not have fragile X syndrome. The XYY syndrome is also associated with autism. Gillberg and Coleman (1992) noted that many of the sex chromosome anomalies associated with the disorder indicate a relative decrease in the influence of the normal "female" chromosomes (XX) and a corresponding increase in the influence of the male (Y) chromosome—with many aspects of autism somehow reflecting aspects of extreme male behavior. Other genetically determined causes of autism that act via metabolic disorder and CNS intoxication include phenylketonuria, lactic acidosis, and errors of purine metabolism.

Environmental factors seem also to be operative, again as in schizophrenia (see Chapter 15). Thus there is evidence of an association with parental exposure to toxins, maternal uterine bleeding especially in the second trimester, maternal viral infections (an excess in the Northern Hemisphere of March and summer births is indicative of peak times for viral infection in the mother, again as in schizophrenia), and perinatal complications.

Gillberg and Coleman (1992) noted that the DA antagonist haloperidol is very effective in treating the stereotypies, withdrawal, hyperactivity, and irritability of autism, whereas DA agonists exacerbate the condition. This observation, linked with evidence of autonomic dysfunctions in the disorder involving heart rate and respiratory control, suggest, they say, an underlying problem in the DAergic brain stem and mesolimbic pathways that project to the mesolimbic cortex. The latter structure includes mesial frontal and mesial temporal lobes and the neostriatum. Consequent temporal lobe anomalies could account for the linguistic and emotional deficits of autism; indeed this structure is itself very vulnerable to perinatal insult, toxins, and viruses. Involvement of the mesial frontal structures would explain the perseveration and asocial behavior so characteristic of autism. There is of course no evidence of gliosis or scarring in autism, although there is often evidence of neurodevelopmental anomalies (Gillberg & Coleman, 1992). Thus autism, just like schizophrenia (see Chapter 15) is best seen as a neurodevelopmental disorder, a second-trimester syndrome.

Around 10% of autistic individuals may exhibit exceptional levels of performance in some specific domain of knowledge. This, known as the savant syndrome (Rimland & Fein, 1988), includes music, visual or verbal memory, mental calculation, and perspective drawing (Mottron & Belleville, 1993). Its neurobiological etiology is indicated by its co-occurrence with epilepsy and various neurological syndromes (Coleman & Gillberg, 1986), and by the fact that neuroimaging and neuropathological studies have evidenced macroscopic and microscopic abnormalities at the level of the cerebellum, limbic system, and cortex (Schopler & Mesibov, 1987). Although neither the exact etiology nor the pathognomic lesion site(s) are yet established, they seem to engender the various cognitive, behavioral, and social manifestations observed in the syndrome (Frith, Morton, & Leslie, 1991).

Moriarty, Ring, and Robertson (1993) noted that the "idiot savant" syndrome, like TS, is more common in males and is associated with autism, with calendrical calculation a strikingly common feature (and see also Treffert, 1988). Calendrical calculation involves an obsessional preoccupation with a limited section of the environment, and a knowledge of rules, regularities, and redundancies that may be tacit, implicit, unformalized, and unconscious. Thus calendrical calculators seem not to be consciously aware of any algorithm that permits the calculation of calendrical information, though their subjectively experienced need for repetitive behaviors may facilitate the tacit acquisition of the requisite skills (Hermelin & O'Connor, 1986). People with OCD, TS, autism, and the idiot savant syndrome all share the same stereotyped, repetitive, obsessive behavior. Casey, Gordon, Mannheim, and Rumsey (1993, and see also Rimland & Fein, 1988) noted that savants have a deficit in disengaging attention due to deficient attentional orienting processes and overselectivity; they display overly efficient gating of external stimuli and overfocusing upon internal processes. Moriarty et al. (1993, pp. 1019–1020) provided the following case history of the savant syndrome in association with TS:

M.C. is a 17-year-old boy who was referred for a neuropsychiatric opinion to the National Hospital, Queen Square, London. He was adopted at the age of 6 weeks. No information about his biological parents was available. There is no history of neurological or psychiatric illness in the adoptive parents. He had normal language development, was emotionally warm and interacted normally with his family, had several friends and engaged normally in play. He gave a fifteen-year history of vocal and motor tics beginning at the age of two, and developed a wide repertoire of motor tics as well as a range of non-verbal vocal tics. In addition, he felt compelled to imitate others' speech and gestures (echolalia and echopraxia) and to repeat the endings of words and phrases (palilalia). The diagnosis of TS was made when M.C. was fifteen. He had a brief trial of haloperidol which diminished the frequency and intensity of his tics but caused intolerable drowsiness. As well as TS, M.C. suffers from asthma and ectopic eczema. He attends a special school for children with learning difficulties where he performs satisfactorily with reading and practical skills but has difficulty with arithmetic. At no time has he been felt to have any abnormal preoccupations or circumscribed interest patterns. He lives at home with his adoptive parents.

From an early age it was noted that M.C. would join in conversations in the family home whenever past events were mentioned by describing a detailed factual recollection of the occasion and knowing the day and date on which it occurred. His descriptions were accurate but not stereotyped. He also became known for his 'party trick' of being able to give the day of the week for any date one chose to mention.

On examination, M.C. had almost continuous severe vocal and motor tics. Neurological examination was otherwise normal. He could suppress the tics voluntarily for a short period of time (less than a minute). His mental state was unremarkable other than the features of TS. Of note, his social interaction was appropriate. When asked to give the day of the week for a series of 17 random dates in 1991, he gave 16 correct answers with a mean speed of response of 5 s. He correctly gave the day of the week for Christmas Day 1985, and 1 April 1980. He was unable to give the day of the week for dates before his own date of birth. When asked to explain how he arrived at the correct day of the week when given a date, he was unable to give any explanation, but he denied 'remembering' the dates and asserted he found others' inability to know the answers to these questions baffling.

Formal neuropsychological examination performed in the Department of Neuropsychology at the National Hospital for Neurology and Neurosurgery showed him to have a verbal IQ of 79 and a performance IQ of 64 on the WAIS-R [Wechsler Adult Intelligence Scale—Revised]. He had a forward digit span of 7, but when asked to repeat digits in the reverse order he could only do so with 3 digits, invariably erring when confronted with 4. His age-scaled score of 5 on the Arithmetic subtest of the WAIS-R lies on the 5th percentile. On the verbal version of the Recognition Memory Test, his score of 48/50 lay at the 75th percentile. However, on the visual version his score of 33/50 showed marked impairment. In addition, he completed the Leyton Obsessional Inventory, and had a symptom score of 13 and trait of 6.

Other Neuropsychiatric Deficits: Aggression and Violence

TS and to a lesser extent HD illustrate how the boundaries between psychiatry and neurology are crumbling. Of course the division itself is relatively recent, and the emerging emphasis upon neuropsychiatric determinants of behavior may prove to be particularly relevant in the context of violence and aggression. As F. Elliott (1992) observed in his overview of a neurological contribution to violence, the latter in the present context may be used to indicate physical interpersonal aggression with the intent to harm or destroy. It is not concerned with collective aggression, or the violent responses of patients to delusions or hallucinations, or with opportunistic premeditated aggression performed for personal profit.

We have already noted (Chapter 10, and see also Elliott, 1992) that the capacity for aggression and its control is vested in an excitatory and inhibitory system situated in the orbitofrontal cortex, septal area, amygdala, caudate, thalamus, and hypothalamus, and elsewhere in the "limbic system," with the neocortex, especially the dorsolateral prefrontal regions, normally exhibiting some degree of inhibitory control (Kötter & Meyer, 1992, critically evaluated the reality of a reified limbic system, and reviewed its empirical foundation). Other recently identified suppressor areas for affective aggression, according to Elliott (1992), include the septal area, the head of the caudate nucleus, the lateral amygdala, and the cortex of the anterior lobe of the cerebellum. Both gamma-aminobutyric acid (GABA) and serotonin inhibit predatory and affective aggression in humans and other species. Low cerebrospinal fluid levels of serotonin metabolites are correlated with impulsivity, aggression, and violently depressive affect. Indeed Brunner, Nelen, Breakefield, Ropers, and van Oost (1993) reported a genetic mutation in males in a large Dutch family, apparently causing periodic outbursts of aggression. A mutation in the gene for the enzyme monoamine oxidase A (MAOA) seems to be involved, allowing the build up of abnormal concentrations of certain neurotransmitters. As none of the women in the family showed evidence of abnormal behavior, this suggests that a recessive gene on the X chromosome is involved.

Elliott (1992) distinguished between three types of violence:

1. *Episodic dyscontrol*, with its intermittent explosive outbursts of primitive rage that may, after an apparently trivial trigger, precipitate unpremeditated attacks, property damage, or even homicide.
2. *Psychopathic aggression*, where shallow affect and cold, callous premeditation is associated with a retardation of social and emotional development, without significant intellectual impairments.
3. *Compulsive violence*, a rare complication of OCD and ADHD and amenable to treatment with serotonin reuptake inhibitors.

Affective Disorders

As observed by Ferrier and Perry (1992), the historical precedent of research into PD resulted in the identification in autopsied brain tissue of a key neuronal deficit and a clinically effective therapy. Similar research in Alzheimer's disease (AD) has as we saw also identified important associations between senile plaques, cholinergic deficits, and cognitive impairment, but not, as yet, an effective therapy. The affective disorders are another collection of neuropsychiatric syndromes where an organic basis is present, though localized changes have generally proved elusive. We shall not therefore deal with the affective disorders in detail, except to note that prefrontal hypometabolism may be found in the chronically depressed (Baxter et al., 1989), that hippocampal degeneration may be involved in some forms of depression, and that there is some evidence (see Ferrier and Perry, 1992) in refractory bipolar disorder of temporal-lobe EEG abnormalities and reduced temporal-lobe volumes, and smaller hippocampal volumes in suicide victims. As Ferrier and Perry (1992) noted, a problem is that patients rarely die during an episode of depression, and if they do they may be unrepresentative of affective disorders in general; drug therapy may further complicate findings. Such autopsy studies as there are in affective disorders have therefore utilized brains from suicide victims, and of individuals who died during an episode of depression from some intercurrent illness. (The neurobiology of suicide may not of course necessarily match that of major depression.)

Bipolar disorder involves episodes of elevated mood state (mania) and lowered mood state (major depression), with disturbances of sleep, appetite, energy, concentration, sexual activity, and self-esteem (Freimer & Reus, 1993). Cyclothymia involves a milder presentation of an otherwise similar disorder. Patients vary enormously in the number, frequency, severity, and regularity of episodes, the ratio of manic to depressive events, and their response to treatment. A genetic component is indicated by family, adoption, and twin studies, though more than one gene is likely to be involved, and because of variability in age of onset, relatives of probands may be wrongly diagnosed as unaffected at time of study, only later becoming ill. Moreover comorbidity of other concurrent psychiatric disorders may modify the expression of affective disorders, leading to misclassification, and bias from assortative mating may further cloud the issue. Indeed, congenital rather than strictly genetic factors could also operate (e.g., from the effects of intrauterine cocaine or alcohol abuse in pregnancy). In any case, genetic heterogeneity is likely, with more than one independent mutation, each accounting for similar behavioral phenotypes (Freimer & Reus, 1993). Lithium salts have proved efficacious in stabilizing mood swings and reducing mania, though their exact mechanism is not understood.

In the affective disorders, disturbed functioning of several monoamines is

implicated. Modern psychopharmacology was born in the 1950s with the introduction of two drugs still widely in use today (Barondes, 1994). Chlorpromazine,
originally an antihistamine, was unexpectedly found to alleviate symptoms of
schizophrenia (see Chapter 15). Then imipramine, initially considered an alternative to chlorpromazine, was found to be efficacious in treating major depression. Its descendants (e.g., fluoxetine, Prozac) selectively block the reuptake
(and thereby prolong the action of) serotonin, without affecting norepinephrine,
and are able to alleviate major depression and also other symptoms previously
considered to be the province of psychotherapy. Such symptoms include excessive sensitivity to criticism, fear of rejection, lack of esteem, and an inability to
experience pleasure (Barondes, 1994). Just as the beneficial effects of chlorpromazine and its DA-blocking derivatives take some time to relieve the symptoms
of schizophrenia, so too the therapeutic effects of these new serotonin-reuptake-
blockers may cause adaptive brain changes that take weeks to develop. In both
drug-free depressives and suicides, serotonergic receptors may be abnormal,
with antidepressants of various kinds countering this abnormality (Ferrier &
Perry, 1992). Moreover, corticotropin-releasing factor (CRF), a neuropeptide
functioning as a neurotransmitter and involved in the mediation of stress responses, is elevated in the cerebrospinal fluid in depression; changes in CRF receptors in the brains of suicide victims have led to the hypothesis that CRF is
overactive in depression (Ferrier & Perry, 1992). Interestingly, there are increasing suggestions of commonalities between OCD and anxiety and depression, and
Bailly et al. (1994) noted increased levels of CRF in individuals suffering from
OCD.

Clearly, neurochemistry is a major determinant of mental illness, and possibly even of such individual personality quirks as hypersensitivity, impulsiveness,
suspicion, introversion, and optimism. Such aspects of personality may lie on a
continuum with mental illness, and be amenable to similar medication and modulation. Of the three major neurotransmitters, DA, norepinephrine, and serotonin, the latter is currently receiving the most attention, and serotonergic
agents are increasingly important with depression, compulsions and obsessions,
anxiety, panic disorder, suicidal ideation, impulse control disorders, substance
abuse, eating and personality disorders, and violence (Stahl, 1993). However,
there are many different serotonin receptor subtypes with, probably, a variety of
separate functions that remain to be elucidated. In any case, serotonergic medication may in the end merely paper over the cracks, in the way that analgesics
help alleviate arthritis without addressing the etiology.

As Bench, Friston, Brown, Frackowiak, & Dolan (1993) observed, the major
depressive disorders are phenomenologically heterogeneous, with diverse clinical
manifestations that include mood disturbances, anxiety, cognitive deficits, and
psychomotor changes. A major problem with a disease-based or lesion approach
in neuropsychiatry is this symptom diversity, coupled with the distributed nature
of the likely neural organization, and with the probable involvement of several

monoamines. Language functions also enjoy a highly distributed neural organization, and in both cases (see e.g., Chapter 3), language and the affective disorders, PET or similar neuroimaging is an optimal tool for simultaneously monitoring, across the whole brain, the ensemble actions of the interacting subsystems. Bench et al. (1993) identified three factors in patients' depressive symptoms, and correlated them with regional cerebral blood flow (rCBF):

1. A factor with a high loading for anxiety and psychomotor agitation. This factor correlated positively with rCBF in the posterior cingulate and inferior parietal lobule bilaterally. The cingulate cortex of course plays an important role in affect, and interconnects with the higher association cortex.

2. A factor with a high loading for psychomotor retardation and depressed mood correlated negatively with rCBF in the dorsolateral prefrontal cortex (DLPFC) and the angular gyrus, on the left. Decreased perfusion of the left DLPFC also occurs with diminished spontaneous speech in aphasia and chronic schizophrenia, the region being important in volitional behavior and willed, intentional activities, and interconnecting with the cingulate. The angular gyrus, apart from an involvement with reading (see Chapter 3), plays an important role in visuospatial orientation and attention (see Chapter 6).

3. A factor with a high loading for cognitive performance correlated positively with rCBF in the left anterior medial prefrontal cortex (i.e., decreased rCBF correlated with increasing cognitive impairment).

From the marriage of new structural and functional imaging systems, we can confidently predict a clearer picture shortly of the systems involved in positive as well as negative affect.

Summary and Conclusions

Extrapyramidal disorders of unwanted movement and posture include other such symptoms as dyskinesia, tremor, myoclonus, tic, ballismus, chorea, athetosis, and dystonia. Writer's cramp (graphospasm) is a model example of dystonia, while the restless-legs syndrome is a very common form of familial, idiopathic dyskinesia, with close parallels with tic disorders and DAergic involvement. Such extrapyramidal disorders further illustrate the close commonalities now emerging between neuropsychiatry and neurology. Other such commonalities include the determinants of aggression and violence, where the prefrontal cortex, limbic system, and striatum are implicated, along with the neurotransmitter serotonin.

Movements associated with tic disorders, although ultimately involuntary in view of their compulsive nature, nevertheless are capable of some degree of voluntary modulation. It is possible that they start as voluntary movements, only later becoming automatic. They are sudden repetitive, nonrhythmic, stereotyped movements, without apparent purpose and caricatures of normal behaviors.

Simple or complex, they may involve face, head, limbs, body, or the articulatory apparatus. Additionally, patients may copy another's words (echolalia) or actions (echopraxia) or engage in involuntary obscenities by word (coprolalia) or action (copropraxia). All of these actions may be disguised or smuggled into other ongoing activities to conceal them. They usually start in childhood, are more common in males, and are exacerbated by both boredom and stress. There are extensive interconnections between the striatal and limbic systems, and electrostimulation of the cingulate can produce complex stereotypies with an urge to move; conversely, cingulotomy may help alleviate tic behavior and compulsions.

TS is an extreme form of tic disorder, the behavioral manifestations of which have been documented for several hundred years. Only recently, with the advent of successful treatment with DA antagonists, has its study moved from the realm of psychoanalysis to that of behavioral neurology and neuropsychiatry. Simple motor tics are the first to appear around 7 years of age, and may indeed be the only symptoms ever to manifest; complex vocal tics appear later if at all. New cases are rare after puberty, when indeed symptoms may abate. Although TS is considerably more common in boys, the closely related OCD is more common in girls, again indicating modulation by sex hormones. Patients often also present with ADHD.

Although patients with TS are generally of approximately normal intelligence, there may be some impairment of visuoperceptual performance, visuomotor skills, reading, and mathematical ability, with some evidence of deficits in executive functions involving planning and sequencing complex behaviors and organizing goal-directed activities. Apart perhaps from some "soft" signs, the neurological examination is typically normal. The disorder is probably inherited as a single autosomal dominant gene with variable penetrance, which may be expressed either as TS or OCD, or possibly as ADHD. However pre- and postnatal environmental factors, including steroidal hormones, clearly also play a major role, and there is evidence of metabolic abnormalities in the striatum, prefrontal, and limbic cortex, which, incidentally, also help explain some of the more florid symptoms. Anomalies of DAergic and serotonergic function are also implicated, though pharmacological intervention may, for example, simultaneously ameliorate OCD while exacerbating tics. Generally, a serotonergic caudate and orbitofrontal syndrome seems implicated in OCD.

Autism shares with OCD many common features. It is probably a consequence of a number of different disease states, possibly working through the DAergic mesial frontal and mesial temporal lobes. More boys than girls are affected, verbal skills more so than performance measures, and the disorder is characterized by restricted, repetitive, obsessive activities, a fascination for a limited range of objects or activities, an insistence on "sameness," solitary tendencies, and a marked lack of social empathy. It is as if the sufferer lacks a "theory of mind." Although the condition typically manifests after early infancy, it appears to be a neurodevelopmental disorder with its origins, possibly viral, in early fetal life. Again like OCD, serotonergic dysfunction also may be implicated. Epilepsy is common later in the disorder's course, and brain imaging often suggests structural anomalies. A familial pattern often is also apparent, though its manifestation may depend upon the underlying etiology.

The affective disorders involve additional neuropsychiatric syndromes where organic bases are present, but localized changes have proved elusive. Much of the available clinical information has come from suicide victims. Again, anomalies of the serotonergic system seem to be implicated, together with CRF. The depressive disorders are phenomenologically heterogeneous, with a highly distributed neural organization. Consequently, neuroimaging methods may well be the way forward; so far, the DLPFC and cingulate seem to be involved, together with the mesial prefrontal cortex.

Further Reading

Chase, T. N., Friedhoff, A. J., & Cohen, D. J. (Eds.). (1992). *Tourette syndrome: Genetics, neurobiology and treatment. Advances in neurology* (Vol. 58). New York: Raven Press.

Comings, D. E. (1990). *Tourette syndrome and human behavior.* Duart, CA: Hope Press.

Frith, U. (Ed.). (1991). *Autism and Asperger syndrome.* Cambridge: Cambridge University Press.

Frith, U. (1993, June). Autism. *Scientific American* 78–84.

Devinsky, O., & Geller, B. D. (1992). Gilles de la Tourette syndrome. In A. B. Joseph & R. R. Young (Eds.), *Movement disorders in neurology and neuropsychiatry* (pp. 471–478). Oxford: Blackwell Scientific.

Elliott, F. A. (1992). Violence: The neurologic contribution, an overview. *Archives of Neurology, 49,* 595–603.

Gillberg, C., & Coleman, M. (1992). *The biology of the autistic syndromes: Clinics in developmental medicine No. 126.* (2nd ed.) Oxford: Mac Keith Press.

Jankovic, J. (1992). Diagnosis and classification of tics and Tourette syndrome. In T. N. Chase, A. J. Friedhoff, & D. J. Cohen (Eds.), *Advances in neurology* (Vol. 58, pp. 7–14). New York: Raven Press.

Kurlan, R. (Ed.) (1992). *Handbook of Tourette's syndrome and related tic and behavioral disorders.* New York: Marcel Dekker.

Mendlewicz, J. (1994). The search for a manic depressive gene: From classical to molecular genetics. In F. Bloom (Ed.), *Progress in Brain Research* (Vol. 100, pp. 255–259). Amsterdam: Elsevier.

Obler, L., & Fein, D. (Eds.) (1988). *The exceptional brain: The neuropsychology of talent.* New York: Guilford.

Shapiro, A. K., Shapiro, E. S., Young, J. G., & Feinberg, T. E. (1988). *Gilles de la Tourette syndrome.* (2nd ed.) New York: Raven Press.

Singer, H. S., & Walkup, J. T. (1991). Tourette syndrome and other tic disorders: Diagnosis, pathophysiology and treatment. *Medicine, 70,* 15–32.

Stefanis, C. N. (1994). Schizophrenia: Neurobiological perspectives. In F. Bloom (Ed.), *Progress in brain research* (Vol. 100, pp. 267–272). Amsterdam: Elsevier.

Suchowersky, O. (1994). Gilles de la Tourette syndrome. *Canadian Journal of Neurological Science, 21,* 48–52.

Chapter 15

A Disorder of Thought

Schizophrenia

$\mathcal{A}$s Waddington (1993a) observed, schizophrenia presents an enormous challenge for the neuroscientist. It is a common disorder, with a lifetime morbid risk of approximately 8 per 1000, and is perhaps the most debilitating disease to affect humankind, occurring often "out of the blue," ruining careers, personal relationships, and lives. Patients are disabled from comparatively early in life, often becoming psychotic even in their late teens, and becoming highly dependent upon society and often disruptive of it. Perhaps twice as many are borderline cases. It begins with males typically in the mid-20s, and with females a little later in the early 30s, though it can commence anywhere between 8 and the 60s, and most patients will have shown subtle signs of psychosis for months or years before clinical care is initiated for florid psychotic symptoms. Indeed Walker, Grimes, Davis, and Smith (1993) reported that abnormal expressions of negative affect are visible on the faces of children, recorded on home movies, who many years later became schizophrenic. The disorder may be more common in higher latitudes and advanced societies, and once diagnosed it is likely to be lifelong. However, a sudden acute onset in an older individual with previously good personal adjustment, prominent affective symptoms, and no previous history of the disorder tends to offer a better prognosis than where the converse situations apply. So varied are its manifestations, that some see it as a cluster of clinical symptoms rather than a precisely defined single entity or disease. As Frith (1992) noted, there is evidence for a genetic component, and also for environmental contributions including viral infections and stress, birth injury, and neurodevelopmental dysfunction. It is increasingly being seen, however, as organic rather than psychosocial (Straube & Oades, 1992), although factors associated with the latter might affect how (and when) it manifests. Its later onset in females suggests a hormonal contribution, though it could merely reflect better social skills in females. However, Waddington (1993b) saw the earlier male onset, poorer premorbid function, and poorer outcome, together with findings that males are more likely to exhibit minor physical anomalies, and antenatally to have experienced obstetric complications, to be in accordance with proposals that the male brain is more vulnerable to neurodevelopmental disruption leading to schizophrenia. Nevertheless, according to Waddington, it is female patients who show not only greater temporal and geographical variations in rate of occurrence, but also a greater susceptibility to first-trimester dietary insufficiency and second-trimester maternal influenza. The latter, we shall see, is itself in accordance with a hypothesis of viral etiology.

Schizophrenia, therefore, is a complex, heterogeneous disorder with multiple, variable symptoms that cannot always be differentiated in terms of the disease process per se, or the effects of drug treatment, or of prolonged institutionalization. There seems to be no single combination of features sufficient to explain all its symptoms, nor any single biological feature that is necessary for disease causation. It remains a diagnosis of exclusion, like Alzheimer's disease (AD) (see Chapter 9), because none of its clinical features are pathognomic, and associated biochemical, neuroradiologic, physiological, and psychological tests lack sufficient sensitivity for decisive diagnosis (Carpenter & Buchanan, 1994). A differential diagnosis must exclude other psychoses with known organic causes (e.g., temporal-lobe epilepsy, metabolic disturbances, toxic substances, or psychoactive drugs). Affective disorders and recurrent mood disturbances must also be eliminated. In schizophrenia, a recurrent pattern of affective change is uncommon, and the psychotic symptoms of schizophrenia tend to be more bizarre and less intuitively understandable than those associated with affective disorders. The gradual development of the disease, incomplete remissions between psychotic episodes, and prolonged social and occupational impairment all contribute to a diagnosis of schizophrenia (Carpenter & Buchanan, 1994).

The symptoms of schizophrenia concern specifically human functions, and it is not clear how an animal, if affected, could manifest the disorder. Among other things there may be a separation of cognitive from emotional functions, hence its appellation, from the Greek σχίξειν φρένα, to divide the mind. Thus inappropriate affect may result in uproarious laughter in discussing the death of a loved one. Information processing and social interaction are disturbed, the personality "fragments," with loss of reality, psychotic episodes, hallucinations (abnormal perceptions, e.g., of voices) and delusions (aberrant beliefs, paranoia, and persecution complexes), incoherent thoughts, disordered memory, confusion, and dementia. Some or all of these problems of course can manifest as a consequence of drugs or epilepsy, degenerative disease, brain tumors, hormonal, or metabolic disturbances. However, to the schizophrenic, the world, instead of being stripped of its meaning as in associative agnosia, may become excessively and terrifyingly rich in semantic possibilities; instead of losing the ability to reason, the schizophrenic develops a reasoning that is so cryptic and obscure that it strikes the observer as incoherent (Heinrichs, 1993).

One Disease or Many?

Like diabetes, there may be a variety of different subtypes of the disorder, perhaps involving different neurotransmitters or clusters of genes. It has long been possible to identify four major categories:

1. *Catatonic*, involving a largely psychomotor deficit
2. *Hebephrenic*, with disorganized behavior and flat inappropriate affect
3. *Paranoid*, with delusions and hallucinations
4. *Undifferentiated*, where some or all of these symptoms manifest.

Over time, patients often progress between these subtypes, more typically perhaps from the paranoid to the hebephrenic than vice versa, and the condition (or conditions) may verge onto bipolar manic–depression, via the "borderline" personality. Moreover, just as no one underlying disease has yet been identified unambiguously as schizophrenia, so too there has been a multitude of alternative classificatory schemes based on checklists of signs and symptoms. Thus in the days when psychosocial factors were deemed to be probably more significant than those involving brain pathology, the disorder was dichotomized into "process" versus "reactive" schizophrenia. The latter subset of patients was seen to have had a normal premorbid adjustment, with rapid disease onset in response to stress, with marked affective symptoms, and a generally better prognosis. Conversely, process schizophrenics, whose illness was seen as the result of reactivity to stress, had a generally poor premorbid adjustment, anhedonia (an inability to experience positive emotions), flattened affect, an insidious disease onset without a clear identifying or precipitating cause, and a poor prognosis with chronicity (Fowles, 1992). We shall shortly review a general dichotomization of symptoms into positive and negative; it is possible to use this dichotomy to classify patients into type 1 (with largely positive symptoms, i.e., extra- or supranormal behaviors), and type 2 (with largely negative symptoms or deficits in behavior.) Type 1 schizophrenics are then seen as having an acute onset, a good premorbid adjustment, and subsequent prognosis, with an episodic course that is amenable to antidopaminergic medication, and little evidence of brain pathology. Type 2 schizophrenics lack the positive symptoms of hallucinations and delusions, exhibiting instead the negative symptoms of poverty of action, speech, and thought, blunted affect, and emotional and social withdrawal (and see Crow, 1980). Their long-term, chronic condition is far less amenable to medication, and may be associated with structural abnormalities of the temporal and frontal lobes. They are less likely to have a family history of the disorder, and more likely to report a raised incidence of perinatal problems.

Yet another approach (for review, see e.g., Carpenter and Buchanan, 1994) divides the psychopathology of schizophrenia into three relatively independent cluster symptoms:

1. *Psychotic symptoms*, with hallucinations and delusions
2. *Cognitive impairment*, with tangentiality of thought, loss of goals, incoherence, and looseness of associations
3. *Negative symptoms*, with diminished affect, poverty of speech, and diminished interests, sense of purpose, and social drives.

Symptoms of course constitute the subjective experiences as reported by the patient, in contrast to *signs*, which make up the objective, observable, abnormal behavior. A *syndrome* (such as schizophrenia) consists of a clustering of signs and symptoms that can distinguish it from another such disorder (e.g., affective disorder). A proposed syndrome may be validated by causality and etiology (if known), response to therapy or medication, or by the natural history, clinical course, and outcome of the disease, such as recovery, relapse, or deterioration.

Kraepelin (1896, cited in Frith, 1992) saw the disorder he termed *dementia praecox* (untimely or premature loss of faculties) to be a functional rather than an organic psychosis like AD, but one that differed from manic–depressive psychosis, another functional disorder, because the latter fluctuated whereas schizophrenia progressed. Bleuler (1911) coined the term *schizophrenia* to capture the idea of a separation between emotion or feeling and thought, cognition, or understanding (Frith, 1992). Psychosis, a break with reality manifesting usually as hallucinations, delusions, or a disruption in thought processes, is of course now central to the classification of schizophrenia. Before the introduction of the dopamine (DA)-blocking neuroleptics (e.g., chlorpromazine) in the 1950s, at least half of all hospital beds in advanced countries were occupied by psychiatric cases; now, with successful drug amelioration, less than a quarter of beds are so occupied, even though half of all mental illness involves schizophrenia. Although there has been a considerable increase in the incidence of the disorder over the last 200 or so years, this may only reflect an increase in life expectancy, as it is a disease of adulthood, just as Huntington's disease (HD) is one of the mature years, and AD is a disorder of the senium. Its increased incidence in lower socioeconomic classes and the unmarried may suggest potentiation by stress, though there is little evidence of war having the same effect—quite the reverse, in fact. Social interaction clearly plays some, possibly comparatively minor, role, rather than stress per se.

The problem with schizophrenia, as we saw, is that there is still no clear, unambiguous, uniform, and characteristic brain abnormality, so that diagnosis is still largely dependent upon symptoms and time course. As Waddington (1993a) observed, it is difficult to study a disorder where there is no known biological marker, and whose diagnostic boundaries derive largely from patients' verbal descriptions of their mental experiences. There are therefore a series of Catch-22 situations; diagnosis of schizophrenia relies first upon defining it, and as yet no one can agree upon its definition, or diagnosis. Moreover, all diagnostic schemes exclude disorders with obvious organic bases, which is itself paradoxical given that nowadays there is increasing evidence for an organic basis, and in fact patients *excluded* from diagnosis of schizophrenia because of obvious organicity may well be the ones who will eventually lead us to the physiological basis of the disorder (Frith, 1992). Other diagnostic problems are the fact that people with affective psychosis, or the old manic–depressive disorder, also display positive symptoms of delusions and hallucinations, whereas people with Parkinson's disease (PD), as we have seen, display poverty of action and other similar (to those of schizophrenia) negative signs. The *DSM-III-R* defines schizophrenia as having the following characteristics:

1. Characteristic psychotic symptoms for at least a week
2. Social functioning below previous levels
3. No major mood changes such as depression or elation
4. Continuous signs of the disturbance for at least 6 months
5. No evidence of organic features

The characteristic psychotic symptoms must include two or more of delusions (e.g., of persecution, grandeur, or of thought broadcasting), hallucinations (e.g., of a voice with content unrelated to the current mood, perhaps telling the patient what to do, commenting negatively on the patient, or talking to someone else), incoherence, catatonic immobility, and flat or grossly inappropriate affect (Frith, 1992).

Course and Progression of Schizophrenia

The course and eventual outcome of schizophrenia is hardly less variable than its initial or early manifestations. Although the patient may oscillate between intense and reduced emotional expression, the former state tends to predominate earlier in the course of the disease, often giving way later to a blunting of affect and emotional flatness. Of the three likely outcomes, recovery, intermittent or partial remission, or chronic progressive deterioration, unfortunately the latter is the most common, with a poor prognosis. Perhaps one-quarter might make a more or less full recovery, one-quarter may remain severely or increasingly disturbed, and the remaining half may experience a more or less severe and fluctuating course. On the other hand, patients presenting with an acute sudden onset tend to have a better prognosis than where the onset is gradual, or auditory hallucinations manifest early. Occasionally a patient may experience a series of psychotic episodes, with comparative normality during the intervening periods, despite perhaps some eccentricity, social isolation, poverty of speech and/or affect, poor attention span, and lack of motivation. Straube and Oades (1992) calculated that full remission may occur in one-quarter of cases, with a similar number maintaining mild residual symptoms, and around half still having moderate to severe symptoms 20 to 40 years after first admission. It should be noted that apart from a tendency to suicide, schizophrenics have normal longevity. The fact that the course of illness is not always one of uniformly progressive deterioration, and that differences may occur between cultures, indicates that ongoing environmental or social factors may play a role. Frequently, patients in later stages exhibit bizarre motor behaviors known as tardive ("late") dyskinesia. It is often claimed that this phenomenon is a consequence of prolonged drug treatment; certainly prolonged exposure to DA antagonists can induce such irreversible effects, though they have been noted in the days before the advent of neuroleptic medication, and they seem to be more common in patients without a family history of the disorder. It is therefore entirely possible that the phenomenon is associated with DAergic dysfunction that includes the basal ganglia (BG) (see Chapter 11). Indeed Caligiuri and Lohr (1994) reviewed voluntary motor disturbances in schizophrenics, which include motor incoordination, disturbed pursuit tracking, difficulty following movement sequences, and desynchronized tapping. They also themselves report, in neuroleptic-naive patients, disturbances in the voluntary control of steady-state force. Degree of force instability, suggesting a motor disturbance resembling tardive dyskinesia, was correlated with positive but not negative symptoms. They concluded that such disturbances in the control of isometric force may represent spontaneous dyskinesia in neuroleptic-naive patients. Where tardive dyskinesia is thought to be drug related, Wirshing and Cummings (1990) noted that it appears after at least 3 months of neuroleptic treatment, it becomes more prevalent in the aged, and may be more common in women, in patients with mood disorders, and in brain-injured patients. Unlike other idiopathic dystonias and dyskinesias (see Chapter 14), tardive dyskinesia tends to involve the lower face and distal limbs, to be increased by rest and distraction and decreased by action, and to be volitionally suppressed. A variety of other related tardive conditions can occur (e.g., chorea, akathisia, tics, and myoclonus), and all can be worsened by anticholinergic agents and improved by DA-blocking and depleting agents. Tardive akathisia (ἀκάθισις, an inability to stay sitting down) is reviewed in detail by Sachdev and Loneragan (1991), and is characterized, as the name implies, by a restless inability to sit still, and is otherwise very similar to the naturally occurring and strongly familial disorder of Ekbom's ("restless legs") syndrome (see Chapter 14). A feeling of unease or discomfort, particularly referable to the legs, commonly follows prolonged neuroleptic treatment. Unlike the tardive dyskinesias, choreas, and myoclonus, however, but similar to "idiopathic" tics and Tourette's syndrome (TS), the patient can voluntarily inhibit the leg movements for a short time. Chronic DAergic changes consequent upon medication in the mesolimbic or striatal pathways or both seem to be implicated.

In the popular mind, schizophrenia tends to be associated with the florid hallucinatory or delusional behavior characterized by positive symptomatology. Hallucinations tend to be auditory in schizophrenia, and are rarely visual, unlike those consequent upon drug misuse. (Occasionally visual hallucinations involving monstrous or terrifying faces—paraprosopia—may be reported, and patients may experience the delusional misidentification of Capgras' syndrome, see Chapter 5). Although prosopagnosic nonrecognition of familiar faces may stem from mesial temporal lobe lesions, disturbances of perceived familiarity may be a consequence, as in schizophrenia, of limbic malfunction (see e.g. Young, Newcombe, de Haan, Small, & Hay, 1993). The similarity of auditory hallucinations to those experienced during temporal lobe epilepsy or under temporal lobe stimulation (Penfield & Perrot, 1963) indicates their likely origin in a defective mesolimbic system. The often-associated unfounded or delusional beliefs of persecution or control clearly derive from the hallucinations, and may be the origin of folk beliefs in witchcraft. Indeed excessive religiosity has itself been linked to temporal lobe malfunction (Schiff, Sabin, Geller, Alexander, & Mark, 1982), especially of the right hemisphere (RH).

False delusional beliefs may include the following (see e.g., Frith, 1992; Straube & Oades, 1992):

1. Thought insertion, as from an outside source into one's head
2. Thought broadcast, from one's own mind to others'
3. Thought withdrawal, from one's own mind by someone else
4. Hearing one's own thoughts spoken aloud
5. Hearing voices either addressing one (in the second person) or commenting about one
6. Delusions of control, where an external force makes one act
7. Delusions of persecution, where someone is out to get one
8. Delusions of reference, where the actions and gestures of strangers are believed to be of special significance

Both hallucinations and delusions may also occur in HD (see Chapter 13), though such patients tend to be more cautious and guarded about them, and less likely to admit to them. Another associated positive symptom is thought disorder; the patient has problems with abstract concepts and may tend to give literal rather than metaphorical interpretations to proverbs, like an RH patient (see Chapter 4), and may produce bizarre or inappropriate associations, and exhibit a general lack of organization. Yet other signs that may be regarded as positive include incoherence of speech with repeated shifts of topic, and a lack of logical connectivity between phrases and sentences. There may be incongruity of affect, with the expressed emotion (e.g., laughter) not fitting the gravity of the topic (e.g., death of a spouse). There may also be stereotyped repetitive movements. Generally, positive symptoms tend to be associated more with the earlier acute phase, and to respond better to medication, than the (often later-occurring) negative symptoms that seem on the whole to be associated more with brain rather than DAergic changes. Nevertheless, auditory hallucinations are present in a majority (maybe two-thirds, according to Straube & Oades, 1992) of patients. They seem to manifest as a form of inner speech that is so intense that it is interpreted as coming from outside. Indeed it may be reduced if the patient is distracted by other external input that demands concentration.

Negative signs that may be associated with schizophrenia include lack of emotional expression (as in PD), a flattened or blunted affect (athymia), apathy (abulia), and inertia. There is typically poor social interaction, with little speech (alogia) or communication, with an incapacity extending to vocal inflections, gesture, and nonverbal communication. Behavior is autistic in nature. Spontaneous movements are greatly reduced, again as in PD. Grooming and hygiene are seriously affected, and there is a genuine lack of interest in recreation and sex, and a reduction in the ability to feel intimacy and to form friendships manifests as anhedonia and asociality. These signs are often associated with cognitive impairment and various brain abnormalities (see below), especially frontal, and respond poorly to medication. However, it is still not unusual for some patients to switch back and forth between predominantly positive and negative symptomatology.

Klimidis, Stuart, Minas, Copolov, and Singh (1993) disputed the simple division of schizophrenic phenomena into positive (productive) symptoms such as delusions, hallucinations, formal thought disorder, and bizarre behavior, and negative (deficit) symptoms such as flat affect, avolition, anhedonia, alogia, and attentional disturbances. Instead, they proposed a tripartite structure of negative symptoms (consisting, separately, of negative signs and social dysfunctions), positive thought disorder, and hallucinations and delusions. The last category bifurcates into hallucinations and delusions, with delusions themselves bifurcating into grandiosity (grandiose delusions, delusions of reference, etc.) and loss of boundary delusions. The latter are seen to be constituted by mind reading, thought insertion, delusions of control, and so on. Clearly, the last word is yet to be said on this issue.

While cognitive impairment may manifest in the presence of negative symptomatology, schizophrenia need not necessarily be associated with a reduction in IQ, or the ability to think logically, deductively, or abstractly. Instead, there may typically be a tendency to deviate therefrom, due maybe to a preoccupation with task-irrelevant information (Straube & Oades, 1992). Eye movements, however, are often abnormal, with irregularities in smooth pursuit, especially when attention is not focussed—a phenomenon claimed to occur at an above-chance level even in otherwise apparently unaffected first-degree relatives. "Counter-saccades" (*deliberate* eye movement in the direction *opposite* to that of a suddenly occurring peripheral signal) are particularly slow. It is probable that these problems reflect involvement in the DAergic oculomotor circuit from cortex to striatum and back to the frontal eye fields (see Chapter 10). Such oculomotor and cognitive anomalies as described above are also reflected in schizophrenic speech. Thus there is a general lack of coherence, a poverty of content, tangentiality, derailment, changes of topic, and illogicality that leads to a loss of goal. The patient typically follows irrelevant associations, which in certain artificial experimental situations such as priming can even be advantageous. There may be verbal perseverations and stereotypies with excessive self-reference, socially inappropriate discourse, confabulation, and delusion, along with outright Wernicke-like neologisms. Clearly temporal and frontal involvement is indicated. Other aspects of schizophrenic conversation include problems in getting across to listeners the appropriate context and viewpoint. There is a lack of referential cues and the listener's perspective is not taken into account. There is no gross memory deficit, but there may be disorganization in working memory and problems in cognitive ordering, grouping, or clustering during the input (encoding) stage; distractibility may also lead to deficits in the automatic organizational processes necessary for efficient memory function (Frith, 1992; Straube & Oades, 1992). Procedural and semantic memory, however, is largely unaffected. Despite popular belief to the contrary, there is no evidence that schizophrenics are any more creative or original in their thought than others, though there is some evidence that this might be true in "unaffected" near relatives. If so, this would suggest that there may be a genetic advantage, in certain instances, for a schizophreniform personality, perhaps accounting for the apparent persistence (see below) of a genetic contribution to the disorder down through the generations, despite the devastating handicap experienced by those who are strongly affected.

Attentional Deficits

Attentional deficits are perhaps one of the oldest problems thought to be associated with schizophrenia. The idea in fact goes back to Kraepelin and Bleuler—that patients are easily derailed, put off, interrupted, or deviated by anything that happens, whether relevant or not, leading to a severe problem in concentration (Straube & Oades, 1992). Similarly, unimportant events may be interpreted as striking, significant, or intense, perhaps because they achieve a deep level of processing and cannot easily be rejected, leading to a problem of "disattending" or distraction (Straube & Oades, 1992). We may therefore invoke an apparent deficit in selective attention, a problem in discriminating signals from noise, leading to an increase in false alarms, especially perhaps in the acute, paranoid subgroup who also have difficulty with simultaneous concurrent tasks and sustained attention (vigilance). With dichotic listening tasks, intrusions occur more at a *semantic* level than where channel selection is based upon *physical* features, (e.g., voice or sex of speaker); again, schizophrenics are more likely to be distracted by irrelevant content than sound. However, chronic patients with predominantly negative signs are less likely to experience distraction; indeed their reduced scanning of the environment may even result in problems in detecting and responding to peripherally occurring events. Attentional anomalies such as those described above may also be reflected in abnormal evoked potentials (EP) to rare, or relevant, events (Straube & Oades, 1992), indicating that schizophrenics may possess a poor cognitive model of stimulus probability. Indeed such patients seem generally to make poor use of advance information, context, redundancy, or stored contingency rules such as are employed by normal individuals to shorten response times.

Feedback and Feed Forward in Schizophrenia

According to Straube and Oades (1992) problems such as those reviewed above can be seen in the context of a high degree of dependence upon external feedback; schizophrenics therefore, like patients with PD (see Chapter 12), may not be able to rely upon their own internally generated feedback. Frith (1992) took these ideas further, suggesting that patients cannot in fact recognize their internal thought processes and subvocal speech as their own, due to a failure in self-monitoring and the feed-forward (reafference or corollary discharge) system (see e.g., von Holst & Mittelstadt, 1950), which comes into play whenever any voluntary activity is undertaken. Thus we normally correct for the effects of self-generated activity when interacting with the environment; we do not see the world swing past us when we voluntarily move our eyes, but only when our eyes are passively moved by an external force. Consequently, according to Frith, positive symptoms occur because brain structures for willed actions no longer send corollary discharges to the posterior regions concerned with perception; the result is that self-generative changes in perception are misinterpreted as having external causes, so accounting for the experienced hallucinations and delusions. With further damage, of course, all response generation will fail, leading to the lack of willed action and negative features of advanced schizophrenia. However, such delusions will be lost with severe end-state schizophrenia or DA blockers because of the reduction in voluntary activity. It is noteworthy that if the electromyogram (EMG) from the speech apparatus is amplified during a hallucinatory episode, a signal is obtained corresponding to actual subvocal speech. Moreover, single photon emission tomography (SPECT) during such episodes reveals activity in Broca's area (McGuire, Shah, & Murray, 1993) and left-hemisphere (LH) temporal regions, indicating involvement of the cortical areas specialized for language and inner speech. Thus schizophrenics seem unable to monitor the generation of their own thoughts and inner speech, which are consequently perceived as alien verbal hallucinations. Delusions similarly would seem to be due to schizophrenics misattributing their own thoughts to others. Faulty monitoring may extend to intentions, especially perhaps in the context of social interactions, with the patient thinking his or her thoughts, emotions, and actions are controlled by someone else. Childhood autism (see Frith, 1991, and Chapter 14) is therefore seen as a complementary disorder, with the child quite unable to infer the mental states of others, rather than, as in schizophrenia, misrepresenting or misattributing his or her own or those of others. Both disorders involve an inadequate or inappropriate "theory of mind" (Whiten and Byrne, 1988). The alien-hand syndrome (Goldberg, Mayer, & Toglia, 1981) may therefore be another related disorder, now involving utilization behavior (Lhermitte, 1983). Here, certain types of frontal patient (see Chapter 7 where the alien-hand syndrome is addressed in more detail) experience stimulus-elicited behavior (e.g., compulsive grasping) without any feeling of effort or intendedness. Indeed, frontal damage associated with a fundamental deficit in the generation of willed action is clearly apparent in three aspects of the schizophrenic's behavior: (1) in poverty of action, speech, and thought, where the problem is one of initiation; (2) in his or her perseverative and stereotyped actions, where the problem is instead one of termination; and (3) in the inappropriate, stimulus-driven actions described here, with loss of inhibition of inappropriate behaviors. Shallice (1988) grouped such deficits under the rubric of failure of a frontal supervisory system. In PD (Chapter 14), willed intentions may be formed but typically cannot easily be connected to action, whereas frontal patients and schizophrenics seem even to lack the ability to *form* the intentions. (Similarly, the expressionless face of the parkinsonian patient, unlike that of the schizophrenic, is only superficial.) We shall shortly review the evidence for frontal lobe damage in schizophrenia, but will close this section with the suggestion of Strange (1992) that the septo-hippocampal system may receive information from the prefrontal cortex relating to planning and goal behavior, and act as a comparator, monitoring any discrepancies between what was intended and what was achieved. Damage in this region could then result in the mislabeling of internally generated behavior as externally generated or controlled.

Brain Anomalies in Schizophrenia

Neuropsychological deficits in schizophrenia seem often to map on to the brain regions (especially prefrontal) thought likely to be damaged by analogy with deficits occurring in known instances of brain damage. A caveat however should be made, in light of schizophrenics' very abnormal ongoing psychological states of anxiety, stress, arousal, and vigilance, which could artifactually determine the neuropsychological deficits. Another caveat is that other clinical patients, with acquired lesions in regions matching those thought to be involved in schizophrenia, generally do not exhibit anything like full-blown schizophrenia; something else must also be happening, and that "something" may well concern limbic structures involved in the coordination of conceptual and significance aspects of reality.

As Frith (1992) noted, it has taken many years to find evidence for brain abnormalities in schizophrenia, the disorder long being known as the graveyard of neuropathology, as the brains of its sufferers are easily recognizable because they looked so normal. However, regions now thought to be affected (prefrontal, parietal, and temporal) are those that are maximally evolved in *Homo sapiens* (Bradshaw & Rogers, 1993).

There has long been a debate as to whether schizophrenia is associated with a disorder of the LH (Flor-Henry, 1989; Waddington, 1993a) or the RH (Cutting, 1990, noted the presence in schizophrenia of such apparently RH phenomena as flattened or abnormal affect, delusional misidentification, and formal thought disorder). It may of course be the case that more withdrawn schizophrenics with predominantly negative symptoms have largely LH deficits, whereas florid individuals with predominantly positive symptomatology experience largely RH dysfunction. Although CT and MRI scans do perhaps indicate a relatively greater LH involvement (Straube & Oades, 1992), frontally with negative and temporally with positive symptoms, perhaps reflecting the auditory–verbal hallucinations so characteristic of schizophrenia (Marengo, 1993), metabolic (PET and blood flow) and electrophysiological (EEG) studies tend to be ambiguous. (Early, Haller, Posner, & Raichle, 1994, however, reported an abnormally high ratio of blood flow in the left globus pallidus (GP) of never-medicated patients, though findings with such techniques may generally be confounded by neuroleptic medication.) Studies involving handedness, visual field, and ear asymmetries are similarly equivocal. The corpus callosum is of course the main channel of interhemispheric communication (see Chapter 7), with fibers connecting homotopically situated cortical regions in the two hemispheres. Woodruff, Pearlson, Geer, Barta, & Chilcoat reviewed reported associations between callosal dysgenesis and schizophrenia, and found in their own study that the middle regions of the corpus callosum, interconnecting the temporal lobes, are reduced in schizophrenia. Although such callosal anomalies could be secondary to primary damage to the medial temporal lobe, they could also themselves then be directly responsible for secondary schizophrenic symptoms. In this context, Waddington (1993a) reviewed evidence for schizophrenic loss of asymmetry in the Sylvian fissure, a major speech-related landmark (see Chapter 2), indicative of a possible neurodevelopmental anomaly in the etiology of the disorder. Seeman (1993) noted that in schizophrenics the size, area, and amount of gray matter are reduced in the left temporal lobe, including the parahippocampal gyrus, with reduced blood flow; patients tend to make leftward turns (indicative of underactivity of the left striatum) and to neglect the right side of space, and where neuroleptics cause asymmetric hemiparkinsonism their right limbs are more affected. Moreover, their putamens have reduced D2 receptor density on the left, their left caudates have less DA than on the right, and there is reduced blood flow in the left GP.

CT and MRI scans show that about one-third of schizophrenics exhibit mild cerebral atrophy, with widening of the lateral ventricles or the cortical sulci or both, especially in frontal and temporal (amygdalar and hippocampal) regions (and see e.g., Beier et al., 1992, and Figure 1), and in patients exhibiting pre-

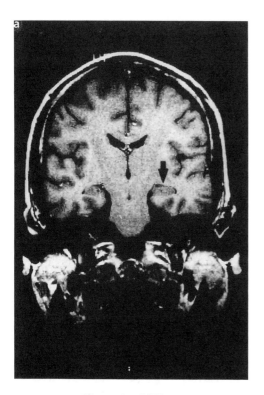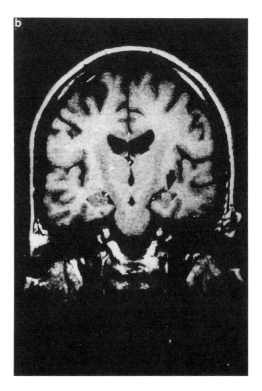

Figure 1   MRI scans of the left and right hippocampus in a normal control (a) and a patient with schizophrenia (b). The shape and size of the hippocampus are altered and the lateral ventricles enlarged in the schizophrenic patient. (From Carpenter & Buchanan, 1994. Represented by permission of the *New England Journal of Medicine*, 330, 685.)

dominantly negative symptoms or chronicity or both. Structural changes in the superior temporal gyrus may be related to positive symptoms (e.g., hallucinations and thought disorder). Again, the fact that the phenomenon appears to be nonprogressive and without gliosis ("scarring"), is indicative of an early neurodevelopmental anomaly rather than a later degenerative disorder. However, these atrophic changes may occur also in people suffering from (nonschizophrenic) affective disorder, HD, and AD. Moreover, they are diffuse and nonspecific, as well as being at most statistical in nature, rather than clearly diagnostic; indeed many normal individuals overlap into the schizophrenic distributions. Straube and Oades (1992) noted that schizophrenic individuals with atrophic changes may exhibit a poor drug response, poor neuropsychological test performance, involuntary movements, and reduced DA metabolism. They may also include cases with a history of birth complications and without a family history of the disorder, though one should not discount the possibility of problems in sampling and retrospective assessment.

Frontal atrophy has been reported from anterior dorsolateral (where "intentionality" may be mediated), orbitofrontal (with its involvement in social behaviors), and cingulate regions, together with the supplementary motor area (SMA). As we saw in Chapter 10, damage to the latter structure may affect our ability to operate from cues that are internally rather than externally derived (where the premotor cortex may be involved instead), a problem also apparent in PD. We have already noted the tendency for schizophrenics to mismanage and misinterpret internal cuing and their own and others' goal-directed intentionality and will to act. Similarly, amphetamine, in addition to provoking or exacerbating schizophreniform behaviors, induces repetitive stereotypic responses and inhibits reversal learning in animals; it causes the balance of response selection now to be largely stimulus (i.e., externally) rather than internally driven (Frith, 1992). Of course, the damage in schizophrenia, as we have already argued, is likely to extend well beyond the frontal lobes alone, and to involve the cortico-cortical reentrant loop via the striatum, pallidum, and thalamus (see Chapter 11), as well as other subcortical and cortical structures. Thus although patients with PD and damage elsewhere (e.g., in the BG) in the above circuit can "intend" but not "perform," schizophrenics can, to a gross approximation, "perform" but not "intend." The frontal damage in schizophrenia would similarly account for their lack or abnormality of affect and asociality, together with their abnormalities of smooth pursuit eye movements and saccadic control noted earlier; here the particular cortico-cortical circuit involving the BG and frontal eye fields presumably is involved. Finally, the usual frontal blood-flow changes occurring during, for example, working memory and Wisconsin Card Sorting Test tasks may be altered or absent in schizophrenics (Berman, Torrey, Daniel, & Weinberger, 1992; Randolf, Goldberg, & Weinberger, 1993). Such "hypofrontality," indicative of deficient "executive" functioning, may be especially apparent in those with evidence of hippocampal involvement (Waddington, 1993b). This temporal limbic structure interconnects with prefrontal regions in the context of working memory tasks (Goldman-Rakic, 1992), and damage here may account for the reports of disrupted working memory and fragmented and disorganized behavior in schizophrenia (see e.g., Fleming, Goldberg, & Gold, 1994). Damage to the hippocampus and associated parahippocampal gyrus (entorhinal cortex) located mesially in the temporal lobe will also account for psychotic, hallucinatory, and disturbed emotional behavior, and alterations in perceived meaningfulness. When we shortly consider possible genetic and environmental etiologies of schizophrenia, we shall note (and see Waddington, 1993a) that observed anomalies in the cytoarchitecture of the hippocampus, parahippocampal gyrus, cingulate, and prefrontal cortices are characteristic of disrupted cortical and other neuronal migration during early to midgestation. The absence of (inflammatory) gliosis indicates damage early rather than late in neurodevelopment. This hypothesis therefore redirects attention away from the idea of factors disturbing previously normal functions at or just before onset of clinical symptoms, and toward very early changes, for whatever reason, resulting in symptoms maybe two or more decades later. Randolf, Goldberg, and Weinberger (1993) in fact proposed that early disruption of dorsolateral prefrontal cortex (DLPFC) results in a premorbid attentional dysfunction, and abnormal development of corticolimbic circuitry. Subsequently, failure of the prefrontal cortex to respond appropriately to stress may result in mesolimbic *hyper*activity, with onset of clinical symptoms.

The first psychotic episode in schizophrenia may often be preceded by such early prodromal signs as social withdrawal or isolation, low self-esteem, impaired role functioning, low social adjustment, odd behavior and ideas, blunted affect and impaired personal hygiene, increased anxiety and feelings of vulnerability, and feelings of alienation and unreality. These and other characteristics such as passivity, a short attention span, and a history of disruptive behavior may even date back to childhood, and indicate a long-standing neuropathology.

Drug Treatment of Schizophrenia: The Dopamine Hypothesis

In the 1950s chlorpromazine, originally used as a major tranquilizer, was found to be efficacious for about 75% of cases, after several weeks of treatment, against the positive and to a lesser extent the negative signs and symptoms of schizophrenia. These effects were apparently unrelated to its efficacy as a tranquilizer. However, it and similar neuroleptics, which produce a blockade of D2 DA receptors, and reduce the activity of mesolimbic DA neurons, may only help about half the patients for any prolonged period. Indeed prolonged treatment may be associated with (reversible, on cessation of medication) parkinsonian akinesia, rigidity, and tremor, and irreversible tardive dyskinesia. The latter involves involuntary movements, especially of the mouth, tongue, and jaw. There is however historical evidence of tardive-dyskinetic movements occurring in chronic schizophrenics even before the advent of such medication, suggesting that DAergic changes in the motor system may in time develop endogenously to the disorder. Paradoxically, carefully titrated amounts of neuroleptics may even help *control* tardive dyskinesia.

The DA hypothesis in the etiology of schizophrenia does not claim that the disorder stems from excess DA per se in the brain; this is certainly not found at autopsy. Rather, supersensitivity of DA receptors are thought responsible, though again one must distinguish between the effects of the disease and of the treatment. The hypothesis is based upon the observation that typical antipsychotic neuroleptics can all block DA receptors in direct proportion to their clinical efficacy in controlling the disorder. Similarly, amphetamine can exacerbate the disorder in patients, and cause psychotic behaviors in previously clinically normal individuals, by promoting the release of DA. Problems with the hypothesis include the observation that although the DA-blocking effect with neuroleptics is rapid, it takes several weeks for symptoms to abate. This phenomenon however is probably due to initial compensatory increases in DA release that take time to overcome. Seeman, Guan, and van Tol (1993) reported that D2 and D3 receptors are only mildly increased in schizophrenia, whereas D4 receptors are greatly increased, and that this is not due to chronic medication. It should therefore be noted that several DAergic pathways are known and are likely to be involved to greater or less extents in the disorder. The *nigrostriatal* pathway of course is the one predominantly involved in the motor and cognitive symptomatology of PD (see Chapter 12), originating in the brain stem substantia nigra (SN), and terminating in the striatum. The ventral tegmentum projects to the nucleus accumbens and amygdala, hippocampus, cingulate, lateral septal nuclei, and entorhinal cortex as part of the *mesolimbic* pathway (see Chapter 11), which is involved in reward, emotions, and memory; the ventral tegmentum also projects to the prefrontal cortex as part of the *mesocortical* tract. Excess of DA receptors in the mesolimbic pathway and deficiency in the mesocortical tract could explain the positive and negative symptoms respectively, and the differential—and delayed—effects of neuroleptics if they act to alter the balance of activity between the different systems. Alternatively, D3 receptors, concentrated in the limbic system, and binding to DA, may tend to suppress behavior rather than stimulate it. Of course, neurotransmitters other than DA are also likely to be involved in schizophrenia, for example, noradrenaline and other biogenic amines such as serotonin (5-HT) according to Larvelle et al. (1993), together with glutamate (excitatory) and gamma-aminobutyric acid (GABA) (inhibitory), as part of a long chain that is faulty in schizophrenia (see e.g., Carpenter & Buchanan, 1994). Thus glutamatergic and DAergic systems are reciprocally interrelated, glutamate regulating DA release; the NMDA (N-methyl-D-aspartate) subtype of glutamate receptor in the hippocampus, which is involved in long-term potentiation and memory functions (see Chapter 8) may be especially vulnerable in schizophrenia.

Schizophrenia as an Inherited Disorder

The etiology of schizophrenia may include some or all of a genetic component, effects of an early gestational viral infection in the mother, perinatal injury, and psychosocial factors. If the latter have been largely discounted as a primary cause, they could still increase the manifestations of the disorder.

Evidence for heritability comes from the strong familial clustering that has long been noted, even with cross-adoption. According to Fowles (1992), if both parents or a monozygotic (MZ) twin have the disorder, the risk is 46%; it is 11% for a first-degree relative or dizygotic (DZ) twin, and 5% for a second-degree relative, over a baseline rate of just under 1%. However, the healthy twin in a discordant MZ pair conveys as much genetic vulnerability to offspring as does the afflicted twin (Gottesmann & Bertelsen, 1989), indicating that the genetic tendency to the disorder may be unexpressed but still passed on. A dominant gene, according to classical Mendelian principles, should give 100% concordance in MZ and 50% concordance in DZ twins. The above observations, and the fact that adopted-out children of schizophrenic parents do not show as high a risk as children staying with their schizophrenic parents, indicate the additional operation of environmental factors (Straube & Oades, 1992); a certain genetic makeup perhaps predisposes one to the disorder, or permits its manifestation when a threshold of environmental influences has been surpassed. Conversely, a considerable number of relatives of schizophrenics show milder schizotypal disorders or borderline personalities, or may even score highly on creativity, and occupy positions of high social status, neither of which apply to those with the disorder itself; such observations indicate the probable operation of a polygenic contribution, with a large number of genes contributing additively, rather than a single gene locus. Indeed the role of environmental factors could well vary inversely with the magnitude of genetic liability (Fowles, 1992). In any case, the heritability of psychotic symptoms may differ from that of negative symptoms (Carpenter & Buchanan, 1994). Waddington (1993a) reviewed recent findings obtained with new molecular genetic techniques. Perhaps not surprisingly, claimed linkages to DNA markers on the long arm of chromosomes 5 and 11 have not been replicated (Wang, Black, Andreasen, & Crowe, 1993). In a very recent study, Arinami et al. (1994) reported that a variant of the DA D2 receptor gene was significantly higher among a group of Japanese schizophrenics than controls, though patients with it showed less severe thought disorder and negative symptoms than those without it. They concluded that this Cys 311 variant of the D2 receptor may be a genetic risk factor for some types of schizophrenia. However, as M. Owen (1992) observed, schizophrenia, as well as constituting a graveyard for neuropathologists, may also prove to be the nemesis of many molecular geneticists.

Environmental Factors

A consensus seems to be emerging (see e.g., Straube & Oades, 1992; Waddington, 1993a) that a genetic susceptibility together with environmental influences underlie schizophrenia. Thus season of birth (late winter to early spring), with associated increased risk of viral (influenza?) infection in the mother during midpregnancy, may account for 5 to 10% of the overall risk. This effect may be relatively larger for those born in large towns, again suggesting an infective origin. Waddington (1993a) drew a parallel with multiple sclerosis where again there are temporal and spatial variations in incidence, the likelihood of an early viral infection and subsequent (auto)immune dysfunction; in the case of schizophrenia such phenomena (a retrovirus altering the genome, an active or inactive viral infection, viral-activated immunopathology, autoimmune dysfunction, and a secondary influence of maternal viral infection) may result in neurodevelopmental disturbances the effects of which may take decades to manifest. R. W. Doty and Davis (1991, and see also R. W. Doty, 1989) in turn proposed that the disorder is caused by perturbed interhemispheric processing due to virally induced damage to the brain stem serotonergic raphe nuclei, thus incorporating earlier claims of abnormal lateralization and callosal anomalies in schizophrenia. Serotonergic and DAergic neurotransmission systems are thought to interact in a modulatory capacity, and hallucinogens are known to act upon the serotonergic system. R. W. Doty and Davis claimed that the raphe system is involved in interhemispheric balancing and integration via the habenular commissural system; the latter receives emotional and limbic input, and is itself asymmetric in many submammalian species (Bradshaw & Rogers, 1993). Moreover, viruses are thought to target the serotonergic system. Indeed, Kerwin (1993) suggested that intrauterine infection or other trauma may disrupt the glutamate-dependent microtubule assembly in developing neurons, due to excess glutamate release. Such cytoskeletal abnormalities would result in developmental anomalies in the medial temporal lobe especially (e.g., the hippocampus, amygdala, and parahippocampal gyrus). Similar excitotoxicity has of course also been proposed (e.g., for Huntington's disease, see Chapter 13).

In conclusion, there is now little evidence for the old idea that schizophrenia is due to faulty parenting or socialization. The latter may exacerbate but cannot cause the disorder. Indeed the incidence of schizophrenia is no greater in war-torn or famine-stricken regions. A disturbed family background may merely reflect the genetic contribution in the family, rather than being directly causal itself. However, an inherited predisposition may be highlighted or potentiated by stress. Obstetric complications may play a part—especially perhaps in males— being associated with younger age of onset; however as Waddington (1993a) observed, the heterogeneity of the complications makes it unlikely that any one particular complication may be the cause of the disorder, perhaps by inducing hypoxia. An early viral insult may induce neurodevelopmental anomalies at a critical period, again perhaps in those genetically or otherwise environmentally susceptible. Schizophrenia, traditionally in the psychiatric domain, should, despite the current criteria ("absence of a clear organic origin") be seen as an organic disorder at the interface between thought and the emotions, and involving neurotransmitters and pathways shared with such unambiguously "neurological" disorders as PD and HD.

Case History

Frith (1992, p. 2) provided the following example (Case 1.2, studied by Fiona MacMillan and Eve Johnstone, and reported in Fiona MacMillan's MD thesis) of a case history of schizophrenia:

SW was a 24-year-old mathematics teacher who was admitted to hospital after a four-week history of increasingly odd behavior. He had left his flat and returned home, only to later leave and return home again. SW took up sports and pursued them excessively, became uncharacteristically irritable and aggressive, unable to tolerate any music being played. He became preoccupied with incomprehensible life difficulties and expressed odd ideas, saying he should go to the police as his rent book was falsified and that he was teaching the wrong syllabus at school.

A few days after admission, SW became floridly psychotic, with grossly disordered speech. His affect fluctuated from tears to elation over minutes. His general manner was distant, absorbed and perplexed. He expressed the beliefs that television and radio referred to him and certain records on the radio were chosen deliberately to remind him of his past life. He was convinced his food was poisoned and felt his head and genitals were compressed as a result of an aeroplane flying overhead. SW described thought insertion and thought echo and heard hallucinatory sounds of a klaxon, and occasionally single words.

There was a slow reduction of these positive psychotic features over a four-month period of treatment with antipsychotic drugs.

Following discharge SW had severe disabling negative symptoms: a lack of spontaneity and volition, and poverty of speech. After a year, he returned to work. He remained free of psychotic features, but his family was well aware of his persisting defect.

Summary and Conclusions

Schizophrenia affects around 1% of the population at some time in their lives, striking in young adulthood, on average slightly earlier in males. However, certain personality characteristics may date back even to childhood. It manifests in so many different ways (e.g., catatonic, hebephrenic, paranoid, and undifferentiated) that some believe that more than one disorder is involved, though the symptoms specifically concern the higher human aspects of personality. As with TS, it is now generally recognized as being organic rather than psychosocial, despite the basic loss of contact with reality, which seems to underlie the disorder. However, social factors and stress may still exacerbate or unmask the condition.

The disorder is still diagnosed essentially from the accompanying signs and symptoms, as no clear, unambiguous, uniform, or characteristic brain abnormality has yet been identified; moreover, its boundaries from other affective disorders have not been clearly established. The characteristic psychotic symptoms must include delusions (of persecution, grandeur, or thought broadcasting, for example), hallucinations (typically of voice, involving inappropriate comment), incoherence, catatonic immobility, and flat or grossly inappropriate affect. A defective mesolimbic system is implicated somewhere in the etiology. In the early stages, there may be intense emotional experience, later perhaps giving way to flatness of affect. Acute, sudden onset is associated with better prognosis than where the onset is gradual, or auditory hallucinations manifest early. Full remission may occur in one-quarter of patients, though the course of the illness is not always uniformly progressive. Positive symptoms (hallucinations, delusions, thought disorder, incoherence, incongruity of affect, and stereotyped, repetitive movements) tend to be associated with the earlier, acute phase, and to respond better to DA antagonists. Negative symptoms (e.g., lack of emotional expression, flattened or blunted affect, apathy, and inertia) tend to be later occurring, less amenable to medication, and associated with imageable brain changes.

Although there may be some cognitive impairment, there is not necessarily any reduction in IQ or logical abilities, except perhaps for a tendency to be distracted or to be preoccupied with task-irrelevant information. These latter aspects may relate to the oculomotor irregularities and tangentiality of speech that is also observed. Although working, procedural, and semantic memories are largely unaffected, there may be a general disorganization and distractibility extending even to "unaffected" close relatives. A deficit in selective attention may therefore pervade many aspects of schizophrenic behavior, with important aspects ignored, and unimportant details seized upon.

We may possibly interpret the above attentional deficits in terms of an unusually high degree of dependence upon external feedback, with a corresponding inability, by schizophrenics, to depend upon their own internally generated feedback, or even to recognize their own thought processes *as* their own. Schizophrenia may therefore involve a failure in self-monitoring and the feedforward system that comes into play whenever a thought, sequence of inner speech, or voluntary act is initiated. Thoughts and inner speech are therefore mistakenly misattributed to others or perceived as alien hallucinations.

Brain scans reveal mild cerebral atrophy, particularly in the prefrontal and temporal regions of schizophrenic brains, though limbic structures are likely also to be involved, as individuals with corresponding cortical damage alone rarely exhibit full-blown schizophreniform symptoms. However, frontal and temporal anomalies can certainly explain at least some of the symptoms of schizophrenia. The absence of gliosis suggests that the phenomenon is nonprogressive, and likely to reflect an early neurodevelopmental anomaly. It is also diffuse and nonspecific, and statistical in nature rather than clearly diagnostic. However, schizophrenic individuals with such atrophic changes tend to have a poorer prognosis and response to medication, and to have a positive history of birth complications and a negative family history for the disorder.

Neuroleptics, which block the action of DA, are efficacious against the positive symptomatology manifested by at least a subset of patients. However, a possible supersensitivity of DA receptors may underlie the disorder rather than an excess per se of DA in the schizophrenic brain. Alternatively, excess DA receptors in the mesolimbic pathway, and a deficiency in the mesocortical tract, could explain the positive and negative symptoms respectively, and the differential effects of neuroleptics. However, neurotransmitters other than DA are likely also to be involved.

The etiology of schizophrenia may include, variously, a genetic component, effects of an early gestational viral infection in the mother, perinatal injury, and psychosocial factors perhaps as a modulating influence. Although a particular genetic makeup certainly predisposes one to the disorder, the contribution is probably polygenic, with a number of genes contributing additively, rather than involving a single gene locus; environmental influences must also operate. A viral infection in the mother during midpregnancy is currently a favored candidate.

Further Reading

David, A., & Cutting, J. (1993). *The neuropsychology of schizophrenia*, Hillsdale, NJ: Lawrence Erlbaum.

Fowles, D. C. (1992). Schizophrenia: Diathesis-stress revisited. *Annual Review of Psychology, 43*, 303–336.

Frith, C. D. (1992). *The cognitive neuropsychology of schizophrenia.* Hillsdale, NJ: Lawrence Erlbaum.

Kavanagh, D. (Ed.) (1992). *Schizophrenia: An overview*. London: Chapman & Hall.

Randolph, C., Goldberg, T. E., & Weinberger, D. R. (1993). The neuropsychology of schizophrenia. In K. M. Heilman & E. Valenstein (Eds.), *Clinical neuropsychology* (3rd ed.) (pp. 499–522). Oxford: Oxford University Press.

Straube, E. R., & Oades, R. D. (1992). *Schizophrenia: Empirical research and findings.* San Diego: Academic Press.

Waddington, J. L. (1993). Schizophrenia: Developmental neuroscience and pathobiology. *The Lancet, 341*, 531–536.

Final Thoughts

The Present and Future Status of Experimental and Clinical Neuropsychology

*I*n Europe, well over one hundred years ago, the tradition was firmly established of correlating acquired deficits in behavior, subsequent to localized brain injury, with the hypothesized function of the damaged underlying structures. The brain was therefore seen as consisting of a series of interconnected centers each playing a special role in various aspects of behavior. Such localization of function, consequent upon lesion mapping, has retained a strong following to the present day, though the emphasis nowadays is more upon the patterns of interconnections in a distributed system than upon relatively isolated centers, and upon how such systems may process information. Nevertheless, there has always been a strongly reactive or countervailing viewpoint. As seen in Chapter 1, this latter viewpoint rejects the approach of the "diagram makers"; it either views the brain as a much more unitary organ, or it eschews any attempt to determine the biological underpinnings, in favor of an emphasis upon the abstract information-processing architecture of the hypothetical system, as constrained by the observable input–output functions. All models of brain function have of course tended to be couched in terms of the contemporary and prevailing technological culture, be it hydraulic or concerned with railroads, telephone networks, or computers. Although digital computers have offered us many useful and illustrative images of possible neural functioning in terms of hardware, the software revolution has greatly advanced the cause of cognitive science and the new hybrid discipline of cognitive neuroscience. Readers of this book will have noted that we have chosen generally to follow a rather more traditional or biological approach of trying to relate lesion (where identifiable) and syndrome (where unambiguous) to brain function. This is not because we reject alternative approaches—indeed we would stress that the brain must be seen as a series of interconnected systems, just as are found with respect to bodily functions. However, we feel that there is still a need to document systematically the multitude of recent clinical findings in a biological context, and to attempt thereby to derive and refine models of how the human brain may operate in health as well as in sickness.

If we are to reject an almost phrenological representation of faculties (speech, memory, spatial cognition, object recognition, etc.) by discrete brain centers, in favor of the operation of a distributed processing system, clearly the exact nature of the deficit is likely to vary depending on where in that system damage occurs. Thus sequencing and bimanual integration may both be affected by damage to the basal ganglia (BG), or to the supplementary motor area (SMA), which receives input from the BG, but there are likely to be subtle differences depending on which component is affected. We may not yet know exactly what those differences are, but they are likely to reflect the particular contributions of the two structures. Some processes however, such as calculation and musical activity, are likely to be far less unitary than others, and to consist of many parts or aspects, differentially mediated. Indeed, the practice of music clearly involves more than the mere sum of its individual parts. Only some of the components need to be disrupted for the patient to be classified as acalculic or amusic, though again the nature of the resultant amusia or acalculia will reflect the kind of damage experienced.

Some disorders lend themselves more than others to a traditional biological approach. Indeed the aphasias have a longer history of systematic empirical observation than almost any other deficit. Other disorders (e.g., schizophrenia) were not identified until much later, and indeed it may be no accident that schizophrenia until recently was addressed in largely psychodynamic terms, given that its first systematic description coincided with the apogee of the psychodynamic tradition. It is interesting, and to us reassuring, that psychiatry (in the form of neuropsychiatry) and neurology, after decades of divorce (as we saw in Chapter 1), are beginning once again to get together. Schizophrenia, autism, and Tourette's syndrome (TS) are amongst their bridesmaids for their second attempt at union. Such a union of course is in large measures a consequence of the success that drug treatment has had in ameliorating some of the more unpleasant manifestations of such disorders; modern neurogenetics (see below) has been another such facilitator. The aphasias and agnosias of course are largely resistant to such pharmacological intervention.

Some of the disorders with which we have dealt involve problems with apparently "hard-wired" aspects of behavior like memory, praxis, movement, and perhaps object recognition. Language, as we saw in Chapter 1, is subject to overriding genetic and mutational control, but nevertheless involves learning the local tongue at certain critical developmental stages. Reading, as we noted in Chapter 3, is a much more recent social phenomenon, and must probably be acquired in a much more formal manner than speech, though apocryphal accounts do describe instances of young children apparently learning to read from watching children's TV programs. Nevertheless, the alexias seem amenable to taxonomic description as readily as the aphasias, and indeed accounts of alexia are grossly parasitic upon treatments of the aphasias, even though certain principles such as direct or mediated access clearly are not common to both sets of disorder. When we turn to the amusias and the acalculias, although we see attempts at borrowing certain principles from the aphasias and the alexias, particularly with respect to input–output variables, the analogies may seem more forced.

Although much effort has gone into studying the aphasias and the alexias, because of the impact they have upon daily life, the same may not be said of the amusias and the acalculias. Indeed many individuals may never seek help for the loss of a musical ability that they never even knew they had. Amusics who *do* seek assistance are likely to come from a self-selected, professional group, an observation that highlights one of the problems in the selection and availability of patients for clinical study, that of differences in premorbid ability and/or training. In fact, inconsistent findings, between studies, may largely reflect inconsistencies in patient selection, methodology, dependent and independent variables, locus, extent, nature, and duration of lesion, whether it occurred in youth or adulthood, whether it was slow or rapidly developing, and so on. Such considerations of course provide ammunition for those who would reject group studies in favor of the single-case design.

At this point we may note that terms such as *aphasia, agnosia, apraxia, amnesia*, employing the privative prefix *a-*, generally are more common in the neurological literature than the prefix *dys-*, as in dysphasia, dyslexia, and so on. Strictly, the latter form should in almost all cases be preferred, as usually there is some preservation of function, except perhaps with global aphasia. However, we have followed convention by generally employing the *a-* form.

In various places in this book we have argued that although a motor response that is not constrained by appropriate sensory or perceptual parameters is likely to fail, perception itself should be viewed at least partially in terms of implicit response processes. Both processes may therefore be viewed as lying at the extremes of a perceptuomotor continuum that is subject to attentional and memorial modulation—attentional so as to limit higher levels of processing to what is currently "relevant," and memorial so that the organism's fund of learned experience may be usefully accessed. Memory therefore seems to be involved in almost everything we do (hence the devastating effects of amnesia and dementia), and agnosia, aphasia, and apraxia are also in a real sense true amnesic syndromes. Nevertheless, attention plays more of a role in the disorders of spatial cognition, of thought and of movement [i.e., Parkinson's (PD), and Huntington's disease (HD), and TS], than in the apraxias, aphasias, alexias, and agnosias. Indeed, in these terms apraxia and agnosia may be conceptualized as lying at opposite ends of the perceptuomotor continuum, both however with little attentional modulation; the aphasias and the alexias are also distributed along the perceptuomotor continuum with little attentional modulation, the movement disorders (HD and PD) are motor-attentional, whereas schizophrenia, autism, TS, and unilateral neglect partake of all three aspects, perceptual, motor, and attentional.

If we can *functionally* distinguish between the disorders in the above terms, can we also make a similar *structural* (e.g., cortical–subcortical) distinctions? As we saw in the contexts of PD, HD, and Alzheimer's disease (AD), dementia has been dichotomized into cortical and subcortical. However as we also saw, such a distinction, while heuristically useful, has proved difficult to maintain both structurally and functionally. Although many of the computations necessary to mediate word or object recognition, word or gesture production, and so on are clearly dependent upon intact neocortical functioning, paleocortical or limbic mechanisms underlie the emotional aspects of behavior. That is not to say, however, that the latter mechanisms alone are at fault in disorders of thought such as schizophrenia, though they may indeed be involved in delusional behavior and disorders of misidentification. As we saw, prosopagnosia may also involve a dissociation between the two systems or mechanisms, cortical and limbic.

Such considerations highlight another emerging distinction, that between explicit and tacit (or implicit) knowledge. Indeed, tacit or implicit knowledge seems all pervasive, both in everyday life (raising questions about the reality, utility, and mechanisms of consciousness, see below) and in clinical pathology. Agnosics, alexics, amnesics, apraxics, and those suffering from unilateral neglect all show remarkable preservation of abilities at an unconscious, automatic level, while unable to perform deliberately and consciously. It is probably not enough just to invoke multiple representation of a function, onion-skin fashion, with loss at the surface, conscious (cortical?) level and preservation at deeper and more automatic levels of processing, though it is far from clear exactly how we *are* to address this problem. One possible lead comes from recent thoughts about the organization of memory and the declarative–procedural distinction, though again such a distinction is clearer in terms of functional architecture than in terms of structural anatomy. That said, cerebellar–striatal mechanisms, as we saw, are clearly more important for automatic, unconscious procedural memory (skills, conditioning) than for conscious, declarative memory where temporal–hippocampal mechanisms may be preeminent.

Inevitably, such considerations return us again to the issue of localization of function. The common co-occurrence of certain syndromes such as disorders of color, face recognition, and route finding suggest common or adjacent mechanisms. The existence of localizing signs provides fundamental support to such concepts, and disconnection syndromes (within the aphasias, and of course in the acallosal syndrome) also imply some degree of localization of function. However, the intriguing phenomenon of category-specific deficits in the aphasias and agnosias *need* not, as we saw, even reflect adjacency of retrieval mechanisms, let alone of storage or of mediation. Indeed, although category-specific deficits are definitely compatible with localization of function, other explanations almost certainly are to be preferred.

New functional imaging techniques (see Chapter 1) are set to provide us with information hitherto undreamt of by clinical neuropsychologists. The *normal* brain can now be "viewed" as it undertakes "normal" (or at least "ecologically valid") "everyday" tasks, in speech, reading, memory, object (or face) classification or recognition, music or even chess playing (Nichelli et al., 1994), to name but a few recent studies. Indeed, the days of clinical neurology and neuropsychology providing a window upon brain functions (as opposed to their purely medical applications) may now be numbered, as neuroimaging throws a clear light upon the *network* functions of the normal, workaday brain. If so, we shall also have neatly sidestepped that other current debate, the relative utility, and validity, of single-case versus group designs. Until that day comes, however, we are firmly convinced of the worth of group designs, wherever possible, while recognizing that odd, rare, and individual cases can still throw much-needed light where sufficient numbers of appropriately (if not approximately) matched individuals are unavailable.

Some disorders are clearly adventitious when, for example, due to stroke, injury, or tumor. Others, as we saw, may have a genetic basis. Mann (1994, and see also Plomin, Owen, & McGuffin, 1994) noted that exactly 70 years previously the German dermatologist Hermann Siemens compared the school performances of identical and fraternal twins; over three-quarters of the (genetically) identical twins received similar grades, compared to only one fifth of the fraternal twins. So was born the technique for estimating the possible extent of a genetic contribution to human behavior. Tighter control of environmental contributions came with cross-fostering studies, and attempts were made to resolve the old nature–nurture debate. For many years such studies were discredited because of their perceived implications for eugenics, and only now is behavioral genetics undergoing rehabilitation. As Mann (1994) observed, molecular geneticists have entered the field, and journals are nowadays filled with claims that heredity plays a role in everything from gregariousness and general cognitive ability to alcoholism, violence, and manic depression. As we have seen, a genetic basis for HD is inescapable, and it almost certainly underlies autism, TS, and AD. Rosenberg (1993) reviewed recent advances in neurogenetics and the progress made by linkage analysis whereby the chromosomal location of a neurological disease is determined, the gene responsible for the disorder is isolated, cloned, and sequenced, and the gene product or protein identified. Such procedures clearly indicate possible avenues of therapeutic intervention. Recently, a genetic underpinning of male homosexuality has also been proposed (Hamer, Hu, Magnuson, Hu, & Pattatucci, 1993), with respect to a small stretch of DNA on the X chromosome. Whether it is a matter of an inherited propensity for violence (Brunner, Nelen, Breakefield, Ropers, & van Oost, 1993) or homosexuality (Byne & Parsons, 1993), or whether it is "simply" a matter of an anatomical difference in the hypothalamus of homosexual and heterosexual men (LeVay, 1991), two kinds of response to such "knowledge" can occur. Either the problems of such people can be said to be their own fault, as they carry "bad" genes, with consequent erosion of human dignity, or the individuals can feel exonerated, and no longer responsible for what is after all their inescapable biology. In spite of the moral, ethical, forensic, and political contentiousness in this field, there is a growing, if grudging, consensus that heredity plays *some* role, maybe only modulatory upon the environmental influences, in many if not most aspects of human behavior ranging from intelligence and sexual preferences to alcoholism, gambling, drug abuse, violence, and artistry. Indeed Plomin et al. (1994) noted that research with diverse twin and adoption designs has found genetic influence on parenting, childhood accidents, TV viewing, classroom environments, peer groups, life events, divorce, drug use, education, and socioeconomic status. Thus, ostensibly, environmental measures seem to assess genetically influenced personality characteristics, and individuals to some extent may create their own experiences for genetic reasons.

We may feel more comfortable with and accepting of the idea of a genetic determinism for alcoholism than for intelligence. Critics, however, of the whole concept of genetic determinism will still argue that statistical treatment is flawed, the selection of cases and controls is biased, and sample sizes are inadequate (Mann, 1994). Nor is it always easy to define the behavioral traits in question. Moreover, apart from some simple exceptions like HD, most behaviors, normal or pathological, are likely to involve multiple genes that interact in complex, nonlinear ways. Simplistic and punitive restraint of individuals with "bad genes" is not a likely scenario, though the possibilities for positive or supportive intervention may become limitless.

Polygenic influences upon a behavior pattern are likely to result in a continuous distribution of the latter, with pathology representing the quantitative extremes of the continuum. The future challenge will be to use molecular genetic techniques to identify the genes and environmental influences in such complex systems.

Although it is clearly a mistake to view evolution as progressing inexorably toward ever more advanced organisms or organizations (Gould, 1991), the human brain has only the great apes as competitors for brain size and complexity (Bradshaw & Rogers, 1993). Nevertheless in the last 10 years possible *similarities* have been sought between ourselves and other species (e.g., with respect to language, tool use, consciousness, and self-awareness), rather than a continued emphasis upon what makes us different or unique. We have earlier dealt with such issues in the context of the evolution of relevant brain structures and functions (Bradshaw & Rogers, 1993). Suffice it to say that philosophical, neuropsychological, and neurological disputes still rage about the definitions and nature of these processes, especially consciousness (Brown, 1991; Dennett, 1991; Humphrey, 1992), which may of course itself encompass a more narrowly defined *self*-awareness. Our own view is that just as language can be realized via alternative channels at both a peripheral and a central level (see Chapter 2 and our discussion on deaf signing), it is sensible to talk about a potential capacity for some form of language in chimpanzees (Savage-Rumbaugh, Sevcik, Brakke, Rumbaugh, & Greenfield, 1990), even though the peripheral and especially the central mechanisms are, in human terms, decidedly nonstandard (Deacon, 1989). Likewise, praxis (in the context of tool use) and self-consciousness (at least as indexed by self-recognition in a mirror) are clearly present in the chimpanzee (Bradshaw & Rogers, 1993). Self-consciousness ranks with the origins of matter and life as one of the three great remaining mysteries. Is it an emergent property of a sufficiently complex information-processing system, or has it been selected for, in evolution, perhaps as an essential for participation within a sufficiently complex society? Would the last person on Earth, if she or he had never encountered a fellow, be *self*-conscious, or merely conscious? Is consciousness the act of monitoring a line of action that has been selected out of a number of competing alternatives? If we cannot be conscious of and successfully monitor more than one course of action *simultaneously* (as opposed to via rapid alternation, and see Chapter 7 and our discussion of independent processing, wills, and streams of consciousness in the commissurotomized patient), then consciousness and selective attention may be fundamentally interdependent.

We opened this final chapter by addressing an issue with which this book has in one way or another been largely concerned: functional localization. We would close by returning to it and making some historical observations. Aristotle (384–322 B.C.) argued that nerves originated in the heart, sensations were appreciated there, and it was the center for thought—and indeed the occupant of the most honorable bodily location, the very center. Although a cardiocentric view dominated Arabic and medieval European thought, Hippocrates (born about 460 B.C.) and Galen (129–199 A.D.) adopted the alternative cephalocentric position:

> Tell me where is fancie bred
> or in the heart or in the head?
> (Shakespeare,
> *Merchant of Venice*,
> Act III, ii.)

Thus in the Hippocratic writings we find:

> Men ought to know that from the brain alone arise our pleasures, joy, laughter and jests as well as our sorrows, pains, griefs, and tears. . . . Eyes, ears, tongue, hands and feet act in accordance with the discernment of the brain . . . it is the brain that is the interpreter of consciousness. Some people say the heart is the organ with which we think, and that it feels pain and anxiety, but it is not so. (Hippocrates, *The Sacred Disease*, XVII.)

Although empirical investigations and dissections of cadavers reached a peak in the Italian Renaissance with Leonardo da Vinci, it was not, as MacMillan (in press) noted, until the early decades of the 19th century that Home proposed the clinical principles of cerebral localization of function in modern (i.e., non-phrenological) form:

> The various attempts which have been made to procure accurate information re-specting the functions that belong to the individual portions of the human brain, having been attended with little success, it occurred to me that were all anatomical surgeons to collect in one view all appearances they had met with, in cases of injury to that organ, and the effects that such injuries produced upon its functions, a body of evidence might be formed, that would materially advance this highly important investigation. (Home, 1814)

However in this context, which forms the theme to this book as a whole, we would invoke the famous dictum attributed to Hippocrates:

ὁ βίος βραχύς, ἡ δὲ τέχνη μακρή, ὁ δὲ καιρὸς ὀξύς, ἡ δὲ πεῖρα σφαλερή, ἡ δὲ κρίσις χαλεπή

Thus, indeed, life is short though the art is long; the opportunity is fleeting, the experiment perilous and the judgment difficult.

References

Aaron, P. G., & Joshi, R. M. (1989). *Reading and writing disorders in different orthographic systems.* Dordrecht: Kluwer Academic.

Adams, R. D., & Salam-Adams, M. (1992). Disorders of movement associated with diseases of the cerebellum, particularly the degenerations. In A. B. Joseph & R. R. Young (Eds.), *Movement disorders in neurology and neuropsychiatry* (pp. 310–318). Oxford: Blackwell Scientific.

Agostino, R., Berardelli, A., Formica, A., Accornero, N., & Manfredi, M. (1992). Sequential arm movements in patients with Parkinson's disease, Huntington's disease and dystonia. *Brain, 115,* 1481–1495.

Akelaitis, A. J. (1943). Studies on the corpus callosum. VII. Study of language function (tactile and visual lexia and graphia) unilaterally following section of the corpus callosum. *Journal of Neuropathology and Experimental Neurology, 2,* 226–262.

Albert, M. L. (1973). A simple test of visual neglect. *Neurology, 23,* 658–664.

Allport, A. (1989). Visual attention. In M. I. Posner (Ed.), *Foundations of cognitive science* (pp. 631–682). Cambridge, MA.: Bradford (MIT Press).

American Psychiatric Association. (1987). *Diagnostic and statistical manual of mental disorders.* (3rd ed.) Washington, DC: Author.

Andersen, R. A., Snyder, L. H., Li, C-S., & Stricanne, B. (1993). Coordinate transformations in the representation of spatial information. *Current Opinion in Neurobiology, 3,* 171–176.

Anderson, D. N. (1988). The delusion of inanimate doubles: Implications for understanding the Capgras phenomenon. *British Journal of Psychiatry, 153,* 694–699.

Anton, G. (1893). Beitrage zur klinischen Beurteilung zur Localisation der Muskelsinnstorungen im Grosshirn. *Zeitschrift Heilk, 14,* 313–348.

Appollonio, I., Grafman, J., Clark, K., Nichelli, P., Zeffiro, T., & Hallett, M. (1994). Implicit and explicit memory in patients with Parkinson's disease with and without dementia. *Archives of Neurology, 51,* 359–367.

Ardila, A. (1993). On the origins of calculation abilities. *Behavioural Neurology, 6,* 89–97.

Arinami, T., Itokawa, M., Enguchi, H., Tagaya, H., Yano, S., Shimizu, H., Hamaguchi, H., & Toru, M. (1994). Association of dopamine D2 receptor molecular variant with schizophrenia. *The Lancet, 343,* 703–704.

Armatas, C., Summers, J. J., & Bradshaw, J. L. (1994). Mirror movements in normal adult subjects. *Journal of Clinical and Experimental Neuropsychology, 16,* 405–413.

Aylward, E. H., Brandt, J., Codori, A. M., Mangus, R. S., Barta, P. E., & Harris, G. J. (1994). Reduced basal ganglia volume associated with the gene for Huntington's disease in asymptomatic at-risk persons. *Neurology, 44,* 823–828.

Baddeley, A. D. (1986). *Working memory*. Oxford: Clarendon Press.

Baddeley, A. D. (1992). Working memory. *Science, 255*, 556–559.

Baily, D., Servant, D., Dewailly, D., Beuscart, R., Racadut, A., Fossati, P., & Parquet, P.-J. (1994). Corticotropin releasing factor stimulation test in obsessive compulsive disorder. *Biological Psychiatry, 35*, 143–146.

Bancaud, J., Brunet-Bourgin, F., Chauvel, P., & Halgren, E. (1994). Anatomical origin of déjà vu and vivid 'memories' in human temporal lobe epilepsy. *Brain, 117*, 71–90.

Barbut, D., & Gazzaniga, M. S. (1987). Disturbances in conceptual space involving language and speech. *Brain, 110*, 1487–1496.

Baron, J. A. (1986). Cigarette smoking and Parkinson's disease. *Neurology, 36*, 1490–1496.

Baron-Cohen, S., Harrison, J., Goldstein, L. H., & Wyke, M. (1993). Coloured speech perception: Is synaesthesia what happens when modularity breaks down. *Perception, 22*, 419–426.

Barondes, S. H. (1994). Thinking about prozac. *Science, 263*, 1102–1103.

Bartlett, C. L. & Pashek, G. V. (1994). Taxonomic theory and practical implications in aphasia classification. *Aphasiology, 8*, 103–126.

Basso, A. (1993). Amusia. In F. Boller & J. Grafman (Eds.), *Handbook of neuropsychology* (Vol. 8, pp. 391–410). Amsterdam: Elsevier.

Battersby, W. S., Bender, M. B., & Pollack, M. (1956). Unilateral spatial agnosia (inattention) in patients with cerebral lesions. *Brain, 79*, 68–93.

Bauer, R. M. (1993). Agnosia. In K. M. Heilman & E. Valenstein (Eds.), *Clinical neuropsychology* (3rd ed.) (pp. 215–278). Oxford: Oxford University Press.

Bauer, R. M., Tobias, B., & Valenstein, E. (1993). Amnesic disorders. In K. M. Heilman & E. Valenstein (Eds.), *Clinical neuropsychology* (3rd ed.) (pp. 523–602). Oxford: Oxford University Press.

Bauman, M. L. (1992). Neuropathology of autism. In A. B. Joseph & R. R. Young (Eds.), *Movement disorders in neurology and neuropsychology* (pp. 662–666). Oxford: Blackwell Scientific.

Baxter, L. R., Schwartz, J. M., Phelps, M. E., Mazziotta, J. C., Guze, B. H., Selin, C. E., Gerner, R. H., & Sumida, R. M. (1989). Reduction of prefrontal cortex glucose metabolism common to three types of depression. *Archives of General Psychiatry, 46*, 243–250.

Baylis, G. C., & Driver, J. (1992). Visual parsing and response competition: The effect of grouping factors. *Perception and Psychophysics, 51*, 145–162.

Baylis, G. C., & Driver, J. (1993). Visual attention and objects: Evidence for hierarchical coding of locations. *Journal of Experimental Psychology: Human Perception and Performance, 19*, 451–470.

Baylis, G. C., Driver, J., & Rafal, R. D. (1993). Visual extinction and stimulus repetition. *Journal of Cognitive Neuroscience, 5*, 453–466.

Baynes, K. (1990). Language and reading in the right hemisphere: Highways or byways of the brain. *Journal of Cognitive Neuroscience, 2*, 159–179.

Bear, D., Schiff, D., Saver, J., Greenberg, M., & Freeman, R. (1986). Quantitative analysis of cerebral asymmetry. *Archives of Neurology, 43*, 598–603.

Beaumont, J. G., & Davidoff, J. B. (1992). Assessment of visuo-perceptual dysfunction. In J. R. Crawford, D. M. Parker, & W. M. McKinlay (Eds.), *A handbook of neuropsychological assessment* (pp. 115–140). Hillsdale, NJ: Lawrence Erlbaum.

Bechara, A., Damasio, A. R., Damasio, H., & Anderson, S. W. (1994). Insensitivity to future consequences following damage to human prefrontal cortex. *Cognition, 50,* 7–15.

Behrmann, M., & Moscovitch, M. (1994). Object-centered neglect in patients with unilateral neglect: effects of left-right coordinates of objects. *Journal of Cognitive Neuroscience, 6,* 1–16.

Behrmann, M., Moscovitch, M., Black, S. E., & Mozer, M. (1990). Perceptual and conceptual mechanisms in neglect dyslexia: two contrasting case studies. *Brain, 113,* 1163–1183.

Behrmann, M., & Tipper, S. P. (1994). Object-based visual attention: Evidence from unilateral neglect. In C. Umiltà & M. Moscovitch (Eds.), *Attention and performance XIV: Conscious and nonconscious processing and cognitive functioning* (pp. 351–375). Cambridge, MA: MIT Press.

Bellas, D. N., Novelly, R. A., Eskenazi, B., & Wasserstein, J. (1988). Unilateral displacement in the olfactory sense: A manifestation of the unilateral neglect syndrome. *Cortex, 24,* 267–275.

Bellugi, U., Bihrle, A., Jernigan, T., Trauner, D., & Doherty, S. (1990). Neuropsychological, neurological and neuroanatomical profile of William's syndrome. *American Journal of Medical Genetics Supplement, 6,* 115–125.

Bench, C. J., Friston, K. J., Brown, R. G., Frackowiak, R. S. J., & Dolan, R. J. (1993). Regional cerebral blood flow in depression measured by positron emission tomography: The relationship with clinical dimensions. *Psychological Medicine, 23,* 579–590.

Bender, M. B., Feldman, M., & Sobin, A. J. (1968). Palinopsia. *Brain, 91,* 321–328.

Benson, D. F. (1993a). Aphasia. In K. M. Heilman & E. Valenstein (Eds.), *Clinical neuropsychology* (3rd ed.) (pp. 17–36). Oxford: Oxford University Press.

Benson, D. F. (1993b). Prefrontal abilities. *Behavioural Neurology, 6,* 75–81.

Benson, D. F., Gardner, H., & Meadows, J. C. (1976). Reduplicative paramnesia. *Neurology, 26,* 147–151.

Benton, A. L. (1977). The amusias. In M. Critchley & R. A. Henson (Eds.), *Music and the brain* (pp. 378–397). London: William Heinemann Medical Books Ltd.

Benton, A. L. (1990). Face recognition. *Cortex, 26,* 491–499.

Benton, A. L., & Tranel, D. (1993). Visuoperceptual, visuospatial, and visuoconstructive disorders. In K. M. Heilman & E. Valenstein (Eds.), *Clinical neuropsychology* (3rd ed.) (pp. 165–214). Oxford: Oxford University Press.

Berciano, J. (1993). Olivopontocerebellar atrophy. In J. Jankovic & E. Tolosa (Eds.), *Parkinson's disease and movement disorders* (pp. 163–189). Baltimore: Williams & Wilkins.

Bergman, H., Wichmann, T., & DeLong, M. R. (1990). Reversal of experimental Parkinsonism by lesions of the subthalamic nucleus. *Science, 249,* 1436–1438.

Berlucchi, G. (1990). Commissurotomy studies in animals. In F. Boller & J. Grafman (Eds.), *Handbook of neuropsychology* (Vol. 4, pp. 9–47). Amsterdam: Elsevier.

Berman, K. F., Torrey, E. F., Daniel, D. G., & Weinberger, D. R. (1992). Regional cerebral blood flow in monozygotic twins discordant and concordant for schizophrenia. *Archives of General Psychiatry, 49,* 927–934.

Berson, R. J. (1983). Capgras' syndrome. *American Journal of Psychiatry, 140,* 969–978.

Berthier, M. L. (1994). Foreign accent syndrome. *Neurology, 44,* 990–991.

Berthier, M. L., & Starkstein, S. E. (1994). Catastrophic reaction in crossed aphasia. *Aphasiology, 8,* 89–95.

Berti, A., & Rizzolatti, G. (1992). Visual processing without awareness: evidence from unilateral neglect. *Journal of Cognitive Neuroscience, 4,* 345–351.

Bever, T. G., & Chiarello, R. J. (1974). Cerebral dominance in musicians and nonmusicians. *Science, 185,* 137–139.

Bhatia, K. P., & Marsden, C. D. (1994). The behavioural and motor consequences of focal lesions of the basal ganglia in man. *Brain, 117,* 859–876.

Bishop, D. V. M. (1983). Linguistic impairment after left hemisphere decortication in infancy. *Brain and Language, 10,* 287–317.

Bisiach, E. (1991). Extinction and neglect: Same or different? In J. Paillard (Ed.), *Brain and space* (pp. 251–257). Oxford: Oxford University Press.

Bisiach, E. (1993). Mental representation in unilateral neglect and related disorders: The Twentieth Bartlett Memorial Lecture. *The Quarterly Journal of Experimental Psychology, 46A,* 435–461.

Bisiach, E., & Berti, A. (1987). Dyschiria. An attempt at its systemic explanation. In M. Jeannerod (Ed.), *Neurophysiological and neuropsychological aspects of spatial neglect* (pp. 183–201). Amsterdam: North-Holland.

Bisiach, E., & Berti, A. (1989). Unilateral misrepresentation of distributed information: Paradoxes and puzzles. In J. W. Brown (Ed.), *Neuropsychology of visual perception* (pp. 145–161). Hillsdale: Erlbaum.

Bisiach, E., Brouchon, M., Poncet, M., & Rusconi, M. L. (1993). Unilateral neglect in route description. *Neuropsychologia, 31,* 1255–1267.

Bisiach, E., Capitani, E., Luzzatti, C., & Perani, D. (1981). Brain and conscious representation of outside reality. *Neuropsychologia, 19,* 543–551.

Bisiach, E., Cornacchia, L., Sterzi, R., & Vallar, G. (1984). Disorders of perceived auditory lateralization after lesions of the right hemisphere. *Brain, 107,* 37–52.

Bisiach, E., Geminiani, G., Berti, A., & Rusconi, M. L. (1990). Perceptual and premotor factors of unilateral neglect. *Neurology, 40,* 1278–1281.

Bisiach, E., & Luzzatti, C. (1978). Unilateral neglect of representational space. *Cortex, 14,* 129–133.

Bisiach, E., Luzzatti, C., & Perani, D. (1979). Unilateral neglect, representational schema and consciousness. *Brain, 102,* 609–618.

Bisiach, E., Perani, D., Vallar, G., & Berti, A. (1986). Unilateral neglect: Personal and extrapersonal. *Neuropsychologia, 24,* 759–767.

Bisiach, E., & Rusconi, M. L. (1990). Break-down of perceptual awareness in unilateral neglect. *Cortex, 26,* 643–649.

Bisiach, E., & Vallar, G. (1988). Hemineglect in humans. In F. Boller & J. Grafman (Eds.), *Handbook of neuropsychology* (Vol. 1, pp. 195–222). Amsterdam: Elsevier.

Bisiach, E., Vallar, G., Perani, D., Papagno, C., & Berti, A. (1986). Unawareness of disease following lesions of the right hemisphere: Anosognosia for hemiplegia and anosognosia for hemianopia. *Neuropsychologia, 24,* 471–482.

Bladin, P. F., & Berkovic, S. F. (1984). Striatocapsular infarction: Large infarcts in the lenticulostriate arterial territory. *Neurology, 34,* 1423–1430.

Bleuler, E. (1911). *Dementia praecox oder die Gruppe der Schizophrenien.* Leipzig: Denticke.

Bodamer, J. (1947). Die prosopagnosie. *Archiv fur Psychiatrie und Nervenkrankheiten, 179*, 6–53.

Bogen, J. E., & Vogel, P. J. (1962). Cerebral commissurotomy in man. *Bulletin of the Los Angeles Neurological Societies, 27*, 169–172.

Bogousslavsky, J., Miklossy, J., Regli, F., Deruaz, J. P., Assal, G., & Delaloye, G. (1988). Subcortical neglect: Neuropsychological, SPECT and neuropathological correlations with anterior choroidal artery infarction. *Annals of Neurology, 23*, 448–452.

Boller, F., & Grafman, J. (1985). Acalculia. In J. A. M. Frederiks (Ed.), *Handbook of clinical neurology*, Vol. 1 (45): *Clinical Neuropsychology* (pp. 473–481). Amsterdam: Elsevier.

Bottini, G., Sterzi, R., & Vallar, G. (1992). Directional hypokinesia in spatial hemineglect: A case study. *Journal of Neurology, Neurosurgery and Psychiatry, 55*, 562–565.

Bradshaw, J. L. (1989). *Hemispheric specialization and psychological function*. Chichester, UK: Wiley.

Bradshaw, J. L. (in press). Gail D.—Poizner, Klima and Bellugi's (1987) deaf agrammatic signer: Form and function in the left cerebral hemisphere for speech and language. In C. Code (Ed.), *Classic cases in neuropsychology*. Hillsdale, NJ: Erlbaum.

Bradshaw, J. L., & Nettleton, N. C. (1981). The nature of hemispheric specialization in man. *Behavioral and Brain Sciences, 4*, 51–63.

Bradshaw, J. L., & Nettleton, N. C. (1983). *Human Cerebral Asymmetry*. Englewood Cliffs, NJ: Prentice Hall.

Bradshaw, J. L., Nettleton, N. C., Pierson, J. M., Wilson, L. E., & Nathan, G. (1987). Co-ordinates of extracorporeal space. In M. Jeannerod (Ed.), *Neurophysiological and neuropsychological aspects of spatial neglect* (pp. 41–68). North-Holland: Elsevier Science Publishers B.V.

Bradshaw, J. L., Phillips, J. G., Dennis, C., Mattingley, J. B., Andrews, D., Chiu, E., Pierson, J. M., & Bradshaw, J. A. (1992). Initiation and execution of movement sequences in those suffering from and at risk of developing Huntington's disease. *Journal of Clinical and Experimental Neuropsychology, 14*, 179–192.

Bradshaw, J. L., & Rogers, L. J. (1993). *The evolution of lateral asymmetries, language, tool use, and intellect*. San Diego: Academic Press.

Bradshaw, J. L., Spataro, J., Harris, M., Nettleton, N. C., & Bradshaw, J. A. (1988). Crossing the midline by four to eight year old children. *Neuropsychologia, 26*, 221–235.

Bradshaw, J. L., Waterfall, M. L., Phillips, J. G., Iansek, R., Mattingley, J. B., & Bradshaw, J. A. (1993). Reorientation of attention in Parkinson's disease: An extension to the vibrotactile modality. *Neuropsychologia, 31*, 51–66.

Brain, W. R. (1941). Visual disorientation with special reference to lesions of the right cerebral hemisphere. *Brain, 64*, 244–271.

Brandt, J. (1991). Cognitive impairments in Huntington's disease: Insights into the neuropsychology of the striatum. In F. Boller & J. Grafman (Eds.), *Handbook of neuropsychology* (Vol. 5, pp. 241–261). Amsterdam: Elsevier.

Bravi, D., Mouradian, M. M., Roberts, J. W., Davis, T. L., Sohn, Y. H., & Chase, T. N. (1994). Wearing-off fluctuations in Parkinson's disease: Contribution of postsynaptic mechanisms. *Annals of Neurology, 36*, 27–31.

Breier, A., Buchanan, R. W., Elkashef, A., Munson, R. C., Kirkpatrick, R., & Gellad, F. (1992). Brain morphology and schizophrenia: A magnetic resonance imaging study of limbic, prefrontal cortex and caudate structures. *Archives of General Psychiatry, 49,* 921–926.

Breitner, J. C. S. (1990). Life table methods and assessment of familial risk in Alzheimer's disease. *Archives of General Psychiatry, 47,* 395–396.

Breitner, J. C. S., Gau, B. A., Welsh, K. A., Plassman, B. L., McDonald, W. M., Helms, M. J., & Anthony, J. C. (1994). Inverse association of anti-inflammatory treatments and Alzheimer's disease. *Neurology, 44,* 227–232.

Brentari, D., & Poizner, H. (1994). A phonological analysis of a deaf parkinsonian signer. *Language and Cognitive Processes, 9,* 69–99.

Bressler, S. L., Coppola, R., & Nakamura, R. (1993). Episodic multiregional cortical coherence at multiple frequencies during visual task performance. *Nature, 366,* 153–156.

Brewer, B. (1992). Unilateral neglect and the objectivity of spatial representation. *Mind and Language, 7,* 222–239.

Broca, P. (1861). Perte de la parole. Ramollissement chronique et destruction partielle du lobe antérieur gauche du cerveau. *Bulletin de la Société d'Anthropologie* (Paris), *2,* 235–238.

Brotchie, P., Iansek, R., & Horne, M. K. (1991a). Motor function of the monkey globus pallidus: 2. Cognitive aspects of movement and phasic neuronal activity. *Brain, 114,* 1685–1702.

Brotchie, P., Iansek, R., & Horne, M. K. (1991b). Motor function of the monkey globus pallidus: 1. Neuronal discharge and parameters of movement. *Brain, 114,* 1667–1683.

Brown, A. L., & Murphy, D. R. (1989). Cryptomnesia: Delineating inadvertent plagiarism. *Journal of Experimental Psychology: Learning, Memory and Cognition, 15,* 432–442.

Brown, J. W. (1991). *Self and process: Brain states of the conscious present.* New York: Springer.

Brown, P. (1994). Pathophysiology of spasticity. *Journal of Neurology, Neurosurgery and Psychiatry, 57,* 773–777.

Brown, V., Schwarz, U., Bowman, E. M., Fuhr, P., Robinson, D. L., & Hallett, M. (1993). Dopamine dependent reaction time deficits in patients with Parkinson's disease are task specific. *Neuropsychologia, 31,* 459–469.

Bruce, V., & Young, A. (1986). Understanding face recognition. *British Journal of Psychology, 77,* 305–327.

Brunner, H. G., Nelen, M., Breakefield, X. O., Ropers, H. H., & van Oost, B. A. (1993). Abnormal behavior associated with the structural gene for monoamine oxidase A. *Science, 262,* 578–580.

Bruun, R. D., & Budman, C. L. (1992). The natural history of Tourette syndrome. In T. N. Chase, A. J. Friedhoff, & D. J. Cohen (Eds.), *Tourette syndrome; Genetics, neurobiology and treatment: advances in neurology* (Vol. 58, pp. 1–5). New York: Raven Press.

Bruyer, R. (1993). Failures of face processing in normal and brain-damaged subjects. In F. Boller & J. Grafman (Eds.), *Handbook of neuropsychology* (Vol. 8, pp. 411–436). Amsterdam: Elsevier.

Burden, V., Bradshaw, J. L., Nettleton, N. C., & Wilson, L. (1985). Hand and hemispace effects in tactual tasks in children. *Neuropsychologia, 23,* 515–525.

Bush, A. I., Pettingell, W. H., Multhaup, G., Paradis, M. d., Vonsattel, J.-P., Gusella, J. F.,

Beyreuther, K., Masters, C. L., & Tanzi, R. E. (1994). Rapid introduction of Alzheimer A amyloid formation by zinc. *Science, 265,* 1464–1466.

Bushnell, M. C., Goldberg, M. E., & Robinson, D. L. (1981). Behavioural enhancement of visual responses in monkey cerebral cortex: I. Modulation in posterior parietal cortex related to selective visual attention. *Journal of Neurophysiology, 46,* 755–771.

Butter, C. M., & Kirsch, N. (1992). Combined and separate effects of eye patching and visual stimulation on unilateral neglect following stroke. *Archives of Physical Medicine and Rehabilitation, 73,* 1133–1139.

Butter, C. M., Kirsch, N. L., & Reeves, G. (1990). The effect of lateralised stimuli on unilateral spatial neglect following right hemisphere lesions. *Restorative Neurology and Neuroscience, 2,* 39–46.

Butter, C. M., Rapcsak, S., Watson, R. T., & Heilman, K. M. (1988). Changes in sensory inattention, directional motor neglect and "release" of the fixation reflex following a unilateral frontal lesion: A case report. *Neuropsychologia, 26,* 533–545.

Butters, N., & Cermak, L. S. (1986). A case study of the forgetting of autobiographical knowledge: Implications for the study of retrograde amnesia. In D. Rubin (Ed.), *Autobiographical memory* (pp. 253–272). Cambridge: Cambridge University Press.

Butters, N., Heindel, W. C., & Salmon, P. (1990). Dissociation of implicit memory in dementia: Neurological implications. *Bulletin of the Psychonomic Society, 28,* 359–366.

Byne, W., & Parsons, B. (1993). Human sexual orientation: The biologic theories reappraised. *Archives of General Psychiatry, 50,* 228–239.

Caligiuri, M. P., & Lohr, J. B. (1994). A disturbance in the control of muscle force in neuroleptic-naive schizophrenic patients. *Biological Psychiatry, 35,* 104–111.

Calne, D. B. (1992). Involuntary movements: An overview. In A. B. Joseph and R. R. Young (Eds.), *Movement Disorders in Neurology and Neuropsychiatry* (pp. 1–2). Oxford: Blackwell Scientific.

Calvanio, R., Petrone, P. N., & Levine, D. N. (1987). Left visual spatial neglect is both environment-centered and body-centered. *Neurology, 37,* 1179–1183.

Campbell, D. C., & Oxbury, J. M. (1976). Recovery from unilateral visuo-spatial neglect? *Cortex, 12,* 303–312.

Caplan, B. (1985). Stimulus effects in unilateral neglect? *Cortex, 21,* 69–80.

Caplan, L. R. (1988). Posterior cerebral artery syndromes. In P. J. Vinken, G. W. Bruyn, & H. L. Klawans (Eds.), *Vascular diseases, Part 1* (pp. 409–415). Amsterdam: Elsevier.

Caplan, L. R. (1992). Intracerebral haemorrhage. *The Lancet, 339,* 656–658.

Caplan, L. R., Schmahmann, J. D., Kase, C. S., Feldmann, E., Baquis, G., Greenberg, J. P., Gorelick, P. B., Helgason, C., & Hier, D. B. (1990). Caudate infarcts. *Archives of Neurology, 47,* 133–143.

Cappa, S. F., Guariglia, C., Messa, C., Pizzamiglio, L., & Zoccolotti, P. (1991). Computed tomography correlates of chronic unilateral neglect. *Neuropsychology, 5,* 195–204.

Cappa, S. F., Sterzi, R., Vallar, G., & Bisiach, E. (1987). Remission of hemineglect and anosognosia during vestibular stimulation. *Neuropsychologia, 25,* 775–782.

Cappa, S. F., & Vallar, G. (1992). Neuropsychological disorders after subcortical lesions: Implications for neural models of language and spatial attention. In G. Vallar, S. F. Cappa, & C-W. Wallesch (Eds.), *Neuropsychological disorders associated with subcortical lesions* (pp. 7–41). Oxford: Oxford University Press.

Caramazza, A. (1986). On drawing inferences about the structure of normal cognitive systems from the analysis of patterns of impaired performance: The case for single-patient studies. *Brain and Cognition, 5,* 41–66.

Caramazza, A. (Ed.) (1988). Methodological problems in cognitive neuropsychology. *Cognitive Neuropsychology, 5,* 517–623.

Caramazza, A., & Badecker, W. (1991). Clinical syndromes are not God's gift to cognitive neuropsychology: A reply to a rebuttal to an answer to a response to the case against syndrome-based research. *Brain and Cognition, 16,* 211–226.

Caramazza, A., & Hillis, A. E. (1990). Spatial representation of words in the brain implied by studies of a unilateral neglect patient. *Nature, 346,* 267–269.

Cardebat, D., Démonet, J.-F., Celsis, P., Puel, M., Viallard, G., & Marc-Vergnes, J.-P. (1994). Right temporal compensatory mechanisms in a deep dysphasic patient: A case report with activation study by SPECT. *Neuropsychologia, 32,* 97–103.

Carpenter, W. T., & Buchanan, R. W. (1994). Schizophrenia. *The New England Journal of Medicine, 330,* 681–689.

Carr, K. (1993). A question of conformation. *Nature, 365,* 386.

Casey, B. J., Gordon, C. T., Mannheim, G. B., & Rumsey, J. M. (1993). Dysfunctional attention in autistic savants. *Journal of Clinical and Experimental Neuropsychology, 15,* 933–946.

Cath, D. C., Hoogduin, C. A. L., van de Wetering, B. J. M., van Woerkom, T. C. A. M., Roos, R. A. C., & Rooymans, H. G. M. (1992). Tourette syndrome and obsessive-compulsive disorder. An analysis of associated phenomena. In T. N. Chase, A. J. Friedhoff, & D. J. Cohen (Eds.), *Advances in neurology* (Vol. 58, pp. 33–41). New York: Raven Press.

Chan, J., Brin, M. F., & Fahn, S. (1991). Idiopathic dystonia: Clinical characteristics. *Movement Disorders, 6,* 119–126.

Chen, H.-C. & Tzeng, O. (1992). *Language processing in chinese.* Amsterdam: Elsevier.

Chieffi, S., Gentilucci, M., Allport, A., Sasso, E., & Rizzolatti, G. (1993). Study of selective reaching and grasping in a patient with unilateral parietal lesion: Dissociation effects of residual spatial neglect. *Brain, 116,* 1119–1137.

Chiu, E. (1994). What's in a name—dementia or dysmentia? *International Journal of Geriatric Psychiatry, 9,* 1–4.

Chomsky, N. (1980). Rules and representations. *Behavioral and Brain Sciences, 3,* 1–61.

Christensen, J. E. J. (1991). New treatment of spasmodic torticollis. *The Lancet, 338,* 572.

Christodoulou, G. N. (1978). Syndrome of subjective doubles. *American Journal of Psychiatry, 135,* 249–251.

Clark, C. R., Geffen, G. M., & Geffen, L. B. (1989). Catecholamines and the covert orientation of attention. *Neuropsychologia, 27,* 131–139.

Clark, J. M., & Campbell, J. I. D. (1991). Integrated versus modular theories of number skills and acalculia. *Brain and Cognition, 17,* 204–239.

Clark, R. F., & Goate, A. M. (1993). Molecular genetics of Alzheimer's disease. *Archives of Neurology, 50,* 1164–1172.

Cohen, A., and Rafal, R. D. (1991). Attention and feature integration: Illusory conjunctions in a patient with a parietal lobe lesion. *Psychological Science, 2,* 106–110.

Cohen, D. J., Bruun, R. D., & Leckman, J. F. (Eds.). (1988). *Tourette's syndrome and tic disorders: Clinical understanding and treatment.* New York: Wiley.

Cohen, F. E., Pan, K.-M., Huang, Z., Baldwin, M., Fletterick, R. J., & Prusiner, S. B. (1994). Structural clues to prion replication. *Science, 264,* 530–531.

Cohen, J. D., Romero, R. D., & Farah, M. J. (1994). Disengaging from the disengage function: The relation of macrostructure to microstructure in parietal attention deficits. *Journal of Cognitive Neuroscience, 6,* 377–387.

Cohen, L., Gray, F., Meyrignac, C., Dehaene, S., & Degos, J.-D. (1994). Selective deficit of visual size perception: Two cases of hemimicropsia. *Journal of Neurology, Neurosurgery and Psychiatry, 57,* 73–78.

Cohen, M. S., & Bookkheimer, S. Y. (1994). Localization of brain function using magnetic resonance imaging. *Trends in Neurosciences, 17,* 268–277.

Coleman, M., & Gillberg, C. (1986). *The biology of autistic syndromes.* New York: Praeger.

Colombo, A., De Renzi, E., & Faglioni, P. (1976). The occurrence of visual neglect in patients with unilateral cerebral disease. *Cortex, 12,* 221–231.

Colombo, A., De Renzi, E., & Gentilini, M. (1982). The time course of visual hemi-inattention. *Archiv für Psychiatrie und Nervenkrankheiten, 231,* 539–546.

Coltheart, M. (1980a). Reading, phonological recoding and deep dyslexia. In M. Coltheart, K. E. Patterson, & J. C. Marshall (Eds.), *Deep dyslexia* (pp. 197–226). London: Routledge & Kegan Paul.

Coltheart, M. (1980b). Deep dyslexia: A right-hemisphere hypothesis. In M. Coltheart, K. E. Patterson, & J. C. Marshall (Eds.), *Deep dyslexia* (pp. 326–380). London: Routledge & Kegan Paul.

Coltheart, M. (1985). Right hemisphere reading revisited. *Behavioral and Brain Sciences, 8,* 363–379.

Coltheart, M., Patterson, K. E., & Marshall, J. C. (Eds.) (1980). *Deep dyslexia.* London: Routledge & Kegan Paul.

Conway, M. A. (1993). Impairments of autobiographical memory. In F. Boller & J. Grafman (Eds.), *Handbook of neuropsychology,* (Vol. 8, pp. 175–192). Amsterdam: Elsevier.

Corbetta, M., Miezin, F. M., Dobmeyer, S., Shulman, G. L., & Petersen, S. E. (1990). Attentional modulation of neural processing of shape, color and velocity in humans. *Science, 248,* 1556–1559.

Corbetta, M., Miezin, F. M., Shulman, G. L., & Petersen, S. E. (1993). A PET study of visuospatial attention. *The Journal of Neuroscience, 13,* 1202–1226.

Corder, E. H., Saunders, A. M., Strittmatter, W. J., Schmechel, D. E., Gaskell, P. C., Small, G. W., Roses, A. D., Haines, J. L., & Pericak-Vance, M. A. (1993). Gene dose of apolipoprotein E type 4 allele and the risk of Alzheimer's disease in late onset families. *Science, 261,* 921–923.

Corkin, S. (1984). Lasting consequences of medial temporal lobectomy: Clinical cause and experimental findings in H. M. *Seminars in Neurology, 4,* 249–259.

Coslett, H. B., & Monsul, N. (1994). Reading with right hemisphere: Evidence from transcranial magnetic stimulation. *Brain and Language, 46,* 198–211.

Coslett, H. B., & Saffran, E. M. (1989). Evidence for preserved reading in "pure alexia." *Brain, 112,* 327–359.

Costa, L. D., Vaughn, H. G. J., Howitz, M., & Ritter, W. (1969). Patterns of behavioural deficits associated with visual spatial neglect. *Journal of Nervous and Mental Disease, 5,* 242–263.

Courbon, P., & Fail, G. (1927). Illusion de Frégoli. *Bulletin de la Société Clinique de Médicine Mentale, 15,* 121–124.

Courbon, P., & Tusques, J. (1932). Illusions d'intermetamorphose et de charme. *Annals Medico-Psychologiques, 90,* 401–405.

Courchesne, E., Townsend, J., & Saitoh, O. (1994). The brain in infantile autism: Posterior fossa structures are abnormal. *Neurology, 44,* 214–223.

Cowell, P. E., Kertesz, A., & Denenberg, V. H. (1993). Multiple dimensions of handedness and the human corpus callosum. *Neurology, 43,* 2353–2357.

Cowey, A., Small, M., & Ellis, S. (1994). Left visuo-spatial neglect can be worse in far than in near space. *Neuropsychologia, 32,* 1059–1066.

Cowey, A., & Stoerig, P. (1992). Reflections on blindsight. In A. D. Milner & M. D. Rugg (Eds.), *The Neuropsychology of Consciousness* (pp. 11–38). San Diego: Academic Press.

Coyle, J. T., & Puttfarcken, P. (1993). Oxidative stress, glutamate, and neurodegenerative disorders. *Science, 262,* 689–695.

Crick, F. (1994). *The astonishing hypothesis.* London: Simon and Schuster.

Critchley, M. (1953). *The parietal lobes.* London: Edward Arnold.

Cronin-Golomb, A. (1986). Subcortical transfer of cognitive information in subjects with complete forebrain commissurotomy. *Cortex, 22,* 499–519.

Crosson, B. (1992). Is the striatum involved in language? In G. Vallar, S. F. Cappa and C.-W. Wallesch (Eds.), *Neuropsychological disorders associated with subcortical lesions* (pp. 268–293). Oxford: Oxford University Press.

Crosson, B., Novack, T. A., & Trenerry, M. R. (1988). Subcortical language mechanisms: Window on a new frontier. In H. A. Whitaker (Ed.), *Phonological processes and brain mechanisms* (pp. 24–58). New York: Springer.

Crow, T. J. (1980). Molecular pathology of schizophrenia: More than one disease process? *British Medical Journal, 280,* 66–68.

Crowne, D. P., Yeo, C. H., & Russell, I. S. (1981). The effects of unilateral frontal eye field lesions in the monkey: Visual-motor guidance and avoidance behavior. *Behavioural Brain Research, 2,* 165–185.

Cummings, J. L. (1990). Introduction. In J. L. Cummings (Ed.), *Subcortical dementia* (pp. 3–16). New York: Oxford University Press.

Cummings, J. L. (1992). Parkinson's disease and parkinsonism. In A. B. Joseph & R. R. Young (Eds.), *Movement disorders in neurology and neuropsychiatry* (pp. 195–203). Oxford: Blackwell Scientific.

Cummings, J. L. (1993). Frontal-subcortical circuits and human behavior. *Archives of Neurology, 50,* 873–879.

Cummings, J. L., & Hegarty, A. (1994). Neurology, psychiatry and neuropsychiatry. *Neurology, 44,* 209–213.

Cummings, J. L., & Huber, S. J. (1992). Visuospatial abnormalities in Parkinson's disease. In S. J. Huber & J. L. Cummings (Eds.), *Parkinson's disease: Neurobehavioral aspects* (pp. 59–73). New York: Oxford University Press.

Cummings, J. L., Syndulko, K., Goldberg, Z., & Treiman, D. M. (1982). Palinopsia reconsidered. *Neurology, 32,* 444–447.

Cunnington, R., Bradshaw, J. L., & Iansek, R. (in press). The role of the supplementary motor area in the control of voluntary movement. *Human Movement Science.*

Cunnington, R., Iansek, R., Bradshaw, J. L., & Phillips, J. G. (in press). Movement-related potentials in Parkinson's disease: Presence and predictability of temporal and spatial cues. *Brain*.

Cutting, J. (1978). Study of anosognosia. *Journal of Neurology, Neurosurgery and Psychiatry, 41,* 548–565.

Cutting, J. (1990). *The right cerebral hemisphere and psychiatric disorder.* Oxford: Oxford University Press.

Cytowic, R. (1989). *Synesthesia: A union of the senses.* New York: Springer.

Daffner, K. R., Ahern, G. L., Weintraub, S., & Mesulam, M.-M. (1990). Dissociated neglect behavior following sequential strokes in the right hemisphere. *Annals of Neurology, 28,* 97–101.

Dalla Barba, G. (1993). Prospective memory: A 'new' memory system? In F. Boller & J. Grafman (Eds.), *Handbook of neuropsychology* (Vol. 8, pp. 239–252). Amsterdam: Elsevier.

Damasio, A. R. (1990). Category-related recognition defects as a clue to the neural substrates of knowledge. *Trends in Neurosciences, 13,* 95–98.

Damasio, A. R. (1991). Signs of aphasia. In M. T. Sarno (Ed.) *Acquired aphasia* (2nd ed.) (pp. 27–44). San Diego: Academic Press.

Damasio, A. R., & Anderson, S. W. (1993). The frontal lobes. In K. M. Heilman & E. Valenstein (Eds.), *Clinical neuropsychology* (3rd ed.) (pp. 409–460). Oxford: Oxford University Press.

Damasio, A. R., Damasio, H., & Chui, H. C. (1980). Neglect following damage to frontal lobe or basal ganglia. *Neuropsychologia, 18,* 123–132.

Damasio, A. R., Damasio, H., & Van Hoesen, G. W. (1982). Prosopagnosia: Anatomic basis and behavioral mechanisms. *Neurology, 32,* 331–341.

Damasio, A. R., & Tranel, D. (1993). Nouns and verbs are retrieved with differently distributed neural systems. *Proceedings of the National Academy of Sciences, 90,* 4957–4960.

Damasio, A. R., Tranel, D., & Damasio, H. (1990). Face agnosia and the neural substrates of memory. *Annual Review of Neuroscience, 13,* 89–109.

Damasio, A. R., Tranel, D., & Damasio, H. (1993). Similarity of structure and the profile of visual recognition defects: A comment on Gaffan and Heywood. *Journal of Cognitive Neuroscience, 5,* 371–372.

Damasio, H. (1991). Neuroanatomical correlates of the aphasias. In M. T. Sarno (Ed.), *Acquired aphasia* (2nd ed.) (pp. 45–72). San Diego: Academic Press.

Darian-Smith, C., & Gilbert, C. D. (1994). Axonal sprouting accompanies functional reorganization in adult cat striate cortex. *Nature, 368,* 737–740.

Davidoff, J. (1991). *Cognition through color.* Cambridge, MA: MIT Press.

Deacon, T. W. (1989). The neural circuitry underlying primate calls and human language. *Human Evolution, 4,* 367–401.

de Graaf, J. B., Sittig, A. C., & Denier van der Gon, J. J. (1991). Misdirections in slow goal-directed arm movements and pointer-setting tasks. *Experimental Brain Research, 84,* 434–438.

Della Sala, S., & Logie, R. H. (1993). When working memory does not work: The role of working memory in neuropsychology. In F. Boller & J. Grafman (Eds.), *Handbook of neuropsychology* (Vol. 8, pp. 1–62). Amsterdam: Elsevier.

DeLong, M. R. (1993). Overview of basal ganglia function. In N. Mano, I. Hamada, & M. R. DeLong (Eds.), *Role of the cerebellum and basal ganglia in voluntary movement* (pp. 65–70). Amsterdam: Elsevier.

Delwaide, P. J., & Gonce, M. (1993). Pathophysiology of Parkinson's signs. In J. Jankovic & E. Tolosa (Eds.), *Parkinson's disease and movement disorders* (2nd ed.) (pp. 77–92). Baltimore: Williams & Wilkins.

Démonet, J.-F., Celsis, P., Puel, M., Cardebat, D., Marc-Vergnes, J. P., & Rascol, A. (1992). Thalamic and non-thalamic subcortical aphasia: A neurolinguistic and SPECT approach. In G. Vallar, S. F. Cappa, & C.-W. Wallesch (Eds.), *Neuropsychological disorders associated with subcortical lesions* (pp. 397–411). Oxford: Oxford University Press.

Denes, G., Semenza, C., Stoppa, E., & Lis, A. (1982). Unilateral spatial neglect and recovery from hemiplegia: A follow-up study. *Brain, 105,* 543–552.

Dennett, D. C. (1991). *Consciousness explained.* Boston: Little, Brown & Co.

Dennis, M. (1988). Language and the young damaged brain. In T. Boll & B. K. Bryant (Eds.), *Clinical neuropsychology and brain function* (pp. 85–123). Washington, DC: American Psychological Association.

De Pauw, K. W., Szulecka, T. K., & Poltock, T. L. (1987). Frégoli syndrome after cerebral infarction. *Journal of Nervous and Mental Disease, 175,* 433–438.

De Renzi, E. (1989). Apraxia. In F. Boller & J. Grafman (Eds.), *Handbook of neuropsychology* (Vol. 2, pp. 245–263). Amsterdam: Elsevier.

De Renzi, E., Faglioni, P., Grossi, D., & Nichelli, P. (1991). Apperceptive and associative forms of prosopagnosia. *Cortex, 27,* 213–221.

De Renzi, E., Gentilini, M., Faglioni, P., & Barbieri, C. (1989). Attentional shift towards the rightmost stimuli in patients with left visual neglect. *Cortex, 25,* 231–237.

De Renzi, E., Gentilini, M., & Pattacini, F. (1984). Auditory extinction following hemisphere damage. *Neuropsychologia, 22,* 733–744.

De Renzi, E., & Lucchelli, F. (1994). Are semantic systems separately represented in the brain? The case of living category impairment. *Cortex, 30,* 3–25.

De Renzi, E., Perani, D., Carlesimo, G. A., Silveri, M. C., & Fazio, F. (1994). Prosopagnosia can be associated with damage confined to the right hemisphere—an MRI and PET study and the review of the literature. *Neuropsychologia, 32,* 893–902.

D'Esposito, M., McGlinchey-Berroth, R., Alexander, M. P., Verfaellie, M., & Milberg, W. P. (1993). Dissociable cognitive and neural mechanisms of unilateral visual neglect. *Neurology, 43,* 2638–2644.

Devinsky, O. (1992). *Behavioural neurology: 100 maxims.* London: Edward Arnold.

Devinsky, O., & Geller, B. D. (1992). Gilles de la Tourette syndrome. In A. B. Joseph & R. R. Young (Eds.), *Movement disorders in neurology and neuropsychiatry* (pp. 471–478). Oxford: Blackwell Scientific.

Devor, E. J. (1990). Untying the Gordian knot: The genetics of Tourette syndrome. *Journal of Nervous and Mental Disease, 178,* 669–679.

DeYeo, E. A., Felleman, D. J., Van Essen, D. C., & McClendon, E. (1994). Multiple processing streams in occipitotemporal cortex. *Nature, 371,* 151–154.

Diamond, S. G., Markham, C. H., Rand, R. W., Becker, D. P., & Treciokas, L. J. (1994). Four-year follow-up of adrenal-to-brain transplants in Parkinson's disease. *Archives of Neurology, 51,* 559–563.

Diamond, S. P., & Bender, M. B. (1965). On auditory extinction and alloacusis. *Transactions of the American Neurological Association, 90*, 154–157.

Dick, J., Wood, R., Bradshaw, J. L., & Bradshaw, J. A. (1987). Programmable visual display for diagnosing, assessing and rehabilitating unilateral neglect. *Medical and Biological Engineering and Computing, 25*, 109–111.

Dick, J. P., Rothwell, J. C., Day, B. L., Cantello, R., Buruma, D., Gioux, M., Benecke, R., Berardelli, A., Thompson, P. D., & Marsden, C. D. (1989). The Bereitschaftspotential is abnormal in Parkinson's disease. *Brain, 112*, 233–244.

Dimond, J. J., & Beaumont, J. G. (1973). Differences in the vigilance performances of the right and left hemispheres. *Cortex, 9*, 259–266.

Di Monte, D. A. (1991). Mitochondrial DNA and Parkinson's disease. *Neurology, 41* (suppl. 2), 38–42.

Domesick, V. B. (1990). Subcortical anatomy: The circuitry of the striatum. In J. L. Cummings (Ed.), *Subcortical dementia* (pp. 31–43). New York: Oxford University Press.

Donnan, G. A. (1992). Investigation of patients with stroke and transient ischemic attacks. *The Lancet, 339*, 473–477.

Doty, R. L., Stern, M. B., Pfeiffer, C., Gollomp, S. M., & Hurtig, H. I. (1992). Bilateral olfactory dysfunction in early stage treated and untreated idiopathic Parkinson's disease. *Journal of Neurology, Neurosurgery and Psychiatry, 55*, 138–142.

Doty, R. W. (1989). Schizophrenia: A disease of interhemispheric processes at forebrain and brainstem levels? *Behavioural Brain Research, 34*, 1–33.

Doty, R. W., & Davis, R. G. (1991). *Retrograde transport from olfactory bulb to brainstem raphe: The etiology of schizophrenia?* Paper presented at the third IBRO World Congress of Neuroscience, Montréal, Québec.

Driver, J., & Baylis, G. C. (1989). Movement and visual attention: The spotlight metaphor breaks down. *Journal of Experimental Psychology: Human Perception and Performance, 15*, 448–456.

Driver, J., Baylis, G. C., & Rafal, R. D. (1992). Preserved figure-ground segregation and symmetry perception in visual neglect. *Nature, 360*, 73–75.

Driver, J., & Halligan, P. W. (1991). Can visual neglect operate in object-centred coordinates? An affirmative single-case study. *Cognitive Neuropsychology, 8*, 475–496.

Dubois, B., Boller, F., Pillon, B., & Agid, Y. (1991). Cognitive deficits in Parkinson's disease. In F. Boller & J. Grafman (Eds.), *Handbook of neuropsychology* (vol. 5, pp. 195–240). Amsterdam: Elsevier.

Dubois, B., & Pillon, B. (1992). Biochemical correlates of cognitive changes and dementia in Parkinson's disease. In S. J. Huber & J. L. Cummings (Eds.), *Parkinson's disease: Neurobehavioral aspects* (pp. 178–198). New York: Oxford University Press.

Dubois, B., Verin, M., Teixeira-Ferreira, C., Sirigu, A., & Pillon, B. (1994). How to study frontal lobe functions in humans. In A.-M. Thierry, J. Glowinski, P. S. Goldman-Rakic and Y. Christen (Eds.), *Motor and cognitive functions of the prefrontal cortex* (pp. 1–16). New York: Springer.

Duffy, J. R., & Duffy, R. J. (1990). The assessment of limb apraxia: The limb apraxia test. In G. R. Hammond (Ed.), *Cerebral control of speech and limb movements* (pp. 503–534). Amsterdam: North Holland.

Duhamel, J-R., Colby, C. L., & Goldberg, M. E. (1992). The updating of the representation of visual space in parietal cortex by intended eye movements. *Science, 255,* 90–92.

Duncan, J. (1984). Selective attention and the organization of visual information. *Journal of Experimental Psychology: General, 113,* 501–517.

Duncombe, M., Bradshaw, J. L., Iansek, R., & Phillips, J. G. (1994). Parkinsonian patients without dementia or depression do not suffer from bradyphrenia as indexed by performance in mental rotation tasks with and without advance information. *Neuropsychologia, 32,* 1383–1396.

Dunnett, S. B. (1991). Neural transplants as a treatment for Alzheimer's disease? *Psychological Medicine, 21,* 825–830.

Early, T. S., Haller, J. W., Posner, M. I., & Raichle, M. (1994). The left striato-pallidal hyperactivity model of schizophrenia. In A. S. David & J. C. Cutting (Eds.), *The neuropsychology of schizophrenia* (pp. 15–37). Hillsdale, NJ: Lawrence Erlbaum.

Eglin, M., Robertson, L. C., & Knight, R. T. (1989). Visual search performance in the neglect syndrome. *Journal of Cognitive Neuroscience, 1,* 372–385.

Eglin, M., Robertson, L. C., Knight, R. T., & Brugger, P. (1994). Search deficits in neglect patients are dependent on size of the visual scene. *Neuropsychology, 8,* 451–463.

Egly, R., Driver, J., & Rafal, R. D. (1994). Shifting visual attention between objects and locations: Evidence from normal and parietal lesion subjects. *Journal of Experimental Psychology: General, 123,* 161–177.

Egly, R., & Homa, D. (1991). Reallocation of visual attention. *Journal of Experimental Psychology: Human Perception and Performance, 17,* 142–159.

Ekman, P. (1993). Facial expression and emotion. *American Psychologist, 48,* 384–392.

Elliott, D., Roy, E. A., Goodman, D., Carson, R. G., Chua, R., & Maraj, B. K. U. (1993). Asymmetries in the preparation and control of manual aiming movements. *Canadian Journal of Experimental Psychology, 47,* 570–589.

Elliott, F. A. (1992). Violence: The neurologic contribution, an overview. *Archives of Neurology, 49,* 595–603.

Ellis, A. W., Flude, B. M., & Young, A. W. (1987). "Neglect dyslexia" and the early visual processing of letters in words and nonwords. *Cognitive Neuropsychology, 4,* 439–464.

Ellis, H. D., & Young, A. W. (1990). Accounting for delusional misidentifications. *British Journal of Psychiatry, 157,* 239–248.

Etcoff, N. L., Freeman, R., & Cave, K. R. (1991). Can we lose memories of faces? Content specificity and awareness in a prosopagnosic. *Journal of Cognitive Neuroscience, 3,* 25–41.

Fahn, S., & Cohen, G. (1992). The oxidant stress hypothesis in Parkinson's disease. *Annals of Neurology, 32,* 804–812.

Fahn, S., Tolosa, E., & Marin, C. (1993). Clinical rating scale for tremor. In J. Jankovic & E. Tolosa (Eds.), *Parkinson's disease and movement disorders* (2nd ed.) (pp. 271–280). Baltimore: Williams & Wilkins.

Fallan, T., & Schwab-Stone, M. (1992). Methodology of epidemiological studies of tic disorders and comorbid psychopathology. In T. N. Chase, A. J. Friedhoff, & D. J. Cohen (Eds.), *Advances in neurology* (Vol. 58, pp. 43–53). New York: Raven Press.

Farah, M. J. (1990). *Visual agnosia: Disorders of object recognition and what they tell us about normal vision.* Cambridge, MA: MIT Press.

Farah, M. J. (1994). Neuropsychological inference with an interactive brain: A critique of the "locality" assumption. *Behavioral and Brain Sciences, 17,* 43–104.

Farah, M. J., Brunn, J. L., Wong, A. B., Wallace, M. A., & Carpenter, P. A. (1990). Frames of reference for allocating attention to space: Evidence from the neglect syndrome. *Neuropsychologia, 28,* 335–347.

Farah, M. J., Monheit, M. A., & Wallace, M. A. (1991). Unconscious perception of "extinguished" visual stimuli: reassessing the evidence. *Neuropsychologia, 29,* 949–958.

Farah, M. J., Wallace, M. A., & Vecera, S. P. (1993). "What" and "where" in visual attention: Evidence from the neglect syndrome. In I. H. Robertson & J. C. Marshall (Eds.), *Unilateral neglect: Clinical and experimental studies* (pp. 123–137). Hove, U.K.: Lawrence Erlbaum Associates.

Feinberg, T. E., & Shapiro, R. M. (1989). Misidentification-reduplication and the right hemisphere. *Neuropsychiatry, Neuropsychology, and Behavioral Neurology, 2,* 39–48.

Ferrier, I. N., & Perry, E. K. (1992). Post-mortem studies in affective disorder. *Psychological Medicine, 22,* 835–838.

Fiorelli, M., Blin, J., Bakchine, S., Laplane, D., & Baron, J. C. (1991). PET studies of cortical diaschisis in patients with motor hemi-neglect. *Journal of the Neurological Sciences, 104,* 135–142.

Fleet, W. S., Valenstein, E., Watson, R. T., & Heilman, K. M. (1987). Dopamine agonist therapy for neglect in humans. *Neurology, 37,* 1765–1771.

Fleming, K., Goldberg, T. E., & Gold, J. M. (1994). Applying working memory constructs to schizophrenic cognitive impairment. In A. S. David & J. C. Cutting (Eds.), *The neuropsychology of schizophrenia* (pp. 197–213). Hillsdale, NJ: Lawrence Erlbaum.

Fletcher, N. A., Harding, A. E., & Marsden, C. D. (1991). The relationship between trauma and idiopathic torsion dystonia. *Journal of Neurology, Neurosurgery and Psychiatry, 54,* 713–717.

Flor-Henry, P. (1989). Interhemispheric relationships and depression in schizophrenia in the perspective of cerebral laterality. In R. Williams & J. T. Dalby (Eds.), *Depression in schizophrenics* (pp. 29–46). New York: Plenum Press.

Fodor, J. A. (1983). *The modularity of the mind: An essay on faculty psychology.* Cambridge, MA: MIT Press.

Folstein, S. E., Brandt, J., & Folstein, M. F. (1990). Huntington's disease. In J. L. Cummings (Ed.), *Subcortical dementia* (pp. 87–107). New York: Oxford University Press.

Fowles, D. C. (1992). Schizophrenia: Diathesis-stress revisited. *Annual Review of Psychology, 43,* 303–336.

Frackowiak, R. S. J. (1994). Functional mapping of verbal memory and language. *Trends in Neurosciences, 17,* 109–115.

Freimer, N. B., & Reus, V. I. (1993). The genetics of bipolar disorder and schizophrenia. In R. N. Rosenberg, S. B. Prusiner, S. DiMauro, R. L. Barchi, & L. M. Kunkel (Eds.), *The molecular and genetic basis of neurological disease* (pp. 951–963). Boston: Butterworth-Heinemann.

Friedman, R., Ween, J. E., & Albert, M. L. (1993). Alexia. In K. M. Heilman & E. Valenstein (Eds.), *Clinical neuropsychology* (3rd ed.) (pp. 37–62). Oxford: Oxford University Press.

Frith, C. D. (1992). *The cognitive neuropsychology of schizophrenia.* Hillsdale, NJ: Lawrence Erlbaum.

Frith, U. (1991). *Autism and asperger syndrome.* Cambridge: Cambridge University Press.

Frith, U., Morton, J., & Leslie, A. M. (1991). The cognitive basis of a biological disorder: Autism. *Trends in Neurosciences, 14,* 433–438.

Fromkin, V., Krashen, S., Curtiss, S., Rigler, D., & Rigler, M. (1974). The development of language in Genie: A case of language acquisition beyond the critical period. *Brain and Language, 1,* 81–107.

Fuller, G. N., Marshall, A., Flint, J., Lewis, S., & Wise, R. J. S. (1993). Migraine madness: Recurrent psychosis after migraine. *Journal of Neurology, Neurosurgery and Psychiatry, 56,* 416–418.

Funahashi, S., Chafee, M. V., & Goldman-Rakic, P. (1993). Prefrontal neuronal activity in rhesus monkeys performing a delayed anti-saccade task. *Nature, 345,* 753–750.

Funnell, E., & Sheridan, J. (1992). Categories of knowledge? Unfamiliar aspects of living and nonliving things. *Cognitive Neuropsychology, 9,* 135–153.

Gainotti, G., D'Erme, P., & Bartolomeo, P. (1991). Early orientation of attention toward the half-space ipsilateral to the lesion in patients with unilateral brain damage. *Journal of Neurology, Neurosurgery and Psychiatry, 54,* 1082–1089.

Galaburda, A. M. (1984). Anatomical asymmetries. In N. Geschwind & A. M. Galaburda (Eds.), *Cerebral dominance: The biological foundations* (pp. 11–25). Cambridge, MA: Harvard University Press.

Galaburda, A. M. (1994). Developmental dyslexia and animal studies: At the interface between cognition and neurology. *Cognition, 50,* 133–149.

Gall, F. (1825). *Sur les fonctions du cerveau et sur celles de chacune de ses parties.* Paris: Baillière.

Galletti, C., Battaglini, P. P., & Fattori, P. (1993). Parietal neurons encoding spatial locations in craniotopic coordinates. *Experimental Brain Research, 96,* 221–229.

Gardner, H., Winner, E., & Rehak, A. (1991). Artistry and aphasia. In M. T. Sarno (Ed.), *Acquired aphasia* (2nd ed.) (pp. 373–404). San Diego: Academic Press.

Garruto, R. M., & Brown, P. (1994). Tau protein, aluminum, and Alzheimer's disease. *The Lancet, 343,* 989.

Gates, A., & Bradshaw, J. L. (1977). The role of the cerebral hemispheres in music. *Brain and Language, 4,* 403–431.

Gazzaniga, M. S. (1985). *The social brain.* New York: Basic Books.

Gazzaniga, M. S. (1987). Perceptual and attentional processes following callosal section in humans. *Neuropsychologia, 25,* 119–134.

Gazzaniga, M. S. (1988). Brain modularity: Towards a philosophy of conscious experience. In A. J. Marcel & E. Bisiach (Eds.), *Consciousness in contemporary science* (pp. 218–238). Oxford: Clarendon Press.

Gazzaniga, M. S., & Hillyard, S. A. (1971). Language and speech capacity of the right hemisphere. *Neuropsychologia, 9,* 273–280.

Gazzaniga, M. S., & Smylie, C. S. (1984). What does language do for a right hemisphere? In M. S. Gazzaniga (Ed.), *Handbook of cognitive neuroscience* (pp. 199–209). New York: Plenum Press.

Geary, D. C. (1993). Mathematical disabilities: Cognitive, neuropsychological, and genetic components. *Psychological Bulletin, 114,* 345–362.

Georgiou, N., Bradshaw, J. L., Iansek, R., Phillips, J. G., Mattingley, J. B., & Bradshaw, J. A. (1994). Reduction in external cues and movement sequencing in Parkinson's disease. *Journal of Neurology, Neurosurgery and Psychiatry, 57,* 368–370.

Georgiou, N., Iansek, R., Bradshaw, J. L., Phillips, J. G., Mattingley, J. B., & Bradshaw, J. A. (1993). An evaluation of the role of internal cues in the pathogenesis of Parkinsonian hypokinesia. *Brain, 116,* 1575–1587.

Georgopoulos, A. P. (1990). Neurophysiology of reaching. In M. Jeannerod (Ed.), *Atten-

tion and performance XIII: Motor representation and control (pp. 227–263). Hillsdale, NJ: Lawrence Erlbaum Associates.

Georgopoulos, A. P., Schwartz, A. B., & Kettner, R. (1986). Neuronal population coding of movement direction. *Science, 233,* 1416–1419.

Geschwind, N. (1965). Disconnexion syndromes in animals and man. *Brain, 88,* 585–644.

Geschwind, N., & Galaburda, A. M. (1985). Cerebral lateralization: Biological mechanisms, associations, and pathology I-III: A hypothesis and program for research. *Archives of Neurology, 42,* 428–459, 521–552, 634–654.

Ghez, C. (1991a). Voluntary movement. In E. R. Kandel, J. H. Schwartz, & T. M. Jessell (Eds.), *Principles of neural science* (3rd ed.) (pp. 609–625). New York: Elsevier.

Ghez, C. (1991b). The cerebellum. In E. R. Kandel, J. H. Schwartz, & T. M. Jessell (Eds.), *Principles of neural science* (3rd ed.) (pp. 626–646). New York: Elsevier.

Ghez, C., Gordon, J., & Hening, W. (1988). Trajectory control in dystonia. In S. Fahn, C. D. Marsden, & D. B. Calne (Eds.), *Dystonia 2. Advances in neurology* (Vol. 50, pp. 141–155). New York: Raven Press.

Gibb, W. R. G. (1993). The neuropathology of Parkinson's disorders. In J. Jankovic & E. Tolosa (Eds.), *Parkinson's disease and movement disorders* (2nd ed.) (pp. 253–270). Baltimore: Williams & Wilkins.

Gillberg, C., & Coleman, M. (1992). *The biology of the autistic syndromes: Clinics in developmental medicine* (No. 126, 2nd ed.) Oxford: Mac Keith Press.

Gilley, D. W. (1993). Behavioral and affective disturbances in Alzheimer's disease. In R. W. Parks, R. F. Zec, & R. S. Wilson (Eds.), *Neuropsychology of Alzheimer's disease and other dementias* (pp. 112–137). Oxford: Oxford University Press.

Gilman, S. (1994). Cerebellar control of movement. *Annals of Neurology, 35,* 3–4.

Girotti, F., Casazza, M., Musicco, M., & Avanzini, G. (1983). Oculomotor disorders in cortical lesions in man: The role of unilateral neglect. *Neuropsychologia, 21,* 543–553.

Gloor, P., Salanova, V., Olivier, A., & Quesney, L. F. (1993). The human dorsal hippocampal commissure. *Brain, 116,* 1249–1273.

Goedert, M. (1993). Tau protein and the neurofibrillary pathology of Alzheimer's disease. *Trends in Neurosciences, 16,* 460–465.

Golbe, L. I., & Davis, P. H. (1993). Progressive supranuclear palsy. In J. Jankovic & E. Tolosa (Eds.), *Parkinson's disease and movement disorders* (pp. 145–161). Baltimore: Williams & Wilkins.

Golbe, L. I., & Langston, J. W. (1993). The etiology of Parkinson's disease: New directions for research. In J. Jankovic & E. Tolosa (Eds.), *Parkinson's disease and movement disorders* (2nd ed.) (pp. 93–102). Baltimore: Williams & Wilkins.

Goldberg, G., Mayer, N. H., & Toglia, J. U. (1981). Medial frontal cortex and the alien hand sign. *Archives of Neurology, 38,* 683–686.

Goldberg, M. E., & Bushnell, M. C. (1981). Behavioural enhancement of visual responses in monkey cerebral cortex: II. Modulation in frontal eye fields specifically related to saccades. *Journal of Neurophysiology, 46,* 773–787.

Goldman-Rakic, P. S. (1988a). Topography of cognition: Parallel distributed networks in primate association cortex. *Annual Review of Neuroscience, 11,* 137–156.

Goldman-Rakic, P. S. (1988b). Changing concepts of cortical connectivity: Parallel distributed cortical networks. In P. Rakic & W. Singer (Eds.), *Neurobiology of neocortex* (pp. 177–202). New York: Wiley.

Goldman-Rakic, P. S. (1992). Working memory and the mind. *Scientific American,* 267(3), 72–79.

Goodale, M. A., & Milner, A. D. (1992). Separate visual pathways for perception and action. *Trends in Neurosciences, 15,* 20–25.

Goodale, M. A., Milner, A. D., Jakobson, L. S., & Carey, D. P. (1990). Kinematic analysis of limb movements in neuropsychological research: Subtle deficits and recovery of function. *Canadian Journal of Psychology, 44,* 180–195.

Goodale, M. A., Milner, A. D., Jakobson, L. S., & Carey, D. P. (1991). A neurological dissociation between perceiving objects and grasping them. *Nature, 349,* 154–156.

Goodglass, H. (1993). *Understanding aphasia.* San Diego: Academic Press.

Gordon, H. W. (1990). Neuropsychological sequelae of partial commissurotomy. In F. Boller & J. Grafman (Eds.), *Handbook of neuropsychology* (Vol. 4, pp. 85–97). Amsterdam: Elsevier.

Gordon, H. W., & Bogen, J. E. (1974). Hemispheric lateralization of singing after intracarotid sodium amylobarbitone. *Journal of Neurology, Neurosurgery and Psychiatry, 37,* 727–738.

Gottesmann, I. I., & Bertelson, A. (1989). Confirming unexpressed genotypes for schizophrenia. *Archives of General Psychiatry, 46,* 867–872.

Gould, S. J. (1991). *Wonderful life: The Burgess shale and the nature of history.* London: Penguin.

Grossi, D., Angelini, R., Pecchinenda, A., & Pizzamiglio, L. (1993). Left imaginal neglect in heminattention: Experimental study with the O'clock Test. *Behavioural Neurology, 6,* 155–158.

Grossi, D., Modafferi, A., Pelosi, L., & Trojano, L. (1989). On the different roles of the cerebral hemispheres in mental imagery: The "O'clock Test" in two clinical cases. *Brain and Cognition, 10,* 18–27.

Growdon, J. H. (1992). Treatment for Alzheimer's disease. *The New England Journal of Medicine, 327,* 1306–1308.

Gruneberg, M. M., Morris, P. E., & Sykes, R. N. (Eds.) (1988). *Practical aspects of memory: Current research and issues. Vol. 1, Memory in everyday life.* Chichester: Wiley.

Grüsser, O.-J., & Landis, T. (1991). *Visual agnosias and other disturbances of visual perception and cognition.* London: MacMillan Press.

Guariglia, C., & Antonucci, G. (1992). Personal and extrapersonal space: A case of neglect dissociation. *Neuropsychologia, 30,* 1001–1009.

Guariglia, C., Padovani, A., Pantano, P., & Pizzamiglio, L. (1993). Unilateral neglect restricted to visual imagery. *Nature, 364,* 235–237.

Gupta, M. (1993). Parkinson's disease pathology: Multineurotransmitter systems defects. In J. S. Schneider & M. Gupta (Eds.), *Current concepts in Parkinson's disease research* (pp. 21–39). Seattle: Hogrefe & Huber.

Guridi, J., Luquin, M. R., Herrero, M. T., & Obeso, J. A. (1993). The subthalamic nucleus: A possible target for stereotaxic surgery in Parkinson's disease. *Movement Disorders, 8,* 421–429.

Gusella, J. F., & MacDonald, M. E. (1994). Huntington's disease and repeating trinucleotides. *The New England Journal of Medicine, 330,* 1450–1451.

Gusella, J. F., MacDonald, M. E., Ambrose, C. M., & Duyao, M. P. (1993). Molecular genetics of Huntington's disease. *Archives of Neurology, 50,* 1157–1163.

Haaland, K. Y., & Harrington, D. L. (1994). Limb-sequencing deficits after left but not right hemisphere damage. *Brain and Cognition, 24*, 104–122.

Haber, R. N. (1969). Eidetic images. *Scientific American, 220*, 36–44.

Hallett, M. (1993). Physiology of basal ganglia disorders: An overview. *The Canadian Journal of Neurological Sciences, 20*, 177–183.

Hallett, M., Berardelli, A., Delwaide, P., Freund, H.-J., Kimura, J., Lücking, C., Rothwell, J. C., & Shahani, B. T. (1994). Central EMG and tests of motor control: Report on an IFCN committee. *Electroencephalography and Clinical Neurophysiology, 90*, 404–432.

Halligan, P. W., Donegan, C. A., & Marshall, J. C. (1992). When is a cue not a cue? On the intractability of visuospatial neglect. *Neuropsychological Rehabilitation, 2*, 283–293.

Halligan, P. W., & Marshall, J. C. (1988). How long is a piece of string? A study of line bisection in a case of visual neglect. *Cortex, 24*, 321–328.

Halligan, P. W., & Marshall, J. C. (1989a). Two techniques for the assessment of line bisection in visuo-spatial neglect: A single case study. *Journal of Neurology, Neurosurgery and Psychiatry, 52*, 1300–1302.

Halligan, P. W., & Marshall, J. C. (1989b). Perceptual cueing and perceptuo-motor compatibility in visuo-spatial neglect: A single case study. *Cognitive Neuropsychology, 6*, 423–435.

Halligan, P. W., & Marshall, J. C. (1991a). Left neglect for near but not far space in man. *Nature, 350*, 498–500.

Halligan, P. W., & Marshall, J. C. (1991b). Figural modulation of visuo-spatial neglect: A case study. *Neuropsychologia, 29*, 619–628.

Halligan, P. W., & Marshall, J. C. (1992). Left visuo-spatial neglect: A meaningless entity? *Cortex, 28*, 525–535.

Halligan, P. W., & Marshall, J. C. (1993a). The history and clinical presentation of neglect. In I. H. Robertson & J. C. Marshall (Eds.), *Unilateral neglect: Clinical and experimental studies* (pp. 3–25). Hove, U.K.: Lawrence Erlbaum Associates.

Halligan, P. W., & Marshall, J. C. (1993b). When two is one: A case study of spatial parsing in visual neglect. *Perception, 22*, 309–312.

Halligan, P. W., & Marshall, J. C. (1994a). Toward a principled explanation of unilateral neglect. *Cognitive Neuropsychology, 11*, 167–206.

Halligan, P. W., & Marshall, J. C. (1994b). Right-sided cueing can ameliorate left neglect. *Neuropsychological Rehabilitation, 4*, 63–73.

Halligan, P. W., & Marshall, J. C. (1994c). Figural perception and parsing in visuospatial neglect. *NeuroReport, 5*, 537–539.

Halligan, P. W., Marshall, J. C., & Wade, D. T. (1990). Do visual field deficits exacerbate visuo-spatial neglect? *Journal of Neurology, Neurosurgery and Psychiatry, 53*, 487–491.

Halligan, P. W., Marshall, J. C., & Wade, D. T. (1992a). Contrapositioning in a case of visual neglect. *Neuropsychological Rehabilitation, 2*, 125–135.

Halligan, P. W., Marshall, J. C., & Wade, D. T. (1992b). Left on right: Allochiria in a case of left visuo-spatial neglect. *Journal of Neurology, Neurosurgery and Psychiatry, 55*, 717–719.

Halsband, U., Ito, N., Tanji, J., & Freund, H.-J. (1993). Role of premotor cortex and the supplementary motor area in the temporal control of movement in man. *Brain, 116,* 243–266.

Hamer, D. H., Hu, S., Magnuson, V. L., Hu, N., & Pattatucci, A. M. L. (1993). A linkage between DNA markers on the chromosome and male sexual orientation. *Science, 261,* 321–327.

Hammond-Tooke, G. D., & Pollock, M. (1992). Depression, dementia, and Parkinson's disease. In A. B. Joseph & R. R. Young (Eds.), *Movement disorders in neurology and neuropsychiatry* (pp. 221–229). Oxford: Blackwell Scientific.

Handley, S. L., & Dursun, S. M. (1992). Serotonin and Tourette's syndrome: Movements such as head shakes and wet-dog shakes may model human tics. *Advances in the Biosciences, 85,* 235–248.

Hardy, J. (1994). ApoE4, amyloid, and Alzheimer's disease. *Science, 263,* 454.

Hardy, J., & Roberts, G. W. (1993). Smoking and neurodegenerative diseases. *The Lancet, 342,* 1238–1239.

Harlow, J. M. (1868). Recovery from passage of an iron bar through the head. *Publications of the Massachusetts Medical Society, 2,* 327–347.

Harrington, C. R., Wischik, C. M., McArthur, F. K., Taylor, G. A., Edwardson, J. A., & Candy, J. M. (1994). Alzheimer's-disease-like changes in tau protein processing: Association with aluminum accumulation in brains of renal dialysis patients. *The Lancet, 343,* 993–997.

Harrington, D. L., & Haaland, K. Y. (1992). Motor sequencing with left hemisphere damage. *Brain, 115,* 857–874.

Harris, G. C., & Aston-Jones, G. (1994). Involvement of D2 dopamine receptors in the nucleus accumbens in the opiate withdrawal syndrome. *Nature, 371,* 155–157.

Harvey, M., Milner, A. D., & Roberts, R. C. Spatial bias in visually-guided reaching and bisection following right cerebral stroke. *Cortex, 30,* 343–350.

Hasselbalch, S. G., Øberg, G., Sørensen, S., Andersen, A. R., Waldemar, G., Schmidt, J. F., Fenger, K., & Paulson, O. B. (1992). Reduced regional cerebral blood flow in Huntington's disease studied by SPECT. *Journal of Neurology, Neurosurgery and Psychiatry, 55,* 1018–1023.

Hauser, R. A., & Olanow, C. W. (1993). Parkinson's disease. In L. Barclay (Ed.), *Clinical geriatric neurology* (pp. 155–169). Philadelphia: Lea & Febiger.

Hebb, D. O. (1949). *The organization of behavior.* New York: Wiley.

Hécaen, H. (1962). Clinical symptomatology in right and left hemispheric lesions. In V. B. Mountcastle (Ed.), *Interhemispheric Relations and Cerebral Dominance* (pp. 215–243). Baltimore: The Johns Hopkins Press.

Heilman, K. M., Bowers, D., Coslett, H. B., Whelan, H., & Watson, R. T. (1985). Directional hypokinesia: Prolonged reaction times for leftward movements in patients with right hemisphere lesions and neglect. *Neurology, 35,* 855–859.

Heilman, K. M., Bowers, D., Valenstein, E., & Watson, R. T. (1987). Hemispace and hemispatial neglect. In M. Jeanerrod (Ed.), *Neurophysiological and neuropsychological aspects of spatial neglect* (pp. 115–150). Amsterdam: North-Holland.

Heilman, K. M., Bowers, D., & Watson, R. T. (1984). Pseudoneglect in a patient with partial callosal disconnection. *Brain, 107,* 519–532.

Heilman, K. M., & Howell, F. (1980). Seizure-induced neglect. *Journal of Neurology, Neurosurgery and Psychiatry, 43,* 1035–1040.

Heilman, K. M., & Rothi, L. J. G. (1993). Apraxia. In K. M. Heilman & E. Valenstein (Eds.), *Clinical Neuropsychology* (3rd ed.) (pp. 141–164). Oxford: Oxford University Press.

Heilman, K. M., & Valenstein, E. (1972). Frontal lobe neglect in man. *Neurology, 22,* 660–664.

Heilman, K. M., & Valenstein, E. (1979). Mechanisms underlying hemispatial neglect. *Annals of Neurology, 5,* 166–170.

Heilman, K. M., Valenstein, E., & Watson, R. T. (1994). The what and the how of neglect. *Neuropsychological Rehabilitation, 4,* 133–139.

Heilman, K. M., & Van Den Abell, T. (1979). Right hemispheric dominance for mediating cerebral activation. *Neuropsychologia, 17,* 315–321.

Heilman, K. M., & Watson, R. T. (1978). Changes in the symptoms of neglect induced by changing task strategy. *Archives of Neurology, 35,* 47–49.

Heilman, K. M., Watson, R. T., & Valenstein, E. (1980). Unilateral gaze deficit associated with hemispatial neglect. *Neurology, 30,* 360.

Heilman, K. M., Watson, R. T., & Valenstein, E. (1993). Neglect and related disorders. In K. M. Heilman & E. Valenstein (Eds.), *Clinical neuropsychology* (3rd ed.) (pp. 279–336). New York: Oxford University Press.

Heinrichs, R. W. (1993). Schizophrenia and the brain. *American Psychologist, 48,* 221–233.

Henderson, V. W. (1990). Alalia, aphemia, and aphasia. *Archives of Neurology, 47,* 85–88.

Henderson, V. W., Paganini-Hill, A., Emanuel, C. K., Dunn, M. E., & Buckwalter, J. G. (1994). Estrogen replacement therapy in older women: Comparisons between Alzheimer's disease cases and nondemented control subjects. *Archives of Neurology, 51,* 896–900.

Henschen, S. E. (1920). *Klinische und Anatomische Beitrage zur Pathologie des Gehirns Teil 5: Ueber Aphasie, Amusie und Akalkulie.* Stockholm: Nordiska Bokhanden.

Henschen, S. E. (1925). Clinical and anatomical contributions on brain pathology. *Archives of Neurology and Psychiatry, 13,* 226—249.

Henson, R. A. (1985). Amusia. In J. A. M. Frederiks (Ed.), *Handbook of clinical neurology, Vol. 1, 45): Clinical neuropsychology* (pp. 483–490). New York: Elsevier.

Hermelin, B., & O'Connor, N. (1986). Idiot-savant calendrical calculations: Rules and regularities. *Psychological Medicine, 16,* 885–893.

Hier, D. B., Davis, K. R., Richardson, E. P., & Mohr, J. P. (1977). Hypertensive putaminal hemorrhage. *Annals of Neurology, 1,* 152–159.

Hier, D. B., Mondlock, J., & Caplan, L. R. (1983). Recovery of behavioural abnormalities after right hemisphere stroke. *Neurology, 33,* 345–350.

Hjaltason, H., & Tegnér, R. (1992). Darkness improves line bisection in unilateral spatial neglect. *Cortex, 28,* 353–358.

Hoehn, M. M., & Yahr, M. D. (1967). Parkinsonism: Onset, progression and mortality. *Neurology, 17,* 427–442.

Holloway, R. L., Anderson, P. J., Defendini, P. J., & Harper, C. (1993). Sexual dimorphism of the human corpus callosum from three independent samples: Relative size of the corpus callosum. *American Journal of Physical Anthropology, 92,* 481–498.

Holmes, G. (1922). The clinical symptoms of cerebellar disease and their interpretation. *The Lancet, 1,* 1177–1182, 1231–1237; 2, 59–65, 111–115.

Holst, von E., & Mittelstaedt, H. (1950). Das Reafferenzprinzip. Wechselwirkungen zwischen Zentralnervensystem und Peripherie. *Naturwissenschaften, 37,* 464–476.

Holtzman, J. D. (1984). Interactions between cortical and subcortical visual areas: Evidence from human commissurotomy patients. *Vision Research, 24,* 801–813.

Home, E. (1814). Observations on the functions of the brain. *Philosophical Transactions of the Royal Society of London, 104,* 469–486.

Horgan, J. (1994, May). Neural eavesdropping. *Scientific American,* 9.

Horwitz, B., Rumsey, J. M., Grady, C. L., & Rapoport, S. I. (1988). The cerebral metabolic landscape in autism: Intercorrelations of regional glucose utilization. *Archives of Neurology, 45,* 749–755.

Howard, D., Patterson, K., Wise, R., Brown, W. D., Friston, K., Weiller, C., & Frackowiak, R. (1992). The cortical localization of the lexicons. *Brain, 115,* 1769–1782.

Howes, D., & Boller, F. (1975). Simple reaction time: Evidence for focal impairment from lesions of the right hemisphere. *Brain, 98,* 317–332.

Huber, S. J., & Shuttleworth, E. C. (1990). Neuropsychological assessment of subcortical dementia. In J. L. Cummings (Ed.), *Subcortical dementia* (pp. 71–86). New York: Oxford University Press.

Humphrey, N. (1992). *A history of the mind: Evolution and birth of consciousness.* London: Chatto & Windus.

Humphreys, G. W., & Riddoch, M. J. (1993). Object agnosias. *Baillière's Clinical Neurology, 2,* 339–359.

Humphreys, G. W., & Riddoch, M. J. (1994). Attention to within-object and between-object spatial representations: Multiple sites for visual selection. *Cognitive Neuropsychology, 11,* 207–241.

Humphreys, G. W., Troscianko, T., Riddoch, M. J., Boucart, M., Donnelly, N., & Harding, G. F. A. (1992). Covert processing in different visual recognition systems. In A. D. Milner & M. D. Rugg (Eds.), *The neuropsychology of consciousness* (pp. 39–68). San Diego: Academic Press.

Hung, T.-P., & Ryu, S.-J. (1988). Anterior cerebral artery syndromes. In P. J. Vinken, G. W. Bruyn, & H. L. Klawans (Eds.), *Vascular diseases, Part 1* (pp. 339–352). Amsterdam: Elsevier.

Huntington, G. (1872). On chorea. *Medicine and Surgery Reporter, 26,* 317–321.

Huppert, F. A. (1991, July). *Neuropsychology in the community: Epidemiological studies of normal aging and dementia.* Paper presented at the International Neuropsychological Society/Australian Society for the Study of Brain Impairment. Pacific Rim Conference, Gold Coast, Australia.

Iacono, R. P., & Lonser, R. R. (1994). Reversal of Parkinson's akinesia by pallidotomy. *The Lancet, 343,* 418–419.

Ikeda, A., Lüders, H., Burgess, R., & Shibasaki, H. (1992). Movement related potentials recorded from supplementary motor area and primary motor area. *Brain, 115,* 1017–1043.

Ikeda, H., Head, G. M., & Ellis, C. J. K. (1994). Electrophysiological signs of retinal dopamine deficiency in recently diagnosed Parkinson's disease and a follow up study. *Vision Research, 34,* 2629–2638.

Ishiai, S., Sugishita, M., Ichikawa, T., Gono, S., & Watabiki, S. (1993). Clock-drawing test and unilateral spatial neglect. *Neurology, 43,* 106–110.

Italian ALS Study Group. (1993). Branched-chain amino acids and amyotrophic lateral sclerosis: A treatment failure? *Neurology, 43,* 2466–2470.

Ivry, R. B., & Diener, H. C. (1991). Impaired velocity perception in patients with lesions of the cerebellum. *Journal of Cognitive Neuroscience, 3,* 355–365.

Jackson, J. H. (1932). Case of large cerebral tumour without optic neuritis and with left hemiplegia and imperception. *Royal Ophthalmological Hospital Reports* (1876, 8, 434–444). Reprinted in I. Taylor (Ed.), *Selected writings of John Hughlings Jackson* (Vol. 2). London: Hodder and Stoughton.

Jacobs, B. L., & Fornal, C. A. (1993). 5-HT and motor control: A hypothesis. *Trends in Neurosciences, 16,* 346–351.

Jacoby, L. L., & Kelley, C. (1992). Unconscious influences of memory: Dissociations and automaticity. In A. D. Milner & M. D. Rugg (Eds.), *The neuropsychology of consciousness* (pp. 201–234). San Diego: Academic Press.

James, W. (1890). *The Principles of Psychology* (Vol. 1). London: Macmillan.

Jankovic, J. (1987). The neurology of tics. In C. D. Marsden & S. Fahn (Eds.), *Movement disorders 2* (pp. 383–406). London: Butterworths.

Jankovic, J. (1992). Diagnosis and classification of tics and Tourette syndrome. In T. N. Chase, A. J. Friedhoff, & D. J. Cohen (Eds.), *Tourette syndrome: Genetics, neurobiology and treatment: Advances in neurology* (Vol. 58, pp. 7–14). New York: Raven Press.

Jankovic, J., Leder, S., Warner, D., & Schwartz, K. (1991). Cervical dystonia: Clinical findings and associated movement disorders. *Neurology, 41,* 1088–1091.

Jason, G. (1990). Disorders of motor function following cortical lesions: Review and theoretical considerations. In G. E. Hammond (Ed.), *Cerebral control of speech and limb movements* (pp. 141–168). Amsterdam: North Holland.

Jason, G. W., & Pajurkova, E. M. (1992). Failure of metacontrol: Breakdown in behavioural unity after lesion of the corpus callosum and inferomedial frontal lobes. *Cortex, 28,* 241–260.

Jeannerod, M., & Rossetti, Y. (1993). Visuomotor coordination as a dissociable visual function: Experimental and clinical evidence. *Baillière's Clinical Neurology, 2,* 439–460.

Jeeves, M. A. (1990). Agenesis of the corpus callosum. In F. Boller & J. Grafman (Eds.), *Handbook of neuropsychology* (Vol. 4, pp. 99–114). Amsterdam: Elsevier.

Jenner, P. (1994). Oxidative damage in neurodegenerative disease. *The Lancet, 344,* 796–798.

Jerison, H. J. (1986). Evolutionary biology of intelligence: The nature of the problem. In H. J. Jerison & I. Jerison (Eds.), *Intelligence and evolutionary biology* (pp. 1–12). New York: Springer.

Joanette, Y., Goulet, P., & Hannequin, D. (1990). *Right hemisphere and verbal communication.* New York: Springer.

Jones, A. W. R., & Richardson, J. S. (1990). Alzheimer's disease: Clinical and pathological characteristics. *International Journal of Neuroscience, 50,* 147–168.

Jones, D. L., Bradshaw, J. L., Phillips, J. G., Iansek, R., Mattingley, J. B., & Bradshaw, J. A. (1994). Allocation of attention to programming of movement sequences in Parkinson's disease. *Journal of Clinical and Experimental Neuropsychology, 16,* 117–128.

Jones, D. L., Phillips, J. G., Bradshaw, J. L., Iansek, R., & Bradshaw, J. A. (1992). Impairments in bilateral alternating movements in Parkinson's disease. *Journal of Neurology, Neurosurgery and Psychiatry, 55,* 503–506.

Jones, G. V. (1983). On double dissociation of function. *Neuropsychologia, 21,* 397–400.

Jonides, J., Smith, E. E., Koeppe, R. A., Awh, E., Minoshima, S., & Mintun, M. A. (1993). Spatial working memory in humans as revealed by PET. *Nature, 363,* 623–626.

Joseph, A. B. (1986). Focal central nervous system abnormalities in patients with misidentification syndrome. *Bibliotheca Psychiatrica, 164*, 68–79.

Joseph, A. B. (1987). Delusional misidentifcation of the Capgras and intermetamorphosis types responding to chlorazepate. *Acta Psychiatrica Scandinavica, 75*, 330–332.

Kahn, H. J. (1991). Prologue to special issue on cognitive and neuropsychological aspects of calculation disorders. *Brain and Cognition, 17*, 97–101.

Kahn, H. J., & Whitaker, H. A. (1991). Acalculia: An historical review of localization. *Brain and Cognition, 17*, 102–115.

Kaiser, J. (1994). Alzheimer's: Could there be a zinc link? *Science, 265*, 1365–1366.

Kamaki, M., Kawamura, M., Moriya, H., & Hirayama, K. (1993). "Crossed homonymous hemianopia" and "crossed left hemispatial neglect" in a case of Marchiafava-Bignami disease. *Journal of Neurology, Neurosurgery, and Psychiatry, 56*, 1027–1032.

Kanner, L. (1943). Autistic disturbances of affective contact. *Nervous Child, 2*, 217–250.

Kaplan, E. (1991). Aphasia-related disorders. In M. T. Sarno (Ed.), *Acquired aphasia* (2nd ed.) (pp. 313–338). San Diego: Academic Press.

Kaplan, J. A., & Gardner, H. (1989). Artistry after unilateral brain disease. In F. Boller & J. Grafman (Eds.), *Handbook of neuropsychology* (Vol. 2, pp. 141–155). Amsterdam: Elsevier.

Kaplan, R. F., Meadows, M.-E., Verfaellie, M., Kwan, E., Ehrenberg, B. L., Bromfield, E. B., & Cohen, R. A. (1994). Lateralization of memory for the visual attributes of objects: Evidence from the posterior cerebral artery amobarbital test. *Neurology, 44*, 1069–1073.

Kaplan, R. F., Verfaellie, M., DeWitt, D., & Caplan, L. R. (1990). Effects of changes in stimulus contingency on visual extinction. *Neurology, 40*, 1299–1301.

Kapur, N., Ellison, D., Smith, M. P., McClellan, D. I., & Burrows, E. H. (1992). Focal retrograde amnesia following bilateral temporal lobe pathology: A neuropsychological and magnetic resonance study. *Brain, 115*, 73–86.

Kapur, N., Turner, A., & King, C. (1988). Reduplicative paramnesia: Possible anatomical and neuropsychological mechanisms. *Journal of Neurology, Neurosurgery and Psychiatry, 51*, 579–581.

Karnath, H.-O. (1988). Deficits of attention in acute and recovered hemi-neglect. *Neuropsychologia, 20*, 27–45.

Karnath, H.-O. (1994a). Spatial limitations of eye movements during ocular exploration of simple line drawings in neglect syndrome. *Cortex, 30*, 319–330.

Karnath, H.-O. (1994b). Subjective body orientation in neglect and the interactive contribution of neck muscle proprioception and vestibular stimulation. *Brain, 117*, 1001–1012.

Karnath, H.-O., Christ, K., & Hartje, W. (1993). Decrease of contralateral neglect by neck muscle vibration and spatial orientation of trunk midline. *Brain, 116*, 383–396.

Karnath, H.-O., & Hartje, W. (1987). Residual information processing in the neglected visual half-field. *Journal of Neurology, 234*, 180–184.

Karnath, H.-O., Schenkel, P., & Fischer, B. (1991). Trunk orientation as the determining factor of the "contralateral" deficit in the neglect syndrome and as the physical anchor of the internal representation of body orientation in space. *Brain, 114*, 1997–2014.

Kartsounis, L. D., & Crewes, H. (1994). Horologagnosia: An impairment of the ability to tell the time. *Journal of Neurology, Neurosurgery and Psychiatry, 57*, 384–385.

Katz, R. B., & Lanzoni, S. M. (1992). Automatic activation of word phonology from print in deep dyslexia. *Quarterly Journal of Experimental Psychology, 45A*, 575–608.

Katzman, R. (1993). Education and the prevalence of dementia and Alzheimer's disease. *Neurology, 43*, 13–20.

Keele, S. W., & Ivry, R. (1990). Does the cerebellum provide a common computation for diverse tasks? A timing hypothesis. *Annals of the New York Academy of Sciences, 608*, 179–207.

Kennedy, J. S., & Whitehouse, P. (1993). Alzheimer's disease. In L. Barclay (Ed.), *Clinical geriatric neurology* (pp. 76–89). Philadelphia: Lea & Febiger.

Kertesz, A. (1991). Language cortex. *Aphasiology, 5*, 207–234.

Kerwin, R. J. (1993). Glutamate receptors, microtubule associated proteins and developmental anomaly in schizophrenia: An hypothesis. *Psychological Medicine, 23*, 547–551.

Kim, S.-G., Ugurbil, K., & Strick, P. L. (1994). Activation of a cerebellar output nucleus during cognitive processing. *Science, 265*, 949–951.

Kimberg, D. Y., & Farah, M. J. (1993). A unified account of cognitive impairments following frontal lobe damage: The role of working memory in complex, organized behavior. *Journal of Experimental Psychology, 122*, 411–428.

Kimura, D. (1983). Sex differences in cerebral organization for speech and praxic functions. *Canadian Journal of Psychology, 37*, 19–35.

Kimura, D. (1987). Are men's and women's brains really different? *Canadian Psychology, 28*, 133–147.

Kinsbourne, M. (1987). Mechanisms of unilateral neglect. In M. Jeannerod (Ed.), *Neurophysiological and neuropsychological aspects of spatial neglect* (pp. 69–86). Amsterdam: North-Holland.

Kinsbourne, M. (1993). Orientational bias model of unilateral neglect: Evidence from attentional gradients within hemispace. In I. H. Robertson & J. C. Marshall (Eds.), *Unilateral neglect: Clinical and experimental studies* (pp. 63–86). Hove, U.K.: Lawrence Erlbaum Associates.

Kinsella, G., & Ford, B. (1985). Hemi-inattention and the recovery patterns of stroke patients. *International Rehabilitation Medicine, 7*, 102–106.

Kirk, A. and Kertesz, A. (1994). Cortical and subcortical aphasias compared. *Aphasiology, 8*, 65–82.

Kish, S. J., El-Awar, M., Schut, L., Leach, L., Oscar-Berman, M., & Freedman, M. (1988). Cognitive deficits in olivopontocerebellar atrophy: Implications for the cholinergic hypothesis of Alzheimer's disease. *Annals of Neurology, 24*, 200–206.

Klimidis, S., Stuart, G. W., Minas, I. H., Copolov, D. L., & Singh, B. S. (1993). Positive and negative symptoms in the psychoses: Re-analysis of published SAPS and SANS global ratings. *Schizophrenia Research, 9*, 11–18.

Kopelman, M. D. (1993). The neuropsychology of remote memory. In F. Boller & J. Grafman (Eds.), *Handbook of neuropsychology* (Vol. 8, pp. 215–238). Amsterdam: Elsevier.

Kornhuber, H. H., Deeke, L., Lang, W., Lang, M., & Kornhuber, A. (1989). Will, volitional actions, attention and cerebral potentials in man. In W. A. Hershberger (Ed.), *Volitional actions: Advances in neurology 62* (pp. 107–168). Amsterdam: Elsevier.

Koroshetz, W. J., Myers, R. H., & Martin, J. (1992). The neurology of Huntington's disease. In A. B. Joseph & R. R. Young (Eds.), *Movement disorders in neurology and neu-*

ropsychiatry (pp. 167–177). Oxford: Blackwell Scientific.

Koroshetz, W. J., Myers, R. H., & Martin, J. (1993). Huntington's disease. In R. N. Rosenberg, S. B. Prusiner, S. DiMauro, R. L. Barchi, & L. M. Kunkel (Eds.), *The molecular and genetic basis of neurological disease* (pp. 737–752). Boston: Butterworth-Heinemann.

Kosslyn, S. M. (1990). Mental imagery. In D. N. Osherson, S. M. Kosslyn, & J. M. Hollerbach (Eds.), *Visual cognition and action, Vol. 2: An invitation to cognitive science* (pp. 73–97). Cambridge, MA: MIT Press.

Kosslyn, S. M. (1994). On cognitive neuroscience. *Journal of Cognitive Neuroscience, 6,* 297–303.

Kosslyn, S. M., Holtzman, J. D., Farah, M. J., & Gazzaniga, M. S. (1985). A computational analysis of mental image generation: Evidence from functional dissociations in split-brain patients. *Journal of Experimental Psychology: General, 114,* 311–341.

Kötter, R., & Meyer, N. (1992). The limbic system: A review of its empirical foundation. *Behavioural Brain Research, 52,* 105–127.

Kramer, A. F., & Jacobson, A. (1991). Perceptual organization and focused attention: The role of objects and proximity in visual processing. *Perception and Psychophysics, 50,* 267–284.

Kremer, B., Goldberg, P., Andrew, S. E., Theilmann, J., Telenius, H., Zeisler, J., Squitieri, F., Lin, B., Bassett, A., Almqvist, E., Bird, T. D., & Haydn, M. R. (1994). A worldwide study of the Huntington's disease mutation: The sensitivity and specificity of measuring CAG repeats. *The New England Journal of Medicine, 330,* 1403–1406.

Kurlan, R. (1992). The pathogenesis of Tourette's syndrome: A possible role for hormonal and excitatory neurotransmitter influences in brain development. *Archives of Neurology, 49,* 874–876.

Làdavas, E. (1987). Is the hemispatial deficit produced by right parietal lobe damage associated with retinal or gravitational coordinates? *Brain, 110,* 167–180.

Làdavas, E., Del Pesce, M., Mangun, G. R., & Gazzaniga, M. S. (1994). Variations in attentional bias of the disconnected cerebral hemispheres. *Cognitive Neuropsychology, 11,* 57–74.

Làdavas, E., Menghini, G., & Umiltà, C. (1994). A rehabilitation study of hemispatial neglect. *Cognitive Neuropsychology, 11,* 75–95.

Làdavas, E., Paladini, R., & Cubelli, R. (1993). Implicit associative priming in a patient with left visual neglect. *Neuropsychologia, 31,* 1307–1320.

Làdavas, E., Petronio, A., & Umiltà, C. (1990). The deployment of visual attention in the intact field of hemineglect patients. *Cortex, 26,* 307–317.

Làdavas, E., Umiltà, C., Ziani, P., Brogi, A., & Minarini, M. (1993). The role of right side objects in left side neglect: A dissociation between perceptual and directional motor neglect. *Neuropsychologia, 31,* 761–773.

Lambert, A. J. (1991). Interhemispheric interaction in the split brain. *Neuropsychologia, 29,* 941–948.

Lambert, A. J. (1993). Attentional interaction in the split-brain: Evidence from negative priming. *Neuropsychologia, 31,* 313–324.

Landsberg, J. P., McDonald, B., & Watt, F. Absence of aluminum in neuritic plaque cores in Alzheimer's disease. *Nature, 360,* 65–68.

Lang, W., Obrig, H., Lindinger, G., Cheyne, D., & Deeke, L. (1990). Supplementary motor area activation while tapping bimanually different rhythms in musicians. *Experi-*

mental Brain Research, 79, 504–514.

Lange, K. W., Robbins, T. W., Marsden, C. D., James, M., Owen, A. M., & Paul, G. M. (1992). L-dopa withdrawal in Parkinson's disease selectively impairs cognitive performance in tests sensitive to frontal lobe dysfunction. *Psychopharmacology, 107,* 394–404.

Laplane, D., & Degos, J. D. (1983). Motor neglect. *Journal of Neurology, Neurosurgery and Psychiatry, 46,* 152–158.

Larvelle, M., Abi-Dargham, A., Casanova, M. F., Toti, R., Weinberger, D. R., & Kleinman, J. E. (1993). Selective abnormalities of prefrontal serotonergic receptors in schizophrenia. *Archives of General Psychiatry, 50,* 810–818.

Lashley, K. S. (1950). In search of the engram. *Symposium of the Society of Experimental Biology, 4,* 454–482.

Lassonde, M., Sauerwein, H., McCabe, N., Laurencelle, L., & Geoffroy, G. (1988). Extent and limits of cerebral adjustment to early section or congenital absence of the corpus callosum. *Behavioural Brain Research, 30,* 165–181.

Lawson, I. R. (1962). Visual-spatial neglect in lesions of the right cerebral hemisphere: A study in recovery. *Neurology, 12,* 23–33.

Leckman, J. F., Pauls, D. L., Peterson, B. S., Riddle, M. A., Anderson, G. M., & Cohen, D. J. (1992). Pathogenesis of Tourette syndrome: Clues from the clinical phenotype and natural history. In T. N. Chase, A. J. Friedhoff, & D. J. Cohen (Eds.), *Advances in neurology,* (Vol. 58, pp. 15–24). New York: Raven Press.

Leckman, J. F., & Peterson, B. S. (1993). The pathogenesis of Tourette's syndrome: Epigenetic factors active in early CNS development. *Biological Psychiatry, 34,* 425–427.

Lecours, A. R., & Joanette, Y. (1980). Linguistic and other psychological aspects of paroxysmal aphasia. *Brain and Language, 10,* 1–23.

LeDoux, J. E. (1994, June). Emotion, memory, and the brain. *Scientific American,* 32–39.

Lees, A. J. (1990). Progressive supranuclear palsy (Steele-Richardson-Olszewski syndrome). In J. L. Cummings (Ed.), *Subcortical dementia* (pp. 123–131). New York: Oxford University Press.

Lees, A. J., & Tolosa, E. (1993). Tics. In J. Jankovic & E. Tolosa (Eds.), *Parkinson's disease and movement disorders* (2nd ed.) (pp. 329–336). Baltimore: Williams & Wilkins.

Le Fever, F. F. (1992). Equivalent operations find equivalent unilateral neglect and search patterns in near and far space: New use for an oral "cancellation" test. *Journal of Clinical and Experimental Neuropsychology, 14,* 85 (abstract).

Leicester, J., Sidman, M., Stoddard, L. T., & Mohr, J. P. (1969). Some determinants of visual neglect. *Journal of Neurology, Neurosurgery and Psychiatry, 32,* 580–587.

Leiner, H. C., Leiner, A. L., & Dow, R. S. (1993a). Cognitive and language functions of the human cerebellum. *Trends in Neurosciences, 16,* 444–447.

Leiner, H. C., Leiner, A. L., & Dow, R. S. (1993b). The role of the cerebellum in the human brain. *Trends in Neurosciences, 16,* 453–454.

LeVay, S. (1991). A difference in hypothalamic structure between heterosexual and homosexual men. *Science, 253,* 1034–1037.

Levin, B. (1990). Spatial cognition in Parkinson's disease. *Alzheimer's Disease and Associated Disorders, 4,* 161–170.

Levin, B. E., Tomer, R., & Rey, G. J. (1992). Clinical correlates of cognitive impairment in Parkinson's disease. In S. J. Huber and J. L. Cummings (Eds.), *Parkinson's disease:*

Neurobehavioral aspects (pp. 97–106). New York: Oxford University Press.

Levin, H. S., Goldstein, F. C., & Spiers, P. A. (1993). Acalculia. In K. M. Heilman & E. Valenstein (Eds.), *Clinical neuropsychology* (3rd ed.) (pp. 91–122). Oxford: Oxford University Press.

Levine, D. N. (1990). Unawareness of visual and sensorimotor defects: A hypothesis. *Brain and Cognition, 13*, 233–281.

Levine, D. N., & Calvanio, R. (1989). Prosopagnosia—a defect in visual configurational processing. *Brain and Cognition, 10*, 149–170.

Levine, D. N., Calvanio, R., & Rinn, W. E. (1991). The pathogenesis of anosognosia for hemiplegia. *Neurology, 41*, 1770–1781.

Levine, D. N., & Grek, A. (1984). The anatomic basis of delusions after right cerebral infarction. *Neurology, 34*, 577–582.

Levine, D. N., Kaufman, K. J., & Mohr, J. P. (1978). Inaccurate reaching associated with superior parietal lobe tumor. *Neurology, 28*, 556–561.

Levine, D. N., Warach, J. D., Benowitz, L., & Calvanio, R. (1986). Left spatial neglect: Effects of lesion size and premorbid brain atrophy on severity and recovery following right cerebral infarction. *Neurology, 36*, 362–366.

Levy, J., & Trevarthen, C. (1976). Metacontrol of hemispheric function in human split-brain patients. *Journal of Experimental Psychology: Human Perception and Performance, 2*, 299–312.

Lewis, J. M. (1993). Jonathan Swift and Alzheimer's disease. *The Lancet, 342*, 504.

Lezak, M. D. (1983). *Neuropsychological assessment* (2nd ed.). New York: Oxford University Press.

Lhermitte, F. (1983). "Utilisation behaviour" and its relation to lesions of the frontal lobes. *Brain, 106*, 237–255.

Liberman, A. M., & Mattingly, I. G. (1985). The motor theory of speech perception revised. *Cognition, 21*, 1–36.

Lichtheim, L. (1885). On aphasia. *Brain, 7*, 433–484.

Lieberman, P. (1989). The origins of some aspects of human language and cognition. In P. Mellars & C. Stringer (Eds.), *The human revolution: Behavioral and biological perspectives on the origins of modern humans* (pp. 391–414). Princeton: Princeton University Press.

Lissauer, H. (1890). Ein Fall von Seelenblindheit nebst einem Beitrage zur Theorie derselben. *Archiv für Psychiatrie und Nervenkrankheiten, 21*, 222–270.

Liu, G. T., Bolton, A. K., Price, B. H., & Weintraub, S. (1992). Dissociated perceptual-sensory and exploratory-motor neglect. *Journal of Neurology, Neurosurgery and Psychiatry, 55*, 701–706.

Livingstone, M., & Hubel, D. (1988). Segregation of form, color, movement and depth: Anatomy, physiology and perception. *Science, 240*, 740–749.

Lohr, J. B., & Wisniewski, A. A. (1987). *Movement disorders: A neuropsychiatric approach.* New York: Guilford Press.

Lovell, J. (1993, December). Musings on aphasia. *Modern Medicine of Australia*, 54–55.

Lu, S. T., Hamalainen, M. S., Hari, R., Ilmoniemi, R. J., Lounasmaa, O. V., Sams, M., & Vilkman, V. (1991). Seeing faces activates three separate areas outside the occipital visual cortex in man. *Neuroscience, 43*, 287–290.

Lueck, C. J., Zeki, S., Friston, K. J., Deiber, M.-P., Cope, P., Cunningham, V. J., Lammertsma, A. A., Kennard, C., & Frackowiak, R. S. J. (1989). The color centre in the cerebral cortex of man. *Nature, 340,* 386–389.

Malapani, C., Pillon, B., Dubois, B., & Agid, Y. (1994). Impaired simultaneous cognitive task performance in Parkinson's disease. *Neurology, 44,* 319–326.

Mann, C. C. (1994). Behavioural genetics in transition. *Science, 264,* 1686–1689.

Manning, L., & Kartsounis, L. D. (1993). Confabulations related to tacit awareness in visual neglect. *Behavioural Neurology, 6,* 211–213.

Manto, M., Godaux, E., & Jacquy, J. (1994). Cerebellar hypermetria is larger when the inertial load is artifically increased. *Annals of Neurology, 35,* 45–52.

Manyam, B. V. (1990). Paralysis agitans and levodopa in "Ayurveda": Ancient Indian medical treatise. *Movement Disorders, 5,* 47–48.

Marengo, J. T. (1993). Disordered thinking and cerebral dysfunction: Laterality effects, language and intellectual functions. *Archives of Clinical Neuropsychology, 8,* 497–509.

Margolin, D. I. (1991). Cognitive neuropsychology: Resolving enigmas about Wernicke's aphasia and other higher cortical disorders. *Archives of Neurology, 48,* 751–765.

Mark, V. W., Kooistra, C. A., & Heilman, K. M. (1988). Hemispatial neglect affected by non-neglected stimuli. *Neurology, 38,* 1207–1211.

Marr, D. (1982). *Vision.* New York: Freeman.

Marsden, C. D. (1994). Parkinson's disease. *Journal of Neurology, Neurosurgery and Psychiatry, 57,* 672–681.

Marsden, C. D., & Obeso, J. A. (1994). The functions of the basal ganglia and the paradox of stereotaxic surgery in Parkinson's disease. *Brain, 117,* 877–897.

Marsh, N. V., & Kersel, D. A. (1993). Screening tests for visual neglect following stroke. *Neuropsychological Rehabilitation, 3,* 245–257.

Marshall, J. C., & Halligan, P. W. (1988). Blindsight and insight in visuospatial neglect. *Nature, 336,* 766–767.

Marshall, J. C., & Halligan, P. W. (1990). Line bisection in a case of visual neglect: psychophysical studies with implications for theory. *Cognitive Neuropsychology, 7,* 107–130.

Marshall, J. C., & Halligan, P. W. (1991). A study of plane bisection in four cases of visual neglect. *Cortex, 27,* 277–284.

Marshall, J. C., & Halligan, P. W. (1993). Visuo-spatial neglect: A new copying test to assess perceptual parsing. *Journal of Neurology, 240,* 37–40.

Marshall, J. C., & Halligan, P. W. (1994). The yin and yang of visuo-spatial neglect: A case study. *Neuropsychologia, 32,* 1037–1057.

Marshall, J. C., & Newcombe, F. (1973). Patterns of paralexia: A psycholinguistic approach. *Journal of Psycholinguistic Research, 2,* 175–199.

Martin, J. B. (1993). Molecular genetics of neurological diseases. *Science, 262,* 674–676.

Martin, R. C. (1993). Short-term memory and sentence processing: Evidence from neuropsychology. *Memory and Cognition, 21,* 176–183.

Martin, R. C., Shelton, J. R., & Yafee, L. S. (1994). Language processing and working memory: Neuropsychological evidence for separate phonological and semantic capacities. *Journal of Memory and Language, 33,* 83–111.

Martyn, C. N. (1992). The epidemiology of Alzheimer's disease in relation to aluminum. In CIBA Foundation symposium No. 169 (ed.), *Aluminum in biology and medicine* (pp. 69–86). New York: Wiley.

Mattingley, J. B. (in press). Paterson and Zangwill's case of unilateral neglect: Insights from 50 years of experimental inquiry. In C. Code & C. W. Wallesch (Eds.), *Classic Cases in Neuropsychology*. Hove: Erlbaum.

Mattingley, J. B., Bradshaw, J. L., & Bradshaw, J. A. (1994). Horizontal visual motion modulates focal attention in left unilateral neglect. *Journal of Neurology, Neurosurgery and Psychiatry, 57*, 1228–1235.

Mattingley, J. B., Bradshaw, J. L., & Bradshaw, J. A. (in press). The effects of unilateral visuospatial neglect on perception of Müller-Lyer illusory figures. *Perception*.

Mattingley, J. B., Bradshaw, J. L., Bradshaw, J. A., & Nettleton, N. C. (1994a). Residual rightward attentional bias after apparent recovery from right hemisphere damage: Implications for a multicomponent model of neglect. *Journal of Neurology, Neurosurgery and Psychiatry, 57*, 597–604.

Mattingley, J. B., Bradshaw, J. L., Bradshaw, J. A., & Nettleton, N. C. (1994b). Recovery from directional hypokinesia and bradykinesia in unilateral neglect. *Journal of Clinical and Experimental Neuropsychology, 16*, 861–876.

Mattingley, J. B., Bradshaw, J. L., Nettleton, N. C., & Bradshaw, J. A. (1994). Can task-specific perceptual bias be distinguished from unilateral neglect? *Neuropsychologia, 32*, 805–817.

Mattingley, J. B., Bradshaw, J. L., & Phillips, J. G. (1992). Impairments of movement initiation and execution in unilateral neglect: Directional hypokinesia and bradykinesia. *Brain, 115*, 1849–1874.

Mattingley, J. B., Bradshaw, J. L., Phillips, J. G., & Bradshaw, J. A. (1993). Reversed perceptual asymmetry for faces in left unilateral neglect. *Brain and Cognition, 23*, 145–165.

Mattingley, J. B., Phillips, J. G., & Bradshaw, J. L. (1994). Impairments of movement execution in unilateral neglect: A kinematic analysis of directional bradykinesia. *Neuropsychologia, 32*, 1111–1134.

Mattingley, J. B., Pierson, J. M., Bradshaw, J. L., Phillips, J. G., & Bradshaw, J. A. (1993). To see or not to see: The effects of visible and invisible cues on line bisection judgments in unilateral neglect. *Neuropsychologia, 31*, 1201–1215.

Mayes, A. R. (1992). Automatic memory processes in amnesia: How are they mediated? In A. D. Milner & M. D. Rugg (Eds.), *The neuropsychology of consciousness* (pp. 235–262). San Diego: Academic Press.

Mayeux, R., Stern, Y., Ottman, R., Tatemichi, T. K., Tang, M.-X., Maestre, G., Ngai, C., Tycko, B., & Ginsberg, H. (1993). The apolipoprotein E4 allele in patients with Alzheimer's disease. *Annals of Neurology, 34*, 752–754.

Mayeux, R., Tang, M.-X., Marder, K., Côté, L. J., & Stern, Y. (1994). Smoking and Parkinson's disease. *Movement Disorders, 9*, 207–212.

Mazoyer, B. M., Tzourio, N., Frak, V., Syrota, A., Murayama, N., Levrier, O., Salamon, G., Dehaene, S., Cohen, L., & Mehler, J. (1993). The cortical representation of speech. *Journal of Cognitive Neuroscience, 5*, 467–479.

Mazzoni, M., Moretti, P., Pardossi, L., Vista, M., Muratorio, A., & Puglioli, M. (1993). A case of music imperception. *Journal of Neurology, Neurosurgery and Psychiatry, 56*, 322–324.

McCarthy, R. A. (1994). Neuropsychology: Going loco? *Behavioral and Brain Sciences,*

17, 73–74.

McCloskey, M. (1993). Theory and evidence in cognitive neuropsychology: A "radical" response to Robertson, Knight, Rafal and Shimamura (1993). *Journal of Experimental Psychology: Learning, Memory and Cognition, 19,* 718–734.

McCloskey, M., Aliminosa, D., & Macaruso, P. (1991). Theory-based assessment of acquired dyscalculia. *Brain and Cognition, 17,* 285–308.

McCloskey, M., Aliminosa, D., & Sokol, S. M. (1991). Facts, rules, and procedures in normal calculation: Evidence from multiple single-patient studies of impaired arithmetic fact retrieval. *Brain and Cognition, 17,* 154–203.

McCloskey, M., Sokol, S. M., Caramazza, A., & Goodman-Schulman, R. A. (1990). Cognitive representations and processes in number production: Evidence from cases of acquired dyscalculia. In A. Caramazza (Ed.), *Cognitive neuropsychology and neurolinguistics: Advances in models of cognitive function and impairment* (pp. 1–32). Hillsdale, NJ: Lawrence Erlbaum.

McDonald, S., Tate, R. L., & Rigby, J. (1994). Error types in ideomotor apraxia: A qualitative analysis. *Brain and Cognition, 25,* 250–270.

McGuire, P. K., Shah, G. M. S., & Murray, R. M. (1993). Increased blood flow in Broca's area during auditory hallucinations in schizophrenia. *The Lancet, 342,* 703–706.

McKenna, P., & Warrington, E. K. (1993). The neuropsychology of semantic memory. In F. Boller & J. Grafman (Eds.), *Handbook of neuropsychology* (Vol. 8, pp. 193–214). Amsterdam: Elsevier.

MacMillan, M. (in press). Experimental and clinical studies of localization before Flourens. *Journal of the History of the Neurosciences.*

McNeil, J. E., Cipolotti, L., & Warrington, E. K. (1994). The accessibility of proper names. *Neuropsychologia, 32,* 193–203.

McNeil, J. E., & Warrington, E. K. (1993). Prosopagnosia: A face-specific disorder. *Quarterly Journal of Experimental Psychology, 46A,* 1–10.

McNeil, J. E., & Warrington, E. K. (1994). A dissociation between addition and subtraction with written calculation. *Neuropsychologia, 32,* 717–728.

Meador, K. J., Allen, M. E., Adams, R. J., & Loring, D. W. (1991). Allochiria vs. allesthesia: Is there a misperception? *Archives of Neurology, 48,* 546–549.

Meador, K. J., Loring, D. W., Baron, M. S., Rogers, O. L., & Kimpel, T. G. (1988). Hemispatial-limb hypometria. *International Journal of Neuroscience, 42,* 71–75.

Meador, K. J., Loring, D. W., Bowers, D., & Heilman, K. M. (1987). Remote memory and neglect syndrome. *Neurology, 37,* 522–526.

Mel, B. W. (1991). A connectionist model may shed light on neural mechanisms for visually guided reaching. *Journal of Cognitive Neuroscience, 3,* 273–292.

Merzenich, M. M. (1987). Dynamic neocortical processes and the origins of higher brain functions. In J.-P. Changeux & M. Konishi (Eds.), *The neural and molecular bases of learning* (pp. 337–358). Chichester: Wiley.

Mestel, R. (1993, November). Can oestrogen fend off Alzheimer's? *New Scientist,* 10.

Mesulam, M.-M. (1981). A cortical network for directed attention and unilateral neglect. *Annals of Neurology, 10,* 309–325.

Mesulam, M.-M. (1985). Attention, confusional states, and neglect. In M-M. Mesulam (Ed.), *Principles of behavioral neurology* (pp. 125–167). Philadelphia: F. A. Davis.

Mesulam, M.-M. (1990). Large-scale neurocognitive networks and distributed processing for attention, language, and memory. *Annals of Neurology, 28,* 597–613.

Metter, E. J. (1992). Role of subcortical structures in aphasia: Evidence from studies of resting cerebral glucose metabolism. In G. Vallar, S. F. Cappa, & C.-W. Wallesch (Eds.), *Neuropsychological disorders associated with subcortical lesions* (pp. 478–500). Oxford: Oxford University Press.

Michel, E. M., & Troost, B. T. (1980). Palinopsia: Cerebral localization with computed tomography. *Neurology, 30,* 887–889.

Milberg, W. P., Hebben, N., & Kaplan, E. (1986). The Boston process approach to neuropsychological assessment. In I. Grant & K. M. Adams (Eds.), *Neuropsychological assessment of neuropsychiatric disorders* (pp. 65–86). New York: Oxford University Press.

Milberg, W., & Blumstein, S. (1981). Lexical decision and aphasia: Evidence for semantic processing. *Brain and Language, 14,* 371–385.

Milders, M. V., & Perrett, D. I. (1993). Recent developments in the neuropsychology and physiology of face processing. *Baillière's Clinical Neurology, 2,* 361–387.

Milner, B., Corkin, S., & Teuber, H.-L. (1968). Further analysis of the hippocampal amnesic syndrome: 14-year follow-up study of H.M. *Neuropsychologia, 6,* 217–234.

Milner, B., & Teuber, H.-L. (1968). Alteration of perception and memory in man: Reflections on methods. In L. Weiskrantz (Ed.), *Analysis of behavioral change* (pp. 268–376). New York: Harper.

Mishkin, M., Ungerleider, L. G., & Macko, K. A. (1983). Object vision and spatial vision: Two cortical pathways. *Trends in Neurosciences, 6,* 414–417.

Miyashita, Y. (1993). Inferior temporal cortex: Where visual perception meets memory. *Annual Review of Neuroscience, 16,* 245–263.

Molloy, R., Brownell, H. H., & Gardner, H. (1990). Discourse comprehension by right hemisphere stroke patients: Deficits of prediction and revision. In Y. Joanette & H. H. Brownell (Eds.), *Discourse ability and brain damage: Theoretical and empirical perspectives* (pp. 113–130). New York: Springer.

Montgomery, S. A. (1992). The diagnostic place of OCD and panic disorder as serotonergic illnesses. *Advances in the Biosciences, 85,* 325–333.

Moore, R. Y. (1990). Subcortical chemical neuroanatomy. In J. L. Cummings (Ed.), *Subcortical dementia* (pp. 44–58). New York: Oxford University Press.

Morecraft, R. J., Geula, C., & Mesulam, M-M. (1993). Architecture of connectivity within a cingulo-fronto-parietal neurocognitive network for directed attention. *Archives of Neurology, 50,* 279–284.

Morgan, M., Bradshaw, J. L., Phillips, J. G., Mattingley, J. B., Iansek, R., & Bradshaw, J. A. (1994). Effects of hand and age upon abductive and adductive movements: A kinematic analysis. *Brain and Cognition, 25,* 194–206.

Moriarty, J., Ring, H. A., & Robertson, M. M. (1993). An idiot savant calendrical calculator with Gilles de la Tourette syndrome: Implications for an understanding of the savant syndrome. *Psychological Medicine, 23,* 1019–1021.

Morris, M. E., Summers, J. J., Matyas, T., & Iansek, R. (1994). Current status of the motor program. *Physical Therapy, 74,* 738–752.

Morrison-Bogorad, M., Rosenberg, R. N., Sparkman, D. R., Weiner, M. F., & White, C. L. (1993). Alzheimer's disease. In R. N. Rosenberg, S. B. Prusiner, S. DiMauro, R. L. Barchi, & L. M. Kunkel (Eds.), *The molecular and genetic basis of neurological disease* (pp. 767–777). Boston: Butterworth-Heinemann.

Morrow, L. A., & Ratcliff, G. (1988). The disengagement of covert attention and the neglect syndrome. *Psychobiology, 16,* 261–269.

Moscovitch, M., Vriezen, E., & Goshen-Gottstein, Y. (1993). Implicit tests of memory in patients with focal lesions or degenerative brain disorders. In F. Boller & J. Grafman (Eds.), *Handbook of neuropsychology* (Vol. 8, pp. 133–174). Amsterdam: Elsevier.

Mottron, L., & Belleville, S. (1993). A study of perceptual analysis in a high-level autistic subject with exceptional graphic abilities. *Brain and Cognition, 23,* 279–309.

Mozer, M. C., & Behrmann, M. (1990). On the interaction of selective attention and lexical knowledge: A connectionist account of neglect dyslexia. *Journal of Cognitive Neuroscience, 2,* 96–123.

Mushiake, H., Inase, M., & Tanji, J. (1991). Neuronal activity in the primate premotor, supplementary and precentral motor cortex during visually guided and internally determined sequential movements. *Journal of Neurophysiology, 66,* 705–718.

Myers, J. J., & Sperry, R. W. (1985). Interhemispheric communication after section of the forebrain commissures. *Cortex, 21,* 249–260.

Myers, R. E., & Sperry, R. W. (1953). Interocular transfer of a visual form discrimination habit in cats after section of the optic chiasm and corpus callosum. *Anatomical Record, 115,* 351.

Narabayashi, H. (1993). Three types of akinesia in the progressive course of Parkinson's disease. *Advances in Neurology, 60,* 18–24.

Navon, D. (1977). Forest before trees: The precedence of global features in visal perception. *Cognitive Psychology, 9,* 353–383.

Nebes, R. D. (1989). Semantic memory in Alzheimer's disease. *Psychological Bulletin, 106,* 377–394.

Nebes, R. D. (1990a). The commissurotomized brain: Introduction. In F. Boller & J. Grafman (Eds.), *Handbook of neuropsychology* (Vol. 4, pp. 3–8). Amsterdam: Elsevier.

Nebes, R. D. (1990b). Semantic memory function and dysfunction in Alzheimer's disease. In T. M. Hess (Ed.), *Aging and cognition* (pp. 265–296). New York: Elsevier.

Neumann, O. (1990). Visual attention and action. In O. Neumann & W. Prinz (Eds.), *Relationships between perception and action* (pp. 227–267). Berlin: Springer.

Nichelli, P., Grafman, J., Pietrini, P., Alway, D., Carton, J. C., & Miletich, R. (1994). Brain activity in chess playing. *Nature, 369,* 191.

Nichelli, P., Rinaldi, M., & Cubelli, R. (1989). Selective spatial attention and length representation in normal subjects and in patients with unilateral spatial neglect. *Brain and Cognition, 9,* 57–70.

Niemann, J., Winker, T., Gerling, J., Landwehrmeyer, B., & Jung, R. (1991). Changes of slow cortical negative DC-potentials during the acquisition of a complex finger motor task. *Experimental Brain Research, 85,* 417–422.

Norman, D. A., & Bobrow, D. G. (1975). On data-limited and resource-limited processes. *Cognitive Psychology, 7,* 44–64.

Obeso, J. A., & Giménez-Roldán, S. (1988). Clinicopathological correlation in symptomatic dystonia. In S. Fahn, C. D. Marsden, & D. B. Calne (Eds.), *Dystonia 2. Advances in neurology* (Vol. 50, pp. 113–122). New York: Raven Press.

Obeso, J. A., Rothwell, J. C., & Marsden, C. D. (1981). Simple tics in Gilles de la Tourette's syndrome are not prefaced by a normal premovement EEG potential. *Journal of Neurology, Neurosurgery and Psychiatry, 44,* 735–738.

Ogden, J. A. (1985a). Anterior-posterior interhemispheric differences in the loci of lesions producing visual hemineglect. *Brain and Cognition, 4*, 59–75.

Ogden, J. A. (1985b). Contralesional neglect of constructed visual images in right and left brain damaged patients. *Neuropsychologia, 23*, 273–277.

Ogden, J. A. (1987). The "neglected" left hemisphere and its contribution to visuo-spatial neglect. In M. Jeannerod (Ed.), *Neurophysiological and neuropsychological aspects of spatial neglect* (pp. 215–233). Amsterdam: North-Holland.

Ogden, J. A. (1988). Onset of motor neglect following a right parietal infarct and its recovery consequent on the removal of a right frontal meningioma. *New Zealand Journal of Psychology, 17*, 24–31.

Ojemann, G. A., Ojemann, J. G., Lettich, E., & Berger, M. (1989). Cortical language localization in left, dominant hemisphere. *Journal of Neurosurgery, 71*, 316–326.

Ojemann, J. G., Ojemann, G. A., & Lettich, E. (1992). Neuronal activity related to faces in human right nondominant temporal cortex. *Brain, 115*, 1–13.

O'Keefe, J., & Nadel, L. (1978). *The hippocampus as a cognitive map.* Oxford: Oxford University Press.

Olanow, C. W. (1990). Oxidation reactions in Parkinson's disease. *Neurology, 40* (suppl. 3), 32–37.

Ollat, H. (1992). Dopaminergic insufficiency reflecting cerebral ageing: Value of a dopaminergic agonist, piribedil. *Journal of Neurology, 239* (suppl. 1), S13–S16.

Orsini, D. L., & Satz, P. (1986). A syndrome of pathological left handedness: Correlates of early left hemisphere injury. *Archives of Neurology, 43*, 333–337.

Ovellette, G. P., & Baum, S. R. (1993). Acoustic analysis of prosodic cues in left- and right-hemisphere damaged patients. *Aphasiology, 8*, 257–283.

Owen, A. M., James, M., Leigh, P. N., Summers, B. A., Marsden, C. D., Quinn, N., Lange, K. W., & Robbins, T. W. (1992). Progression of fronto-striatal cognitive deficits in Parkinson's disease. *Brain, 115*, 1727–1751.

Owen, A. M., Roberts, A. C., Hodges, J. R., Summers, B. A., Polkey, C. E., & Robbins, T. W. (1993). Contrasting mechanisms of impaired attentional set-shifting in patients with frontal lobe damage or Parkinson's disease. *Brain, 116*, 1159–1175.

Owen, M. J. (1992). Will schizophrenia become a graveyard for molecular geneticists? *Psychological Medicine, 22*, 289–293.

Pandya, D. N., & Rosene, D. L. (1985). Some observations on trajectories and topography of commissural fibers. In A. G. Reeves (Ed.), *Epilepsy and the corpus callosum* (pp. 21–39). New York: Plenum.

Park, S., Como, P. G., Cui, L., & Kurlan, R. (1993). The early course of the Tourette's syndrome clinical spectrum. *Neurology, 43*, 1712–1715.

Parkin, A. J., & Leng, R. C. (1993). *Neuropsychology of the amnesic syndrome.* Hove, UK: Lawrence Erlbaum.

Parkinson, J. (1817). *Essay on the shaking palsy.* London: Sherwood, Neely and Jones.

Parks, R. W., Long, D. L., Levine, D. S., Crockett, D. J., McGeer, E. G., McGeer, P. L., Dalton, I. E., Zec, R. F., Becker, R. E., Coburn, K. L., Siler, G., Nelson, M. E., & Bower, J. M. (1991). Parallel distributed processing and neural networks: Origins, methodology and cognitive functions. *International Journal of Neuroscience, 60*, 195–214.

Parry-Jones, B. (1992). A bulimic ruminator? The case of Dr. Samuel Johnson. *Psycholog-*

ical Medicine, 22, 851–862.

Pasqual-Leone, A., Gomez-Tortosa, E., Grafman, J., Alway, D., Nichelli, P., & Hallett, M. (1994). Induction of visual extinction by rapid-rate transcranial magnetic stimulation of parietal lobe. *Neurology, 44,* 494–498.

Paterson, A., & Zangwill, O. L. (1944). Disorders of visual space perception associated with lesions of the right hemisphere. *Brain, 67,* 331–358.

Paterson, A., & Zangwill, O. L. (1945). A case of topographical disorientation associated with unilateral cerebral lesion. *Brain, 68,* 188–212.

Patten, B. M. (1993). Wilson's disease. In J. Jankovic & E. Tolosa (Eds.), *Parkinson's disease and movement disorders* (pp. 217–233). Baltimore: Williams & Wilkins.

Patterson, K. E., & Besner, D. (1984). Is the right hemisphere literate? *Cognitive Neuropsychology, 1,* 315–341.

Patterson, K. E., Marshall, J. C., & Coltheart, M. (Eds.) (1985). *Surface dyslexia: Neuropsychological and cognitive studies of phonological reading.* London: Lawrence Erlbaum.

Paulesu, E., Frith, C. D., & Frackowiak, R. S. J. (1993). The neural correlates of the verbal component of working memory. *Nature, 362,* 342–345.

Paulin, M. G. (1993). The role of the cerebellum in motor control and perception. *Brain, Behaviour and Evolution, 41,* 39–50.

Penfield, W., & Perot, P. (1963). The brain's record of auditory and visual experience: A final summary and discussion. *Brain, 86,* 595–696.

Penney, J. B., & Young, A. B. (1993). Huntington's disease. In J. Jankovic & E. Tolosa (Eds.), *Parkinson's disease and movement disorders* (2nd ed.) (pp. 205–216). Baltimore: Williams & Wilkins.

Perani, D., Vallar, G., Cappa, S., Messa, C., & Fazio, F. (1987). Aphasia and neglect after subcortical stroke: A clinical/cerebral perfusion correlation study. *Brain, 110,* 1211–1229.

Perani, D., Vallar, G., Paulesu, E., Aberoni, M., & Fazio, F. (1993). Left and right hemisphere contribution to recovery from neglect after right hemisphere damage—an ^{18}F FDG PET study of two cases. *Neuropsychologia, 31,* 115–125.

Perenin, M. T., & Vighetto, A. (1983). Optic ataxia: A specific disorder in visuomotor coordination. In A. Hein & M. Jeannerod (Eds.), *Spatially oriented behavior* (pp. 305–326). New York: Springer.

Peretz, I. (1990). Processing of local and global musical information by unilateral brain-damaged patients. *Brain, 113,* 1185–1205.

Petersen, S. E., & Fiez, J. A. (1993). The processing of single words studied with positron emmission tomography. *Annual Review of Neuroscience, 16,* 509–530.

Petersen, S. E., Fox, P. T., Posner, M. I., Mintun, M., & Raichle, M. E. (1989). Positron emission tomographic studies of the processing of single words. *Journal of Cognitive Neuroscience, 1,* 153–170.

Petersen, S. E., Fox, P. T., Snyder, A., & Raichle, M. E. (1990). Activation of extrastriate and frontal cortical areas by visual words and word-like stimuli. *Science, 249,* 1041–1044.

Peterson, B., Riddle, M. A., Cohen, D. J., Katz, L. D., Smith, J. C., Hardin, M. T., & Leckman, J. F. (1993). Reduced basal ganglia volumes in Tourette's syndrome using three-dimensional reconstruction techniques from magnetic resonance images. *Neurology, 43,* 941–949.

Petri, H. L., & Mishkin, M. (1994). Behaviorism, cognitivism, and the neuropsychology of memory. *American Scientist, 82*, 30–37.

Phillips, J. G., Bradshaw, J. L., Chiu, E., & Bradshaw, J. A. (1994). Characteristics of handwriting of patients with Huntington's disease. *Movement Disorders, 9*, 521–530.

Phillips, J. G., Bradshaw, J. L., Iansek, R., & Chiu, E. (1993). Motor functions of the basal ganglia. *Psychological Research, 55*, 175–181.

Pick, A. (1903). Clinical studies III: On reduplicative paramnesia. *Brain, 26*, 260–267.

Pierson-Savage, J. M., Bradshaw, J. L., Bradshaw, J. A., & Nettleton, N. C. (1988). Vibro-tactile reaction times in unilateral neglect. *Brain, 111*, 1531–1545.

Pirozzolo, F. J., Swihart, A. A., Rey, G. J., Mahurin, R., & Jankovic, J. (1993). Cognitive impairments associated with Parkinson's disease and other movement disorders. In J. Jankovic & E. Tolosa (Eds.), *Parkinson's disease and movement disorders* (2nd ed.) (pp. 493–510). Baltimore: Williams & Wilkins.

Pizzamiglio, L., Cappa, S., Vallar, G., Zoccolotti, P., Bottini, G., Ciurli, P., Guariglia, C., & Antonucci, G. (1989). Visual neglect for far and near extra-personal space in humans. *Cortex, 25*, 471–477.

Pizzamiglio, L., Frasca, R., Guariglia, C., Inoccia, C., & Antonucci, G. (1990). Effect of optokinetic stimulation in patients with visual neglect. *Cortex, 26*, 535–540.

Plomin, R., Owen, M. J., & McGuffin, P. (1994). The genetic basis of complex human behaviors. *Science, 264*, 1733–1739.

Plourde, G., & Sperry, R. W. (1982). Left hemisphere involvement in left spatial neglect from right-sided lesions: A commissurotomy study. *Brain, 107*, 95–106.

Poizner, H., Bellugi, U., & Klima, E. S. (1990). Biological foundations of language: Clues from sign language. *Annual Review of Neuroscience, 13*, 283–307.

Polinsky, R. J. (1993). Shy-Drager syndrome. In J. Jankovic & E. Tolosa (Eds.), *Parkinson's disease and movement disorders* (pp. 191–203). Baltimore: Williams & Wilkins.

Pons, T. P., Garraghty, P. E., Ommaya, A. K., Kaas, J. H., Taub, E., & Mishkin, M. (1991). Massive cortical reorganization after sensory deafferentation in adult macaques. *Science, 252*, 1857–1860.

Posner, M. I. (1978). *Chronometric explorations of the mind.* Englewood Cliffs, NJ: Lawrence Erlbaum.

Posner, M. I. (1980). Orienting of attention. *The Quarterly Journal of Experimental Psychology, 32*, 3–25.

Posner, M. I. (1993). Seeing the mind. *Science, 262*, 673–674.

Posner, M. I. (1994). Local and distributed processes in attentional orienting. *Behavioral and Brain Sciences, 17*, 78–79.

Posner, M. I., & Dehaene, S. (1994). Attentional networks. *Trends in Neurosciences, 17*, 75–79.

Posner, M. I., Inhoff, A., Friedrich, F. J., & Cohen, A. (1987). Isolating attentional systems: A cognitive-anatomical analysis. *Psychobiology, 15*, 107–121.

Posner, M. I., & Petersen, S. E. (1990). The attention system of the human brain. *Annual Review of Neuroscience, 13*, 25–42.

Posner, M. I., Petersen, S. E., Fox, P. T., & Raichle, M. E. (1988). Localization of cognitive operations in the human brain. *Science, 240*, 1627–1631.

Posner, M. I., Snyder, C. R. R., & Davidson, B. J. (1980). Attention and the detection of signals. *Journal of Experimental Psychology: General, 109*, 160–174.

Posner, M. I., Walker, J. A., Friedrich, F. J., & Rafal, R. D. (1984). Effects of parietal injury on covert orienting of attention. *The Journal of Neuroscience, 4,* 1863–1874.

Posner, M. I., Walker, J. A., Friedrich, F. J., & Rafal, R. D. (1987). How do the parietal lobes direct covert attention? *Neuropsychologia, 25,* 135–145.

Price, B. H., & Mesulam, M. (1985). Psychiatric manifestations of right hemisphere infarctions. *Journal of Neurology, Neurosurgery and Psychiatry, 173,* 610–614.

Prichard, J. W., & Brass, L. M. (1992). New anatomical and functional imaging methods. *Annals of Neurology, 32,* 395–400.

Prusiner, S. B. (1993). Genetic and infectious prion diseases. *Archives of Neurology, 50,* 1129–1153.

Puccetti, R. (1985). Experiencing two selves: The history of a mistake. *Behavioral and Brain Sciences, 8,* 646–647.

Pulsinelli, W. (1992). Pathophysiology of acute ischaemic stroke. *The Lancet, 339,* 533–536.

Raichle, M. E. (1993). The scratch pad of the mind. *Nature, 363,* 583–584.

Raichle, M. E. (1994, April). Visualizing the mind. *Scientific American,* 36–42.

Rajan, R., Irvine, D. R. F., Wise, L. Z., & Heil, P. (1993). Effect of unilateral partial cochlear lesions in adult cats on the representation of lesioned and unlesioned cochleas in primary auditory cortex. *Journal of Comparative Neurology, 338,* 17–49.

Ramachandran, V. S., Cronin-Golomb, A., & Myers, J. J. (1986). Perception of apparent motion by commissurotomy patients. *Nature, 320,* 358–359.

Ramón y Cajal, S. (1989). *Recollections of my life.* Cambridge, MA: MIT Press.

Randolph, C., Goldberg, T. E., & Weinberger, D. R. (1993). The neuropsychology of schizophrenia. In K. M. Heilman & E. Valenstein (Eds.), *Clinical neuropsychology* (3rd ed.) (pp. 499–522). Oxford: Oxford University Press.

Rapcsak, S. Z., Ochipa, C., Beeson, P. M., & Rubens, A. B. (1993). Praxis and the right hemisphere. *Brain and Cognition, 23,* 181–202.

Rapcsak, S. Z., Watson, R. T., & Heilman, K. M. (1987). Hemispace-visual field interactions in visual extinction. *Journal of Neurology, Neurosurgery and Psychiatry, 50,* 1117–1124.

Rapoport, J. L. (1990). Obsessive compulsive disorder and basal ganglia dysfunction. *Psychological Medicine, 20,* 465–469.

Ratcliff, G., & Davies-Jones, G. A. B. (1972). Defective visual localization in focal brain wounds. *Brain, 95,* 49–60.

Rauch, S. L., Jenike, M. A., Alpert, N. M., Baer, L., Breiter, H. C. R., Savage, C. R., & Fischman, A. J. (1994). Regional cerebral blood flow measured during symptom provocation in obsessive compulsive disorder using oxygen 15-labelled carbon dioxide and positron emission tomography. *Archives of General Psychology, 51,* 67–70.

Regard, M., Cook, N. D., Wieser, H. G., & Landis, T. (1994). The dynamics of cerebral dominance during unilateral limbic seizures. *Brain, 117,* 91–104.

Regland, B., & Gottfries, C.-G. (1992). The role of amyloid β-protein in Alzheimer's disease. *The Lancet, 340,* 467–468.

Rentschler, I., Treutwein, B., & Landis, T. (1994). Dissociation of local and global processing in visual agnosia. *Vision Research, 34,* 963–971.

Riddoch, M. J., & Humphreys, G. W. (1983). The effect of cueing on unilateral neglect. *Neuropsychologia, 21,* 589–599.

Riggs, J. E. (1993). Smoking and Alzheimer's disease: Protective effect or differential survival bias? *The Lancet, 342*, 793–794.

Rimland, B., & Fein, D. (1988). Special talents of autistic savants. In L. K. Obler & D. Fein (Eds.), *The exceptional brain* (pp. 474–492). New York: Guilford.

Rizzo, M., Hurtig, R., & Damasio, A. R. (1987). The role of scanpaths in focal learning and recognition. *Annals of Neurology, 22*, 41–45.

Rizzolatti, G., & Berti, A. (1990). Neglect as a neural representation deficit. *Revue Neurologique, 146*, 626–634.

Rizzolatti, G., & Berti, A. (1993). Neural mechanisms of spatial neglect. In I. H. Robertson & J. C. Marshall (Eds.), *Unilateral neglect: Clinical and experimental studies* (pp. 87–106). Hove, U.K.: Erlbaum.

Rizzolatti, G., & Camarda, R. (1987). Neural circuits for spatial attention and unilateral neglect. In M. Jeannerod (Ed.), *Neurophysiological and neuropsychological aspects of spatial neglect* (pp. 289–313). Amsterdam: North-Holland.

Rizzolatti, G., & Gallese, V. (1988). Mechanisms and theories of spatial neglect. In F. Boller & J. Grafman (Eds.), *Handbook of neuropsychology* (Vol. 1, pp. 223–246). Amsterdam: Elsevier.

Robbins, T. W. (1994). Neural substrates of neglect: Speculations and animal models. *Neuropsychological Rehabilitation, 4*, 189–191.

Roberts, G. W., Gentleman, S. M., Lynch, A., Murray, L., Landon, M., & Graham, D. I. (1994). β Amyloid protein deposition in the brain after severe head injury: Implications for the pathogenesis of Alzheimer's disease. *Journal of Neurology, Neurosurgery and Psychiatry, 57*, 419–425.

Robertson, I. H. (1989). Anomalies in the laterality of omissions in unilateral left visual neglect: Implications for an attentional theory of neglect. *Neuropsychologia, 27*, 157–165.

Robertson, I. H. (1993). The relationship between lateralised and non-lateralised attentional deficits in unilateral neglect. In I. H. Robertson & J. C. Marshall (Eds.), *Unilateral neglect: Clinical and experimental studies* (pp. 257–275). Hove, U.K.: Lawrence Erlbaum Associates.

Robertson, I. H., Gray, J., Pentland, B., & Waite, L. (1990). A randomised controlled trial of computer-based cognitive rehabilitation for unilateral left visual neglect. *Archives of Physical Medicine and Rehabilitation, 71*, 663–668.

Robertson, I. H., Halligan, P. W., & Marshall, J. C. (1993). Prospects for the rehabilitation of unilateral neglect. In I. H. Robertson & J. C. Marshall (Eds.), *Unilateral neglect: Clinical and experimental studies* (pp. 279–292). Hove, U.K.: Lawrence Erlbaum Associates.

Robertson, I. H., & North, N. T. (1992). Spatio-motor cueing in unilateral left neglect: The role of hemispace, hand and motor activation. *Neuropsychologia, 30*, 553–563.

Robertson, I. H., & North, N. T. (1993). Fatigue versus disengagement in unilateral neglect. *Journal of Neurology, Neurosurgery and Psychiatry, 56*, 717–719.

Robertson, I. H., North, N. T., & Geggie, C. (1992). Spatiomotor cueing in unilateral left neglect: Three case studies of its therapeutic effects. *Journal of Neurology, Neurosurgery and Psychiatry, 55*, 799–805.

Robertson, L. C., Knight, R. T., Rafal, R., & Shimamura, A. P. (1993). Cognitive neuropsychology is more than single-case studies. *Journal of Experimental Psychology: Learning, Memory and Cognition, 19*, 710–717.

Robertson, L. C., & Lamb, M. R. (1991). Neuropsychological contributions to part-whole organization. *Cognitive Psychology, 23,* 299–330.

Robin, D. A., Tranel, D., & Damasio, H. (1990). Auditory perception of temporal and spectral events in patients with focal left and right cerebral lesions. *Brain and Language, 39,* 539–555.

Rock, I. (1975). *An introduction to perception.* New York: MacMillan.

Rock, I. (1981). Anorthoscopic perception. *Scientific American, 244,* 145–153.

Rode, G., & Perenin, M. T. (1994). Temporary remission of representational hemineglect through vestibular stimulation. *NeuroReport, 5,* 869–872.

Roeltgen, D. P. (1987). Loss of deep dyslexic reading ability from a second left-hemispheric lesion. *Archives of Neurology, 44,* 346–348.

Roland, P. E., Larsen, B., Lassen, N. A., & Skinhøj, E. (1980). Supplementary motor area and other cortical areas in organization of voluntary movements in man. *Journal of Neurophysiology, 43,* 118–136.

Rondot, P. (1991). The shadow of movement. *Journal of Neurology, 238,* 411–419.

Rose, S. (1992). *The making of memory. From molecules to mind.* New York: Anchor Books/Doubleday.

Rosenberg, R. N. (1993). An introduction to the molecular genetics of neurological disease. *Archives of Neurology, 50,* 1123–1128.

Roses, A. D., Strittmatter, W. J., Pericak-Vance, M. A., Corder, E. H., Saunders, A. M., & Schmechel, D. E. (1994). Clinical application of apolipoprotein E genotyping to Alzheimer's disease. *The Lancet, 343,* 1564–1565.

Ross, E. (1981). The aprosodias: Functional-anatomical organization of the affective components of language in the right hemisphere. *Archives of Neurology, 38,* 561–569.

Ross, E. D. (1988). Fact vs. fancy or is it all just semantics? *Archives of Neurology, 45,* 338–339.

Rossi, P. W., Kheyfets, S., & Reding, M. J. (1990). Fresnel prisms improve visual perception in stroke patients with homonymous hemianopia or unilateral visual neglect. *Neurology, 40,* 1597–1599.

Rothwell, J. (1994). *Control of human voluntary movement.* (2nd ed.). London: Chapman & Hall.

Rothwell, J. C., & Obeso, J. A. (1987). The anatomical and physiological basis of torsion dystonia. In C. D. Marsden & S. Fahn (Eds.), *Movement Disorders 2* (pp. 313–331). London: Butterworths.

Roy, E. A., & Square-Storer, P. A. (1990). Evidence for common expressions of apraxia. In G. E. Hammond (Ed.), *Cerebral control of speech and limb movements* (pp. 478–502). Amsterdam: North Holland.

Rubens, A. B. (1985). Caloric stimulation and unilateral visual neglect. *Neurology, 35,* 1019–1024.

Rumelhart, D. E., & McClelland, J. L. (1986). *Parallel distributed processing* (Vol. 1). Cambridge, MA: MIT Press.

Sachdev, P., & Loneragan, C. (1991). The present status of akathisia. *The Journal of Nervous and Mental Disease, 179,* 381–391.

Sackeim, H. A., Nobler, M. S., Prudic, J., Devanand, D. P., McElhinney, M., Coleman, E., Settembrino, J., & Maddatu, V. (1992). Acute effects of electroconvulsive therapy on hemispatial neglect. *Neuropsychiatry, Neuropsychology, and Behavioral Neurology, 5,* 151–160.

Sacks, O. (1992). Tourette's syndrome and creativity: Exploiting the ticcy witticisms and witty ticcicisms. *British Medical Journal, 305,* Dec. 19/26, 1515–1516.

Sagar, H. J. (1990). Aging and age-related neurological disease: Remote memory. In F. Boller & J. Grafman (Eds.), *Handbook of Neuropsychology* (Vol. 4, pp. 311–324). Amsterdam: Elsevier.

Saint-Cyr, J. A., and Taylor, A. E. (1993). Cognitive dysfunction in Parkinson's disease. In J. S. Schneider & M. Gupta (Eds.), *Current concepts in Parkinson's disease research* (pp. 41–58). Seattle: Hogrefe & Huber.

Sakashita, Y. (1991). Visual attentional disturbance with unilateral lesions in the basal ganglia and deep white matter. *Annals of Neurology, 30,* 673–677.

Sams, M., Hari, R., Rif, J., & Knuutila, J. (1993). The human auditory sensory memory trace persists about 10 sec: Neuromagnetic evidence. *Journal of Cognitive Neuroscience, 5,* 363–370.

Sano, M., & Mayeux, R. (1992). Biochemistry of depression in Parkinson's disease. In S. J. Huber & J. L. Cummings (Eds.), *Parkinson's disease: Neurobehavioral aspects* (pp. 229–239). New York: Oxford University Press.

Santamaria, J., & Tolosa, E. (1992). Clinical subtypes of Parkinson's disease and depression. In S. J. Huber & J. L. Cummings (Eds.), *Parkinson's disease: Neurobehavioral aspects* (pp. 217–228). New York: Oxford University Press.

Sartori, B., & Job, R. (1988). The oyster with four legs: A neuropsychological study on the interaction of visual and semantic information. *Cognitive Neuropsychology, 5,* 677–709.

Savage-Rumbaugh, E. S., Sevcik, R. A., Brakke, K. E., Rumbaugh, D. M., & Greenfield, P. M. (1990). Symbols: Their communicative use, comprehension, and combination by bonobos (Pan paniscus). In C. Rovee-Collier & L. P. Lipsitt (Eds.), *Advances in Infancy Research* (Vol. 6, pp. 221–278). Norwood, NJ: Ablex.

Schacter, D. L. (1992). Consciousness and awareness in memory and amnesia: Critical issues. In A. D. Milner & M. D. Rugg (Eds.), *The neuropsychology of consciousness* (pp. 180–200). San Diego: Academic Press.

Schacter, D. L., Chiu, C.-Y. P., & Ochsner, K. N. (1993). Implicit memory. A selective review. *Annual Review of Neuroscience, 16,* 159–182.

Schapira, A. H. V. (1994). Evidence for mitochondrial dysfunction in Parkinson's disease—a critical appraisal. *Movement Disorders, 9,* 125–138.

Schiff, H. B., Sabin, T. D., Geller, A., Alexander, L., & Mark, V. (1982). Lithium in aggressive behavior. *American Journal of Psychiatry, 139,* 1346–1348.

Schmidt, R. A. (1988). *Motor control and learning: A behavioral emphasis.* Champaign, IL: Human Kinetics Publishers.

Schnabel, J. (1993). New Alzheimer's therapy suggested. *Science, 260,* 1719–1720.

Schneider, G. E. (1969). Two visual systems. *Science, 163,* 895–902.

Schneider, G. E., & Shiffrin, R. M. (1977). Controlled and automatic human information processing. I. Detection, search and attention. *Psychological Review, 84,* 1–66.

Schnider, A., Benson, D. F., Alexander, D. N., & Schnider-Klaus, A. (1994). Nonverbal environmental sound recognition after unilateral hemispheric stroke. *Brain, 117,* 281–287.

Schopler, E., & Mesibov, G. B. (1987). *Neurobiological issues in autism.* New York: Plenum.

Schulteis, G., & Koob, G. (1994). Dark side of drug dependence. *Nature, 371,* 108–109.

Schwartz, A. S., Marchok, P. L., Kreinick, C. J., & Flynn, R. E. (1979). The asymmetric lateralization of tactile extinction in patients with unilateral cerebral dysfunction. *Brain, 102,* 669–684.

Scoville, W. B., & Milner, B. (1957). Loss of recent memory after bilateral hippocampal lesions. *Journal of Neurology, Neurosurgery and Psychiatry, 20,* 11–21.

Seeman, P. (1993). Schizophrenia as a brain disease. *Archives of Neurology, 50,* 1092–1095.

Seeman, P., Guan, H.-C., & Van Tol, H. H. M. (1993). Dopamine D4 receptors elevated in schizophrenia. *Nature, 365,* 441–445.

Seignot, M. J. N. (1961). *Un cas de maladie des tics de Gilles de la Tourette guéri par le R-1625. Annals Medico-Psychologiques, 119,* 578–579.

Selkoe, D. J. (1991, November). Amyloid protein and Alzheimer's disease. *Scientific American,* 40–47.

Serby, M., Larson, P., & Kalkstein, D. (1991). The nature and course of olfactory deficits in Alzheimer's disease. *American Journal of Psychiatry, 148,* 357–360.

Sergent, J. (1982). The cerebral balance of power: Confrontation or co-operation? *Journal of Experimental Psychology: Human Perception and Performance, 8,* 253–272.

Sergent, J. (1986). Subcortical coordination of hemisphere activity in commissurotomized patients. *Brain, 109,* 357–369.

Sergent, J. (1987). A new look at the human split brain. *Brain, 110,* 1375–1392.

Sergent, J. (1990). Furtive incursions into bicameral minds: Integrative and coordinating role of subcortical structures. *Brain, 113,* 537–568.

Sergent, J. (1993). Music, the brain and Ravel. *Trends in Neurosciences, 16,* 168–172.

Sergent, J. (1994). Brain-imaging studies of cognitive functions. *Trends in Neurosciences, 17,* 221–227.

Sergent, J., & Signoret, J.-L. (1992). Functional and anatomical decomposition of face processing: Evidence from prosopagnosia and PET study of normal subjects. *Philosophical Transactions of the Royal Society of London B, 335,* 55–62.

Sergent, J., Zuck, E., Terriah, S., & MacDonald, B. (1992). Distributed neural network underlying musical sight-reading and keyboard performance. *Science, 257,* 106–109.

Seymour, S. E., Reuter-Lorenz, P. A., & Gazzaniga, M. S. (1994). The disconnection syndrome: Basic findings reaffirmed. *Brain, 117,* 105–115.

Shallice, T. (1988). *From neuropsychology to mental structure.* Cambridge: Cambridge University Press.

Shallice, T., & Burgess, P. W. (1991). Deficits in strategy application following frontal damage in man. *Brain, 114,* 727–741.

Shallice, T., Fletcher, P., Frith, C. D., Grasby, P., Frackowiak, R. S. J., & Dolan, R. J. (1994). Brain regions associated with acquisition and retrieval of verbal episodic memory. *Nature, 368,* 633–635.

Sheehy, M. P., Rothwell, J. C., & Marsden, C. D. (1988). Writer's cramp. In S. Fahn, C. D. Marsden, & D. B. Calne (Eds.), *Dystonia 2. Advances in neurology* (Vol. 50, pp. 457–472). New York: Raven Press.

Shimuzu, N., & Okiyama, R. (1993). Bereitschaftspotential preceding voluntary saccades is abnormal in patients with Parkinson's disease. *Advances in Neurology, 60,* 398–402.

Shoulson, I., & Kurlan, R. (1993). Inherited disorders of the basal ganglia. In R. N.

Rosenberg, S. B. Prusiner, S. DiMauro, R. L. Barchi, & L. M. Kunkel (Eds.), *The molecular and genetic basis of neurological disease* (pp. 753–763). Boston: Butterworth-Heinemann.

Sidtis, J. J. (1985). Bilateral language and commissurotomy: Interactions between the hemispheres with and without the corpus callosum. In A. Reeves (Ed.), *Epilepsy and the corpus callosum* (pp. 369–380). New York: Plenum Press.

Singer, H. S., Reiss, A. L., Brown, J. E., Aylward, E. H., Shih, B., Chee, E., Harris, E. L., Reader, M. J., Chase, G. A., Bryan, R. N., & Denckla, M. B. (1993). Volumetric MRI changes in basal ganglia of children with Tourette's syndrome. *Neurology, 43,* 950–956.

Singer, H. S., & Walkup, J. T. (1991). Tourette syndrome and other tic disorders: Diagnosis, pathophysiology and treatment. *Medicine, 70,* 15–32.

Singer, W. (1994). A new job for the thalamus. *Nature, 369,* 444–445.

Snow, B. J., & Calne, D. B. (1992). The etiology of Parkinson's disease. In A. B. Joseph & R. R. Young (Eds.), *Movement disorders in neurology and neuropsychiatry* (pp. 230–235). Oxford: Blackwells Scientific.

Soderfeldt, B., Rönnberg, J., & Risberg, J. (1994). Regional cerebral blood flow in sign language users. *Brain and Language, 46,* 59–68.

Soliveri, P., Brown, R. G., Jahanshahi, M., & Marsden, C. D. (1992). Effect of practice on performance of a skilled motor task in patients with Parkinson's disease. *Journal of Neurology, Neurosurgery and Psychiatry, 55,* 454–460.

Speedie, L. J., Wertman, E., Ta'ir, J., & Heilman, K. M. (1993). Disruption of automatic speech following a right basal ganglia lesion. *Neurology, 43,* 1768–1774.

Sperry, R. W. (1961). Cerebral organization and behavior. *Science, 133,* 1749–1757.

Sperry, R. W. (1968). *Mental unity following surgical disconnection of the cerebral hemispheres. The Harvey lectures, Series 62.* New York: Academic Press.

Sperry, R. W., Zaidel, E., & Zaidel, D. (1979). Self-recognition and social awareness in the deconnected minor hemisphere. *Neuropsychologia, 17,* 153–166.

Spiers, P. A., Schomer, D. L., Blume, H. W., Kleefield, J., O'Reilly, G., Weintraub, S., Osborne-Shaefer, P., & Mesulam, M.-M. (1990). Visual neglect during intracarotid amobarbital testing. *Neurology, 40,* 1600–1606.

Spoont, M. R. (1992). Modulatory role of serotonin in neural information processing: Implications for human psychopathology. *Psychological Bulletin, 112,* 330–350.

Squire, L. R., Knowlton, B., & Musen, G. (1993). The structure and organization of memory. *Annual Review of Psychology, 44,* 453–495.

Stacy, M., & Jankovic, J. (1992). Clinical and neurobiological aspects of Parkinson's disease. In S. J. Huber & J. L. Cummings (Eds.), *Parkinson's disease: Neurobehavioral aspects* (pp. 10–31). New York: Oxford University Press.

Stahl, S. M. (1993). Serotonergic mechanisms and the new antidepressants. *Psychological Medicine, 23,* 281–285.

Stam, C. J., Visser, S. L., Op de Coul, A. A. W., De Sonneville, L. M. J., Schellens, R. L. L. A., Brunia, C. H. M., de Smet, J. S., & Gielen, G. (1993). Disturbed frontal regulation of attention in Parkinson's disease. *Brain, 116,* 1139–1158.

Staton, R. D., Brumback, R. A., & Wilson, H. (1982). Reduplicative paramnesia, a disconnection syndrome of memory. *Cortex, 18,* 23–26.

Steg, G., & Johnels, B. (1993). Physiological mechanisms and assessment of motor disorders in Parkinson's disease. In H. Narabayashi, T. Nagatsu, N. Yanagisawa, & Y. Mizuno (Eds.), *Advances in neurology* (Vol. 60, pp. 358–365). New York: Raven Press.

Stein, J. F. (1991). Space and the parietal association areas. In J. Paillard (Ed.), *Brain and space* (pp. 185–222). Oxford: Oxford University Press.

Stern, M. B., Doty, R. L., Dotti, M., Coreoron, P., Crawford, D., McKeown, D. A., Adler, C., Gollomp, S., & Hurtig, H. (1994). Olfactory function in Parkinson's disease subtypes. *Neurology, 44,* 266–268.

Stoel-Gammon, C., & Otomo, K. (1986). Babbling development of hearing-impaired and normally hearing subjects. *Journal of Speech and Hearing Disorders, 51,* 33–41.

Stoetter, B., Braun, A. R., Randolph, C., Gernert, J., Carson, R. E., Herscovitch, P., & Chase, T. N. (1992). Functional neuroanatomy of Tourette syndrome: Limbic-motor interactions studied with FDG PET. In T. N. Chase, A. J. Friedhoff, & D. J. Cohen (Eds.), *Advances in neurology* (Vol. 58, pp. 213–243). New York: Raven Press.

Stone, S. P., Patel, P., Greenwood, R. J., & Halligan, P. W. (1992). Measuring visual neglect in acute stroke and predicting its recovery: The visual neglect recovery index. *Journal of Neurology, Neurosurgery and Psychiatry, 55,* 431–436.

Stone, S. P., Wilson, B., Wroot, A., Halligan, P. W., Lange, L. S., Marshall, J. C., & Greenwood, R. J. (1991). The assessment of visuo-spatial neglect after acute stroke. *Journal of Neurology, Neurosurgery and Psychiatry, 54,* 345–350.

Storey, E., & Beal, M. F. (1993). Neurochemical substrates of rigidity and chorea in Huntington's disease. *Brain, 116,* 1201–1222.

Strange, P. G. (1992). *Brain biochemistry and brain disorders.* Oxford: Oxford University Press.

Straube, E. R., & Oades, R. D. (1992). *Schizophrenia: Empirical research and findings.* San Diego: Academic Press.

Strittmatter, W. J., Weisgraber, K. H., Goedert, M., Saunders, A. M., Huang, D., Corder, E. H., Dong, L.-M., Jakes, R., Alberts, M. J., Gilbert, J. R., Han, S.-H., Hulette, C., Einstein, G., Schmechel, D. E., Pericak-Vance, M. A., & Roses, A. D. (1994). Hypothesis: Microtubule instability and paired helical filament formation in the Alzheimer disease brain are related to apolipoprotein E genotype. *Experimental Neurology, 125,* 163–171.

Stuss, D. T., & Benson, D. F. (1986). *The frontal lobes.* New York: Raven Press.

Suchowersky, O. (1994). Gilles de la Tourette syndrome. *The Canadian Journal of Neurological Sciences, 21,* 48–52.

Sutherland, G. R., & Richards, R. I. (1994). Dynamic mutations. *American Scientist, 82,* 157–163.

Takayama, Y., Sugishita, M., Akiguchi, I., & Kimura, J. (1994). Isolated acalculia due to left parietal lesion. *Archives of Neurology, 51,* 286–291.

Tanaka, K. (1993). Neuronal mechanisms of object recognition. *Science, 262,* 685–688.

Tarsy, D. (1992). Restless legs syndrome. In A. B. Joseph & R. R. Young (Eds.), *Movement disorders in neurology and neuropsychiatry* (pp. 397–400). Oxford: Blackwell Scientific.

Taylor, A. E., & Saint-Cyr, J. A. (1992). Executive function. In S. J. Huber & J. L. Cummings (Eds.), *Parkinson's disease: Neurobehavioral aspects* (pp. 74–85). New York: Oxford University Press.

Tegnér, R., & Levander, M. (1991a). Through a looking glass. A new technique to demonstrate directional hypokinesia in unilateral neglect. *Brain, 114*, 1943–1951.

Tegnér, R., & Levander, M. (1991b). The influences of stimulus properties on visual neglect. *Journal of Neurology, Neurosurgery and Psychiatry, 54*, 882–887.

Teuber, H.-L. (1955). Physiological psychology. *Annual Review of Psychology, 9*, 267–276.

The Huntington's Disease Collaborative Research Group. (1993). A novel gene containing a trinucleotide repeat that is expanded and unstable on HD chromosomes. *Cell, 72*, 971–983.

Theeuwes, J. (1993). Visual selective attention: A theoretical analysis. *Acta Psychologica, 83*, 93–154.

Thompson, P. D., Berardelli, A., Rothwell, J. C., Day, B. L., Dick, J. P. R., Benecke, R., & Marsden, C. D. (1988). The coexistence of bradykinesia and chorea in Huntington's disease and its implications for theories of basal ganglia control of movement. *Brain, 111*, 223–244.

Tovée, M. J., & Cohen-Tovée, E. M. (1993). The neural substrates of face processing models: A review. *Cognitive Neuropsychology, 10*, 505–528.

Treffert, D. A. (1988). The idiot savant: A review of the syndrome. *American Journal of Psychiatry, 145*, 563–572.

Treisman, A. M. (1986). Properties, parts and objects. In K. R. Boff, L. Kaufmann, & J. P. Thomas (Eds.), *Handbook of perception and human performance: Vol. II, Cognitive processes and performance* (pp. 35:1–35:70). New York: Wiley.

Treisman, A. M., & Gelade, G. (1980). A feature-integration theory of attention. *Cognitive Psychology, 12*, 97–136.

Trevarthen, C. (1970). Experimental evidence for a brainstem contribution to visual perception in man. *Brain, Behaviour and Evolution, 3*, 338–352.

Trevarthen, C. (1990). Integrative functions of the cerebral commissures. In F. Boller & J. Grafman (Eds.), *Handbook of neuropsychology* (Vol. 4, pp. 49–83). Amsterdam: Elsevier.

Trevarthen, C., & Sperry, R. W. (1973). Perceptual unity of the ambient visual field in human commissurotomy patients. *Brain, 96*, 547–570.

Trojano, L., Crisci, C., Lanzillo, B., Elfante, R., & Caruso, G. (1993). How many alien hand syndromes? Follow-up of a case. *Neurology, 43*, 2710–2712.

Tulving, E. (1989). Remembering and knowing the past. *American Scientist, 77*, 361–367.

Tulving, E., Kapur, S., Craik, F. I. M., Moscovitch, M., & Houle, S. (1994). Hemispheric encoding/retrieval asymmetry in episodic memory: Positron emission tomography findings. *Proceedings of the National Academy of Sciences, 91*, 2016–2020.

Tulving, E., Kapur, S., Markowitsch, H. J., Craik, F. I. M., Habib, R., & Houle, S. (1994). Neuroanatomical correlates of retrieval in episodic memory: Auditory sentence recognition. *Proceedings of the National Academy of Sciences, 91*, 2012–2015.

Turjanski, N., Sawle, G. V., Playford, E. D., Weeks, R., Lammerstma, A. A., Lees, A. J., & Brooks, D. J. (1994). PET studies of the presynaptic and postsynaptic dopaminergic system in Tourette's syndrome. *Journal of Neurology, Neurosurgery and Psychiatry, 57*, 688–692.

Turner, A. M., & Greenough, W. T. (1985). Differential rearing effects on rat visual cortex synapses: 1. Synaptic and neuronal density and synapses per neuron. *Brain Research, 329*, 195–203.

Tyler, L. K. (1992). The distinction between implicit and explicit language function: Evidence from aphasia. In A. D. Milner & M. D. Rugg (Eds.), *The neuropsychology of consciousness* (pp. 159–179). San Diego: Academic Press.

Tyszka, J. M., Grafton, S. T., Chew, W., Woods, R. P., & Colletti, P. M. (1994). Parceling of mesial frontal motor areas during ideation and movement using functional magnetic resonance imaging at 1.5 tesla. *Annals of Neurology, 35,* 746–749.

Uitti, R. J., Ahlskog, J. E., Maraganore, D. M., Muenter, M. D., Atkinson, E. J., Cha, R. H., & O'Brien, P. C. (1993). Levodopa therapy and survival in idiopathic Parkinson's disease. *Neurology, 43,* 1918–1926.

Valenstein, E., & Heilman, K. M. (1981). Unilateral hypokinesia and motor extinction. *Neurology, 31,* 445–448.

Valenstein, E., Van Den Abell, T., Watson, R. T., & Heilman, K. M. (1982). Nonsensory neglect from parietotemporal lesions in monkeys. *Neurology, 32,* 1198–1201.

Vallar, G. (1991). Current methodological issues in human neuropsychology. In F. Boller & J. Grafman (Eds.), *Handbook of neuropsychology* (Vol. 5, pp. 343–378). Amsterdam: Elsevier.

Vallar, G. (1993). The anatomical basis of spatial hemineglect in humans. In I. H. Robertson & J. C. Marshall (Eds.), *Unilateral neglect: Clinical and experimental studies* (pp. 27–59). Hove, U.K.: Lawrence Erlbaum Associates.

Vallar, G., Bottini, G., Rusconi, M. L., & Sterzi, R. (1993). Exploring somatosensory hemineglect by vestibular stimulation. *Brain, 116,* 71–86.

Vallar, G., & Perani, D. (1986). The anatomy of unilateral neglect after right-hemisphere stroke lesions. A clinical/CT-scan correlation study in man. *Neuropsychologia, 24,* 609–622.

Vallar, G., Perani, D., Cappa, S. F., Messa, C., Lenzi, G. L., & Fazio, F. (1988). Recovery from aphasia and neglect after subcortical stroke: Neuropsychological and cerebral perfusion study. *Journal of Neurology, Neurosurgery and Psychiatry, 51,* 1269–1276.

Vallar, G., Rusconi, M. L., Bignamini, L., Geminiani, G., & Perani, D. (1994). Anatomical correlates of visual and tactile extinction in humans: A clinical CT scan study. *Journal of Neurology, Neurosurgery and Psychiatry, 57,* 464–470.

Vallar, G., Rusconi, M. L., & Bisiach, E. (1994). Awareness of contralesional information in unilateral neglect: Effects of verbal cueing, tracing, and vestibular stimulation. In C. Umiltà & M. Moscovitch (Eds.), *Attention and performance XV: Conscious and nonconscious information processing* (pp. 377–391). Cambridge, MA: MIT Press.

Vallar, G., Sandroni, P., Rusconi, M. L., & Barbieri, S. (1991). Hemianopia, hemianesthesia, and spatial neglect: A study with evoked potentials. *Neurology, 41,* 1918–1922.

van Gijn, J. (1992). Subarachnoid haemorrhage. *The Lancet, 339,* 653–655.

Vargha-Khadem, F., Isaacs, E., & Mishkin, M. (1994). Agnosia, alexia and a remarkable form of amnesia in an adolescent boy. *Brain, 117,* 683–703.

Ventris, M., & Chadwick, J. (1956). *Documents in Mycenaean Greek.* Cambridge: Cambridge University Press.

Vilkki, J. (1984). Hemi-inattention after ventrolateral thalamotomy. *Neuropsychologia, 22,* 399–408.

Villardita, C., Smirni, P., & Zappalà, G. (1983). Visual neglect in Parkinson's disease. *Archives of Neurology, 40,* 737–739.

Volpe, B. T., & Hirst, W. (1983). Amnesia following rupture and repair of an anterior communicating artery aneurysm. *Journal of Neurology, Neurosurgery and Psychiatry, 46,* 704–709.

Volpe, B. T., LeDoux, J. E., & Gazzaniga, M. S. (1979). Information processing of visual stimuli in an "extinguished" visual field. *Nature, 282,* 722–724.

von Economo, C. (1931). *Encephalitis lethargica: Its sequelae and treatment.* Oxford: Oxford University Press.

von Monakow, C. (1914). *Die Lokalisation in Grosshirn und der Abbau der Funktion durch Korticale Herde.* Wiesbaden: Bergmann.

Vonsattel, J. P. (1992). Neuropathology of Huntington's disease. In A. B. Joseph & R. R. Young (Eds.), *Movement disorders in neurology and neuropsychiatry* (pp. 186–194). Oxford: Blackwell Scientific.

Wada, J., & Rasmussen, T. (1960). Intracarotid injection of sodium amytal for the lateralization of cerebral speech dominance. Experimental and clinical observations. *Journal of Neurosurgery, 17,* 266–282.

Waddington, J. L. (1993a). Schizophrenia: Developmental neuroscience and pathobiology. *The Lancet, 341,* 531–536.

Waddington, J. L. (1993b). Sight and insight: "Visualisation" of auditory hallucinations in schizophrenia? *The Lancet, 342,* 692–693.

Wade, D. T., Wood, V. A., & Hewer, R. L. (1988). Recovery of cognitive function soon after stroke; A study of visual neglect, attention span and verbal recall. *Journal of Neurology, Neurosurgery and Psychiatry, 51,* 10–13.

Walker, E. F., Grimes, K. E., Davis, D. M., & Smith, A. J. (1993). Childhood precursors of schizophrenia: Facial expressions of emotion. *American Journal of Psychiatry, 150,* 1654–1660.

Wallace, W., Ahlers, S. T., Gotlib, J., Bragin, V., Sugar, J., Gluck, R., Shea, P. A., Davis, K. L., & Haroutunian, V. (1993). Amyloid precursor protein in the cerebral cortex is rapidly and persistently induced by loss of subcortical innervation. *Proceedings of the National Academy of Sciences, 90,* 8712–8716.

Walsh, K. W. (1985). *Understanding brain damage.* New York: Churchill Livingstone.

Walters, A. S., Hening, W. A., & Chokroverty, S. (1991). Review and videotape recognition of idiopathic restless legs syndrome. *Movement Disorders, 6,* 105–110.

Wang, Z. W., Black, D., Andreasen, N. C., & Crowe, R. R. (1993). A linkage study of chromosome 11q in schizophrenia. *Archives of General Psychiatry, 50,* 212–216.

Ward, R., Goodrich, S. J., & Driver, J. (in press). Grouping reduces visual extinction: Neuropsychological evidence for weight-linkage in visual selection. *Visual Cognition.*

Warrington, E. K. (1982). Neuropsychological studies of object recognition. *Philosophical Transactions of the Royal Society of London B, 298,* 15–33.

Warrington, E. K. (1985). Agnosia: The impairment of object recognition. In J. A. M. Frederiks (Ed.), *Handbook of clinical neurology, Vol. 1 (45): Clinical neuropsychology* (Gen. Eds. P. Vinken, G. Bruyn, & G. Klawans) (pp. 333–349). Amsterdam: Elsevier.

Warrington, E. K., & Langdon, D. (1994). Spelling dyslexia: A deficit of the visual word-form. *Journal of Neurology, Neurosurgery and Psychiatry, 57,* 211–216.

Warrington, E. K., & McCarthy, R. A. (1987). Categories of knowledge: Further fractionations and an attempted integration. *Brain, 111,* 1273–1296.

Watson, R. T., & Heilman, K. M. (1979). Thalamic neglect. *Neurology, 29,* 690–694.

Watson, R. T., Miller, B. D., & Heilman, K. M. (1978). Nonsensory neglect. *Annals of Neurology, 3,* 505–508.

Watson, R. T., Rothi, L. J. G., & Heilman, K. M. (1992). Apraxia: A disorder of motor programming. In A. B. Joseph & R. R. Young (Eds.), *Movement disorders in neurology and neuropsychiatry* (pp. 681–690). Oxford: Blackwell Scientific.

Webster, J. S., Rapport, L. J., Godlewski, M. C., & Abadee, P. S. (1994). Effect of attentional bias to right space on wheelchair mobility. *Journal of Clinical and Experimental Neuropsychology, 16,* 129–137.

Weiller, C., Willmes, K., Reiche, W., Thron, A., Isensee, C., Buell, U., & Ringelstein, E. B. (1993). The case of aphasia or neglect after striatocapsular infarction. *Brain, 116,* 1509–1525.

Weinberg, J., Diller, L., Gordon, W. A., Gerstman, L., Lieberman, A., Lakin, P., Hodges, G., & Ezrachi, O. (1977). Visual scanning training effect on reading-related tasks in acquired right brain damage. *Archives of Physical Medicine and Rehabilitation, 58,* 479–486.

Weinberg, J., Diller, L., Gordon, W. A., Gerstman, L., Lieberman, A., Lakin, P., Hodges, G., & Ezrachi, O. (1979). Training sensory awareness and spatial organization in people with right brain damage. *Archives of Physical Medicine and Rehabilitation, 60,* 491–496.

Weintraub, S., & Mesulam, M.-M. (1988). Visual hemispatial inattention: Stimulus parameters and exploratory strategies. *Journal of Neurology, Neurosurgery and Psychiatry, 51,* 1481–1488.

Wernicke, C. (1874). *Der Aphasiche Symptomenkomplex: Eine Psychologische Studie auf Anatomischer Basis.* Breslau: Max Cohn & Weigert.

Wernicke, T. F., & Reischies, F. M. (1994). Prevalence of dementia in old age: Clinical diagnoses in subjects aged 95 years and older. *Neurology, 44,* 250–253.

West, M. J., Coleman, P. D., Flood, D. G., & Troncoso, J. C. (1994). Differences in the pattern of hippocampal neuronal loss in normal ageing and Alzheimer's disease. *The Lancet, 344,* 769–772.

Wexler, N. (1993, November). *Repeating the past, expanding the present, changing the future.* Paper presented at the Sunderland Lecture Theatre, Medical School, University of Melbourne.

Whitehouse, P. J., Lerner, A., & Hedera, P. (1993). Dementia. In K. M. Heilman & E. Valenstein (Eds.), *Clinical neuropsychology* (3rd ed.) (pp. 603–646). Oxford: Oxford University Press.

Whiten, A., & Byrne, R. W. (1988). The Machiavellian intelligence hypothesis. In R. W. Byrne & A. Whiten (Eds.), *Machiavellian intelligence* (pp. 1–9). Oxford: Clarendon Press.

WHO MONICA Project, Principal Investigators. (1988). The World Health Organization MONICA Project (monitoring trends and determinants in cardiovascular disease): A major international collaboration. *Journal of Clinical Epidemiology, 41,* 105–114.

Wichmann, T., & DeLong, M. R. (1993). Pathophysiology of Parkinsonian motor abnormalities. In H. Narabayashi, T. Nagatsu, N. Yanagisawa, & Y. Mizuno (Eds.), *Advances in neurology* (Vol. 60, pp. 53–61). New York: Raven Press.

Wigan, A. L. (1844). *The duality of the mind.* London: Longman, Brown & Green.

Wilson, F. A. W., O'Scalaidhe, S. P., & Goldman-Rakic, P. S. (1993). Dissociation of object and spatial processing domains in primate prefrontal cortex. *Science, 260,* 1955–1958.

Wilson, M. A., & McNaughton, B. L. (1993). Dynamics of the hippocampal ensemble code for space. *Science, 261,* 1055–1057.

Wirshing, W. C., & Cummings, J. L. (1990). Tardive movement disorders. *Neuropsychiatry, Neuropsychology, and Behavioral Neurology, 3,* 23–35.

Witelson, S. F. (1989). Hand and sex differences in the isthmus and genu of the corpus callosum. *Brain, 112,* 799–832.

Witelson, S. F. (1993). Clinical neurology as data for basic neuroscience: Tourette's syndrome and the human motor system. *Neurology, 43,* 859–861.

Woodcock, J. H. (1992). Behavioral aspects of Huntington's disease. In A. B. Joseph & R. R. Young (Eds.), *Movement disorders in neurology and neuropsychiatry* (pp. 178–185). Oxford: Blackwell Scientific.

Woodruff, P. W. R., Pearlson, G. D., Geer, M. J., Barta, P. E., & Chilcoat, H. D. (1993). A computerized magnetic resonance imaging study of corpus callosum morphology in schizophrenia. *Psychological Medicine, 23,* 45–56.

Wright, S., Young, A. W., & Hellawell, D. J. (1993). Frégoli delusion and erotomania. *Journal of Neurology, Neurosurgery and Psychiatry, 56,* 322–323.

Yahr, M. D. (1993). Parkinson's disease. The L-dopa era. In H. Narabayashi, T. Nagatsu, N. Yanagisawa, & Y. Mizuno (Eds.), *Advances in neurology* (Vol. 60, pp. 11–17). New York: Raven Press.

Yang, T. T., Gallen, C., Schwartz, B., Bloom, F. E., Ramachandran, V. S., & Cobb, S. (1994). Sensory maps in the human brain. *Nature, 368,* 592–593.

Young, A. B., & Penney, J. B. (1993). Biochemical and functional organization of the basal ganglia. In J. Jankovic & E. Tolosa (Eds.), *Parkinson's disease and movement disorders* (2nd ed.) (pp. 1–12). Baltimore: Williams & Wilkins.

Young, A. W., Flude, B. M., & Ellis, A. W. (1991). Delusional misidentification incident in a right hemisphere stroke patient. *Behavioural Neurology, 4,* 81–87.

Young, A. W., Newcombe, F., de Haan, E. H. F., Small, M., & Hay, D. C. (1993). Face perception after brain injury: Selective impairments affecting identity and expression. *Brain, 116,* 941–959.

Zaidel, E. (1990a). The sage of right-hemisphere reading. In C. Trevarthen (Ed.), *Brain circuits and functions of the mind. Essays in honor of Roger W. Sperry* (pp. 304–319). Cambridge: Cambridge University Press.

Zaidel, E. (1990b). Language functions in the two hemispheres following complete cerebral commissurotomy and hemispherectomy. In F. Boller & J. Grafman (Eds.), *Handbook of neuropsychology* (Vol. 4, pp. 115–150). Amsterdam: Elsevier.

Zaidel, E., Clarke, J., & Suyenobu, B. (1990). Hemispheric independence: A paradigm case for cognitive neuroscience. In A. B. Scheibel & A. F. Wechsler (Eds.), *Neurobiology of higher cognitive function* (pp. 297–352). New York: Guilford.

Zaidel, E., & Schweiger, A. (1984). On wrong hypotheses about the right hemisphere: Commentary on K. Patterson & D. Besner: Is the right hemisphere literate? *Cognitive Neuropsychology, 1,* 351–364.

Zangwill, O. L. (1976). Thought and the brain. *British Journal of Psychology, 67,* 301–314.

Zarit, S. H., & Kahn, R. L. (1974). Impairment and adaptation in chronic disabilities: Spatial inattention. *The Journal of Nervous and Mental Disease, 159,* 63–72.

Zatorre, R. J. (1984). Music perception and cerebral function: A critical review. *Music Perception, 2,* 196–221.

Zec, R. F. (1993). Neuropsychological functioning in Alzheimer's disease. In R. W. Parks, R. F. Zec, & R. S. Wilson (Eds.), *Neuropsychology of Alzheimer's disease and other dementias* (pp. 3–80). Oxford: Oxford University Press.

Zipser, D., & Andersen, R. A. (1988). A back-propagation programmed network that stimulates response properties of a subset of posterior parietal neurons. *Nature, 331,* 679–684.

Zoccolotti, P., Antonucci, G., Judica, A., Montenero, P., Pizzamiglio, L., & Razzano, C. (1989). Incidence and evolution of the hemineglect disorder in chronic patients with unilateral right brain damage. *International Journal of Neuroscience, 47,* 209–216.

Zola-Morgan, S., & Squire, L. R. (1993). Neuroanatomy of memory. *Annual Review of Neuroscience, 16,* 547–563.

Zülch, K.-J., & Hossmann, V. (1988). Patterns of cerebral infarctions. In P. J. Vinken, G. W. Bruyn, & H. L. Klawans (Eds.), *Vascular diseases, Part 1* (pp. 175–198). Amsterdam: Elsevier.

Zurif, E., Swinney, D., & Fodor, J. A. (1991). An evaluation of assumptions underlying the single-patient-only position in neuropsychological research: A reply. *Brain and Cognition, 16,* 198–210.

Author Index

M

Subject Index